CONSERVATION
GENETICS

CASE HISTORIES
FROM NATURE

CONSERVATION
GENETICS

CASE HISTORIES
FROM NATURE

Edited by
John C. Avise
James L. Hamrick

CHAPMAN & HALL

New York • Albany • Bonn • Boston • Cincinnati • Detroit • London • Madrid • Melbourne
Mexico City • Pacific Grove • Paris • San Francisco • Singapore • Tokyo • Toronto • Washington

Cover design and art direction: John C. Avise and Andrea Meyer, emDASH inc.
Illustrations: Dorset W. Trapnell

Copyright © 1996
Chapman & Hall

Printed in the United States of America

For more information, contact:

Chapman & Hall
115 Fifth Avenue
New York, NY 10003

Thomas Nelson Australia
102 Dodds Street
South Melbourne, 3205
Victoria, Australia

Nelson Canada
1120 Birchmount Road
Scarborough, Ontario
Canada, M1K 5G4

International Thomson Editores
Campos Eliseos 385, Piso 7
Col. Polanco
11560 Mexico D.F. Mexico

Chapman & Hall
2-6 Boundary Row
London SE1 8HN
England

Chapman & Hall GmbH
Postfach 100 263
D-69442 Weinheim
Germany

International Thomson Publishing Asia
221 Henderson Road #05-10
Henderson Building
Singapore 0315

International Thomson Publishing-Japan
Hirakawacho-cho Kyowa Building, 3F
1-2-1 Hirakawacho-cho
Chiyoda-ku, 102 Tokyo
Japan

1 2 3 4 5 6 7 8 9 10 XXX 01 00 99 97 96 95

Library of Congress Cataloging-in-Publication Data

Conservation genetics : case histories from nature / John C. Avise and J.L. Hamrick (eds.).
     p. cm.
    Includes bibliographical references and index.
    ISBN 0-412-05581-3
    1. Germplasm resources—Case studies. 2. Population genetics—Case studies.
I. Avise, John C. II. Hamrick, J.L. (James Lewis).
QH75.C6635 1995   95-5253
639.9'01'5751—dc20   CIP

British Library Cataloguing in Publication Data available

To order for this or any other Chapman & Hall book, please contact **International Thomson Publishing, 7625 Empire Drive, Florence, KY 41042.** Phone: (606) 525-6600.
Fax: (606) 525-7778, e-mail: order@chaphall.com.

For a complete listing of Chapman & Hall's titles, send your requests to
Chapman & Hall, Dept. BC, 115 Fifth Avenue, New York, NY 10003.

To our children's children and, we hope, to theirs

CONTENTS

Part II.   Case Histories
that Involve a Regional or Ecosystem Perspective,
or a Pattern and Process Orientation

CONTRIBUTORS

Fred W. Allendorf, Department of Biology, University of Montana, Missoula, MT 59812

John C. Avise, Department of Genetics, University of Georgia, Athens, GA 30602

C. Scott Baker, School of Biological Sciences, University of Aukland, Aukland, New Zealand

Brian W. Bowen, BEECS Genetic Analysis Core, University of Florida, Gainesville, FL 32611

Nicholas J. Georgiadis, New York Zoological Society and Wildlife Conservation Society, Bronx, NY 10640

Mary Jo W. Godt, Department of Botany, University of Georgia, Athens, GA 30602

John E. Graves, Virginia Institute of Marine Science, College of William and Mary, Gloucester Point, VA 23062

Susan M. Haig, National Biological Service, Oregon State University, Corvallis, OR 97331

J. L. Hamrick, Departments of Botany and Genetics, University of Georgia, Athens, GA 30602

Michael Lynch, Department of Biology, University of Oregon, Eugene, OR 97403

Stephen J. O'Brien and Collaborators *, National Cancer Institute, Frederick Cancer Research Facility, Frederick, MD 21701

Stephen R. Palumbi, Department of Zoology and Kewalo Marine Laboratory, University of Hawaii, Honolulu, HI 96816

Theresa R. Pope, Department of Biological Anthropology and Anatomy, Duke University, Durham, NC 27705

Loren H. Rieseberg, Biology Department, Indiana University, Bloomington, IN 47405

* Janice S. Martenson, Sriyanie Miththapala, Diane Janczewski, Jill Pecon-Slattery, Warren Johnson, Dennis A. Gilbert, and Melody Roelke, National Cancer Institute, Frederick, MD; Craig Packer, Department of Ecology and Behavior Biology, University of Minnesota, St. Paul, MN; Mitchell Bush and David E. Wildt, National Zoological Park, Washington DC.

Susan M. Swensen, Biology Department, Indiana University, Bloomington, IN 47405

Alan R. Templeton, Department of Biology, Washington University, St. Louis, MO 63130

Robert C. Vrijenhoek, Center for Theoretical and Applied Genetics, Cook College, Rutgers University, New Brunswick, NJ 08903

Robin S. Waples, Northwest Fisheries Center, Seattle, WA 98112

Robert K. Wayne, Department of Biology, University of California, Los Angeles, CA 90024

Prologue

A nearly inevitable aspect of science is that researchers come to view their own discipline with a special fondness, and seek to identify a broader significance for its perspectives. So it has been in conservation biology, where the fields of ecology, systematics, demography, genetics, ethology, and natural history (and subdisciplines within all of these) sometimes appear to vie for preeminence. This book is about genetics in the context of conservation biology, but emphatically, the intent is not to promote genetic approaches to the exclusion of others. Rather, our motivation has been to illustrate (by empirical examples) how various genetic perspectives can sometimes mesh with those of other biological disciplines in the service of conservation concerns.

Producing this book has been somewhat of a cathartic exercise for us. We both were attracted to biology by a love of nature tracing to early childhood. Our formal college trainings began in applied ecology (JCA's first degree was in fishery biology, JLH's in forestry), and only later did genetics become a focus of our respective research programs. Genetics as an intellectual discipline is endlessly challenging and fascinating (and has provided us with immensely enjoyable careers), but we have not lost sight of our natural-history roots. Neither could we be blind to the fact that the decline of global biodiversity has entered a critical stage, under the collective weight of human population growth. If this book promotes a cause, let it be the broader goals of conservation awareness.

This volume presents a collection of substantial case histories in which genetic approaches have been applied to conservation-related issues involving natural (as opposed to captive) populations. Rationales for this format are as follows: (1) in the last decade, the first large-scale and multifaceted genetic research programs directed at threatened or endangered taxa in nature were initiated; (2) results from these studies provide empirical object lessons of what can (and cannot) be learned from genetic approaches; and (3) by summarizing several of these case histories between two covers, an accessible guide for reference and teaching is made available from a scattered primary literature.

This case-history format also has the acknowledged limitation that the approach is inherently illustrative rather than comprehensive, in at least two regards. First, it was well beyond the current scope to include extensive background on population genetics theory, systematics theory, methods of genetic data analysis, or molecular laboratory techniques (and in any event, such information is readily available elsewhere—e.g., Nei, 1987; Hillis and Moritz, 1990; Weir, 1990; Miyamoto and Cracraft, 1991; Avise, 1994). Nonetheless, the chapters have been written at technical and conceptual levels appropriate for biologists with even an introductory knowledge of genetics. Second, the treatment is taxonomically unbalanced in the sense that some important organismal groups (e.g., invertebrate animals) are neglected. However, the book's content primarily reflects research directions taken by the conservation genetics community, rather than editorial capriciousness. Indeed, to the extent possible, we endeavored to include significant case histories from a variety of animal and plant groups.

The book is divided into two major parts. Chapters in Part I center on particular taxonomic assemblages of threatened or endangered species for which molecular genetic approaches have been employed in a conservation context. In most chapters, an introductory Box summarizes the formal conservation status of the organisms in question, as reflected in two recently updated lists: (1) "Endangered and Threatened Wildlife and Plants," 1993 (a joint production of the U.S. Fish and Wildlife Service and the National Marine Fisheries Service, which lists all organisms currently afforded protection under the U.S. Endangered Species Act); and (2) the "1994 IUCN Red List of Threatened Animals" (a publication of the World Conservation Union [1993], and the most authoritative compendium of jeopardized species worldwide). Other Boxes provide additional background information relevant to conservation interests. Summarized in Part I are many of the major research programs conducted to date in conservation genetics, with the various chapters covering organisms ranging from whales and dolphins (Chapter 2), to large cats (Chapter 3), wolves and allies (Chapter 4), primates (Chapter 5), birds (Chapter 6), marine turtles (Chapter 7), and salmonid fishes (Chapter 8).

Chapters in Part II move increasingly toward a regional or ecosystem perspective, and/or involve a "pattern and process" orientation. Although empirical genetic studies ineluctably deal with particular taxa, chapters in this section tend to involve comparisons of population genetic features observed among diverse organisms inhabiting particular environmental regimes, or sharing particular natural histories. Examples include endemic floras on continents and islands (Chapters 9 and 10, respectively), fishes in pelagic marine environments and in deserts (Chapters 11 and 12, respectively), bovid mammals in African landscapes (Chapter 13), and diverse faunas in the Southeastern United States (Chapter 14). The taxa included in these studies were not necessarily of imperiled status, nor were they invariably allied taxonomically, but the comparative genetic patterns (and the ecological and evolutionary processes thereby inferred) provide important general lessons for conservation biology. A concluding treatment (Chapter 15) deals with the important but largely neglected topic of quantitative genetics in conservation biology, and offers perspectives that sometimes differ considerably from those of earlier chapters.

This book is an outgrowth of a presidential symposium of the Society for the Study of Evolution, held in Athens, Georgia in June 1994, and funded by a grant from the University of Georgia. We wish to thank the speakers at that symposium and the other authors who have joined them in contributing the case histories and perspectives presented here. We also wish to give special thanks to DeEtte Walker and Bill Nelson, who meticulously copyedited and proofed the entire volume (although any errors of course remain our responsibility). We hope that any furtherance of conservation sensibilities promoted by this book may serve as some measure of repayment for the personal satisfaction and inspiration that nature has added to our lives.

JOHN C. AVISE
J. L. HAMRICK
Athens, Georgia, June 1994

REFERENCES

AVISE, J.C. 1994. *Molecular Markers, Natural History and Evolution.* Chapman & Hall, New York.

HILLIS, D.M. and C. MORITZ (eds.). 1990. *Molecular Systematics.* Sinauer, Sunderland, MA.

MIYAMOTO, M.M. and J. CRACRAFT (eds.). 1991. *Phylogenetic Analysis of DNA Sequences.* Oxford University Press, New York.

NEI, M. 1987. *Molecular Evolutionary Genetics.* Columbia University Press, New York.

WEIR, B. 1990. *Genetic Data Analysis.* Sinauer, Sunderland, MA.

WORLD CONSERVATION UNION. 1993. *1994 IUCN Red List of Threatened Animals.* World Conservation Union, Gland, Switzerland.

CONSERVATION GENETICS

CASE HISTORIES FROM NATURE

INTRODUCTION: THE SCOPE
OF CONSERVATION GENETICS

John C. Avise

It has been slightly more than a decade since publication of *Genetics and Conservation* (Schonewald-Cox et al., 1983), one of the seminal compilations of genetic perspectives in conservation biology. Diverse roles for genetic analyses were emphasized, ranging from studies of inbreeding in small populations, to assessment of spatial patterns of genetic variation in nature, to problems of gene flow, hybridization, and systematics. This eclectic view of genetics will also be apparent in this volume.

An important empirical development over the past three decades has been the introduction of various laboratory assays that permit molecular-level characterizations of genes and their protein products. No longer need geneticists restrict attention to bacteria, maize, fruit flies, or other short-generation species that breed readily under controlled conditions, nor need attention be confined to short-term changes in population genetic composition observable directly. Molecular methods have opened the entire biological world for genetic scrutiny, as well as the full spectrum of ecological and evolutionary timescales of genetic differentiation. In the final analysis, biodiversity (the ultimate subject of conservation interest) *is* genetic diversity, and the magnitudes and patterns of this diversity can now be examined in

Table 1.1. Examples of the Diversity of Empirical and Conceptual Roles for Genetics in Conservation Biology[1]

---

1. Issues of "heterozygosity" (within-population genetic variability)

   Examples: Is genetic variation reduced in endangered species?
   Is this a cause for ecological or evolutionary concern?
   Should populations be managed for increased variation?
   Is a certain class of genes of special fitness significance?
   How serious a problem is inbreeding depression?

2. Issues of parentage and kinship

   Examples: Who has bred with whom in captivity and nature?
   How does this illuminate breeding and social structure?
   What is the impact of the mating system on effective population size (N_e), inbreeding, gene flow, etc.?

3. Issues of population structure and intraspecific phylogeny

   Examples: What are the levels of gene flow between populations?
   Can gender-biased dispersal be documented?
   How does gene flow relate to demographic connectedness?
   Are there significant historical partitions within species?
   What are the "stock" structures within harvested species?

4. Issues of species boundaries and hybridization phenomena

   Examples: How distinct genetically are "endangered species"?
   Do phylogenetic perspectives alter species concepts?
   How widespread are hybridization and introgression?
   Do findings affect the legal status of endangered species?
   What forensic applications do genetic markers provide?

5. Issues of species phylogenies and macroevolution

   Examples: What are the phylogenetic relationships of species and higher taxa?

---

[1] For most of the questions posed, a follow-up query is "How might the outcome influence conservation or management strategies?"

any organisms, and at hierarchical levels ranging from micro- to macro-evolutionary (Table 1.1).

Some genetic studies in conservation biology involve rare or endangered taxa. Others involve non-threatened forms from which general lessons of conservation relevance may emerge. The empirical data of genetics (as revealed by molecular techniques or other methods) are interpreted using the analytical tools of population genetics and systematics, which permit estimation of parameters that frequently carry management implications, such as levels of genetic diversity within populations, patterns of mate choice, biological parentage and kinship, magnitudes of gene flow (and demographic connectedness) between populations, levels of introgressive exchange between hybridizing species, the nature of species boundaries, and phylogenetic relationships among higher taxa (Avise, 1994).

It is important to emphasize the multifaceted nature of genetic input into conservation biology because not always has genetics been perceived so broadly. For example, when Lande (1988) concluded that demographic and behavioral considerations should be of greater importance than genetic considerations in endangered-species programs, he implicitly was equating "genetics" with heterozygosity assessment (i.e., estimation of within-population variation). So too were Caro and Laurenson (1994) when they concluded that ecological issues are more important than genetic issues for endangered species. Such treatments of genetics are disturbing not only because they define the field too narrowly (see Table 1.1), but also because they tend to portray genetic perspectives as being inherently in opposition to other conservation outlooks. On the contrary, as several chapters in this book attest, genetic discoveries often provide novel, conservation-relevant insights into organismal ecology, demography, behavior, biogeography, and evolutionary history.

To cite but one example, a popular approach in recent years has been to characterize the matriarchal phylogenies of conspecific organisms using molecular data from maternally transmitted mitochondrial (mt) DNA. The intraspecific mtDNA gene trees are themselves functions of the historical demographies, ecologies, and behaviors of females (e.g., means and variances among females in reproductive success, population structures as influenced by ecological and biogeographic factors, and historical female-dispersal patterns). Furthermore, because of the usual spatial association of females with newly produced young, significant phylogeographic structure in a mtDNA gene tree implies a considerable degree of *demographic* autonomy among populations (at least over ecological timescales, and regardless of male dispersal behavior), a finding which in turn can be relevant to population management (Avise, 1995). The perceived dichotomy between genetics versus other disciplines has been overstated in the literature of conservation biology—in actuality, these "alternative" perspectives more often are intertwined or at least mutually illuminating.

It is equally important to emphasize that genetic findings considered alone seldom translate automatically into definitive conservation strategies. Three hypothetical scenarios (from different levels in the evolutionary hierarchy) exemplify this point. First, suppose that an endangered population is shown to harbor low levels of genetic variation due to historical inbreeding. Under the "dominance" view of inbreeding depression, low genetic variability in the inbred population has little predictive value of that deme's genetic "health" because the population already may have been purged of deleterious recessive alleles. However, under the "overdominance" view, low genetic variation predicts poor genetic health because inbreeding depression (or, perhaps, inability to respond genetically to ecological challenges such as disease) is assumed to result from low heterozygosity (for two recent examples of inbreeding depression in nature, see Jiménez et al., 1994 and Keller et al., 1994). These competing views of the genetic ramifications of inbreeding have been debated for decades (Charlesworth and Charlesworth, 1987), generally remain unresolved (Lacy, 1992), but carry different implications for management of endangered taxa.

In a second hypothetical scenario, suppose that an endangered population is demonstrated by genetic analysis to have received foreign alleles via introgression from a related but more common taxon. Under one recent interpretation (the "hybrid policy") of the U.S. Endangered Species Act, the introgressed population no longer should qualify for legal protection because the individuals are not "purebred." However, further biological considerations might yield just the opposite recommendation, because (1) rarity itself might increase the chances of hybridization with a related species, such that the small populations most likely to benefit from legal protection could be the ones least likely to receive aegis under the hybrid policy; (2) most species of plants and animals probably hybridize at least occasionally in nature, such that few taxa would then strictly qualify for legal protection; and (3) a limited input of foreign genes might in some cases actually improve the reproductive success of a highly inbred recipient population (O'Brien and Mayr, 1991; Grant and Grant, 1992).

A third hypothetical scenario involves two species being considered for limited management resources. Suppose one species is rare and is shown genetically to be ancient and phylogenetically unique, whereas the other is a somewhat more common representative of a recent, species-rich clade. Vane-Wright et al. (1991; see also Faith, 1992) argue that the former species should be given conservation priority under the rationale that it makes a disproportionate contribution to the world's current evolutionary diversity. However, alternative reasoning by Erwin (1991) suggests just the reverse—the common member of the rapidly speciating clade is to be prized because it carries the greater promise for continued evolution and the generation of future biodiversity.

So, genetic data provide merely one starting point for conservation decisions, and many additional scientific considerations (as well as value judgments) normally must be factored in. The sphere of potentially competing viewpoints enlarges further when human social and economic considerations are added (e.g., when preservation of the northern spotted owl becomes pitted against continuance, for even a few more years, of the old-growth timber industry). Solutions to the overall "calculus of biodiversity" have only recently been initiated (May, 1990), and genetic considerations are merely one part of the equations.

Most debates over management strategies result (directly or indirectly) from the triage nature of the conservation effort (Ehrlich, 1988; Crozier, 1992; Eiswerth and Haney, 1992)—limited time and resources are available to confront seemingly limitless problems. Not all extant species can or will be saved in the coming decades, so hard choices present themselves. Conservation decisions should of course include the best available scientific evidence, but how best to gather and utilize this information is itself a matter of debate. For example, comprehensive inventories of biodiversity have been called for to strengthen the scientific underpinnings of conservation strategies (Box 1.1; see also Eldredge, 1992; Alberch, 1993; Macilwain, 1994). However, Renner and Ricklefs (1994) suggest that massive inventories "will produce data of uneven quality and undemonstrated utility for conservation purposes, divert attention from more constructive approaches to conservation, and further weaken systematics and other collection-based inquiry as scientific endeavors." It might be countered that an organized program to describe and map the Earth's biotic diversity (much of which remains poorly known, particularly in biodiversity hotspots such as tropical rainforests and the deep sea) should provide a powerful focus for conservation efforts, rejuvenate primary-purpose systematics, help redress the current funding imbalance in favor of the reductionist and laboratory sciences (Ehrlich, 1994), and in general help biology regain a closer touch with what remains of the natural world (Wilson, 1988).

Box 1.1. A Biological Survey for the Nation

In the United States, a broad initiative known as the National Biological Service (NBS) recently was organized to "gather, analyze, and disseminate the information necessary for the wise stewardship of our Nation's natural resources, and to foster an understanding of our biological systems and the benefits they provide to society" (Secretary of Interior Bruce Babbitt, as quoted in National Research Council, 1993). The NBS is administered by the Department of Interior, and has responsibilities for inventorying, mapping, and monitoring biological resources, performing basic and applied research, and providing scientific support and technical assistance for management and policy decisions. In 1993, a committee organized by the National Research Council (an advisory arm of the National Academies of Sciences and Engineering) recommended that many public and private agencies be recruited to the effort in a "National Partnership for Biological Survey." The contemplated scope of this biological inventory is unprecedented. If successful, the effort should provide an understanding of the nation's biotic resources analogous to that provided for the country's geography by the U.S. Geological Survey, which began in 1879. As evidenced by several of the chapters in this book, genetic approaches can make important empirical and conceptual contributions to the NBS.

Problems of finite resources become amplified in conservation genetics, where even modest appraisals of molecular diversity are time-consuming and relatively expensive. Genetic approaches in conservation biology could be counterproductive if they unduly drained resources from competing conservation programs, or if other conservation initiatives were delayed or dismantled solely on the grounds that available genetic information was incomplete (as will always be true). Molecular genetic approaches will contribute most intelligently to biodiversity assessment when they address controversial or otherwise interesting problems beyond the purview of nonmolecular methods, or when the findings are likely to provide object lessons of broader relevance to the field. The examples presented in this book are cases in point. Most of the authors explicitly convey their personal interpretations (uncensored by the editors) as to the broader conservation relevance of particular genetic observations.

In the 12 years since publication of *Genetics and Conservation*, numerous species have gone extinct (Ehrlich and Wilson, 1991), many more have been placed in jeopardy, and the Earth's biodiversity has come under increased pressure from the direct and indirect effects of explosive growth in human numbers. Neither genetic studies nor those of any other scientific discipline will be of any significant avail in protecting biodiversity over the critical next century unless the unbridled growth in the human population is arrested (Neel, 1994). Preferably, this will occur by a humane and sensible decrease in birthrates, rather than the inevitable increase in death rates that otherwise *will* take place from crowding and disease, starvation, wars over limited resources, or by other horrific means. Even if human societies survive intact into the 22nd century, many other species will not. An environmentally conscious human being in the year 2150 might well look back to the current time and ponder what our generation could possibly have been thinking when, through lack of self-restraint on population growth and environmental degradation, we thoroughly fouled our planet and precipitated one of the greatest mass extinctions the Earth had ever known: "Those societies could not have been completely ignorant—they did develop DNA technologies."

As we conduct our focused genetic (or other) studies of rare and endangered species, let us remain cognizant of the nearsighted nature of this effort in isolation. The underlying causes of the biodiversity crisis must also be seen with a broader vision. Societies must find ways not only to preserve extant genetic diversity, but also to preserve sustainable environments for life in which the ecological and evolutionary processes fostering biodiversity are maintained (Thompson, 1994; Tilman et al., 1994). This will necessitate a curbing of negative human impacts on the environment, which in turn will require intelligent means of managing resources and bridling human population growth (Box 1.2). As stated by Morowitz (1991), "...as human population goes up, biological species diversity goes down. We might be able to moderate the rate of decline, but we cannot fend off the inevitable."

Like other biological disciplines, genetics fundamentally is an attempt to know life, and therein lies its greatest value to conservation biology. "To the degree that we come to understand other organisms, we will place a greater value on them" (Wilson, 1984), and increasingly cherish their presence and evolutionary continuance. As the chapters in this book attest, genetic perspectives truly have enriched our understanding of nature. Let us relish these important contributions, but at the same time retain sight of the underlying root of the conservation predicament (Box 1.2; Figure 1.1).

Box 1.2. The Ultimate Source of the Biodiversity Crisis

In the autumn of 1993, representatives of the National Academies of Science from 58 countries throughout the world met in New Delhi, India, in a "Science Summit on World Population." This unprecedented collaborative activity was motivated by an urgent concern about the expansion of the global human population. The following are excerpts from the joint statement to which these academies were signatory.

"It took hundreds of thousands of years for our species to reach a population level of 10 million, only 10,000 years ago. This number grew to 100 million people about 2,000 years ago and to 2.5 billion by 1950. Within less than the span of a single lifetime, it has more than doubled to 5.5 billion in 1993 ... [Providing] fertility declines to no lower than 2.4 children per woman [the] global population would grow to 19 billion by the year 2100, and to 28 billion by 2150 ... If current predictions of population growth prove accurate and patterns of human activity on the planet remain unchanged, science and technology may not be able to prevent irreversible degradation of the natural environment and continued poverty for much of the world ... Humanity is approaching a crisis point with respect to the interlocking issues of population, environment, and development ... In our judgment, humanity's ability to deal successfully with its social, economic, and environmental problems will require the achievement of zero population growth within the lifetime of our children."

These sobering statements by the world's leading scientific academies highlight the ultimate futility of conservation efforts in the absence of immediate and pronounced brakes on the growth in human numbers. Perhaps the United Nations International Conference on Population Development, which took place in Cairo, Egypt during September 1994, has begun to identify ways to translate these concerns about human population growth into effective action. Only in such hope could the National Academies of Science find an uplifting note: "Let 1994 be remembered as the year when the people of the world decided to act together for the benefit of future generations."

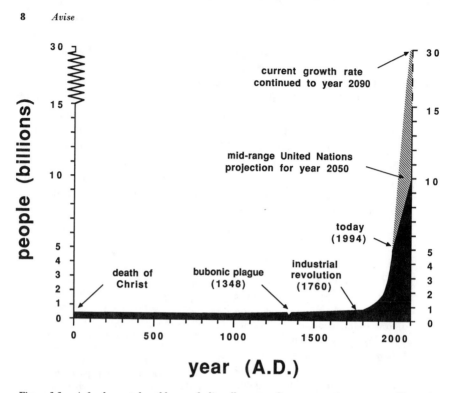

Figure 1.1.   A fundamental problem underlies all present-day conservation concerns. Shown is the chilling growth in the human population in the last two centuries, following all prior millennia in which human numbers were relatively stable and vastly smaller than at present (adapted from Peterson, 1988; Primack, 1993). Also shown are reasonable (in the sense of *realistic*, not *sensible*) population projections over the next century. Barring a dramatic increase in overall death rate (or, far preferably, a dramatic and immediate lowering in birthrate), between 10 and 30 billion people will crowd the Earth in the 21st century.

REFERENCES

ALBERCH, P. 1993. Museums, collections and biodiversity inventories. *Trends Ecol. Evol.* 8:372–375.

AVISE, J.C. 1994. *Molecular Markers, Natural History and Evolution.* Chapman & Hall, New York.

AVISE, J.C. 1995. Mitochondrial DNA polymorphism and a connection between genetics and demography of relevance to conservation. *Conserv. Biol.* 9:686–690.

CARO, T.M. and M.K. LAURENSON. 1994. Ecological and genetic factors in conservation: a cautionary tale. *Science* 263:485–486.

CHARLESWORTH, D. and B. CHARLESWORTH. 1987. Inbreeding depression and its evolutionary consequences. *Annu. Rev. Ecol. Syst.* 18:237–268.

CROZIER, R.H. 1992. Genetic diversity and the agony of choice. *Biol. Conserv.* 61:11–15.

EHRLICH, P.R. 1988. The loss of diversity: causes and consequences. In *Biodiversity*, ed. E.O. Wilson, pp. 21–27. National Academy Press, Washington, DC.

EHRLICH, P.R. 1994. Enhancing the status of population biology. *Trends Ecol. Evol.* 9:157.

EHRLICH, P.R. and E.O. WILSON. 1991. Biodiversity studies: science and policy. *Science* 253:758–762.

EISWERTH, M.E. and J.C. HANEY. 1992. Allocating conservation expenditures across habitats: accounting for inter-species genetic distinctiveness. *Ecol. Econ.* 5:235–249.

ELDREDGE, N. (ed.). 1992. *Systematics, Ecology and the Biodiversity Crisis*. Columbia University Press, New York.

ERWIN, T.L. 1991. An evolutionary basis for conservation strategies. *Science* 253:750–752.

FAITH, F.P. 1992. Conservation evaluation and phylogenetic diversity. *Biol. Conserv.* 61:1–10.

GRANT, P.R. and B.R. GRANT. 1992. Hybridization and bird species. *Science* 256:193–197.

JIMÉNEZ, J.A., K.A. HUGHES, G. ALAKS, L. GRAHAM, and R.C. LACY. 1994. An experimental study of inbreeding depression in a natural habitat. *Science* 266:271–273.

KELLER, L.F., P. ARCESE, J.N.M. SMITH, W.M. HOCHACHKA, and S.C. STEARNS. 1994. Selection against inbred song sparrows during a natural population bottleneck. *Nature* 372:356–357.

LACY, R.C. 1992. The effects of inbreeding on isolated populations: are minimum viable population sizes predictable? In *Conservation Biology*, eds. P.L. Fiedler and S.K. Jain, pp. 277–296. Chapman & Hall, New York.

LANDE, R. 1988. Genetics and demography in biological conservation. *Science* 241:1455–1460.

MACILWAIN, C. 1994. Global effort is launched to create taxonomic map of living organisms. *Nature* 368:3.

MAY, R.M. 1990. Taxonomy as destiny. *Nature* 347:129–130.

MOROWITZ, H.J. 1991. Balancing species preservation and economic considerations. *Science* 253:752–754.

NATIONAL RESEARCH COUNCIL. 1993. *A Bioiogical Survey for the Nation*. National Academy Press, Washington, DC.

NEEL, J.V. 1994. *Physician to the Gene Pool: Genetic Lessons and Other Stories*. Wiley, New York.

O'BRIEN, S.J. and E. MAYR. 1991. Bureaucratic mischief: recognizing endangered species and subspecies. *Science* 251:1187–1188.

PETERSON, R.W. 1988. An Earth ethic: our choices. In: *Earth '88: Changing Geographic Perspectives*, ed. H.J. de Blij, pp. 360–377. National Geographic Society, Washington, DC.

PRIMACK, R.B. 1993. *Essentials of Conservation Biology*. Sinauer, Sunderland, MA.

RENNER, S.S. and R.E. RICKLEFS. 1994. Systematics and biodiversity. *Trends Ecol. Evol.* 9:78.

SCHONEWALD-COX, C.M., S.M. CHAMBERS, B. MACBRYDE, and L. THOMAS (eds.). 1983. *Genetics and Conservation*. Benjamin/Cummings, Menlo Park, CA.

THOMPSON, J.N. 1994. *The Coevolutionary Process*. University of Chicago Press, Chicago, IL.

TILMAN, D., R.M. MAY, C.L. LEHMAN, and M.A. NOWAK. 1994. Habitat destruction and the extinction debt. *Nature* 371:65–66.

VANE-WRIGHT, R.I., C.J. HUMPHRIES, and P.H. WILLIAMS. 1991. What to protect—systematics and the agony of choice. *Biol. Conserv.* 55:235–254.

WILSON, E.O. 1984. *Biophelia*. Harvard University Press, Cambridge, MA.

WILSON, E.O. 1988. The current state of biological diversity. In *Biodiversity*, ed. E.O. Wilson, pp. 3–18. National Academy Press, Washington, DC.

2

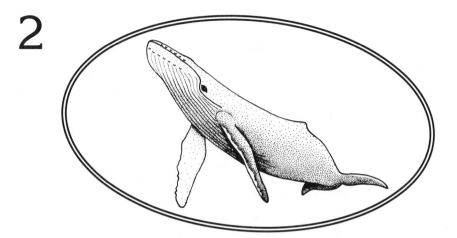

POPULATION STRUCTURE, MOLECULAR SYSTEMATICS, AND FORENSIC IDENTIFICATION OF WHALES AND DOLPHINS

C. Scott Baker and Stephen R. Palumbi

INTRODUCTION

Mysticete whales have been systematically hunted to near extinction by commercial whaling nations. Using only sail power and handheld harpoons, "Yankee" whaling ships severely depleted Northern right, bowhead, and gray whale populations before the end of the 19th century (Townsend, 1935). The invention of the exploding harpoon and the advent of steam-powered vessels at the beginning of the 20th century allowed modern commercial-whaling nations to expand their harvest to the swifter rorquals or fin whales found throughout the oceans of the world. In the Southern Hemisphere alone, modern whalers killed over one million animals between 1920 and 1986 (Allen, 1980; Evans, 1987), driving first the humpback and then the blue, fin, and sei whales to their current depleted status (Box 2.1).

The legacy of commercial whaling has provided us with a natural experiment in conservation biology on a grand, if tragic, scale. At present, how-

Box 2.1. Taxonomic and Conservation Status of Baleen Whales

The order Cetacea is divided into two suborders: the Odontoceti (toothed-whales, dolphins and porpoises), and the Mysticeti (baleen whales). The Mysticeti, the primary subject of this report, are generally considered to include the following species, although recent reports (see text) suggest that the taxonomy may be in need of revision:

1. Family Balaenidae
   bowhead, *Balaena mysticetus*
   northern right, *Eubalaena glacialis*
   southern right, *Eubalaena australis*
2. Family Neobalaenidae
   pygmy right, *Capera marginata*
3. Family Eschrichtiidae
   gray whale, *Eschrichtius robustus*
4. Family Balaenopteridae
   minke, *Balaenoptera acutorostrata*, (*B. bonaerensis*)
   sei, *B. borealis*

Bryde's, *B. edeni*
fin, *B. physalus*
blue, *B. musculus*
humpback, *Megaptera novaeangliae*

All of these species are protected in U.S. waters by the Marine Mammal Protection Act of 1972 and the Endangered Species Act of 1973. (The National Marine Fisheries Service has recently removed the California gray whale from the list of endangered species.) All species of mysticete whales are also considered in Appendix I of the Convention on International Trade in Endangered Species, and (with the exception of the gray whale) are listed as threatened in the 1994 IUCN Red List (World Conservation Union, 1993). The Red List also includes a total of more than 60 species of Odontoceti.

---

ever, the genetic consequences of exploitation in mysticetes as a group and the potentially mitigating differences in the demography and population structure of each species remain largely undescribed. With few exceptions (e.g., Arnason, 1974; Wada, 1983; Arnason et al., 1985), the genetics of Cetacea received little attention until recently. With the advent of new molecular techniques and the ease of sample collection by biopsy darting (Box 2.2), however, interest in cetacean populations has exploded (Hoelzel, 1991b, 1994; see also Table 2.1).

Because most whales have global distributions but are divided into oceanic and regional populations, a complete picture of their genetic structure requires collections from throughout the world's oceans. To date, the most comprehensive study of geographic variation involves mitochondrial (mt) DNA in the humpback whale (e.g., Baker et al., 1990; 1993b; 1994a). This once abundant species was severely depleted by commercial hunting (Box 2.3), but small populations remain in the North Atlantic, North Pacific, and Southern Oceans (Figure 2.1). Results from studies of humpback whales are of interest in their own right, and may provide broader

Box 2.2. Biopsy Darting

Until recently, it was possible to conduct genetic studies of cetaceans only by collecting organ samples from carcasses of commercial catches, or from beachcast or stranded specimens. As commercial whaling declined during the 1970s and ended officially by moratorium in 1986, tissue samples from species of large whales became even more difficult to obtain for most scientists. After about 1986, however, advances in molecular techniques revolutionized approaches to the study of genetic variation and structure in wild populations by allowing the use of small samples of almost any tissue (e.g., Vigilant et al., 1989). Collecting even small tissue samples from whales, however, remained problematic until the refinement of the biopsy dart (Winn et al., 1973; Arnason et al., 1985; Mathews, 1986; Lambertsen, 1987; Amos and Hoelzel, 1990; Baker et al., 1990; Lambertsen et al., 1994).

The biopsy dart is a small cylindrical punch that can be shot from a crossbow, longbow, or modified capture gun. A metal flange prevents deep penetration into the skin and blubber and provides recoil to dislodge the dart. The dart may be tethered for retrieval or may be free floating. Thorough behavioral studies have shown that biopsy darting results in little or no observable change in the behavior of whales in most cases (Brown et al., 1991; Weinrich et al., 1991, 1992; Lambertsen et al., 1994). With this non-lethal approach, it is now possible to conduct a full range of genetic analyses on any cetacean species, from identification of sex (Baker et al., 1991) to sequence analyses of mitochondrial (Baker et al., 1993b) and nuclear DNA (Palumbi and Baker, 1994).

Note: Ironically, despite the advances in molecular genetic techniques and biopsy sampling, lethal takes under scientific whaling permits are still demanded by some nations because of a reported need to collect organs for stock identification by traditional biochemical analyses such as allozymes (Anonymous, 1995).

---

insights into global gene flow in other organisms with high dispersal abilities, large-scale distributions, and complex social behaviors. Studies of allozyme and mtDNA variation have been performed for other species of whales and dolphins as well (see Hoelzel, 1991b, for review), and these provide a comparative backdrop for generalizing the results from humpback whales.

Due to its maternal mode of inheritance and absence of recombination (Brown, 1983; Wilson et al., 1985; Avise et al., 1987; Moritz et al., 1987), mtDNA has been a favored genetic system for analysis of population structure. In general, mtDNA offers two important distinctions from nuclear genetic markers. First, the phylogenetic relationship of mtDNA types reflects the history of maternal lineages within a population or species. Second, all else being equal, the effective population size of mtDNA genomes is one fourth that of autosomal nuclear genes, leading to a higher rate of local

Table 2.1. Molecular Population Genetic Studies of Whales and Dolphins

Species	Method	Populations compared[1]	Reference
Baleen whales			
Fin	Allozymes	Iceland ≠ Spain (between-year variation in Iceland)	Danielsdottir et al., 1991
Fin	Allozymes	Antarctic ≠ N. Pacific ≠ Spanish coast	Wada and Numachi, 1991
Fin	MHC	Iceland = Spain (low variability)	Trowsdale et al., 1989
Minke	mtDNA	Southern Ocean (Antarctic IV = V) ≠ N. Pacific ≠ N. Atlantic	Hoelzel and Dover, 1991c
Minke	Fingerprints	N. Atlantic ≠ N. Pacific	van Pijlen et al., 1991
Minke	Allozymes	West Greenland ≠ Iceland ≠ Norway	Danielsdottir et al., 1992
Minke	Allozymes	Antarctic = Brazil ≠ Pacific coast of Japan ≠ Sea of Japan, Korean coast	Wada and Numachi, 1991
Sei	Allozymes	Antarctic ≠ N. Pacific	Wada and Numachi, 1991
Bryde's	Allozymes	Madagascar ≠ Indonesia ≠ Fiji ≠ Peru ≠ N. Pacific	Wada and Numachi, 1991
Humpback whale	mtDNA	N. Atlantic (Maine = Newfoundland, possible multiple stocks on wintering grounds) ≠ N. Pacific (Hawaii = Alaska ≠ California ≠ Mexico, heterogeneity among Mexican regions) ≠ Southern Ocean (Antarctic IV ≠ Antarctic V)	Baker et al., 1993b; 1990; 1994a; this paper; Medrano-G., 1993
Humpback whale	Fingerprints	N. Atlantic ≠ N. Pacific (Hawaii = Alaska = California)	Baker et al., 1993a
Right whale	mtDNA	N. Atlantic (calving grounds different and genetic variability low) ≠ S. Atlantic	Schaeff et al., 1991
Toothed whales, dolphins, and porpoises			
Spinner dolphin	mtDNA	E. tropical Pacific (morphotypes not different) ≠ Timor Sea, Australia	Dizon et al., 1991
Bottlenose dolphin	Allozymes mtDNA	Florida (Tampa Bay ≠ Sarasota Bay, 20 km distant, prelim.)	Duffield and Wells, 1991
Bottlenose dolphin	mtDNA	N. Atlantic ≠ Gulf of Mexico ≠ Gulf of California	Dowling and Brown, 1993

Table 2.1. *(cont.)*

Species	Method	Populations compared[1]	Reference
Dall's porpoise	Allozymes	N. Pacific = Bering Sea	Winans and Jones, 1988
Pilot whale	Allozymes	Faeroe Islands (9 populations identical)	Anderson, 1988
Pilot whale	Fingerprints Microsatellites	Faeroe Islands (pods different)	Amos et al., 1991; Amos et al., 1993
Pilot whale	Allozymes	Pacific Coast of Japan (2 populations identical)	Wada, 1988
Harbor porpoise	mtDNA	N. Pacific (California = Br. Columbia) ≠ N. Atlantic (Newfoundland = Maine) ≠ Black Sea	Rosel, 1992; Rosel et al., 1995; Wang, 1993
Common dolphins	mtDNA	Species level differences between sympatric morphotypes	Rosel et al., 1994
Killer whale	Fingerprints mtDNA	N. Pacific (sympatric pods different) ≠ S. Atlantic ≠ N. Atlantic	Hoelzel and Dover, 1991b; Stevens et al., 1989.

[1] ≠ indicates that the populations compared differed significantly at the genetic character investigated.

differentiation by random drift (Birky et al., 1983; Neigel and Avise, 1986). The ability to detect local differentiation may also be enhanced by the rapid pace of mtDNA evolution, which is generally considered to be five to ten times higher than nuclear DNA in most species of mammals (Brown, 1983; Wilson et al., 1985); however, the rate of mtDNA evolution may be lower for some cetaceans (Hoelzel and Dover, 1991a; Baker et al., 1993b; Martin and Palumbi, 1993), as it is for turtles (Avise et al., 1992).

Descriptions of the distribution of maternal lineages may be particularly insightful in mammalian species such as whales and dolphins in which social organization is often structured around matrilines and dispersal is often sex-biased. In humpback whales, for example, individuals undertake seasonal migrations between summer feeding grounds and winter breeding and calving grounds, and field observations of migrating individuals suggest that fidelity to a specific migratory destination may be maternally directed (Martin et al., 1984; Baker and Herman, 1987; Clapham and Mayo, 1987). In the so-called "resident" killer whales (*Orcinus orca*) of the Northwest Pacific, individuals appear to remain for life with their natal group and spend much of the year in predictable home ranges (Bigg et al., 1990). Coastal bottlenose dolphins (*Tursiops truncatus*) of the Gulf of Mexico (e.g., Duffield and Wells, 1991; Wells, 1991) and Hector's dolphins (*Cephalorhynchus hectori*) of New Zealand (Slooten and Dawson, 1994) also occupy restricted home ranges,

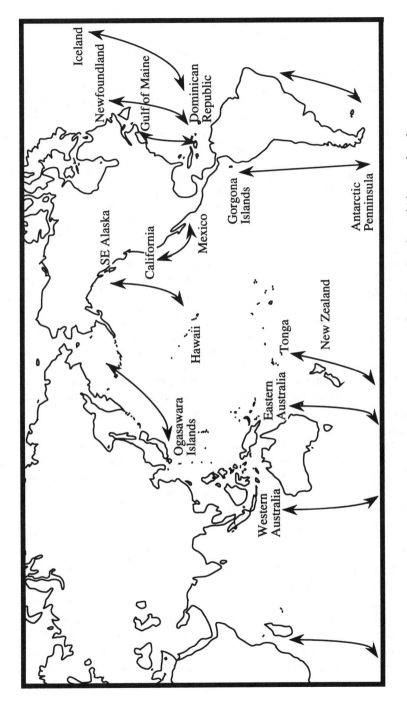

Figure 2.1. Inferred primary migratory connections between regions from which examples of humpback whales were collected for genetic analysis.

Box 2.3. Abundance and Exploitation of Humpback Whales

Although estimates are uncertain, the global population of humpback whales prior to human exploitation probably numbered well over 120,000 individuals, with perhaps 80% found in the oceans of the Southern Hemisphere. Following commercial hunting, which began as early as 1611 (Stone et al., 1987) and greatly intensified during the 20th century, the global population may have declined by more than 95%, and several "stocks" were driven close to extinction. In eastern Australia, for example, a preexploitation stock of 10,000 individuals was reduced by commercial whaling to a low of about 200–500 individuals (Chittleborough, 1965). Following international protection from commercial hunting for this species in 1966, some stocks (e.g., in the western North Atlantic) recovered reasonably well, whereas others (e.g., in the Southern Ocean and the North Pacific) are still represented by only a small fraction of their former abundance (Baker and Herman, 1987; Paterson and Paterson, 1989; Bryden et al., 1990; Best, 1993).

Recent revelations of massive unreported Soviet catches of humpback whales offer some insight into a possible source of the observed differences in population recovery and existing genetic variation (Yablokov, 1994). The total number of unreported catches from four Soviet factory ships operating in the Southern Hemisphere is now thought to have exceeded 45,000 humpback whales following World War II (Anonymous, 1995). This is nearly three times more than the combined reported catches from all nations for this period. Knowledge of the historical demography of this species must be considered incomplete until the details of pirate whaling are fully known.

although their social organizations are more fluid than that of the killer whale. In pelagic species such as pilot whales (*Globicephala melas*), geographic ranges may be large but genetic structure may also be maintained by maternal fidelity to social groups (Amos et al., 1993). These life-history traits provide a behavioral mechanism for the cultural transmission of migratory destinations, or for the inheritance of local home ranges and the possible formation of population subdivisions.

POPULATION AND STOCK STRUCTURE IN HUMPBACK WHALES

Continental landmasses currently separate humpback whales in the Northern Hemisphere into two oceanic populations, the North Atlantic and the North Pacific (Mackintosh, 1965), whereas humpback whales are distributed more or less continuously throughout the oceans of the Southern Hemisphere. Within each oceanic basin, humpback whales undertake annual migrations, averaging 10,000 kilometers roundtrip, between summer feeding grounds in high latitudes and winter breeding and calving grounds in low latitude waters (Figure 2.1). Effective interchange between Northern and Southern Hemisphere populations is limited by the seasonal opposition of this migratory cycle. It has long been assumed that the three major oceanic populations are reproductively isolated from each other, although regions of potential migratory overlap near the equator were noted by Kellogg (1929).

Within each major oceanic population, discontinuous patterns of seasonal distribution and observations of migratory movement by marked individuals suggest that humpback whales form relatively discrete subpopulations that are not separated by obvious geographic barriers (e.g., Kellogg, 1929; Mackintosh, 1965). As elsewhere, the definition of these *population units* or *stocks* is controversial for humpback whales and other cetaceans (Box 2.4). Here, in keeping with its traditional definition, we will use the term *stock* to refer to a subpopulation, defined by geography or demographic evidence, that is assumed to remain more or less isolated from other subpopulations. Our approach has been to test these demographic or geographic stock divisions for evidence of genetic divisions by comparing various combinations of regional or seasonal distributions of mtDNA haplotypes as defined by control-region sequences or restriction fragment length polymorphisms (RFLPs). Within a hypothetical stock, haplotype frequencies from regional or seasonal habitats were also compared, where available, for evidence of non-random assortment or geographic structure. In general, the terms *feeding ground*, *feeding herd*, or *feeding aggregation* are used synonymously to refer to population divisions observed primarily during summer months in high latitude waters; the term *wintering ground* refers to divisions observed in low latitude waters. The term *migratory corridor* refers to regions, such as the eastern and western coasts of Australia, that are used primarily for migration between wintering and summering grounds.

The North Pacific

Based on his examination of logbooks from 19th-century whalers, Kellogg (1929) suggested that humpback whales in the North Pacific were divided into two stocks: (1) an Asian stock that winters in tropical waters south of Japan and travels north to feeding areas in the Sea of Okhotsk and along the Kamchatka Peninsula; and (2) an American stock thought to breed off the western coast of Mexico and to travel northward along the coast of North America to feeding grounds in the Gulf of Alaska, the Bering Sea, and near the Aleutian Islands (see Figure 2.1; further discussion to follow). Although Rice (1974) suggested that whales from the Hawaiian wintering

Box 2.4. Stock Concepts in Cetacean Biology

The concept of *stocks*, or population subdivisions, has a long history of controversy in the management of fisheries and marine mammals (e.g., Chapman, 1974; Donovan, 1991; Dizon et al., 1992). Nevertheless, the assumption of biologically significant subdivisions within oceanic populations has been, and continues to be, fundamental to management schemes of exploited (Donovan, 1991) and protected (National Marine Fisheries Service, 1991) species. For exploited species, an understanding of stock boundaries is critical for estimating abundance, setting catch limits, and interpreting catch statistics and life history parameters. For protected species, stock boundaries are important for assessing population changes, establishing territorial jurisdiction, delineating critical habitats, and verifying catch or trade records.

The question of stock divisions in baleen whales is likely to become more acute now that the IWC has adopted the Revised Management Procedure, a step widely viewed as leading to a return of commercial whaling. Under the conditions of this procedure, quotas for individual stocks are set by a Catch Limit Algorithm (CLA) that uses historical catch information and current estimates of abundance. Although the CLA is generally considered conservative and robust to errors in catch histories and estimates of abundance, it assumes that this information represents a single stock of whales. Uncertainty about stock structure is taken into account through a series of computer simulations, referred to as implementation trials, in which

various plausible scenarios of stock structure are considered. In general, simulations that assume a number of small or localized stocks result in small allowable catch quotas (unless estimates of abundance are very accurate, and they seldom are), whereas simulations that assume a few widely distributed stocks provide larger catch quotas. In simulations in which a number of localized stocks are assumed and the estimates of abundance are variable, an implementation trial can provide catch quotas of zero. Consequently, the criteria for designating the range of "plausible" stock structures and the genetic or demographic evidence to reject or support a given scenario are highly contentious.

Donovan (1991) provides a working definition of a stock as "a relatively homogeneous and self contained population whose losses by emigration and accessions by immigration, if any, are negligible in relationship to the rates of growth and mortality." Given this demographic definition, it is clear that significant genetic differences between population subunits should be considered strong evidence for stocks, and that these population subdivisions should be afforded independent management status.

On the other hand, the absence of significant genetic differences (nuclear or mtDNA) should not be considered conclusive proof of demographic homogeneity (Palumbi et al., 1991; Dizon et al., 1992). Indirect genetic estimates of interchange represent an effective rate averaged over many hundreds of generations. Such estimates are limited by the

Box 2.4. *(cont.)*

rates of mutational change in the chosen genetic marker, and the resolution of the analytic technique. For new mutations revealed by conventional restriction fragment length analyses of mtDNA, this minimum resolution is likely to be on the order of 50,000 years or more (Wilson et al., 1985). For existing polymorphisms, a gene flow parameter (Nm) can be estimated from interlocality variances in allelic frequencies or from frequencies of private alleles (see Chapter 9), and represents the product of population size (N) and the proportion of individuals exchanged between populations (m). Thus, long-term rates of migratory exchange as low as a few females per generation are likely to maintain relatively homogeneous frequencies of mtDNA haplotypes among populations at equilibrium, regardless of population sizes (Birky et al., 1983; Slatkin, 1987). In very small populations, such migratory exchange would be demographically significant. However, among local populations of even moderate size (e.g., 1,000 individuals each), migratory rates an order of magnitude greater than this would be negligible in relation to annual birth- and mortality rates (which are thought

to be approximately 4% for most baleen whale populations). Thus, different stocks as defined by demographic criteria could remain genetically homogeneous indefinitely.

Finally, we should note that the stock concept has generally involved the assumption of extrinsic reproductive isolation. Although analysis of mtDNA variation may prove to be more relevant for many management decisions because of its greater sensitivity to demographic processes, it cannot provide conclusive evidence of reproductive isolation. Further investigation of nuclear DNA markers is necessary to address questions concerning the mating system of cetaceans and the possibility of reproductive isolation between stocks and oceanic populations. As the description of nuclear and mitochondrial DNA variation in the humpback whale has shown, different genetic structures can coexist within a single species (Palumbi and Baker, 1994). Other cetaceans, with their variable and sometimes complex social organization and ability to undertake long-distance migration, are likely to have evolved similarly complex population genetic structures.

grounds are part of an extended American stock, Kellogg (1929) did not consider this group and may have been unaware of its existence (Herman, 1979).

Observations of naturally marked individuals demonstrate that individuals from Alaskan feeding grounds migrate primarily to wintering grounds around the windward islands of Hawaii (Darling and Jurasz, 1983; Darling and McSweeney, 1985; Baker et al., 1986). In the eastern North Pacific, individuals from the central California feeding grounds are thought to winter primarily in the coastal waters of Mexico, including the southeastern coast of Baja California, coastal mainland Mexico, and the nearshore Islas Tres

Marias and Isabel (Urban and Aguayo, 1987; Calambokidis et al., 1990). Although the majority of data on migratory movement of naturally marked individuals suggests a demographic division between the central and eastern component of the North Pacific population (Perry et al., 1990), important exceptions have been noted. A few whales from Alaska have been observed in Mexico, and one whale from central California was observed in Hawaii (Baker et al., 1986). Infrequent movement by individual whales between the Hawaiian and Mexican wintering grounds and between Hawaii and the Ogasawara Islands of Japan has also been documented (Darling and McSweeney, 1985; Baker et al., 1986; Darling and Mori, 1993). Finally, the winter "song" of the humpback whale, a presumed male mating display, changes from year to year but remains similar on the Hawaiian and Mexican wintering grounds (Payne and Guinee, 1983), indicating some acoustical contact between these groups.

To evaluate the genetic significance of these population subdivisions, differences in mtDNA haplotype frequencies were tested among seasonal habitats and between the central and eastern stocks of the North Pacific population, using samples from 95 humpback whales (Baker et al., 1994a); western North Pacific populations were not examined. MtDNA haplotype frequencies differed among the four regional samples from the central and eastern North Pacific, showing a fixed distinction between the southeastern Alaska and central California feeding grounds and a strong connection between southeastern Alaska and the Hawaiian wintering grounds (Figure 2.2A). Surprisingly, the coastal Mexican wintering ground is dominated by haplotypes common to both feeding areas. Differences between the Hawaiian and Mexican wintering grounds were significant but considerably less marked than differences between feeding grounds.

Differences between the central and eastern component of the North Pacific population were tested by comparing combined samples from Hawaii and Alaska to those from central California and Mexico (Figure 2.2). Despite the intermingling of haplotypes in Mexico, this test supports the division of the North Pacific into a central stock that feeds in Alaskan waters and winters predominantly in Hawaii, and an eastern or "American" stock that migrates between feeding grounds along the coast of California and wintering grounds along the coast of Mexico. A more detailed study of mtDNA sequence variation in humpbacks from Mexican wintering grounds supports this conclusion and suggests a further (though weaker) segregation between the offshore Revilla Gigedo Islands and the coastal mainland (Medrano-G., 1993).

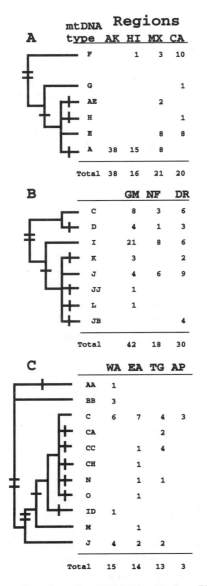

Figure 2.2. Genealogy and numbers of mtDNA haplotypes from RFLP analysis in samples of humpback whales from: (A) central California (CA), Mexico (MX), Hawaii (HI), and southeastern Alaska (AK), all within the central and eastern North Pacific; (B) southern Gulf of Maine (GM), Newfoundland (NF), and Dominican Republic (DR), all within the western North Atlantic; and (C) Western Australia (WA), Eastern Australia (EA), Tonga (TG), and Antarctic Peninsula (AP), all within the Southern Ocean. Bars along the tree indicate inferred restriction site changes (adapted from Baker et al., 1994a).

The North Atlantic

Humpback whales in the western North Atlantic congregate each winter to give birth and presumably to breed in the shallow waters over submerged banks and along the coastlines of the West Indies (Whitehead, 1982; Martin et al., 1984; Mattila et al., 1989; Katona and Beard, 1990). During summer months, individual whales predictably return to only one of four geographically distinct coastal feeding grounds (Katona and Beard, 1990): Iceland, western Greenland, Newfoundland (including the coast of Labrador), and the southern Gulf of Maine. Movement between feeding grounds is restricted and fidelity to a particular feeding ground is high. An eastern North Atlantic stock of humpback whales is thought to migrate from feeding grounds along the coast of northern Europe (including Norway) to winter grounds near the Cape Verde Islands, but little is known of the current status of this group (Bannister et al., 1984).

We tested differences in mtDNA haplotype frequencies between two western North Atlantic feeding grounds and between feeding and wintering grounds using samples from 90 humpback whales (Figure 2.2B). No significant differences were found between haplotype frequencies in the Gulf of Maine and Newfoundland, although there was a surprising degree of heterogeneity between two distinct sample sets collected within the Gulf of Maine (Box 2.5). We then pooled samples from the two feeding grounds for comparison to samples from the Dominican Republic in the West Indies wintering ground. In this comparison, significant heterogeneity suggests that some component of the wintering population is not represented at the two feeding grounds sampled. Because the Caribbean wintering congregation includes individuals from all known feeding grounds (Mattila et al., 1989; Katona and Beard, 1990), the observed heterogeneity may reflect differences between the westerly continental feeding grounds sampled here, and insular eastern feeding grounds (e.g., Greenland, Iceland, and Norway).

The Southern Ocean

Humpback whales in the Southern Hemisphere are thought to form five or six distinct feeding aggregations, referred to as Groups I–VI, distributed discontinuously around the Antarctic continent during the austral summer feeding season (Mackintosh, 1965; Gambell, 1976). The longitudinal boundaries of these feeding grounds are largely coincidental with the division of Antarctic waters into six broad zones which, based on the summer distributions of blue and humpback whales, functioned as political units in the apportionment of catch quotas for commercial whaling in the Antarctic (Tonnessen and Johnsen, 1982). Whales from these feeding aggregations

Box 2.5. Group Mortality

The demographic consequences of mass mortality among marine mammals are known to be severe. Population losses of more than 50%, attributable to parasites, infectious disease, or toxic algal blooms, have been reported in recent years (Harwood and Hall, 1990). The genetic impact of these events, however, has received little attention. A local group mortality of 14 humpback whales in Cape Cod Bay, Massachusetts, during the late fall of 1987 and early winter of 1988, offered a unique opportunity to investigate this problem (Baker et al., 1994b). Postmortem analysis indicated that the whales were feeding on Atlantic mackerel containing elevated levels of saxotoxin, a dinoflagellate neurotoxin responsible for paralytic shellfish poisoning in humans (Geraci et al., 1989). Although the number of deaths was small in comparison to recent widespread mortalities of seals and dolphins or to the mass strandings common among pilot whales or sperm whales, group mortalities of any size are extremely uncommon among baleen whales.

Our analysis of the mtDNA from the dead whales indicated that these individuals were a non-random sample of the larger Gulf of Maine subpopulation. Two maternal lineages represented 80% of the 10 dead whales available for analysis but only 12.5% of the reference population of 32 individuals sampled by biopsy darting the following year.

The circumstances of the group mortality offered several possible explanations for the observed association between mtDNA haplotypes and susceptibility to this environmental perturbation. The deaths occurred late in the feeding season, after the departure of most summer residents, suggesting that the poisoned whales were part of a socially or ecologically distinct subgroup within the larger regional subpopulation. A second possibility is that the whales were feeding independently on the mackerel—an unusual prey for humpback whales in this region (Payne et al., 1990)—but each may have learned this prey preference or idiosyncratic feeding strategy from related individuals. In general, if ecological specialization is culturally inherited from mother to offspring, a correlation between mtDNA haplotypes and feeding preferences, habitat utilization, or site fidelity may develop. Regardless of the specific processes underlying the observed pattern of non-random mortality, these results point to a surprising degree of fine-scale genetic differentiation among whale populations.

migrate to discrete wintering grounds along continental and insular coastlines or over shallow submerged banks in tropical latitudes. Unlike Northern Hemisphere stocks that may represent congregations of seasonally distinct subpopulation units, Southern Hemisphere stocks are thought to remain segregated year-round, migrating between wintering and summering grounds as individual units.

Arguably, the two most extensively studied stocks of humpback whales in the world are those of Antarctic Groups IV and V (Chittleborough, 1965; Dawbin, 1966). Group IV humpback whales migrate along the western coast of Australia to wintering grounds off the northwest coast of Australia. Group V humpbacks segregate along two major corridors during migration to wintering areas in tropical latitudes. The eastern component migrates along the coastline of New Zealand and is presumed to winter primarily near islands in the southwest Pacific, including the Tongan archipelago (Townsend, 1935; Dawbin, 1966). The western component of the Group V stock migrates along the eastern coast of Australia and is thought to winter in coastal waters inshore of the Great Barrier Reef (Figure 2.1; Paterson and Paterson, 1989).

We tested differences in haplotype frequencies between the eastern and western migratory components of Group V and between Groups IV and V, using samples from 45 individuals collected in the Southern Ocean (Figure 2.2C). The potential for genetic divisions or heterogeneity within Group V (i.e., between eastern Australia and Tonga) was tested first because of the historic inclusion of these two geographically distant regions as a single stock. No significant genetic differentiation was found between these two regions. Samples were then pooled and compared to those from Western Australia (Group IV). Here, the historic descriptions of stock differentiation between Groups IV and V (Chittleborough, 1965; Mackintosh, 1965; Dawbin, 1966) were supported by the distribution of mtDNA haplotypes of humpback whales migrating past Western Australia (Group IV), eastern Australia (Group V, western component) and the Tongan archipelago (Group V, eastern component).

Levels of mtDNA Variability

Attempts to reconstruct the historical demography of endangered species from genetic data are generally complicated by the absence of independent estimates of variation prior to periods of exploitation or population bottlenecks (Menotti-Raymond and O'Brien, 1993). The problem is to find an appropriate system for comparison of current variation in order to judge the extent of genetic variation that may have been lost. Although the amplification and sequencing of DNA from museum specimens collected prior to exploitation or population decline offers a powerful direct approach to this problem, adequate samples, particularly for regional populations, are not always available.

In the absence of pre-exploitation samples, we tested the possibility that commercial hunting may have reduced genetic variation in intensively exploited species such as the humpback whale by using the variability of the worldwide population as a baseline for comparison to oceanic or regional populations (Baker et al., 1993b). A worldwide phylogeny of the humpback whale mtDNA sequences (Figure 2.3) confirmed that oceanic populations of this species are independent demographic units with long-term migration rates of less than one female per generation. The mtDNA phylogeny supported the division of maternal lineages into three major clades, referred to as "AE", "CD" and "IJ" to indicate their relationship to haplotypes previously defined by RFLP analysis (see Baker et al., 1990). We found that each oceanic population is dominated, in terms of frequency and diversity, by a different mtDNA clade, although two clades have transoceanic distributions.

Average variation in humpback whale mtDNA control-region sequences was high in both the North Atlantic and Southern Oceans relative to worldwide variation, with each oceanic population retaining 82% and 76%, respectively, of this total (Baker et al., 1993b). High levels of genetic variation were also observed within some stocks or regional habitats relative to their respective oceanic populations. Genetic evidence consistent with a population bottleneck was found only in the central North Pacific (i.e., Hawaii and southeastern Alaska), where mtDNA nucleotide diversity was zero. This low genetic variation may reflect the results of exploitation during two periods of intense hunting in this century (Rice, 1978), the illegal hunt of the Soviet fleet (Yablokov, 1994), or a recent colonization of the Hawaiian wintering ground by a limited number of female humpback whales (Baker et al., 1993b).

Conservation Implications

Given the tremendous mobility of humpback whales and the apparent absence of geographic barriers within oceanic basins, the formation of significant genetic divisions between oceanic stocks suggests strong individual-female fidelity to migratory destinations. Unlike the green turtle (Chapter 7), however, the ability of individual whales to visit alternate wintering grounds and, on occasion, to move between feeding grounds (e.g., Baker et al., 1986) argues against a strict behavioral imprinting underlying this fidelity. Instead, the life history of humpback whales suggests that migratory fidelity develops when a calf first accompanies its mother on her annual migration (Martin et al., 1984; Baker et al., 1987; Clapham and Mayo, 1987).

In the central and eastern North Pacific, where a complete hierarchical analysis of feeding and wintering grounds was possible, segregation of haplotypes was strongest between the two feeding grounds. The two wintering grounds, where calves are presumed to be born, showed genetic evidence of migratory interchange. These results suggest that populations on feeding grounds are segregated more "cleanly" than stocks on wintering grounds. A

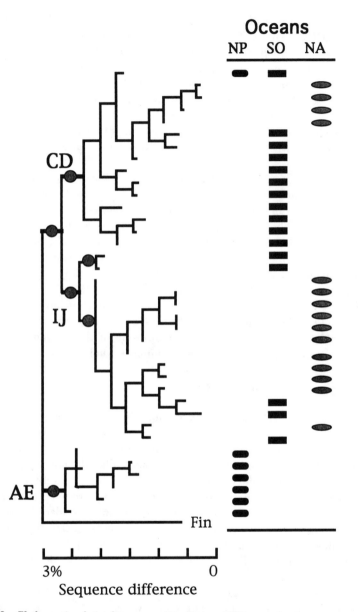

Figure 2.3. Phylogenetic relationships among 37 unique mtDNA control-region sequences from 90 humpback whales from the North Pacific (NP), North Atlantic (NA), and Southern Ocean (SO), rooted by the outgroup method with the fin whale. Shaded circles indicate 85–100% agreement in consensus trees from 500 heuristic searches using randomized entry order (adapted from Baker et al., 1993b). Major clades, as described in text, are indicated by CD, IJ, and AE.

considerably different conclusion about the biological significance of genetic divisions between stocks might have been reached if samples had been collected on the feeding or breeding grounds alone.

The regional segregation of distinct maternal lineages during migration is a strong argument for independent regional management or protection of humpback whales. Were the southeastern Alaska feeding "herd" to be driven to extinction by hunting or habitat destruction, for example, replacement by immigration from California would probably require hundreds of generations. Although it could be argued that the genetic diversity of the oceanic population might suffer only a small loss if a regional lineage was exterminated, this offers little consolation for regional managers and their constituents.

How have most populations of humpback whales survived near-extinction without major losses of mtDNA variation? Models of mtDNA lineage extinction suggest that the decay of genetic variation is slowed if the duration of a population bottleneck is short and the rate of population increase high following the bottleneck (Nei et al., 1975; Neigel and Avise, 1986). Intensive commercial hunting of humpback whales has been recent and relatively short-lived, especially in the North Pacific and Southern Oceans, and some humpback subpopulations recently have increased in abundance (Box 2.3).

Genetic models also show that loss of genetic variation is slower in species with long generation times and age-structured populations (Nei et al., 1975). Humpback whales may reach sexual maturity at age 4–6 years (Chittleborough, 1965; Clapham and Mayo, 1990), suggesting an average generation time on the order of 5–10 years. Thus, there have been only three to six overlapping generations since humpback whales were at their lowest population size. Given the potential life span of humpback whales (Chittleborough, 1959), some individuals that survived commercial hunting during the early 1960s are probably alive today. As Chittleborough (1965) predicted shortly after the collapse of the Southern Ocean stocks of humpback whales, full recovery is likely to take more than 50 years. Our analysis of samples collected halfway through this predicted period can be considered a baseline for measuring further changes in genetic variation as populations recover.

GENETIC PATTERNS IN OTHER WHALES AND DOLPHINS

Other Mysticete Species

Geographic or temporal patterns of mtDNA variation are being used to examine the question of population structure or stocks in several other species of whales, although published information is available for only the Southern Hemisphere minke and Northern right whales (e.g., Hoelzel, 1991b). Minke whales have received considerable attention because of interest in renewed hunting of this species. Preliminary examination of small samples from adjacent Southern Hemisphere locations failed to reveal significant genetic differences (Hoelzel and Dover, 1991c; Wada et al., 1991), but a recent temporal and geographic analysis of mtDNA haplotypes from 1,257 minke whales did suggest heterogeneity within the region, perhaps due to assortment of lineages during seasonal migration from the poorly described breeding grounds (Pastene et al., 1994b). More dramatic genetic differences have emerged from comparisons between oceanic populations of minke whales and between morphological variants within the Southern Hemisphere population (Hoelzel and Dover, 1991c; Wada et al., 1991; Pastene et al., 1994a). These results have implications for management of whale species and will be discussed in detail later.

The Northern right whale is among the most endangered of cetacean species, with an estimated population size of less than 350 individuals (Crone and Kraus, 1990; Schaeff et al., 1993). Preliminary evidence from mtDNA RFLP analysis suggests that the North Atlantic form (considered a taxonomic species; see Box 2.1) is genetically distinct from the Southern right whale. The analysis of mtDNA also suggests heterogeneity of habitat use within the North Atlantic. The Bay of Fundy is currently the only known nursery area for the North Atlantic population, but observations of migratory movement by naturally marked individuals suggest that not all known females visit this area. The frequency and temporal distribution of three mtDNA haplotypes support the hypothesis that females segregate in a second, as yet undescribed, nursery area (Schaeff et al., 1993). In contrast to this inferred site fidelity by mature females, males may move between the two nursery grounds. Given the low reproductive rates of right whales and the high incidental mortality due to human activity (i.e, collision with vessels), locating the hypothesized second nursery area is a management priority.

In addition to studies of population level differences using mtDNA, baleen whales have been examined using allozymes, and a few studies have employed major histocompatibility complex (MHC) polymorphisms or DNA fingerprinting (Table 2.1). These studies generally have uncovered genetic differences between conspecific populations in different oceans. However, outcomes within ocean basins have varied, with some studies showing genetic structure over scales of a few thousand kilometers and other studies of the same populations failing to identify significant structure. In some cases, these discrepancies are probably due to differences in resolving power of different assays. In other cases, contrasting genetic patterns from the same populations might also reflect different evolutionary patterns at different genetic loci. These patterns, in turn, provide insight into cetacean biology and behavior.

Recently, nuclear DNA variability in humpback whale populations was examined using sequences from an intron in a gene for the muscle and cytoskeletal protein actin (Palumbi and Baker, 1994). Universal polymerase chain reaction (PCR) primers that flank this intron were used to amplify a 1,500 base pair fragment from humpback whales sampled from California and Hawaii. In contrast to the pattern revealed by mtDNA, RFLP and sequence analysis of the actin intron showed that the Hawaiian population has the same diversity of intron variants as does the California population.

These differences in the distributions of allelic variants have two possible explanations. First, female movement between the California–Mexico and Alaska–Hawaii migratory routes may be far lower than that of males. Because mtDNA is maternally inherited, evidence of male-mediated gene flow between populations is lost from one generation to the next. By contrast, analysis of introns or other nuclear characters can, in principle, detect such gene flow. Male movement between Hawaii and Mexico is also suggested by the similarity of male vocalizations in these two localities (Payne and Guinee, 1983). A second explanation is that genetic differences between Hawaii and California have developed at the mitochondrial locus, but are not yet evident at nuclear loci, despite low gene flow between these populations at both types of genes. As a result of its maternal inheritance, ancestral mtDNA polymorphisms may drift to fixation more quickly, and *de novo* mutational differences arise more rapidly between populations than is true for most autosomal loci. In our example, the Hawaiian population may be so recent (Herman, 1979) that mtDNA variation, but not nuclear variation, has had time to diverge from the California population. Studies of more nuclear loci are required to test this possibility.

Odontocete Species

Small cetacean populations throughout the world are under direct threat from human activities including hunting, competition through fishing, incidental mortality during fishing, and modification of inshore habitat (Donoghue and Wheeler, 1990). The demographic impact of these threats, however, can only be assessed with information about the size and structure of the populations affected. Unlike most baleen whales, many small cetaceans are extremely localized in distribution and habitat use. Thus, they are more likely to suffer local extinction or depletion due to environmental or human induced perturbations.

Populations of the bottlenose dolphin along the northern Gulf of Mexico and the eastern Atlantic coastlines of the United States recently have experienced high levels of mortality, as evidenced by beachcasts or strandings (Box 2.5). Between July 1987 and March 1988, for example, documented deaths along the U.S. Atlantic coast included at least 750 dolphins—about 50% of the known coastal population (Harwood and Hall, 1990)! The proximate cause of these deaths remains controversial, but has been attributed to dinoflagellate toxins (Hansen, 1992). Despite this dramatic die-off, it is not possible to formally declare this regional population depleted or threatened under current U.S. law without evidence that it represents a distinct stock (Box 2.4). Genetic studies of bottlenose dolphins are now providing this required evidence of population structure. Clear differentiation of mtDNA haplotypes exists between populations in the Gulf of Mexico and the Atlantic coast (Dowling and Brown, 1993), and perhaps between even localized groups separated by less than 20 km (Duffield and Wells, 1991).

The harbor porpoise (*Phocoena phocoena*) is subject to a high level of mortality due to interactions with fisheries along the Pacific and Atlantic coasts of the United States. In an attempt to assess the relative impact of these interactions on local populations, Rosel (1992) examined geographic variation in the mtDNA control region of animals killed as bycatch in British Colombia, California, the North Atlantic, and the Black Sea. No haplotypes were shared between oceans and average sequence differences were substantial (5.5–5.8%). Similarly, Wang (1993) showed that there were fixed differences in haplotypes between harbor porpoises from the eastern North Pacific and western North Atlantic, and an average divergence of 2.4% for the whole mtDNA genome between these two oceanic populations. On the other hand, geographic samples within an ocean basin showed no clear genetic differences. Thus, haplotypes were in similar frequency between British Colombia and California populations (Rosel, 1992; Rosel et al., 1995), and also for harbor porpoises from Newfoundland, the Gulf of St. Lawrence, the Bay of Fundy, and the Gulf of Maine (Wang, 1993). Rosel (1992) offered two possible interpretations of the apparent lack of concordance between intraoceanic genetic and geographic differences in harbor porpoise: (1) Harbor porpoises in nearby locales do not conform to the definition of a fixed geographic stock but may be seasonally sympatric due to migratory movement (i.e., genetic structure is "hidden"); (2) Two historically distinct populations may have undergone secondary contact and admixture, perhaps during recent periods of glaciation.

The problem of "hidden" genetic structure in cetacean populations is highlighted by studies of mtDNA variation in two sympatric forms of killer whales found along the northwestern Pacific coasts of the United States and Canada (Stevens et al., 1989; Hoelzel, 1991a; Hoelzel and Dover, 1991b).

Observations of naturally marked individuals during the past 15–20 years have led to the characterization of two distinct ecotypes: the so-called "resident" pods that feed primarily on fish, and "transient" pods that regularly feed on marine mammals (Bigg et al., 1990). Based on complete sequences of the mtDNA control region from one individual representing each of two resident pods and one transient pod from the Puget Sound and Vancouver Island area, Hoelzel (1991a) found little difference between the residents but comparatively large differences between the resident and transient pods. Additional comparisons of mtDNA sequences and RFLPs from populations elsewhere in the world showed that the sympatric resident and transient pods were as similar as allopatric populations separated by large distances (Hoelzel and Dover, 1991b). These results, although based on few samples, suggest that genetic variation in this species may be highly structured into small local demes defined by maternal relationships and cultural differences, but not closely constrained by geography or distribution (Hoelzel, 1991a). As a result, analyses of genetic patterns based strictly on the geography of collection locales could fail to reveal important aspects of the biology of this species.

The absence of geographic partitioning of genetic variance in killer whales, like the seasonal intermingling and segregation of humpback whale maternal lineages during migration (Baker et al., 1990, 1994a), cautions against an unconsidered interpretation of phylogeographic patterns in cetaceans. The presence of hidden "herd" structure among species as mobile as most cetaceans may confound interpretations of genetic variation from samples collected from beachcast, bycatch, or direct harvest. Genetic analyses are likely to be most informative when interpreted within a demographic, behavioral, or comparative framework.

MOLECULAR SYSTEMATICS OF BALEEN WHALES

Surprisingly, the molecular systematics of the Mysticeti have received little attention until recently. Although generally thought to include only 10–12 species, questions remain about the systematic relationships and species-level designations within this suborder. Here we attempt to address some of this uncertainty by estimating the phylogenetic relationship among mysticete whales using mtDNA control-region sequences. The current analysis involves a total of 600 bp from 14 individuals representing eight species, including representatives of five of the six recognized genera and all four recognized families (Box 2.1). Where possible, sequences from more than one individual were used to evaluate the significance of intraspecific variation and to assure the accuracy of species-specific sequence data. Three species of toothed cetaceans (suborder Odontoceti, family Delphinidae) were chosen as outgroups—the killer whale (*Orcinus orca*), Commerson's dolphin (*Cephalorhynchus commersonii*), and Hector's dolphin (*C. hectori*).

Sequence Divergence Within and Between Species

Observed sequence differences suggest that levels of intraspecific nucleotide variability in the mtDNA control region are low relative to most between-species values. For example, two California gray whales differed by 2.3%, a North Atlantic and a North Pacific humpback whale by 2.0%, a North Atlantic and a North Pacific blue whale by less than 1.0%, and two North Atlantic fin whales by only 0.2%. By contrast, interspecific sequence differences ranged from 4.0% to 28.5% in our total of 55 comparisons among the various mysticete and odontocete species.

The most similar mysticete sequences were from fin and humpback whales (sequence divergence 6.4%). Particularly surprising was the small genetic distance (7.7%) between the gray whale and the blue whale (species traditionally placed in separate taxonomic families), relative to the difference (8.1%) between the blue and the fin (two congeners in *Balaenoptera* reported to produce viable hybrids in the wild; Arnason and Gullberg, 1993). Observed sequence differences between the Balaenidae and the Balaenopteridae/Eschrichtiidae were large (13.8–20.0%) relative to those among species within the latter group (6.2–14.6%) and pairwise differences between the odontocetes and mysticetes were larger still (> 21.0%).

Phylogenetic Reconstructions

A reconstructed phylogeny for the mysticete whales based on the mtDNA control-region sequences is presented in Figure 2.4, from which can be suggested the following: (1) an ancestral placement of the bowhead and pygmy right whales relative to the Eschrichtiidae (gray whale) and the Balaenopteridae; (2) placement of the gray whale next to the blue whale within the Balaenopteridae; (3) placement of the humpback with the fin whale, or intermediate between the fin, the blue, and the gray whale; and (4) an intermediate placement of the minke whale between the Balaenidae/Neobalaenidae and the other representatives of the Balaenopteridae/Eschrichtiidae.

Compared to the conventional systematic ordering of the mysticetes into four families, the most striking feature of this phylogeny is the placement of the gray whale within Balaenopteridae. Support for a monophyletic grouping of the gray and balaenopterid whales based on mtDNA control-

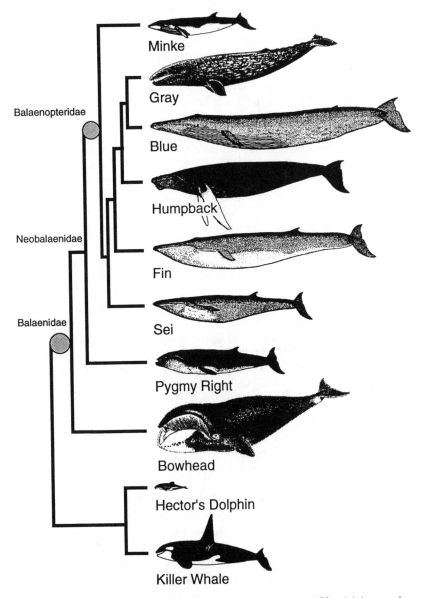

Figure 2.4.   Phylogenetic relationships of baleen whales based on neighbor-joining, parsimony, and maximum likelihood analyses of mtDNA control-region sequences, rooted by the outgroup method with the Hector's dolphin and killer whale. Shaded circles indicate family level groupings supported at > 80% in bootstrap replicates (illustrations adapted from Baker (1983) by V. Ward).

region sequences was tested by a variety of statistical procedures (Kishino and Hasegawa, 1989) and permutation tests (Faith, 1991), and found to be robust. Thus, this molecular phylogeny argues strongly against the generally held hypothesis that the gray whale represents an ancestral lineage distinct from the Balaenopteridae. In an analysis of mtDNA control-region and cytochrome *b* sequences, Arnason and colleagues also noted the close similarity of the gray whales to the other balaenopterids (Arnason et al., 1993; Arnason and Gullberg, 1994).

In addition to its implications for the systematic ordering of the gray whale, this revision provides insight into the evolution of feeding mechanisms in the mysticete whales. Unlike the Balaenidae, which use their long baleen to "skim" the water's surface for small planktonic prey, or the Balaenopteridae, which use their vast pleated throats to "engulf" dense concentrations of plankton or schooling fish (Nemoto, 1970), gray whales use the lower jaw like a shovel to scoop up benthic invertebrates, forcing the mud and water through their coarse baleen plates (Nerini, 1984). This adaptation allows California gray whales to exploit the rich benthic fauna of the Bering and Chukchi Sea during the summer months, fueling their annual coastal migration to the breeding and calving lagoons of Baja California (Rice and Wolman, 1971). Contrary to early descriptions of the gray whale as a "living fossil" (Andrews, 1914), it now appears that these unique morphological characteristics are recently derived adaptations to a specialized benthic feeding ecology. This specialized feeding ecology seems to have been accompanied by a rapid morphological divergence from the common Balaenopteridae *bauplan*.

Conservation Implications

Investigations into the molecular systematics of baleen and toothed whales indicate the need for reconsideration of relationships at several taxonomic levels. In a recent analysis of cetacean mtDNA ribosomal genes, Milinkovitch and others (1993) suggest that the sperm whale is more closely related to the mysticete whales than it is to other toothed whales. If this phylogenetic reconstruction is correct, the suborder Odontoceti, which currently includes all of the toothed whales, dolphins, and porpoises (Box 2.1) cannot be considered a natural or "holophyletic" grouping.

Our phylogenetic reconstruction of the mysticetes calls into question the current family-level status of the gray whale. The finding that the morphologically distinct gray whale may be as closely related to the blue whale as the blue whale is to the fin whale provides further caution against considering only morphological characteristics to distinguish "evolutionarily significant units" among cetaceans (Dizon et al., 1992).

A striking example of a taxonomic problem at the species level that is especially relevant to conservation biology involves the minke whale complex, the group now most likely to be the target of renewed commercial whaling. According to the International Whaling Commission (IWC) schedule, minke whales (*Balaenoptera acutorostrata* and *B. bonaerensis*) are defined as "any whale known as lesser rorqual, little piked whale, minke whale, pike-headed whale or sharp-headed finner" (Anonymous, 1994a; see also Dizon et al., 1992). However, genetic differences between some minke whales appear to be greater than those observed between the sei and Bryde's whales, two species distinguishable morphologically (Wada et al., 1991; Wada and Numachi, 1991; Arnason et al., 1993). Data from mtDNA analyses as well as protein electrophoresis now support the division of the minke whale complex into four distinct genetic units (Hoelzel and Dover, 1991c; Wada et al., 1991; Wada and Numachi, 1991; Arnason et al., 1993; Pastene et al., 1994a, 1994b; Hori et al., 1994): a North Atlantic type, a North Pacific type, a "normal" Southern Hemisphere type, and a "dwarf" or diminutive type sympatric with the Southern Hemisphere form.

Although attempts to manage future hunting of minke whales may not be greatly affected by the observed genetic division between oceans, the presence of genetically distinct sympatric forms in the Southern Hemisphere is cause for concern. The "dwarf" form appears to be rarer and more restricted in its distribution in the Southern Hemisphere (Best, 1985). It is not clear, however, if the dwarf form can be distinguished easily from the common type by shipboard observers and, consequently, how estimates of abundance derived from vessel surveys might be influenced. Obviously, the full taxonomic and demographic status of these two forms must be evaluated in order to set meaningful catch limits (Anonymous, 1994a).

A similar problem is encountered with the "pygmy" form or subspecies of the blue whale in the Southern Hemisphere (Ichihara, 1966). Although distinguishable from the "true" blue whale by a shorter tail, broader baleen plates, and smaller body length at sexual maturity (Ichihara, 1966), these differences may not be obvious to shipboard observers (Kato et al., 1995). Current attempts to reconstruct catch records and to estimate abundance are dependent on the seasonal and latitudinal segregation assumed to be maintained by the two forms (Anonymous, 1995). There is no available evidence of the genetic differences between the pygmy and true blue whales.

Clearly, issues concerning what constitutes a population, a stock, a subspecies, a species, a family, and even a suborder of cetaceans will continue to be active areas of inquiry and debate. We expect that both mitochondrial and nuclear DNA sequences will make important contributions to the eventual resolution of many of these questions.

FORENSIC IDENTIFICATION
OF CETACEAN PRODUCTS

For species threatened by overexploitation, molecular genetics provides yet another type of conservation tool—the forensic identification of commercial products from endangered species. Such products include ivory, horn, shell, meat, feathers, dried leaves, and a host of other commercial items derived from animal or plant materials (e.g., Baskin, 1991; Milner-Gulland and Mace, 1991). Although these products may be impossible to classify on the basis of appearance, they often contain DNA that can be amplified and compared to known or "type" samples. Molecular research on cetaceans provides a good example of the utility of this approach and, in addition, shows the applied value of basic research into species-level phylogenetic patterns.

A global moratorium on commercial whaling was established by the International Whaling Commission, and took effect in 1985 and 1986. However, whaling never actually stopped. IWC member nations have continued to hunt some species of whales for scientific research or for aboriginal and subsistence use. Whales killed for scientific research (largely for allozyme analysis of liver tissue to augment sample sizes already in the thousands; Wada et al., 1991) can legally be sold to consumers, thereby sustaining a commercial market for meat, skin, blubber, and other whale products. When some species are protected by an international prohibition against hunting but similar species are not, it is crucial to identify the origin of products that are actually sold in retail markets. Are the available products exclusively from species hunted legally under international agreements? Or, does the low level of whaling that has persisted since the moratorium serve as an umbrella for the sale of illegal whale products? Furthermore, because whale species can be hunted legally in some regions but not others, can the geographic origin of a whale product be ascertained?

Molecular genetics is uniquely suited to help in these situations. PCR can be used to amplify DNA from very small amounts of tissue. Research efforts to isolate amplifiable DNA from fossil and other ancient tissues have led to a variety of methods of obtaining DNA data from bone and dried tissue (Pääbo, 1988; Cooper et al., 1992; Janczewski et al., 1992). By targeting short sections of highly variable DNA (like the vertebrate mtDNA control region), the probable geographic origins of a specimen can be identified if a sufficient collection of regional samples is available for that species. Statistical methods developed for phylogenetic analyses can then be used to evaluate the reliability of an identification to species or geographic region.

Molecular Genetic Identification
of Whale Products

At the request of Earthtrust, an international conservation organization, we have begun development of a system for monitoring trade in whale and dolphin products using molecular genetic methods. To test this system, we identified the species and, in some cases, the probable geographic sources of 16 samples of whale products purchased throughout the main island of Japan from February to April, 1993. The products ranged from dried and salted strips of meat (marinated in sesame oil and soy sauce) to raw sliced meat, skin, and blubber sold for "sashimi." We used a portable PCR laboratory to amplify a variable portion of the mtDNA control region from each sample. The amplified control region was later sequenced and compared to 24 "type" sequences from 16 toothed or baleen whale species generated by our own research or found by searching international genetic databases (Baker and Palumbi, 1994).

Phylogenetic reconstruction using parsimony and bootstrap simulations unambiguously aligned 14 of the test samples with a sequence from our collection of "type" sequences (Figure 2.5). Eight of these samples grouped with the minke whales, whereas four grouped with fin whales. One sample of mixed strips of marinated meat yielded two identifiable sequences, one of which grouped with the "type" minke whales and the other with the "type" humpback whales. Two samples were placed unambiguously (by bootstrap criteria) within the family Delphinidae. The final sample was intermediate between the sperm whale (Physeteridae) and the harbor porpoise (Phocoenidae), differing from each by > 30% sequence difference.

Further inspection of the test sequences revealed that the minke whale products probably came from two different oceanic populations. Eight samples were similar in sequence to animals sampled from the Queensland coast of Australia and from Antarctica (Figure 2.5). The ninth sample was most similar to a Northern Hemisphere minke whale, differing by 7.7% from the Southern Ocean type sequences. Because minke whales in separate hemispheres are known to be genetically different (Hoelzel, 1991a; Wada et al., 1991; Arnason et al., 1993), the hemisphere of origin of these samples can be ascribed with some confidence, although more information is needed worldwide on minke whale population genetic structure.

Among the fin whale test sequences, one was identical to fin whales sampled near Iceland (Arnason et al., 1991) and in the western Mediterranean, suggesting that the origin of this sample was the North Atlantic. Four other fin whale products, however, differed by 1.6-2.9% from the North Atlantic type sequences. These animals might represent a genetically distinct stock in the North Atlantic, or they might be from a different ocean. Without a more

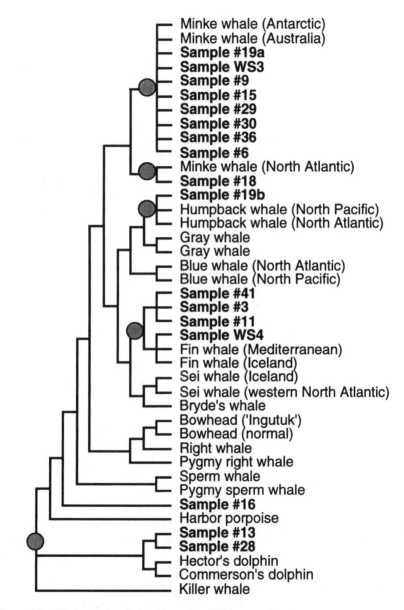

Figure 2.5. Phylogenetic relationships of mtDNA control-region sequences from "test" samples (bold print) of whale products from the Japanese retail market and "type" samples of whales and dolphins (after Baker and Palumbi, 1994). Phylogenetic reconstruction of type and test sequences was based on parsimony analysis (Swofford, 1993). Branches allowing unambiguous identification of test samples (bootstrap > 90%) are shown as shaded circles.

comprehensive genetic database on fin whales, we cannot evaluate the possibility of origins outside of the North Atlantic for these samples.

Comparison of Market Identifications and Catch Records

The phylogenetic framework within which these whale products were analyzed leads to statistically reliable conclusions about the species represented in the retail samples, and in some cases, their geographic origins. Thus, this approach has great power for determining the nature of whale products actually for sale on the retail market. Although the initial sample size collected by Earthtrust agents was small, we can ask whether the distribution of species found on the market matches expectations from international agreements about the capture, sale, and trade of whales.

The hunting of small cetaceans (dolphins, porpoises, and small-toothed whales) in coastal Japanese waters is extensive (Kishiro and Kasuya, 1993), but there is no international system for regulating or documenting these hunts. Our results show that dolphins or other toothed whales are often sold as whale meat (3 of 16 test samples). Whether this meat was misrepresented depends on the interpretation of the Japanese generic term for whale (*kujira*) and on the exact identity of the species (e.g., pilot whales but not bottlenose dolphins might be considered *kujira*). When sold as whale meat, however, a dolphin represents a valuable commodity, up to $2,500 each on the wholesale markets of Korea (Anonymous, 1994b). Such an economic gain is not likely to go unnoticed by local fisherman, who can be expected to target dolphins, porpoises, and other toothed whales for capture, or to send the animals taken in local bycatch to market.

In contrast to the lack of international management of small-toothed whales, there are extensive international agreements about the hunting and international trade of baleen whale products. A review of the catch reports by member nations of the IWC shows that several hundred Southern Hemisphere minke whales were taken annually by Japan under scientific permit in 1991 and 1992. From 1986 to 1992, North Atlantic minke whales were taken only as aboriginal catches. According to IWC resolutions, whales killed under subsistence or aboriginal permits are for local consumption only. Ninety-five North Atlantic minke whales were taken by Norway in 1992 under scientific permit, but a national policy bans the commercial export of these products. The last recorded commercial export of minke whales from Norway was in 1986 (Larson, 1994).

Except for aboriginal catches by Greenland/Denmark, North Atlantic fin whales have not been hunted since 1989, when Iceland killed 68 under scientific permit. Fin whales from oceans other than the North Atlantic have not been hunted legally since the 1986 moratorium. Hunting of humpback whales has been prohibited by international agreement since 1966 (Rice, 1974). No infractions, exceptions, or aboriginal hunting of humpback whales are noted in recent IWC annual reports.

The eight Southern Hemisphere minke whales in our sample are consistent with the reported scientific catch by Japan in recent years. The North Atlantic minke whale could only be part of a recorded IWC hunt and international shipment if it had been in storage outside of Norway since 1986. The fin whales in our sample would have to have been in storage since 1989, and the humpback sample since 1966.

Given that these storage intervals are unlikely, we conclude that whale products currently available on the retail market include species that have been imported illegally and others that may have been hunted or processed illegally. The fraction of potentially illegal samples was high: six of the 13 baleen whale products (46%) are not consistent with recent catch records. The two identified dolphins or small cetaceans may also have been misrepresented to the domestic consumer as whale.

Evidence of Unauthorized Hunting or Trade of Whales

Revelations of the extent of "pirate whaling" by the Soviet fleet in the Southern Hemisphere from the 1950s to the 1970s emphasize the magnitude of illegal whaling that can go unnoticed by the international community (see Box 2.3). More recent incidents confirm that active international and domestic trade in illegal products continues. Early in 1994, the Russian Ministry of the Environment blocked an attempt to export 238 metric tons of whale meat from Russia to Japan (Anonymous, 1994c). In October 1993, an air cargo handler in Oslo, Norway, discovered that 3.5 metric tons of Norwegian "shrimp" bound for export to South Korea via Frankfort was actually whale meat. Reports of large amounts of whale meat in storage in Norway suggest that much of the local catch may be bound for international markets (Mulvaney, 1993). Baleen whales from bycatch of coastal Japanese fisheries are reportedly processed and sold in domestic markets without permission of government agencies (Anonymous, 1993). Korean fishermen recently admitted to large-scale illegal whaling in the North Pacific (Anonymous, 1994b). Japan is presumably the main market for this meat, although some meat is also consumed locally.

Management Implications

Whaling is a contentious and emotional issue. Whereas some people see the hunting of whales as primarily an ethical problem, others have approached it strictly in terms of the management of a sustainable resource. Even when viewed primarily as a management issue, the conditions for a return to commercial whaling have been the subject of much scientific debate, and considerable uncertainty remains about the status and methods for monitoring whale populations.

The ease of molecular testing of whale products, and the ability to pinpoint geographic origins of the samples in many cases, suggest that a comprehensive management plan could be implemented whereby commercial markets are tested regularly for the products actually sold. Inclusion of data from the marketplace into models that estimate the effect of hunting pressure on individual species could then allow a more accurate estimate of population growth of species that are, and are not, being considered for harvest.

Arguments about sustainable hunting of abundant stocks are based on the tacit assumption that depleted or endangered populations and species will continue to enjoy protection. Without an adequate system of monitoring and verification of catches, however, no species can be considered safe.

SUMMARY

1. As judged by geographic distributions of mtDNA lineages, oceanic populations of humpback whales are divided into regional populations or *stocks* that are not separated by obvious geographic barriers. An analysis of samples from around the world shows significant genetic divisions between stocks in each of the three major oceanic populations. A hierarchical analysis of haplotypes shows that population structure can be complex, with greater segregation on summering grounds than on wintering grounds following seasonal migration. The population structure of mtDNA may be attributed to maternally directed (although not necessarily natal) fidelity to migratory destinations.

2. Despite intensive hunting during much of this century, most humpback whale populations appear to have retained high levels of mtDNA diversity. A possible exception is the central North Pacific stock, which is considerably reduced in variation relative to other regional populations.

3. Numerous additional cetacean species have been examined with varying levels of effort for magnitudes of genetic variation and patterns of population subdivisions, using molecular markers from allozyme genes, mitochondrial DNA, and various hypervariable nuclear DNA loci. Results have varied but, not uncommonly, have revealed previously unsuspected "stock" structures in several species. These population subdivisions may reflect physical impediments to gene flow (as between ocean basins), behaviorally and socially mediated movement patterns of relatives within ocean basins, or both. Results have ramifications for management and conservation plans for harvested as well as protected cetaceans.

4. Phylogenetic reconstructions of relationships between baleen and other whales, using sequence information from mtDNA, suggest the need for reconsideration of the current systematic ordering of cetaceans. For example, a family-level placement of the gray whale is not supported, nor are previous descriptions of this species as "primitive." Instead, the gray whale appears to be a recently diverged member of the family Balaenopteridae, most closely related to the humpback and blue whales. Additional management problems relating to the systematic ordering of whales are discussed, including the status of the several forms of minke whales and the two forms of blue whales.

5. Molecular genetic techniques have a demonstrated utility for the identification of the species and geographic origin of whale products purchased on the retail market. A spot check of the Japanese retail market suggests that a substantial proportion of whale products currently for sale is illegal or misrepresented. Systematic molecular genetic testing would provide a useful management tool for discouraging pirate whaling and for interdicting illegal trade in whale products.

REFERENCES

ALLEN, K.R. 1980. *Conservation and Management of Whales.* University of Washington Press, Seattle.

AMOS, B., J. BARRETT, and G.A. DOVER. 1991. Breeding behaviour of pilot whales revealed by DNA fingerprinting. *Heredity* 67:49–55.

AMOS, B., C. SCHLOTTERER, and D. TAUTZ. 1993. Social structure of pilot whales revealed by analytical DNA profiling. *Science* 30:670–672.

AMOS, W. and A.R. HOELZEL. 1990. DNA fingerprinting cetacean biopsy samples for individual identification. *Rep. Int. Whal. Comm.,* Special Issue 12:79–86.

ANDERSON, L.W. 1988. Electrophoretic differentiation among local populations of the long-finned pilot whale, *Globicephala melaena,* at the Faroe Islands. *Can. J. Zool.* 66:1884–1892.

ANDREWS, R.C. 1914. Monographs of the Pacific Cetacea. I. The California gray whale (*Rhachianectess glaucus* Cope). *Mem. Amer. Mus. Nat. Hist.* 1:229–287.

ANONYMOUS. 1993. Six more whales flensed and sold illicitly. Yomiuri Shinbun, Tokyo, April 30.

ANONYMOUS. 1994a. Report of the Scientific Committee. *Rep. Int. Whal. Comm.* 44:41–204.

ANONYMOUS. 1994b. South Koreans caught in the act. *New Sci.* 142:4.

ANONYMOUS. 1994c. Whale of a problem. *New Sci.* 141:11.

ANONYMOUS. 1995. Report of the Scientific Committee. *Rep. Int. Whal. Comm.* 45: *in press.*

ARNASON, U. 1974. Comparative chromosome studies in Cetacea. *Hereditas* 77:1–36.

ARNASON, U., H. BELLAMY, P. EYTHORSSON, R. LUTLEY, J. SIGURJONSSON, and B. WIDEGREN. 1985. Conventionally stained and C-banded karyotypes of a female blue whale. *Hereditas* 102:251–253.

ARNASON, U. and A. GULLBERG. 1993. Comparison between the complete mtDNA sequences of the blue and fin whale, two species that can hybridize in nature. *J. Molec. Evol.* 37:312–322.

ARNASON, U. and A. GULLBERG. 1994. Relationship of baleen whales established by mitochondrial cytochrome *b* sequence comparison. *Nature* 367:726–728.

ARNASON, U., A. GULLBERG, and B. WIDEGREN. 1991. The complete nucleotide sequence of the mitochondrial DNA of the fin whale, *Balaenoptera physalus*. *J. Molec. Evol.* 33:556–568.

ARNASON, U., A. GULLBERG, and B. WIDEGREN. 1993. Cetacean mitochondrial DNA control region: sequences of all extant baleen whales and two sperm whale species. *Molec. Biol. Evol.* 10:960–970.

AVISE, J.C., J. ARNOLD, R.M. BALL, E. BERMINGHAM, T. LAMB, J.E. NEIGEL, C.A. REEB, and N.C. SAUNDERS. 1987. Intraspecific phylogeography: The mitochondrial DNA bridge between population genetics and systematics. *Annu. Rev. Ecol. Syst.* 18:489–522.

AVISE, J.C., B.W. BOWEN, T. LAMB, A.B. MEYLAN, and E. BERMINGHAM. 1992. Mitochondrial DNA evolution at a turtle's pace: evidence for low genetic variability and reduced microevolutionary rate in the Testudines. *Molec. Biol. Evol.* 9:457–473.

BAKER, A.N. 1983. *Whales and Dolphins of New Zealand and Australia*. Victoria University Press, Wellington, New Zealand.

BAKER, C.S., D.A. GILBERT, M.T. WEINRICH, R.H. LAMBERTSEN, J. CALAMBOKIDIS, B. MCARDLE, G.K. CHAMBERS, and S.J. O'BRIEN. 1993a. Population characteristics of DNA fingerprints in humpback whales (*Megaptera novaeangliae*). *J. Hered.* 84:281–290.

BAKER, C.S. and L.M. HERMAN. 1987. Alternate population estimates of humpback whales (*Megaptera novaeangliae*) in Hawaiian waters. *Can. J. Zool.* 65:2818–2821.

BAKER, C.S., L.M. HERMAN, A. PERRY, W.S. LAWTON, J.M. STRALEY, A.A. WOLMAN, G.D. KAUFMAN, H.E. WINN, J.D. HALL, J.M. REINKE, and J. OSTMAN. 1986. Migratory movement and population structure of humpback whales (*Megaptera novaeangliae*) in the central and eastern North Pacific. *Mar. Ecol. Prog. Ser.* 31:105–119.

BAKER, C.S., R.H. LAMBERTSEN, M.T. WEINRICH, J. CALAMBOKIDIS, G. EARLY, and S.J. O'BRIEN. 1991. Molecular genetic identification of the sex of humpback whales (*Megaptera novaeangliae*). *Rep. Int. Whal. Comm.*, Special Issue 13:105–111.

BAKER, C.S. and S.R. PALUMBI. 1994. Which whales are hunted? A molecular genetic approach to monitoring whaling. *Science* 265:1538–1539.

BAKER, C.S., S.R. PALUMBI, R.H. LAMBERTSEN, M.T. WEINRICH, J. CALAMBOKIDIS, and S.J. O'BRIEN. 1990. The influence of seasonal migration on the distribution of mitochondrial DNA haplotypes in humpback whales. *Nature* 344:238–240.

BAKER, C.S., A. PERRY, J.L. BANNISTER, M.T. WEINRICH, R.B. ABERNETHY, J. CALAMBOKIDIS, J. LIEN, R.H. LAMBERTSEN, J. URBAN-RAMIREZ, O. VASQUEZ, P.J. CLAPHAM, A. ALLING, S.J. O'BRIEN, and S.R. PALUMBI. 1993b. Abundant mitochondrial DNA variation and world-wide population structure in humpback whales. *Proc. Natl. Acad. Sci. USA* 90:8239–8243.

BAKER, C.S., A. PERRY, and L.M. HERMAN. 1987. Reproductive histories of female humpback whales *Megaptera novaeangliae* in the North Pacific. *Mar. Ecol. Prog. Ser.* 41:103–114.

BAKER, C.S., R.W. SLADE, J.L. BANNISTER, R.B. ABERNETHY, M.T. WEINRICH, J. LIEN, J. URBAN, P. CORKERON, J. CALAMBOKIDIS, O. VASQUEZ, and S.R. PALUMBI. 1994a. Hierarchical structure of mitochondrial DNA gene flow among humpback whales *Megaptera novaeangliae*, world-wide. *Molec. Ecol.* 3:313–327.

BAKER, C.S., M.T. WEINRICH, G. EARLY, and S.R. PALUMBI. 1994b. Genetic impact of an unusual group mortality among humpback whales. *J. Hered.* 85:52–54.

BANNISTER, J.L., E.D. MITCHELL, K.C. BALCOMB, S.G. BROWN, and A.R. MARTIN. 1984. Report of the subgroup on the North Atlantic humpback whale boundaries. *Rep. Int. Whal. Comm.* 34:181.

BASKIN, Y. 1991. Archaeologist lends a technique to rhino protection. *BioScience* 41:532–534.

BEST, P.B. 1985. External characteristics of southern minke whales and the existence of a diminutive form. *Sci. Rep. Whales Res. Inst. Tokyo* 36:1–33.

BEST, P.B. 1993. Increase rate in severely depleted stocks of baleen whales. *ICES J. Mar. Sci.* 50:169–186.

BIGG, M.A., P.F. OLESIUK, G.M. ELLIS, J.K.B. FORD, and K.C. BALCOMB. 1990. Social organization and genealogy of resident killer whales (*Orcinus orca*) in the coastal waters of British Colombia and Washington State. *Rep. Int. Whal. Comm.*, Special Issue 12:383–406.

BIRKY, C.W., T. MARUYAMA, and P. FUERST. 1983. An approach to population and evolutionary genetic theory for genes in mitochondria and chloroplasts and some results. *Genetics* 103:513–527.

BROWN, M.W., S.D. KRAUS, and D.E. GASKIN. 1991. Reaction of North Atlantic right whales (*Eubalaena glacialis*) to skin biopsy sampling for genetic and pollutant analysis. *Rep. Int. Whal. Comm.*, Special Issue 13:81–90.

BROWN, W.M. 1983. Evolution of animal mitochondrial DNA. In *Evolution of Genes and Proteins*, eds. M. Nei and R.K. Koehn, pp. 62–88. Sinauer, Sunderland, MA.

BRYDEN, M.M., G.P. KIRKWOOD, and R.W. SLADE. 1990. Humpback whales, Area V. An increase in numbers off Australia's east coast. In *Antarctic Ecosystems: Ecological Change and Conservation*, eds. K.R. Kerry and G. Hempel, pp. 271–277. Springer-Verlag, Berlin.

CALAMBOKIDIS, J., J.C. CUBBAGE, G.H. STEIGER, K.C. BALCOMB, and P. BLOEDEL. 1990. Population estimates of humpback whales in the Gulf of the Farallones. *Rep. Int. Whal. Comm.*, Special Issue 12:325–334.

CHAPMAN, D.G. 1974. Status of Antarctic rorqual stocks. In *The Whale Problem*, ed. W.E. Schevill, pp. 218–238. Harvard University Press, Cambridge, MA.

CHITTLEBOROUGH, R.G. 1959. Determination of age in the humpback whale, *Megaptera nodosa* (Bonnaterre). *Aust. J. Mar. Freshw. Res.* 10:125–143.

CHITTLEBOROUGH, R.G. 1965. Dynamics of two populations of humpback whales, *Megaptera novaeangliae* (Borowski). *Aust. J. Mar. Freshw. Res.* 16:33–128.

CLAPHAM, P.J. and C.A. MAYO. 1987. Reproduction and recruitment in individually identified humpback whales (*Megaptera novaeangliae*) observed in Massachusetts Bay: 1979–1985. *Can. J. Zool.* 65:2853–2863.

CLAPHAM, P.J. and C.A. MAYO. 1990. Reproduction of humpback whales (*Megaptera novaeangliae*) observed in the Gulf of Maine. *Rep. Int. Whal. Comm.*, Special Issue 12:171–175.

COOPER, A., C. MOURER-CHAUVIRE, G.K. CHAMBERS, A.V. HAESELER, A.C. WILSON, and S. PÄÄBO. 1992. Independent origins of New Zealand moas and kiwis. *Proc. Natl. Acad. Sci. USA* 89:8741–8744.

CRONE, M.J. and S.D. KRAUS (eds.). 1990. *Right whales* (Eubalaena glacialis) *in the Western North Atlantic: a catalog of identified individuals*. New England Aquarium, Boston, MA.

DANIELSDOTTIR, A.K., E.J. DUKE, and A. ARNASON. 1992. Genetic variation at enzyme loci in North Atlantic minke whales, *Balaenoptera acutorostrata*. *Biochem. Genet.* 30:189–202.

DANIELSDOTTIR, A.K., E.J. DUKE, and P. JOYCE. 1991. Preliminary studies on genetic variation at enzyme loci in fin whales (*Balaenoptera physalus*) and sei whales (*Balaenoptera borealis*) from the North Atlantic. *Rep. Int. Whal. Comm.*, Special Issue 13:115–124.

DARLING, J.D. and C.M. JURASZ. 1983. Migratory destinations of North Pacific humpback whales (*Megaptera novaeangliae*). In *Communication and Behavior of Whales*, ed. R. Payne, pp. 359–368. Westview Press, Boulder, CO.

DARLING, J.D. and D.J. McSWEENEY. 1985. Observations on the migrations of North Pacific humpback whales (*Megaptera novaeangliae*). *Can. J. Zool.* 63:308–314.

DARLING, J.D. and K. MORI. 1993. Recent observations of humpback whales (*Megaptera novaeangliae*) in Japanese waters off Ogasawara and Okinawa. *Can. J. Zool.* 71:325–333.

DAWBIN, W.H. 1966. The seasonal migratory cycle of humpback whales. In *Whales, Dolphins and Porpoises*, ed. K.S. Norris, pp. 145–171. University of California Press, Berkeley.

DIZON, A.E., C. LOCKYER, W.F. PERRIN, D.P. DEMASTERS, and J. SISSON. 1992. Rethinking the stock concept: a phylogenetic approach. *Conserv. Biol.* 6:24–36.

DIZON, A.E., S.O. SOUTHERN, and W.F. PERRIN. 1991. Molecular analysis of mtDNA types in exploited populations of spinner dolphins (*Stenella longirostris*). *Rep. Int. Whal. Comm.*, Special Issue 13:183–202.

DONOGHUE, M. and A. WHEELER. 1990. *Save the Dolphins*. David Bateman, Auckland, New Zealand.

DONOVAN, G.P. 1991. A review of IWC stock boundaries. *Rep. Int. Whal. Comm.*, Special Issue 13:39–68.

DOWLING, T. and W.M. BROWN. 1993. Population structure of the bottlenose dolphins as determined by restriction endonuclease analysis of mitochondrial DNA. *Mar. Mamm. Sci.* 9:138–155.

DUFFIELD, D.A. and R.S. WELLS. 1991. The combined application of chromosome, protein and molecular data from the investigation of social unit structure and dynamics in *Tursiops truncatus. Rep. Int. Whal. Comm.*, Special Issue 13:155–170.

EVANS, P.G.H. 1987. *The Natural History of Whales and Dolphins*. Facts on File Publ., New York.

FAITH, D.P. 1991. Cladistic permutation tests for monophyly and nonmonophyly. *Syst. Zool.* 40:366–375.

GAMBELL, R. 1976. World whale stocks. *Mamm. Rev.* 5:41–53.

GERACI, J.R., D.M. ANDERSON, R.J. TIMPERI, G.A. EARLY, J.H. PRESCOTT, and C.A. MAYO. 1989. Humpback whales (*Megaptera novaeangliae*) fatally poisoned by dinoflagellate toxin. *J. Fish. Res. Bd. Can.* 46:1895–1898.

HANSEN, L.J. 1992. *Report on Investigation of 1990 Gulf of Mexico Bottlenose Dolphin Strandings*. Southeast Fisheries Science Center, National Marine Fisheries Service, Miami, FL.

HARWOOD, J. and A. HALL. 1990. Mass mortality in marine mammals: its implications for population dynamics and genetics. *Trends Ecol. Evol.* 5:254–256.

HERMAN, L.M. 1979. Humpback whales in Hawaiian waters: a study in historical ecology. *Pac. Sci.* 33:1–15.

HOELZEL, A.R. 1991a. Analysis of regional mitochondrial DNA variation in the killer whale: implications for cetacean conservation. *Rep. Int. Whal. Comm.*, Special Issue 13:225–233.

HOELZEL, A.R. (ed.). 1991b. *Genetic Ecology of Whales and Dolphins*. International Whaling Commission, Cambridge, UK.

HOELZEL, A.R. 1994. Genetics and ecology of whales and dolphins. *Annu. Rev. Ecol. Syst.* 25:377–399.

HOELZEL, A.R. and G.A. Dover. 1991a. Evolution of the cetacean mitochondrial D-loop region. *Molec. Biol. Evol.* 8:475–493.

HOELZEL, A.R. and G.A. DOVER. 1991b. Genetic differences between sympatric killer whale populations. *Heredity* 66:191-196.

HOELZEL, A.R. and G.A. Dover. 1991c. Mitochondrial D-loop DNA variation within and between populations of the minke whale (*Balaenoptera acutorostrata*). *Rep. Int. Whal. Comm.*, Special Issue 13:171–181.

HORI, H., Y. BESSHO, R. KAWABATA, I. WATANABE, A. KOGA, and L.A. PASTENE. 1994. World-wide population structure of minke whales deduced from mitochondrial DNA control region sequences. Paper SC/46/SH14 presented to the IWC Scientific Committee, May, 1994 (unpublished).

ICHIHARA, T. 1966. The blue whale, *Balaenoptera musculus brevicauda*, a new subspecies from the Antarctic. In *Whales, Dolphins and Porpoises*, ed. K.S. NORRIS, pp. 79–113. University of California Press, Berkeley.

JANCZEWSKI, D.N., N. YUHKI, D.A. GILBERT, G.T. JEFFERSON, and S.J. O'BRIEN. 1992. Molecular phylogenetic inference from saber-toothed cat fossils of Rancho La Brea. *Proc. Natl. Acad. Sci. USA* 89:9769–9773.

KATO, H., T. MIYASHITA, and H. SHIMADA. 1995. Segregation of the two subspecies of blue whales in the Southern Hemisphere. *Rep. Int. Whal. Comm.*, in press.

KATONA, S.K. and J.A. BEARD. 1990. Population size, migration and substock structure of the humpback whale (*Megaptera novaeangliae*) in the western North Atlantic Ocean. *Rep. Int. Whal. Comm.*, Special Issue 12:295–306.

KELLOGG, R. 1929. What is known of the migration of some of the whalebone whales. *Smithsonian Inst. Annu. Rep.* 1928:467–494.

KISHINO, H. and M. HASEGAWA. 1989. Evaluation of the maximum likelihood estimate of the evolutionary tree topologies from DNA sequence data and the branching order of the Hominoidae. *J. Molec. Evol.* 29:170–179.

KISHIRO, T. and T. KASUYA. 1993. Review of the Japanese dolphin drive fisheries and their status. *Rep. Int. Whal. Comm.* 43:439–452.

LAMBERTSEN, R.H. 1987. A biopsy system for large whales and its use for cytogenetics. *J. Mamm.* 68:443–445.

LAMBERTSEN, R.H., C.S. BAKER, M. WEINRICH, and W.S. MODI. 1994. An improved whale biopsy system designed for multidisciplinary research. In *Nondestructive Biomarkers in Vertebrates*, eds. M.C. Fossi and C. Leonzio, pp. 219–224. Lewis Publishers, London.

LARSON, E. 1994. *Norwegian Banning of Export of Whale Products*. Royal Norwegian Embassy, Canberra.

MACKINTOSH, N.A. 1965. *The Stocks of Whales*. Fishing News (Books), London.

MARTIN, A.P. and S.R. PALUMBI. 1993. Body size, metabolic rate, generation time, and the molecular clock. *Proc. Natl. Acad. Sci. USA* 90:4087–4091.

MARTIN, A.R., S.K. KATONA, D. MATILLA, D. HEMBREE, and T.D. WATERS. 1984. Migration of humpback whales between the Caribbean and Iceland. *J. Mamm.* 65:330–333.

MATHEWS, E.A. 1986. Multiple use of skin biopsies collected from free-ranging gray whales. MS thesis, University of California, Santa Cruz.

MATTILA, D.K., P.J. CLAPHAM, S.K. KATONA, and G.S. STONE. 1989. Humpback whales on Silver Bank, 1984: population composition and habitat use. *Can. J. Zool.* 67:281–285.

MEDRANO-G., L. 1993. Estudio genetico del rorcual jorobado en el Pacifico Mexicano. PhD thesis, Universitas Nacional Autonoma de Mexico.

MENOTTI-RAYMOND, M. and S.J. O'BRIEN. 1993. Dating the genetic bottleneck of the African cheetah. *Proc. Natl. Acad. Sci. USA* 90:3172–3176.

MILINKOVITCH, M.C., G. ORTI, and A. MEYER. 1993. Revised phylogeny of whales suggested by mitochondrial ribosomal DNA sequences. *Nature* 361:346–348.

MILNER-GULLAND, E.J. and R. MACE. 1991. The impact of the ivory trade on the African elephant *Loxodonta africana* population as assessed by data from the trade. *Biol. Conserv.* 55:215–229.

MORITZ, C., T.E. DOWLING, and W.M. BROWN. 1987. Evolution of animal mitochondrial DNA: relevance for population biology and systematics. *Annu. Rev. Ecol. Syst.* 18:269–292.

MULVANEY, K. 1993. Norway caught flogging minke. *BBC Wildlife* 111:62.

NATIONAL MARINE FISHERIES SERVICE. 1991. *Recovery Plan for the Humpback Whale* (Megaptera novaeangliae). National Marine Fisheries Service, Silver Spring, MD.

NEI, M., T. MARUYAMA, and R. CHAKRABORTY. 1975. The bottleneck effect and genetic variability in populations. *Evolution* 29:1–10.

NEIGEL, J.E. and J.C. AVISE. 1986. Phylogenetic relationships of mitochondrial DNA under various demographic models of speciation. In *Evolutionary Processes and Theory*, eds. E. Nevo and S. Karlin, pp. 513–534. Academic Press, London.

NEMOTO, T. 1970. Feeding patterns of baleen whales in the ocean. In *Marine Food Chains*, ed. J.H. Steele, pp. 241–252. Oliver and Boyd, Edinburgh.

NERINI, M. 1984. A review of gray whale feeding ecology. In *The Gray Whale*, eds. M.L. Jones, S.L. Swartz, and S. Leatherwood, pp. 423–448. Academic Press, Orlando, FL.

PÄÄBO, S. 1988. Ancient DNA: Extraction, characterization, molecular cloning, and enzymatic amplification. *Proc. Natl. Acad. Sci. USA* 86:1939–1943.

PALUMBI, S.R. and C.S. BAKER. 1994. Opposing views of humpback whale population structure using mitochondrial and nuclear DNA sequences. *Molec. Biol. Evol.* 11:426–435.

PALUMBI, S.R., A.P. MARTIN, B. KESSING, and W.O. MCMILLAN. 1991. Detecting population structure using mitochondrial DNA. *Rep. Int. Whal. Comm.*, Special Issue 13:271–278.

PASTENE, L.A., Y. FUJISE, and K. NUMACHI. 1994a. Differentiation of mitochondrial DNA between ordinary and dwarf forms of southern minke whales. *Rep. Int. Whal. Comm.* 44:277–282.

PASTENE, L.A., M. GOTO, Y. FUJISE, and K. NUMACHI. 1994b. Further analysis on the spatial and temporal heterogeneity in mitochondrial DNA haplotype distribution in minke whales from Antarctic Areas IV and V. Paper SC/46/SH13 presented to the IWC Scientific Committee (unpublished).

PATERSON, R. and P. PATERSON. 1989. The status of the recovering stock of humpback whales *Megaptera novaeangliae* in east Australian waters. *Biol. Conserv.* 47:33–48.

PAYNE, P.N., D. WILEY, S. YOUNG, S. PITTMAN, P.J. CLAPHAM, and J.W. JOSSI. 1990. Recent fluctuations in the abundance of baleen whales in the southern Gulf of Maine in relation to changes in prey abundance. *Fish. Bull.* 88:687–696.

PAYNE, R. and L.N. GUINEE. 1983. Humpback whale (*Megaptera novaeangliae*) songs as an indicator of "stocks." In *Communication and Behavior of Whales*, ed. R. Payne, pp. 333–358. Westview Press, Boulder, CO.

PERRY, A., C.S. BAKER, and L.M. HERMAN. 1990. Population characteristics of individually identified humpback whales in the central and eastern North Pacific: a summary and critique. *Rep. Int. Whal. Comm.*, Special Issue 12:307–318.

RICE, D.W. 1974. Whales and whale research in the eastern North Pacific. In *The Whale Problem*, ed. W.E. Schevill, pp. 218–238. Harvard University Press, Cambridge, MA.

RICE, D.W. 1978. The humpback whale in the North Pacific: distribution, exploitation, and numbers. In *Report on a Workshop on Problems Related to Humpback Whales* (Megaptera novaeangliae) *in Hawaii*, eds. K.S. Norris and R. Reeves, pp. 29–44. U.S. Marine Mammal Commission, Washington, DC.

RICE, D.W. and A. A. WOLMAN. 1971. The life history and ecology of the gray whale (*Eschrichtius robustus*). *Amer. Soc. Mamm.*, Special Issue 3:1–142.

ROSEL, P.E. 1992. Genetic population structure and systematics of some small cetaceans inferred from mitochondrial DNA sequence variation. PhD thesis, University of California, San Diego.

ROSEL, P.E., A.E. DIZON, and M.G. HAYGOOD. 1995. Variability of the mitochondrial control region in populations of the harbour porpoise, *Phocoena phocoena*, on interoceanic and regional scales. *Can. J. Fish. Aquat. Sci.* 52: in press.

ROSEL, P.E., A.E. DIZON, and J.E. HEYNING. 1994. Genetic analysis of sympatric morphotypes of common dolphins (genus *Delphinus*). *Mar. Biol.* 119:159–167.

SCHAEFF, C., S. KRAUS, M. BROWN, J. PERKINS, R. PAYNE, D. GASKIN, P. BOAG, and B. WHITE. 1991. Preliminary analysis of mitochondrial DNA variation within and between the right whale species *Eubalaena glacialis* and *Eubalaena australis*. *Rep. Int. Whal. Comm.*, Special Issue 13:217–223.

SCHAEFF, C.M., S.D. KRAUS, M.W. BROWN, and B.N. WHITE. 1993. Assessment of the population structure of western North Atlantic right whales (*Eubalaena glacialis*) based on sighting and mtDNA data. *Can. J. Zool.* 71:339–345.

SLATKIN, M. 1987. Gene flow and the geographic structure of natural populations. *Science* 236:787–792.

SLOOTEN, E. and S.M. Dawson. 1994. Hector's dolphin *Cephalorhynchus hectori* (van Beneden 1881). In *Handbook of Marine Mammals: Vol. V, The First Book of Dolphins*, eds. S.H. Ridgway and R. Harrison, pp. 311–322. Academic Press, New York.

STEVENS, T.A., D.A. DUFFIELD, E.D. ASPER, K.G. HEWLETT, A. BOLZ, L.J. GAGE, and G.D. BOSSART. 1989. Preliminary findings of restriction fragment differences in mitochondrial DNA among killer whales (*Orcinus orca*). *Can. J. Zool.* 67:2592–2595.

STONE, G.S., S.K. KATONA, and E.B. TUCKER. 1987. History, migration and present status of humpback whales, *Megaptera novaeangliae*, at Bermuda. *Biol. Conserv.* 42:122–145.

SWOFFORD, D.L. 1993. *PAUP: Phylogenetic Analysis Using Parsimony, 3.1.1.* Illinois Natural History Survey, Champaign.

TONNESSEN, J.N. and A.O. JOHNSEN. 1982. *The History of Modern Whaling.* University of California Press, Berkeley.

TOWNSEND, C.H. 1935. The distribution of certain whales as shown by logbook records of American whaleships. *Zoologica* 19:1–50.

TROWSDALE, J., V. GROVES, and A. ARNASON. 1989. Limited MHC polymorphism in whales. *Immunogenetics* 29:19–24.

URBAN R.J. and A. AGUAYO. 1987. Spatial and seasonal distribution of the humpback whale, *Megaptera novaeangliae*, in the Mexican Pacific. *Mar. Mamm. Sci.* 3:333–344.

VAN PIJLEN, I., W. AMOS, and G.A. DOVER. 1991. Multilocus DNA fingerprinting applied to population studies of the minke whale *Balaenoptera acutorostrata*. *Rep. Int. Whal. Comm.*, Special Issue 13:245–254.

VIGILANT, L., R. PENNINGTON, H. HARPENDING, T.D. KOCHER, and A.C. WILSON. 1989. Mitochondrial DNA sequences in single hairs from a southern African population. *Proc. Natl. Acad. Sci. USA* 86:9350–9354.

WADA, S. 1983. Genetic structure and taxonomic status of minke whales in the coastal waters of Japan. *Rep. Int. Whal. Comm.* 33:361–363.

WADA, S. 1988. Genetic differentiation between two forms of short-finned pilot whales off the Pacific coast of Japan. *Sci. Rep. Whales Res. Inst. Tokyo* 39:91–101.

WADA, S., T. KOBAYASHI, and K. NUMACHI. 1991. Genetic variability and differentiation of mitochondrial DNA in minke whales. *Rep. Int. Whal. Comm.*, Special Issue 13:203–215.

WADA, S. and K.I. NUMACHI. 1991. Allozyme analyses of genetic differentiation among populations and species of *Balaenoptera*. *Rep. Int. Whal. Comm.* 41:125–154.

WANG, J.Y.-C. 1993. Mitochondrial DNA analysis of the harbour porpoise. MS thesis, University of Guelph, Guelph, Ontario, Canada.

WEINRICH, M.T., R.H. LAMBERTSEN, C.S. BAKER, M.R. SCHILLING, and C.R. BELT. 1991. Behavioural responses of humpback whales (*Megaptera novaeangliae*) in the southern Gulf of Maine to biopsy sampling. *Rep. Int. Whal. Comm.*, Special Issue 13:91–98.

WEINRICH, M.T., R.H. LAMBERTSEN, C.R. BELT, M.R. SCHILLING, H.J. IKEN, and S.E. SYRJAL. 1992. Behavioral reactions of humpback whales *Megaptera novaeangliae* to biopsy procedures. *Fish. Bull.* 90:588–598.

WELLS, R.S. 1991. The role of long-term study in understanding the social structure of a bottlenose dolphin community. In *Dolphin Societies*, eds. K. Pryor and K.S. Norris, pp. 199–226. University of California Press, Berkeley.

WHITEHEAD, H. 1982. Populations of humpback whales in the Northwest Atlantic. *Rep. Int. Whal. Comm.* 32:345–353.

WILSON, A.C., R.L CANN, S.M. CARR, M. GEORGE, U.B. GYLLENSTEN, K.M. HELM-BYCHOWSKI, R.G. HIGUCHI, S.R. PALUMBI, E.M. PRAGER, R.D. SAGE, and M. STONEKING. 1985. Mitochondrial DNA and two perspectives on evolutionary genetics. *Biol. J. Linn. Soc.* 26:375-400.

WINANS, G.A. and L.L. JONES. 1988. Electrophoretic variability in Dall's porpoise (*Phocoenoides dalli*) in the North Pacific ocean and Bering Sea. *J. Mamm.* 69:14–21.

WINN, H.E., W.L. BISCHOFF, and A.G. TARUSKI. 1973. Cytological sexing of Cetacea. *Mar. Biol.* 23:343–346.

WORLD CONSERVATION UNION. 1993. 1994 *IUCN Red List of Threatened Animals*. World Conservation Union, Gland, Switzerland.

YABLOKOV, A.V. 1994. Validity of whaling data. *Nature* 367:108.

3

CONSERVATION GENETICS
OF THE FELIDAE

Stephen J. O'Brien and Collaborators *

INTRODUCTION

The Felidae family comprises 38 living species that have fascinated natural-
ists, biologists, artists, children, and nearly everyone else for thousands of
years (Guggisberg, 1975; Nowak, 1991; Seidensticker and Lumpkin, 1991).
Because of their agility, speed, and efficient specialization as killer-carni-
vores, felids occupy the top of the trophic food chain in most habitats on
all continents. Yet their ferociousness and predatory elegance have also
instilled fear in humans, and persecution of felids has been an integral
component of human civilizations, particularly agrarian societies. Even today, many of the big cats are central targets of "problem animal" control
programs throughout the world—from Missoula, Montana (pumas), to
Windhoek, Namibia (cheetahs), to Sasan-Gir, India (lions), to Fos du
Iguazu, Brazil (jaguars). Man's antagonism toward large predators and the
constant pressures on natural habitats have led to the situation in which 37

* Janice S. Martenson, Sriyanie Miththapala, Dianne Janczewski, Jill Pecon-Slattery, Warren Johnson, Dennis A. Gilbert, Melody Roelke, Craig Packer, Mitchell Bush, and David E.
Wildt.

Box 3.1. Taxonomic and Conservation Status in the Family Felidae

Species (number of subspecies)	U.S. Fish and Wildlife Service listing	IUCN listing
Pantherine lineage		
Cheetah, *Acinonyx jubatus* (5)	Endangered	Vulnerable (1 subsp. Endangered)
Lion, *Panthera leo* (8)	Endangered (1 subsp. only)	Endangered (1 subsp. only)
Tiger, *Panthera tigris* (5)	Endangered	Endangered
Leopard, *Panthera pardus* (27)	Endangered	Vulnerable (2 subsp. Endangered)
Jaguar, *Panthera onca* (8)	Endangered	Vulnerable
Snow leopard, *Panthera uncia* (0)	Endangered	Endangered
Clouded leopard, *Neofelis nebulosa* (4)	Endangered	Vulnerable
Marbled cat, *Pardofelis marmorata* (2)	Endangered	
North American lynx, *Lynx canadensis* (2)	Threatened	
Bobcat, *Lynx rufus* (12)	Endangered (1 subsp. only)	
Eurasian lynx, *Lynx lynx* (7)	Threatened	
Spanish lynx, *Lynx pardina* (0)	Endangered	Vulnerable
Caracal, *Caracal caracal* (9)	Threatened	
Serval, *Leptailurus serval* (15)	Endangered (1 subsp. only)	
African golden cat, *Profelis aurata* (2)	Threatened	
Asian golden cat, *Profelis temmincki* (3)	Endangered	
Leopard cat, *Prionailurus bengalensis* (11)	Endangered (1 subsp. only)	
Fishing cat, *Prionailurus viverrina* (2)	Threatened	
Flat-headed cat, *Ictailurus planiceps* (0)	Endangered	
Rusty-spotted cat, *Prionailurus rubiginosa* (2)	Threatened	
Bay cat or Bornean red cat, *Profelis badia* (0)	Threatened	Rare
Iriomote cat, *Mayailurus iriomotensis* (0)	Endangered	Endangered
Jaguarundi, *Herpailurus yagouaroundi* (8)	Endangered (4 subsp. only)	
Puma, *Puma concolor* (30)	Endangered (3 subsp. only)	Endangered (2 subsp. only)

Box 3.1. *(cont.)*

Species (number of subspecies)	U.S. Fish and Wildlife Service listing	IUCN listing
Ocelot lineage		
Ocelot, *Leopardus pardalis* (11)	Endangered	Vulnerable
Margay, *Leopardus wiedii* (11)	Endangered	Vulnerable
Tigrina, *Leopardus tigrina* (4)	Endangered	Vulnerable
Kodkod, *Oncifelis guigna* (2)	Threatened	
Geoffroy's cat, *Oncifelis geoffroyi* (4)	Threatened	
Andean mountain cat, *Oreailurus jacobita* (2)	Endangered	Rare
Pampas cat, *Lynchailurus colocolo* (7)	Threatened	
Domestic cat lineage		
Wild cat, *Felis silvestris* (36) (includes *F.s. libyca*)	Threatened	Vulnerable
Pallas cat, *Otocolobus manul* (3)	Threatened	
Jungle cat, *Felis chaus* (10)	Threatened	
Black-footed cat, *Felis nigripes* (2)	Endangered	
Sand cat, *Felis margarita* (4)	Endangered (1 subsp. only)	Endangered (1 subsp. only)
Chinese desert cat, *Felis bieti* (3)	Threatened	

of the 38 species of extant felids are listed as either endangered or threatened by the international monitors of endangered species (Box 3.1). The only non-endangered felid is the one we do not fear, the domestic cat *Felis catus*, which numbers in the hundreds of millions of individuals worldwide and, ironically, has become a pest and threat to native wildlife in many areas.

Humankind's fascination with cats has led to countless scientific and non-scientific descriptions of their appearance, behavior, and relationships with man. They have been the subject of mythology, art, and even theology. Furthermore, both big and small cats have been captured and trained for hunting or circus performance, as well as displayed in zoos and bred in captivity. These experiences have given insights into the biology and behavior of cats, and have provided empirical opportunities for study of their reproduction, clinical health, and nutrition. This rich informational background, from rigorous science to mythical anecdote, provides a substantial assemblage of explanations and hypotheses about the intrinsic and extrinsic threats faced by felids. Unfortunately, much of what has been written of felid natural history is also unsubstantiated. The advent of molecular genetic technologies has begun to change this situation, in part, by providing the means of examining patterns of genomic variation in the cat family and in particular cat species. The data are revealing phylogenetic relationships of relevance to taxonomy, as well as elucidating characteristics of population structure that may directly affect species survival. By highlighting examples from particular species of Felidae, this chapter summarizes findings that relate to conservation of a group of organisms that some consider among evolution's most charismatic creations.

The methods of analysis will be familiar to readers of this book, since they have been applied to many endangered species (Schonewald–Cox et al., 1983; Avise, 1994; Li and Graur, 1991; O'Brien, 1994 a, 1994b). The molecular methods are the tools of molecular biology: a variety of procedures for assessing the extent and character of genomic variation in DNA sequences of individuals. Technologies from human, mouse, *Drosophila*, and bacterial and plant genetics have been used to track DNA differences in the felids, and various phylogenetic algorithms that reflect different philosophies for relating gene characters in an evolutionary or systematics sense have been applied (Weir, 1990; Li and Graur, 1991; Avise, 1994). Genetic partitions, evidence of inbreeding, and historical migration events have been interpreted against a framework of population genetic theory applied to free-ranging populations (O'Brien, 1994 a, 1994b). Finally, in a synthesis of species characteristics that assess the present and future disposition of felids, medical and ecological disciplines have been recruited to better describe population status on topics ranging from prey base to infectious disease to reproductive fitness.

Because many of the felid populations studied are endangered (e.g., African cheetah, Asiatic lion, Sumatran tiger), we do not have the scientific luxury of designing explicit experiments to test hypotheses. Rather, correlation and multidisciplinary inference are relied upon to draw conclusions. This approach is reminiscent of human genetic analysis, which is restrained from direct experimentation by ethical considerations. Despite these limitations, much has been (and remains to be) learned from descriptive applications of molecular methods to the ecological and evolutionary genetics of felids.

PHYLOGENY AS A BASIS FOR SPECIES RECOGNITION AND PROTECTION

Taxonomy, the systematic classification of plants and animals, had little relevance beyond academic institutions before the mid-1970s. Species were grouped according to morphological types into genera, genera into families, families into orders, and so on. Systematic uncertainties had little relevance to everyday life, and taxonomic resolution was limited. When taxonomic distinctions became the basis for legal protection afforded by the Endangered Species Act of 1973, this innocence was lost forever (U.S. Fish and Wildlife Service, 1973). Disagreements over taxonomic status fueled legal assaults on the Act, and misclassifications led to inappropriate conservation measures resulting in losses of some species. Errors of "oversplitting" and "overlumping" based on guesswork have led to mistaken legal judgments retrospectively revealed by molecular approaches (Avise and Nelson, 1989; Daugherty et al., 1990; O'Brien and Mayr, 1991; Wayne and Jenks, 1991; Geist, 1992). Even today, with vastly improved molecular methods for discriminating taxonomic groups, there remains considerable confusion about the units of conservation that the Endangered Species Act was designed to protect. Finally, because it is unlikely that all endangered species will be afforded equal protection and recourse, proposed bases for priority ranking of endangered species have been advanced. For example, one such criterion for conservation priority involved taxonomic distinctiveness, or depth of phylogenetic divergence (May, 1988). According to this philosophy, an aardvark, which has no close relatives, would rank higher for conservation concern than would a lion, which has several close (less than two million years away) phylogenetic relatives.

Despite the wide popularity of cats as research objects, there remains considerable confusion as to the evolutionary relationships among the living species (Hemmer, 1978; Leyhausen, 1979; Neff, 1982; Collier and O'Brien, 1985; Kitchener, 1991; Nowak, 1991; Salles, 1992). Cat taxonomy based on morphological and behavioral criteria produced earlier classification schemes that grouped cats into as few as two or as many as 19 genera (Nowak, 1991). A major reason for this confusion seems to be a rather recent adaptive radiation that has produced 37 distinct felid species all within the last 6–12 million years (Savage and Russell, 1983; Wayne et al., 1989; Nowak, 1991). Each species displays specific adaptations that play an important role in its ecological balance. For example, 10 African cat species are sympatric but occupy distinctive niches. The same is true for seven South American felids, whose common ancestor invaded the continent only after the formation of the Panama land bridge about 2–3 million years ago (Pecon–Slattery et al., 1994), yet which now exhibit a diverse array of ecological adaptations.

Molecular Phylogeny of the Felidae

Our research group has applied four molecular techniques (protein electrophoresis, microcomplement fixation, DNA–DNA hybridization and Gbanded karyology) to estimate phylogenetic relationships within the Felidae (Figure 3.1; Collier and O'Brien, 1985; O'Brien, 1986; O'Brien et al., 1987a; Wayne et al., 1989). These results did not resolve all relationships, but provided evidence for the occurrence of three primary radiations within the felids. Under the assumption of a molecular clock calibrated with paleontological fossil dates, the data suggest that the earliest separation occurred approximately 10 million years ago, when the ancestor to the South American ocelot lineage diverged. Seven to nine million years ago, the domestic cat lineage was formed, followed by a gradual divergence of the remaining large cats forming the pantherine lineage. The most recently derived species are a group of six large cats forming the *Panthera* genus (lion, tiger, jaguar, leopard, snow leopard, and clouded leopard). In agreement with molecular data (Collier and O'Brien, 1985; O'Brien, 1986), a study of 44 cranial characters in extant felid species (Salles, 1992) resolved three primary clades corresponding to *Leopardus* (ocelot), *Felis* (domestic cat), and the pantherine lineages. However, the morphological data differed by placing several of the pantherine cats (notably Asian golden and marbled cats) outside of the pantherine lineage.

Until recently, further resolution of the phylogenetic topology of the Felidae has been difficult or equivocal with both molecular and morphological approaches. Sequence analysis of mitochondrial (mt) DNA (12S RNA and cytochrome *b* genes) has given additional insight into the pantherine lineage (Janczewski et al., 1995). As illustrated in the majority-rule consensus topology presented in Figure 3.2, the mitochondrial genes analyzed with phenetic, maximum parsimony, and likelihood methods provide additional support for: (1) inclusion of clouded leopard (*Neofelis nebulosa*) in the *Panthera* genus; (2) monophyletic association of puma (*Puma concolor*) and cheetah (*Acinonyx jubatus*); and (3) polyphyletic origins of two golden cat species, *Profelis temmincki* and *P. aurata*. In the ocelot lineage, phylogenetic reconstruction based on high resolution 2DE gel proteins (Figure 3.3) and isozyme markers ($N = 40$ loci) indicated that a major split occurred approximately 5–6 million years ago, leading eventually to three phylogenetic groups (Pecon–Slattery et al., 1994). The earliest divergence led to *Leopardus tigrina*, followed by a split between an ancestor of an unresolved trichotomy of three species (*Oncifelis guigna, O. geoffroyi*, and *Lynchailurus colocolo*) and a recent common ancestor of *Leopardus pardalis* and *L. wiedii*.

The South American Radiation

The ocelot lineage provides an opportunity to study more closely the pattern of a recent monophyletic radiation, because it is known when the radiation likely began. Before the formation of the Panama land bridge

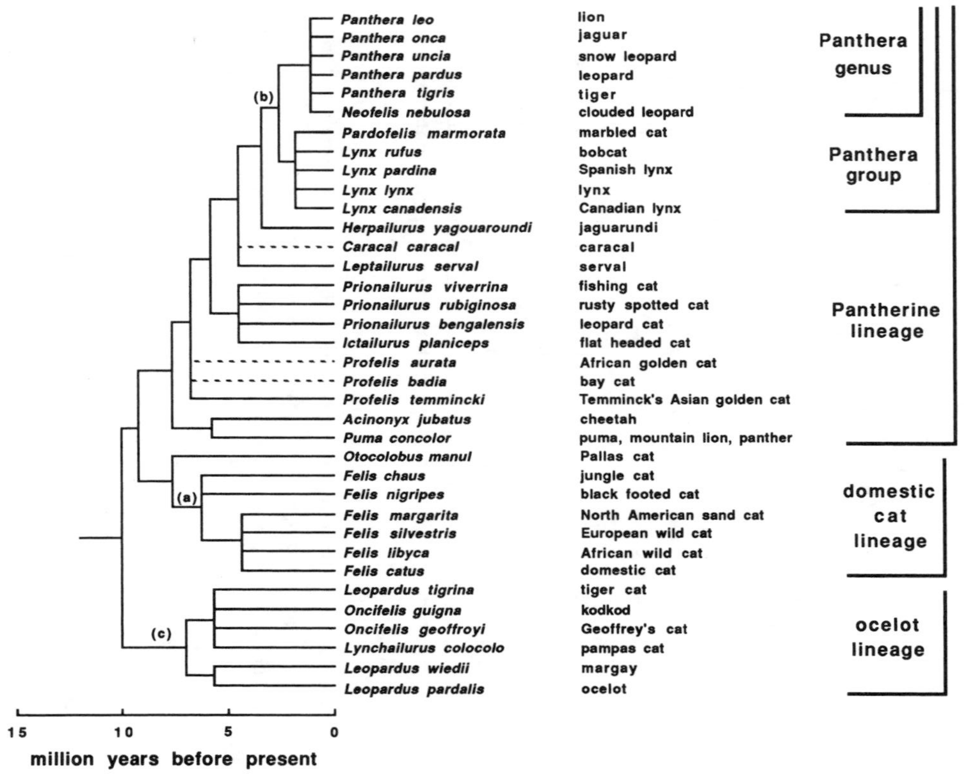

Figure 3.1. Phylogenetic relationships of Felidae based on immunological distances (Collier and O'Brien, 1985), isozyme electrophoresis (O'Brien et al., 1987c), karyology (Wurster-Hill and Gray, 1975; Wurster-Hill and Centerwall, 1982; Modi and O'Brien, 1988), and endogenous retroviruses (Benveniste, 1985): (a) indicates entry of endogenous retroviral families feline leukemia virus and RD-114; (b) indicates members of *Panthera* group with identical karyotypes; (c) indicates South American ocelot lineage, all species of which have 36 chromosomes and share a unique metacentric chromosome C3. Dashed lines indicate tentative placements of species. Classification used as per Kitchener (1991) modified from O'Brien (1986). (From Janczewski et al., 1995).

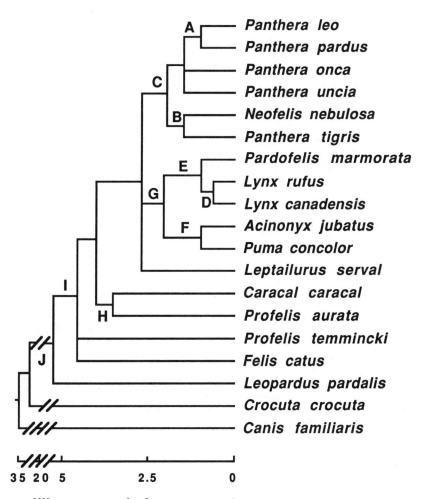

million years before present

Figure 3.2. Phylogeny of the Felidae based on maximum parsimony analysis of mtDNA sequences at the 12S RNA and cytochrome *b* genes, calibrated using the divergence from the hyena (*Crocuta crocuta*) at 20 million years ago (Hunt, 1989). Also included is another outgroup pecies, the dog *Canis familiaris*. Species associations shown here (A–J) represent those observed in a majority of the trees constructed from these genetic data, and are supported by isozymes, albumin immunological distance, karyology, DNA–DNA hybridization, endogenous retrovirus identification, and morphological data (Benveniste, 1985; Collier and O'Brien, 1985; O'Brien et al., 1987c; Modi and O'Brien, 1988). (From Janczewski et al., 1995).

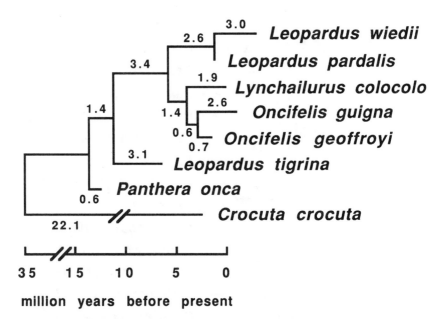

Figure 3.3. Phylogenetic reconstruction of South American felids (with the exception of *Oreailurus jacobita*) based on two-dimensional electrophoretic (2DE) data ($N = 548$ protein loci). Shown is a phenetic summary from the neighbor-joining method (in the Phylip 3.5 computer package) as applied to a distance matrix of Nei's (1978) unbiased minimum genetic distance (Pecon-Slattery et al., 1994). The hyena (*Crocuta crocuta*) is included as an outgroup. The observed inconsistency by which the estimated divergence times (5–10 MYA) of the extant South American species predates the formation of the Panama land bridge (2–3 MYA), can be attributed to either: (a) unequal rates of evolution in the 2DE metric among the different cat species, leading to an overestimation of the divergence times within South America; or (b) divergence of some of the species lineages prior to migration into South America.

between North and South America 2–3 million years ago, there were no carnivores (except marsupials) in South America (Stehli and Webb, 1985). The phylogenetic separation of the ocelot lineage (which was to lead to seven extant species) may have coincided with this event, and if so was very recent (but see Figure 3.3). We have initiated a collection of living "voucher specimens" (blood and skin fibroblast cell culture) from throughout the range of each species (Figure 3.4), with the initial intent of further examining the extent of intrinsic genetic diversity within and between species as assessed from rapidly evolving mitochondrial and nuclear microsatellite loci. Such information should soon permit a coupling of molecular variation with geographic isolation and evolutionary divergence. The major goal is to define precisely the "units" of conservation for each species based on present and historic genetic structure.

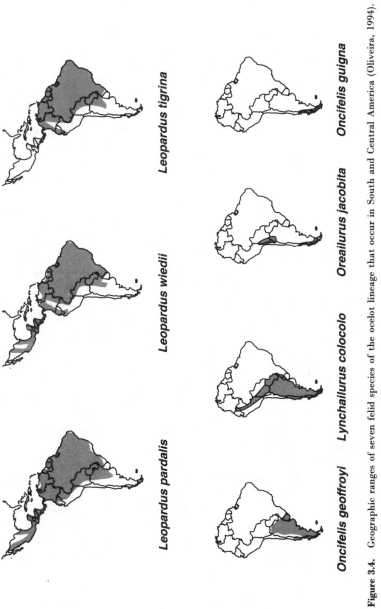

Leopardus tigrina

Leopardus wiedii

Leopardus pardalis

Oncifelis guigna

Oreailurus jacobita

Lynchailurus colocolo

Oncifelis geoffroyi

Figure 3.4. Geographic ranges of seven felid species of the ocelot lineage that occur in South and Central America (Oliveira, 1994).

RECOGNIZING CONSERVATION UNITS WITHIN SPECIES: THE SUBSPECIES QUESTION

The subspecies designation for differentiated, geographically isolated populations has been one of the most controversial topics in the biological sciences. Darwin (1859) was well aware of recognizable biotic divisions below the species level, and called these *races* or *incipient species* because they appeared to be preludes to species development. Mayr (1963) defined a *subspecies* as "an aggregate of local populations of a species, inhabiting a geographic subdivision of the range of the species, and differing taxonomically from other populations of the species." Dobzhansky (1937) termed *subspecies* as "any populations sufficiently distinct to merit a Latin name." More recently, Avise and Ball (1990) attempted to provide objective criteria for subspecies recognition by recommending that subspecies designation should be reserved for populations displaying concordant distinctions in multiple, independent, genetically based traits. O'Brien and Mayr (1991) synthesized the subspecies classification to include "individual populations that share a unique geographic range or habitat, a group of phylogenetically concordant phenotypic characters and a unique natural history relative to other subdivisions of the species."

Many felid species have been extensively subdivided taxonomically, largely by 19th century mammalogists who named populations with little more criteria than a hide or skeleton from a particular geographic locale. As an extreme, there are 27 named subspecies of leopard (Figure 3.5), 30 subspecies of puma, and five subspecies of tiger. The bases of these designations are soft. However, contrary to some authors (Ehrlich and Raven, 1969; Cracraft, 1989), we believe that a formally demonstrated subspecies category is an important unit of conservation because of two factors (O'Brien and Mayr, 1991). The first is the potential to become a new species given sufficient time (Darwin's incipient species). The second is the acquisition of ecologically relevant adaptations during allopatric separation. Of course, it is difficult to say which particular subspecies will realize either or both criteria, but every subspecies has at least the potential.

To provide greater historical genetic rigor to subspecies recognition, the disposition of molecular markers (isozymes, mtDNA restriction fragments, nuclear DNA fingerprints) has been examined in a group of leopard subspecies (Miththapala, 1992). By examining population genetic structure of 90 leopards from 14 named subspecies (see Figure 3.5), the molecular data were sufficient to resolve eight subspecies partitions, and to permit the recognition of several identified groups as distinct populations. For example, all African subspecies were virtually indistinguishable, whereas island populations from Sri Lanka and Java shared several distinctive molecular genetic

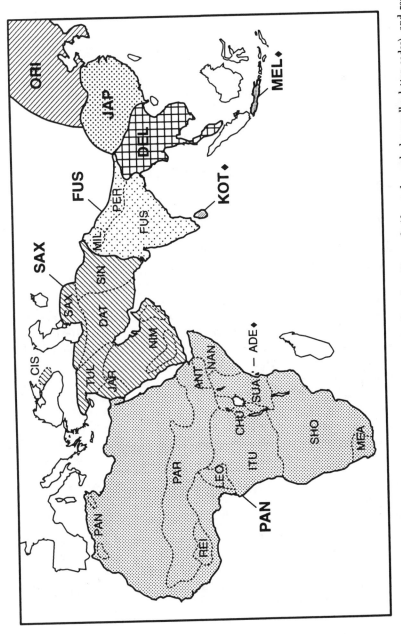

Figure 3.5. Described geographic ranges of 27 previously recognized subspecies of leopards (denoted mostly by smaller letter codes), and recent revision to eight verifiable subspecies partitions (larger, bold letter codes) based on a combined molecular and morphological analysis (Miththapala, 1992). Different hatchings and patterns indicate geographic ranges of the revised subspecies.

61

characters. A parallel morphological analysis of metric cranial characters in leopards affirmed the conclusions of the molecular study, adding credence to the recommendation that the 27 original subspecies be subsumed into eight verifiable subspecies. A similar analysis with puma and tiger subspecies, but utilizing more powerful genetic assays (of mtDNA control-region sequences, microsatellites, and major histocompatibility complex (MHC) Class II sequences) is underway. These results should lead to formal definitions of feline subspecies, which in turn should provide a more solid basis for conservation efforts.

WHEN ENDANGERED SPECIES HYBRIDIZE IN NATURE

An area that has led to confusion and to legal assaults on protection involves the question of *in situ* hybridization of endangered species or subspecies. Historically, the U.S. Fish and Wildlife Service had interpreted that "hybrids" between taxa listed by the Endangered Species Act would not be eligible for protection, a directive largely intended to concentrate management responses on "pure" endangered species (O'Brien and Mayr, 1991). Thus, when molecular genetics revealed natural hybridization involving the Florida panther (Box 3.2), or geographically restricted hybridization between wolves and coyotes (Lehman et al., 1991; Chapter 4), litigious challenges to the protection of these endangered species were based on the precedent of the so-called "hybrid policy." Fortunately, the hybrid policy was suspended after O'Brien and Mayr (1991) argued that these sorts of hybrid events were natural outcomes of evolution, and that the species should not be penalized due to a bureaucratic precedent that did not anticipate the resolving power of molecular genetics.

THE BIG CATS ILLUSTRATE THE COST OF INBREEDING

More than a decade ago, the first results were obtained indicating that African cheetahs (*Acinonyx jubatus*) had significantly less genomic variation than did other felid or mammal species (based on a survey of allozymes, and cellular proteins resolved by two-dimensional polyacrylamide gel electrophoresis [PAGE; O'Brien et al., 1983]). In subsequent studies, additional molecular genetic data confirmed the genetic uniformity of this species (see Table 3.1), and led to the conclusion that the cheetah's ancestors experienced a severe demographic reduction and extreme inbreeding several thou-

Box 3.2. The Florida Panther: A Lesson about Subspecies Hybridization in Nature

The Florida panther is a small relict population of mountain lion (also called cougar or puma) descended from the subspecies *Puma concolor coryi* that ranged throughout the southern United States in the 19th century (O'Brien et al., 1990; Roelke et al., 1993). Less than 50 animals survive today in the Big Cypress Swamp and Everglades ecosystem in southern Florida. The remaining panthers show physiological impairments as a consequence of inbreeding depression, including a 90° kink in the tail vertebrae, a cowlick in the middorsal back, and reproductive defects (see text). The Florida panther was listed as endangered by the U.S. Fish and Wildlife Service in 1967.

In the late 1980s, the existence of two highly distinct genetic stocks was discovered when two family groups appeared in the Everglades. The animals seemed different from the larger population in the adjacent Big Cypress Swamp because they lacked the cowlick and tail kink diagnostic for the subspecies. An allozyme and mtDNA–RFLP analysis showed that the Everglades pumas were indeed distinct from the Big Cypress animals, as well as from pumas in the western U.S. (O'Brien et al., 1990). Phylogenetically, they were closer to puma subspecies from South and Central America. Retrospective inspection of the archives of the Everglades National Park revealed an explanation. Between the years 1957 and 1967, in cooperation with National Park Service officials, seven animals from a captive stock were released into the Everglades (and promptly forgotten). This stock was a mixture of authentic North American *P. concolor coryi* and South American founders. Today, the Everglades population contains a mixture of these two distinct puma lineages.

The genetic advantages of introducing some additional genetic material into a population suffering from inbreeding would have been comforting except for one detail. Three independent opinions from the Solicitor's Office (the legal counsel of the U.S. Fish and Wildlife Service) have ruled with the force of precedent that hybrids between endangered taxa (species, subspecies, or populations) cannot be protected. Their opinions, known as the "Hybrid Policy," concluded that protection of hybrids would not serve to recover listed species, and would likely jeopardize that species' continued existence. My collaborators and I were in an untenable position, wherein publication of the new data would threaten the Florida panther's endangered species status. The solution came when U.S. Fish and Wildlife suspended the hybrid policy indefinitely, largely in response to this "catch-22" of the Florida panther (O'Brien and Mayr, 1991).

sand years ago (likely at the end of the Pleistocene, the time of the most recent Northern Hemisphere glaciation; O'Brien et al., 1983, 1985, 1987d; Wayne et al., 1986; Yuhki and O'Brien, 1990; Menotti-Raymond and O'Brien, 1993). Overall, genetic variability of modern cheetahs was reduced by one to two orders of magnitude compared to other large cat species,

Table 3.1. Evidence for Genetic Uniformity in Cheetahs, and Observed Physiological Correlates

A. Methods indicating reduced genetic variation

	References
1. Allozymes—52 loci	O' Brien et al., 1983, 1987d
2. Two-dimensional PAGE—155 loci	O'Brien et al., 1983
3. Allogeneic skin graft accepted	O'Brien et al., 1985
4. MHC-RFLPs—(six restriction enzymes)	Yuhki & O'Brien, 1990
5. mtDNA—RFLPs	Menotti-Raymond & O'Brien, 1993
6. Microsatellite loci	Menotti-Raymond & O'Brien, 1995
7. Increased fluctuating asymmetry of skeletal measurement	Wayne et al., 1986

B. Physiological correlates

	References
1. Diminished sperm count	O'Brien et al., 1983; Wildt et al., 1983, 1987b
2. Elevated frequency of morphological abnormalities in sperm development (~70%)	O'Brien et al., 1983; Wildt et al., 1983, 1987b
3. Low fecundity in captive breeding attemps	Marker & O'Brien, 1989; Marker-Kraus & Grisham, 1993
4. Captive population is not self-sustaining	Marker & O'Brien, 1989; Marker-Kraus & Grisham, 1993
5. Relatively high incidence (~30%) juvenile mortality even among unrelated parents	O'Brien et al., 1985; Marker-Kraus & O'Brien, 1989
6. Increased population vulnerability to infectious disease outbreaks, notably feline infectious peritonitis	O'Brien et al., 1985; Heeney et al., 1990

Table 3.2. Correlation of Genetic Variation and Reproductive Parameters in Three Lion Populations[1]

Parameter	Serengeti, Tanzania	Ngorongoro Crater, Tanzania	Gir Forest, India
Genetic properties			
Allozyme heterozygosity (%)	3.1	1.5	0.0
Mean % difference			
MHC-RFLPs	21.8	8.0	0.0
DNA fingerprints	48.1	43.5	2.8
Reproductive measures			
Sperm count (X 10^{-6})	34.4 ± 12.8	25.8 ± 11.0	3.3 ± 2.8
% sperm abnormality	24.8 ± 4.0	50.5 ± 6.8	66.2 ± 3.6
No. motile sperm per ejaculate (X 10^{-6})	228.5 ± 65.5	236.0 ± 93.0	45.3 ± 9.9
Testosterone, ng/ml	1.3 − 1.7	0.5 − 0.6	0.1 − 0.3

[1] O'Brien et al. (1987b, 1987c), Wildt et al. (1987a), Yuhki and O'Brien (1990), and Gilbert et al. (1991). When indicated, data are the mean ± standard error of mean.

Table 3.3. Evidence for Genetic Uniformity in the Florida Panther [*Felis (Puma) concolor coryi*] and Observed Physiological Correlates[1]

A. Methods indicating reduced genomic variation

   1. Allozymes—41 loci
   2. mtDNA—RFLPs
   3. DNA fingerprints—feline minisatellites
   4. Aberrant morphological characters
     a. Kink tail
     b. Cowlick

B. Physiological correlates relative to other puma populations

   1. Spermatazoal abnormalities
     a. Diminished sperm count
     b. Low motile sperm per ejaculate
     c. Elevated frequency of morphological abnormalities in sperm development (~ 95%)
     d. 48% abnormal acrosomes in sperm development
   2. Rapid rise to 80% cryptorchidism
   3. Heart murmurs and fatal atrial septal defects
   4. Multiple and widespread exposure to feline pathogenic infectious agents

[1] O'Brien et al. (1990), O'Brien and Mayr (1991), Roelke et al. (1993), and Barone et al. (1994).

Box 3.3. Conservation Concerns for the Asiatic Lion

As recently as 200 years ago, the Asiatic lion (*Panthera leo persica*) occupied a wide range extending from Syria to northern India. The advance of agriculture, the increased use of firearms, and other familiar companions of human population pressure brought the subspecies to extinction in Syria, Iraq, Iran, Afghanistan, and Pakistan in the latter part of the 19th century. The last remaining population consists of about 250 individuals in a 1,400 km^2 area in the Gir Forest in Gujarat State of western India. Published estimates of fewer than 20 lions in the early 1900s led to a complete prohibition of hunting in 1900. The tiny population survives today, but is severely limited by available habitat and by proximity to human agricultural settlements. In the last decade, over 100 lion attacks on humans have occurred, resulting in twelve fatalities. Man-eaters are captured, held in the Sakkarburg Zoo in Junagadh, and used for captive breeding and for research.

A captive breeding program under the Species Survival Plan (SSP) in U.S. zoos was initiated from five founders in 1981. The program was highly successful in producing lions, but the disappearance of phenotypic traits found in free-ranging Asian lions (reduced mane and belly fold) led to a molecular genetic study. The results of an allozyme analysis showed that two of the original founder animals were actually African lions, and that the SSP lions probably were doing so well because of hybrid vigor (O'Brien et al., 1987b, 1987c). The Asiatic lions thus provided an unexpected natural example of the direct improvement that can be achieved by hybridization between subspecies, particularly when one subspecies had a history of inbreeding and associated inbreeding depression (Wildt et al., 1987a).

---

suggesting that the proposed bottleneck likely persisted for several generations or perhaps occurred several times (Nei et al., 1975; O'Brien et al., 1987d). In addition, cheetahs underwent a significant range retraction whereby they disappeared from North America, Europe, and parts of Asia at the end of the Pleistocene.

Correlated with the genetic uniformity of cheetahs are a number of physiological impairments that influence reproduction and contribute to a difficulty in establishing self-sustaining captive populations (Table 3.1). Sperm abnormalities are observed in both free-ranging and captive cheetahs and likely play a key role in the difficulty in achieving a self-sustaining captive population (Marker and O'Brien, 1989; Marker-Kraus and Grisham, 1993). Furthermore, an extreme morbidity and mortality of cheetahs from outbreaks of a nearly benign domestic cat virus (feline infectious peritonitis virus) was interpreted to be a consequence of the homogeneous state of genes that mediate immune defenses (O'Brien et al., 1985; Heeney et al., 1990). Several of these immunological loci, particularly the MHC, are highly variable in other feline and mammalian species. The evolutionary explanation for high variation among immune response loci would be that they offer a "moving target" for microbial pathogens that themselves rapidly evolve genetic adaptations that override the immune defenses of individuals (O'Brien and Evermann, 1988).

The cheetah example has served as a paradigm of the potential for hidden perils that threaten small populations from within. The interpretations have been reinforced by the variety of genetic observations that support the bottleneck hypothesis. Evidence of the cheetah's genetic uniformity was obtained with seven distinct measures of genomic variation (Table 3.1). The three genomic assays that did reveal modest variation in cheetahs (mtDNA–RFLPs, DNA fingerprints, and microsatellites) involve rapidly evolving DNA sequences that likely accumulated variation by mutation after the hypothesized bottleneck (Menotti-Raymond and O'Brien, 1993, 1995). The concordant genetic diminution of multiple estimators of genomic variation lent strong support to the hypothesis of close inbreeding resulting from demographic collapse(s) in the cheetah's history.

The developing legacy of the cheetah led to investigations of the genetic structure of several other endangered felid species. Three other big cat populations are now documented to have suffered severe historical population contractions, leading to inbreeding and probable physiological consequences: lions in the Ngorongoro Crater in Tanzania, Asiatic lions from the Gir Forest in India (Table 3.2; Box 3.3), and pumas in Florida (Table 3.3). Because estimated levels of genomic variation within species are relative measures, the populations of Florida puma and the Ngorongoro and Gir Forest lions provided the equivalent of case-controlled studies for the consequences of genetic depletion (O'Brien et al., 1987b, 1987c, 1990; Wildt et al., 1987a; Roelke et al., 1993). The two lion populations had dramatically reduced variation in allozymes, MHC loci, and DNA minisatellite fingerprints relative to a larger outbred lion population living in the Serengeti (Figure 3.6; O'Brien et al., 1987c; Yuhki and O'Brien, 1990; Gilbert et al., 1991; Packer et al., 1991). Both lion populations also displayed elevated sperm abnormalities and a large reduction in circulating testosterone concentration relative to their outbred Serengeti lion counterparts (O'Brien et al., 1987b; Wildt et al., 1987a). Reproduction by the most genetically deficient lion population (the Asiatic lion) is known to be severely compromised in captive settings (O'Brien et al., 1987b). Furthermore, when Asiatic lions were inadvertently bred to African lion subspecies in North America, the fecundity, reproductive success, and spermatozoal development improved dramatically (Box 3.3; O'Brien et al., 1987b). Combined with the results for the cheetah, it seems clear that inbreeding, at least in the Felidae, can have a direct effect on reproductive performance.

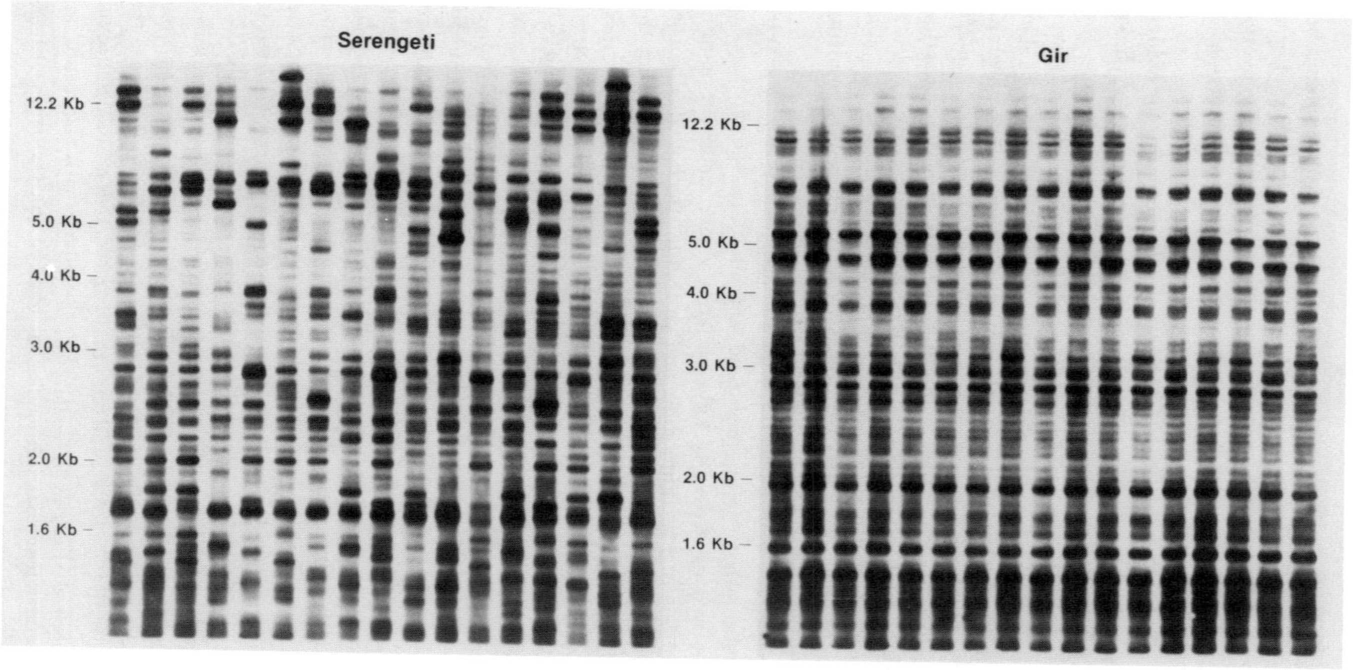

Figure 3.6. DNA fingerprint patterns of unrelated lions from the Serengeti National Park, Tanzania, and from the Gir Forest Sanctuary, India. Both Southern blots contain DNA digested with *Msp*I and hybridized with *FC28* feline-specific minisatellite probe (Gilbert et al., 1991). The genetic uniformity of the Gir Forest population relative to the Serengeti population is evident (especially so in the original gels).

68

A compelling addition to this inference came from the dramatic story of the Florida panther (*Puma concolor coryi*), a native American subspecies of puma (O'Brien et al., 1990; O'Brien and Mayr, 1991; Roelke et al., 1993). Human depredation spurred principally by fear and legends of ferociousness toward livestock and mankind, plus the imposition of bounties, reduced the panther's range from the entire American Southeast to hardwood swamps and adjoining ecosystems in the Everglades National Park and Big Cypress Reserve of south Florida, the only habitat east of the Mississippi occupied by wild pumas today. The major threats to the Florida panther were thought to have been mortality factors, with road kills and illegal hunting accounting for 63% of documented mortalities since 1973.

Genetic studies (Table 3.3) revealed that the Florida panther retained less genomic variation than any puma subspecies from North or South America; also, several cases of incestuous (father/daughter) matings were documented *in situ* (Roelke et al., 1993). The cost of inbreeding in this population was dramatic. Florida panthers have the poorest sperm observed in any felid species; about 95% of the sperm in each ejaculate are malformed (Roelke et al., 1993; Barone et al., 1994). The incidence of cryptorchidism, a rare heritable defect that causes either one or both testicles to remain unde-scended, has risen from 0% to 80% in males born in the last 15 years. In addition, a new fatal, congenital cardiac abnormality has recently appeared in four panther kittens. Finally, Florida panthers are riddled with patho-logical viruses, bacteria, and parasites, and these represent a time bomb waiting to explode as the animals develop debilitating disease. One of the viruses endemic in Florida panthers is a close relative of the feline version of the human AIDS virus—feline immunodeficiency virus (FIV; Olmsted et al., 1992). FIV causes severe immunodeficiency in domestic cats, but whether it causes disease in panthers is not yet certain.

The lessons learned from these studies of population genetic variability in the felids (Tables 3.1–3.3) are several. First, there can be hidden genetic perils, not so apparent from traditional ecological observations, that threaten natural populations. Second, when populations drop to very low numbers (as most endangered species by definition do), the associated gen-etic depletion from genetic drift and close inbreeding carries the risk of inbreeding depression and the expression of congenital abnormalities result-ing from homozygosity of rare deleterious genes. These genes can affect any aspect of development, survivorship, or reproduction in an unpredictable manner. Third, in addition to these heritable defects, inbreeding homogen-izes variation at abundantly polymorphic genes that mediate the immune response, thereby increasing the population's risk of extinction from patho-gens that abrogate the immune defenses of an individual.

CONCLUSIONS

The last decade has seen the emergence of a field that applies the principles and methods of population genetics to species conservation. As for other areas of molecular biotechnology, conservation genetics is an applied science with one important goal being to explicitly describe the composite genomes of small endangered populations. By comparison to well-studied examples, such as those reviewed here, one can make realistic approximations of the recent natural history, present status, and future prognosis of endangered populations. When combined with data from other disciplines (e.g., reproductive biology, epidemiology and the study of infectious disease, and field ecology), the synthesis offers some valuable insights that can be applied directly to species management plans.

The cats comprise but a small fraction of the nearly 5,000 described species of mammals. Close examinations of their molecular genetic structure, and integration of the genetic information with ecological, reproductive, medical, and natural history data have added to an understanding of factors that should be considered in efficacious management plans for these as well as other endangered species groups. Not all species will be saved in the coming decades; indeed, extinction has been a natural process since long before the accelerating influence of humans. However, equipped with the technologies and accumulated knowledge already collected on many of these endangered felids, it is now possible to identify and to address some of the many threats to their continued survival.

SUMMARY

1. All of the world's 37 extant wild species of Felidae are threatened or endangered, often as a direct result of persecution by humans. Ironically, as impressive top-level carnivores, these cat species also hold a special fascination for humans, making them especially important subjects for conservation efforts. Here, an overview is provided of how molecular genetic methods have been employed to ascertain the evolutionary histories, present conservation genetic status, and prospects for the future survival of felid species.

2. A variety of molecular genetic assays, including one- and two-dimensional protein electrophoresis, microcomplement fixation, DNA–DNA hybridization, and direct DNA sequencing, have been employed to examine phylogenetic relationships among felid species. Although these molecular phylogenetic analyses have identified certain evolutionary lineages (e.g., the ocelot lineage, domestic cat lineage, and pantherine lineage), several unresol-

ved clusters of species remain. The lack of resolution is presumably related to the relatively recent and rapid radiation of extant felid species, which has taken place within the last 6–12 million years.

3. Below the taxonomic level of species, most felids were probably "oversplit" into numerous subspecies by 19th century mammalogists. For example, more than 25 subspecies of leopards were traditionally recognized but often poorly defined. Molecular assays provide opportunities for reexamining geographic variation in multiple attributes with known genetic basis, and thereby can yield more robust information for the development of intraspecific taxonomies of relevance to conservation and management. In leopards, for example, such molecular studies indicate that only about eight geographic subspecies probably warrant formal recognition.

4. The genetic and conservation consequences of hybridization between subspecies are discussed, with particular reference to the Florida panther. In this isolated and endangered population in southern Florida, an infusion of genes from South American panthers (following the release of some individuals from a captive stock) has been documented using molecular markers. In this case, the hybridization may actually be of fitness benefit to the recipient population, which was highly inbred. In any event, the genetic findings led to a reinterpretation of an earlier legal directive (the "hybrid policy") under the Endangered Species Act.

5. The severe cost of inbreeding in small populations is well-evidenced by several endangered felid species, including the cheetah, Asiatic lion, and Florida panther. In these populations, a severe depletion of genetic variability (as estimated by several molecular genetic assays) is associated with diminished performance in a number of physiological and reproductive measures related to fitness.

REFERENCES

AVISE, J.C. 1991. Ten unorthodox perspectives on evolution prompted by comparative population genetic findings on mitochondrial DNA. *Annu. Rev. Genet.* 25:45–69.

AVISE, J.C. 1994. *Molecular Markers, Natural History and Evolution.* Chapman & Hall, New York.

AVISE, J.C. and R.M. BALL, Jr. 1990. Principles of genealogical concordance in species concepts and biological taxonomy. *Oxford Surv. Evol. Biol.* 7:45–67.

AVISE, J.C. and W.S. NELSON. 1989. Molecular genetic relationships of the extinct dusky seaside sparrow. *Science* 243:646–648.

BARONE, M.A., M.E. ROELKE, J.G. HOWARD, J.L. BROWN, A.E. ANDERSON, and D.E. WILDT. 1994. Reproductive characteristics of male Florida panthers: comparative studies from Florida, Texas, Colorado, Latin America, and North American zoos. *J. Mamm.* 75:150–162.

BENVENISTE, R.E. 1985. The contributions of retroviruses to the study of mammalian evolution. In *Molecular Evolutionary Genetics* (Monographs in Evolutionary Biology Series), ed. R. MacIntyre, pp. 359–417. Plenum, New York.

COLLIER, G.E. and S.J. O'BRIEN. 1985. A molecular phylogeny of the Felidae: immunological distance. *Evolution* 39:473–487.

CRACRAFT, J. 1989. Speciation and its ontology: The empirical consequences of alternative species concepts for understanding patterns and processes of differentiation. In *Speciation and Its Consequences*, eds. D. Otte and J.A. Endler, pp. 28–59. Sinauer, Sunderland, MA.

DARWIN, C.R. 1859 (reprinted 1964). *On the Origin of Species by Means of Natural Selection.* Harvard University Press, Cambridge, MA.

DAUGHERTY, C.H., A. CREE, J.M. HAY, and M.B. THOMPSON. 1990. Neglected taxonomy and continuing extinctions of tuatara (*Sphenodon*). *Nature* 347:177–179.

DOBZHANSKY, T. 1937. *Genetics and the Origin of Species.* Columbia University Press, New York.

EHRLICH, P. and P. RAVEN. 1969. Differentiation of populations. *Science* 165:1228–1232.

GEIST, V. 1992. Endangered species and the law. *Nature* 357:274–276.

GILBERT, D.A., C. PACKER, A.E. PUSEY, J.C. STEPHENS, and S.J. O'BRIEN. 1991. Analytical DNA fingerprinting in lions: parentage, genetic diversity, and kinship. *J. Hered.* 82:378–386.

GUGGISBERG, C.A. 1975. *Wild Cats of the World.* Taplinger, New York.

HEENEY, J.L., J.F. EVERMANN, A.J. McKEIRNAN, L. MARKER-KRAUS, M.E. ROELKE, M. BUSH, D.E. WILDT, D.G. MELTZER, L. COLLY, J. LUCAS, V.J. MANTON, T. CARO, and S.J. O'BRIEN. 1990. Prevalence and implications of feline coronavirus infections of captive and free-ranging cheetahs (*Acinonyx jubatus*). *J. Virol.* 64:1964–1972.

HEMMER, H. 1978. The evolutionary systematics of living Felidae: present status and current problems. *Carnivore* 1:71–79.

HUNT, R.M. 1989. Evolution of the Aeleuroid Carnivora: significance of the ventral promotorial process of the petrosal, and the origin of basicranial patterns in the living families. *Amer. Mus. Novit.* 2930:1–32.

JANCZEWSKI, D.N., W.S. MODI, J.C. STEPHENS, and S.J. O'BRIEN. 1995. Molecular evolution of mitochondrial 12S RNA and cytochrome b sequences in the pantherine lineage of Felidae. *Molec. Biol. Evol.* 12:690–707.

KITCHENER, A. 1991. *The Natural History of the Wild Cats.* Cornell University Press, Ithaca, NY.

LEHMAN, N., A. EISENHAWER, K. HANSEN, L.D. MECH, R.O. PETERSON, P.J.P. GOGAN, and R.K. WAYNE. 1991. Introgression of coyote mitochondrial DNA into sympatric North American gray wolf populations. *Evolution* 45:104–119.

LEYHAUSEN, P. 1979. *Cat Behavior.* Garland STPM Press, New York.

LI, W.-H. and D. GRAUR. 1991. *Fundamentals of Molecular Evolution*, Sinauer, Sunderland, MA.

MARKER, L. and S.J. O'BRIEN. 1989. Captive breeding of the cheetah (*Acinonyx jubatus*) in North American zoos. *Zoo Biol.* 8:3–16.

MARKER-KRAUS, L. and J. GRISHAM. 1993. Captive breeding of cheetahs in North American zoos: 1987–1991. *Zoo Biol.* 12:5–18.

MAY, R.T. 1988. Conservation and disease. *Conserv. Biol.* 2:28–30.

MAYR, E. 1963. *Animal Species and Evolution.* McGraw-Hill, New York.

MENOTTI-RAYMOND, M. and S.J. O'BRIEN. 1993. Dating the genetic bottleneck of the African cheetah. *Proc. Natl. Acad. Sci. USA* 90:3172–3176.

MENOTTI-RAYMOND, M.A. and S.J. O'BRIEN. 1995. Evolutionary conservation of ten microsatellite loci in four species of Felidae. *J. Hered.* 86:319–322.

MITHTHAPALA, S. 1992. Genetic and morphological variation in the leopard (*Panthera pardus*): A geographically widespread species. PhD dissertation, University of Florida, Gainesville.

MODI, W.S. and S.J. O'BRIEN. 1988. Quantitative cladistic analyses of chromosomal banding data among species in three orders of mammals: Hominoid primates, felids and arvicolid rodents. In *Chromosome Structure and Function*, eds. J.P. Gustafson and R. Appels, pp. 215–242. Plenum, New York.

NEFF, N.A. 1982. *The Big Cats: The Paintings of Guy Coheleach.* Abrams, New York.

NEI, M. 1978. Estimation of average heterozygosity and genetic distance from a small number of individuals. *Genetics* 23:341–369.

NEI, M., T. MARUYAMA, and R. CHAKRABORTY. 1975. The bottleneck effect and genetic variability in populations. *Evolution* 29:1–10.

NOWAK, R.M. 1991. *Walker's Mammals of the World*, 5th ed. Johns Hopkins University Press, Baltimore, MD.

O'BRIEN, S.J. 1986. Molecular genetics in the domestic cat and its relatives. *Trends Genet.* 2:137–142.

O'BRIEN, S.J. 1994a. A role for molecular genetics in biological conservation. *Proc. Natl. Acad. Sci. USA* 91:5748–5755.

O'BRIEN, S.J. 1994b. Genetic and phylogenetic analyses of endangered species. *Annu. Rev. Genet.* 28:467–489.

O'BRIEN, S.J., G.E. COLLIER, R.E. BENVENISTE, W.G. NASH, A.K. NEWMAN, J.M. SIMONSON, M.A. EICHELBERGER, U.S. SEAL, M. BUSH, and D.E. WILDT. 1987a. Setting the molecular clock in Felidae: The great cats, *Panthera*. In *Tigers of the World: The Biology, Biopolitics, Management and Conservation of an Endangered Species*, eds. R.L. Tilson and U.S. Seal, pp. 10–27. Noyes, Park Ridge, NJ.

O'BRIEN, S.J. and J.F. EVERMANN. 1988. Interactive influence of infective disease and genetic diversity in natural populations. *Trends Ecol. Evol.* 3:254–259.

O'BRIEN, S.J., P. JOSLIN, G.L. SMITH III, R. WOLFE, N. SCHAFFER, E. HEATH, J. OTT-JOSLIN, P.P. RAWAL, K.K. BHATTACHERJEE, and J.S. MARTENSON. 1987b. Evidence for African origins of founders of the Asiatic lion species survival plan. *Zoo Biol.* 6:99–116.

O'BRIEN, S.J., J.S. MARTENSON, C. PACKER, L. HERBST, V. DE VOS, P. JOSLIN, J. OTT-JOSLIN, D.E. WILDT, and M. BUSH. 1987c. Biochemical genetic variation in geographic isolates of African and Asiatic lions. *Natl. Geog. Res.* 3:114–124.

O'BRIEN, S.J. and E. MAYR. 1991. Bureaucratic mischief: Recognizing endangered species and subspecies. *Science* 251:1187–1188.

O'BRIEN, S.J., M.E. ROELKE, L. MARKER, A. NEWMAN, C.A. WINKLER, D. MELTZER, L. COLLY, J.F. EVERMANN, M. BUSH, and D.E. WILDT. 1985. Genetic basis for species vulnerability in the cheetah. *Science* 227:1428–1434.

O'BRIEN, S.J., M.E. ROELKE, N. YUHKI, K.W. RICHARDS, W.E. JOHNSON, W.L. FRANKLIN, A.E. ANDERSON, O.L. BASS, Jr, R.C. BELDEN, and J.S. MARTENSON. 1990. Genetic introgression within the Florida panther *Felis concolor coryi*. *Natl. Geog. Res.* 6:485–494.

O'BRIEN, S.J., D.E. WILDT, M. BUSH, T.M. CARO, C. FITZGIBBON, I. AGGUNDEY, and R.E. LEAKEY. 1987d. East African cheetahs: evidence for two population bottlenecks. *Proc. Natl. Acad. Sci. USA* 84:508–511.

O'BRIEN, S.J., D.E. WILDT, D. GOLDMAN, C.R. MERRIL, and M. BUSH. 1983. The cheetah is depauperate in genetic variation. *Science* 221:459–462.

OLIVEIRA, T.G. 1994. *Neotropical Cats: Ecology and Conservation.* Press of the Universidade Federal de Maranhao, Sao Luis, Brazil.

OLMSTED, R.A., R. LANGLEY, M.E. ROELKE, R.M. GOEKEN, D. ADGER–JOHNSON, J.P. GOFF, J.P. ALBERT, C. PACKER, M.K. LAURENSON, T.M. CARO, L. SCHEEPERS, D.E. WILDT, M. BUSH, J.S. MARTENSON, and S.J. O'BRIEN. 1992. Worldwide prevalence of lentivirus infection in wild feline species: epidemiologic and phylogenetic aspects. *J. Virol.* 66:6008–6018.

PACKER, C., D.A. GILBERT, A.E. PUSEY, and S.J. O'BRIEN. 1991. Kinship, cooperation and inbreeding in African lions: A molecular genetic analysis. *Nature* 351:562–565.

PECON-SLATTERY, J., W.E. JOHNSON, D. GOLDMAN, and S.J. O'BRIEN. 1994. Phylogenetic reconstruction of South American felids defined by protein electrophoresis. *J. Molec. Evol.* 39:296–305.

ROELKE, M.E., J.S. MARTENSON, and S.J. O'BRIEN. 1993. The consequences of demographic reduction and genetic depletion in the endangered Florida panther. *Curr. Biol.* 3:340–350.

SALLES, L.O. 1992. Felid phylogenetics: extant taxa and skull morphology (Felidae, Aeluroidae). *Amer. Mus. Novit.* 3047:1–67.

SAVAGE, D.E. and D.E. RUSSELL. 1983. *Mammalian Paleofaunas of the World.* Addison-Wesley, London.

SCHONEWALD-COX, C.M., S.M. CHAMBERS, B. MACBRYDE, and L. THOMAS (eds.). 1983. *Genetics and Conservation: A Reference for Managing Wild Animal and Plant Populations.* Benjamin/Cummings, Menlo Park, CA.

SEIDENSTICKER, J. and S. LUMPKIN (eds.). 1991. *Great Cats: Majestic Creatures of the Wild,* pp. 1–240. Weldon Owen, Sydney, Australia.

STEHLI, F.G. and S.D. WEBB. 1985. *The Great American Biotic Interchange.* Plenum, New York.

U.S. FISH & WILDLIFE SERVICE. 1973. U.S. Endangered Species Act, FWS-F-037.

WAYNE, R.K., R.E. BENVENISTE, D.N. JANCZEWSKI, and S.J. O'BRIEN. 1989. Molecular and biochemical evolution of the Carnivora. In *Carnivore Behavior, Ecology and Evolution,* ed. J.L. GITTLEMAN, pp. 465–494. Cornell University Press, Ithaca, NY.

WAYNE, R.K. and S.M. JENKS. 1991. Mitochondrial DNA analysis implying extensive hybridization of the endangered red wolf *Canis rufus. Nature* 351:565–568.

WAYNE, R.K., W.S. MODI, and S.J. O'BRIEN. 1986. Morphological variability in the cheetah (*Acinonyx jubatus*), a genetically uniform species. *Evolution* 40:78–85.

WEIR, B.S. 1990. *Genetic Data Analysis.* Sinauer, Sunderland, MA.

WILDT, D.E., M. BUSH, K.L. GOODROWE, C. PACKER, A.E. PUSEY, J.L. BROWN, P. JOSLIN, and S.J. O'BRIEN. 1987a. Reproductive and genetic consequences of founding isolated lion populations. *Nature* 329:328–331.

WILDT, D.E., M. BUSH, J.G. HOWARD, S.J. O'BRIEN, D. MELTZER, A. VAN DYK, H. EBEDES, and D.J. BRAND. 1983. Unique seminal quality in the South African cheetah and a comparative evaluation in the domestic cat. *Biol. Reprod.* 29:1019–1025.

WILDT, D.E., S.J. O'BRIEN, J.G. HOWARD, T.M. CARO, M.E. ROELKE, J.L. BROWN, and M. BUSH. 1987b. Similarity in ejaculate-endocrine characteristics in captive versus free-ranging cheetahs of two subspecies. *Biol. Reprod.* 36:351–360.

WURSTER-HILL, D.H. and W.R. CENTERWALL. 1982. The interrelationships of chromosome banding patterns in canids, mustelids, hyena, and felids. *Cytogenet. Cell Genet.* 34:178–192.

WURSTER-HILL, D.H. and C.W. GRAY. 1975. The interrelationship of chromosome banding patterns in procyonids, viverrids, and felids. *Cytogenet. Cell Genet.* 15:306–331.

YUHKI, N. and S.J. O'BRIEN. 1990. DNA variation of the mammalian major histocompatibility complex reflects genomic diversity and population history. *Proc. Natl. Acad. Sci. USA* 87:836–840.

4

CONSERVATION GENETICS
IN THE CANIDAE

Robert K. Wayne

INTRODUCTION

Species of Canidae (wolves, foxes, and relatives) exist in a wide variety of habitats and, in general, are excellent dispersers not strongly limited by topographic or habitat barriers. For example, a single gray wolf may disperse several hundred kilometers during its lifetime (Mech, 1987), and common species such as the red fox, arctic fox, and gray wolf have circumpolar distributions. Some canids are great habitat opportunists. An example is the coyote, which has tripled its geographic range in the past 150 years as its larger competitor, the gray wolf, diminished in numbers (Nowak, 1979; Voigt and Berg, 1987; Moore and Parker, 1992). Many canids also have large population sizes in parts of their geographic range (Ginsberg and Macdonald, 1990). However, over the past few hundred years, predator control measures and habitat fragmentation caused by human activities have dramatically altered the distribution and abundance of canids, such that many species now exist as a patchwork of small, highly unstable, isolated populations (Ginsberg and Macdonald, 1990). As a result, a number of canid species are considered threatened or endangered (Box 4.1).

This chapter focuses on genetic issues that are relevant to the conserva-

Box 4.1. Taxonomic and Conservation Status of Canids

About 35 species of Canidae are recognized worldwide (Stains, 1975; Nowak and Paradiso, 1983; Sheldon, 1992), of which 18 are listed as threatened in the 1994 IUCN Red List (World Conservation Union, 1993). Seven species or subspecies also are listed for protection under the U.S. Endangered Species Act of 1973. The following species are mentioned in the text. Species that have been the primary subjects of molecular genetic assays are shown in bold type, and those indicated by an asterisk are considered threatened by the IUCN.

Species	Range
* 1. *Canis simensis* (Simien jackal)	Ethiopian highlands
2. *Canis mesomelas* (black-backed jackal)	East and South Africa
3. *Canis aureus* (golden jackal)	Eurasia, North and East Africa
4. *Canis adustus* (side-striped jackal)	Sub-Saharan Africa
* 5. *Canis lupus* (gray wolf)	Holarctic
* 6. *Canis rufus* (red wolf)	Southcentral U.S.
7. *Canis latrans* (coyote)	North America
* 8. *Lycaon pictus* (African wild dog)	Sub-Saharan Africa
9. *Vulpes macrotis* (kit fox)	Western U.S.
10. *Vulpes velox* (swift fox)	Central U.S.
11. *Vulpes vulpes* (red fox)	Holarctic
12. *Alopex lagopus* (arctic fox)	Holarctic
* 13. *Urocyon littoralis* (Channel Island fox)	California Channel Islands
14. *Urocyon cinereoargenteus* (gray fox)	North America
* 15. *Dusicyon fulvipes* (Darwin's fox)	Central Chile
16. *Dusicyon griseus* (South American gray fox)	Chile, Argentina
17. *Dusicyon culpaeus* (culpeo fox)	Chile, Argentina
18. *Fennecus zerda* (fennec fox)	Sahara, Middle East

tion of endangered or rare canid populations, and how these issues can be addressed using molecular genetic techniques. Conservation genetics is an emerging field that has at its core questions concerning systematic distinction, the quantity and substructure of genetic variation in natural populations, and the effects of habitat changes and fragmentation on levels of gene flow and hybridization (Avise and Nelson, 1989; Wayne and Jenks, 1991; Avise, 1992, 1994).

Molecular genetic studies of both large and small endangered canids demonstrate how differences in dispersal ability and ecological constraints influence the degree of genetic substructure within canid species. Results from a new class of hypervariable nuclear markers (microsatellites or simple sequence repeats) will be introduced and compared to older studies involving mitochondrial (mt) DNA polymorphisms. Attempts will be made to show how these molecular genetic approaches can be used to understand the phylogenetic histories and levels of distinction of canid species, and to expose the genetic units within each species. Moreover, mitochondrial and nuclear genetic results will be reviewed that document interspecific hybridization between several canid species in disturbed areas. Throughout, genetic observations will be connected to recommendations for management of canids in captivity and the wild.

RATIONALES FOR MOLECULAR GENETIC APPROACHES

Canids generally show only low to moderate levels of intraspecific allozyme polymorphism (Fisher et al., 1976; Ferrell et al., 1980; Wayne and O'Brien, 1987; Kennedy et al., 1991; Wayne et al., 1991a, 1991b). Consequently, recent studies have focused on the mitochondrial genome, because in mammals its sequence generally evolves much faster than that of most nuclear genes (Brown, 1986). Moreover, because mtDNA is maternally inherited in a clonal fashion without recombination, within-species analyses provide a history of maternal lineages that avoids the reticulation caused by recombination. This in turn may allow a more precise reconstruction of historical demographic events such as colonization and gene flow (e.g., Avise et al., 1987; Slatkin and Maddison, 1989; Avise, 1991, 1994). Both restriction fragment analysis, and more recently, mtDNA sequencing via the polymerase chain reaction (PCR), have been applied to large population samples of canids. Examples of estimated levels of observed mtDNA variation are summarized in Table 4.1.

However, mtDNA analysis provides only one perspective on genetic variation. Levels of mtDNA variation are more severely affected than those of nuclear loci by changes in population size. Furthermore, phylogenetic trees based on mtDNA record the history of only a single linked set of genes (Avise et al., 1984; Pamilo and Nei, 1988; Avise, 1991). Until recently, nuclear genes with equivalent evolutionary rates had not been identified or were difficult to survey in large population samples. As an alternative, allele frequencies at hypervariable nuclear loci have been monitored. Some of these assays involved conventional multilocus DNA fingerprinting procedures (Bruford and Burke, 1991), wherein complex DNA banding patterns are produced on gels because several unspecified minisatellite loci are examined simultaneously. Other assays involve microsatellite loci, which are composed of tandem repeats of short sequences 2–5 base pairs (bp) in length. These

Table 4.1. Estimates of mtDNA Variability in 10 Species of Canids (see Box 4.1) as Assayed by Multienzyme RFLP Approaches.

Species (sample size)	Number of genotypes	Max. sequence divergence (%)	Max. distance between locales (km)	Mean no. genotypes per locale	Reference[1]
Simien jackal (54)	1	0.0	150	1.0	1
Black-backed jackal (64)	4	8.4	400	2.8	2
Golden jackal (20)	2	0.1	400	2.0	2
Side-striped jackal (7)	2	0.2	400	2.0	2
Coyote (327)	32	2.5	4,800	4.8	3
Gray wolf (276)	9	0.8	15,000	2.1	4
Kit fox (256)	24	1.5	2,000	3.1	5
Red fox (4)	3	1.2	12,000	–	6
Channel Island fox (150)	5	1.8	250	1.8	7
African wild dog (104)	6	0.9	3,000	2.2	8

[1] 1) Gottelli et al., 1994; 2) Wayne et al., 1990; 3) Lehman and Wayne, 1991; 4) Wayne et al., 1992; 5) Mercure et al., 1993; 6) Geffen et al., 1992; 7) Wayne et al., 1991b; 8) Girman et al., 1993.

simple sequence repeats can be easily amplified by the PCR and separated on acrylamide gels. This procedure allows characterization of alleles at individual loci (assayed one at a time) that are highly polymorphic in large population samples. The PCR basis of microsatellite analysis also permits the use of atypical source material including bones, hair, and feces (Hagelberg et al., 1991; Hoss et al., 1992; Roy et al., 1994a, 1994b). A panel of a dozen or fewer microsatellite loci may be sufficient to accurately quantify components of variation within and among populations and to study individual relatedness within social groups (Amos et al., 1993; Queller et al., 1993; Gottelli et al., 1994; Roy et al., 1994b; Taylor et al., 1994).

SIMIEN JACKAL

Canis simensis is probably the most endangered canid (Ginsberg and Macdonald, 1990; Gottelli and Sillero-Zubiri, 1992), with fewer than 500 individuals remaining in small and highly isolated populations restricted to the Ethiopian highlands above 3,000 meters (Figure 4.1, Gottelli and Sillero-Zubiri, 1992). Simien jackals are more than twice the size of other African jackals, are wolf-like in morphology, and have an organized gray wolf–like pack structure. Indeed, phylogenetic analysis of mtDNA sequences showed that the closest living relatives of Simien jackals are probably the non-African gray wolves and coyotes (Figure 4.2). An evolutionary hypothesis consistent with these results is that Simien jackals are a relict form remaining from a Pleistocene invasion of a wolf-like progenitor into East Africa. The current extent of Ethiopian high-altitude moorland habitats is only 5% of the area existing after the last Ice Age (Yalden, 1983; Kingdon, 1990; Gottelli et al., 1994). Consequently, the geographic range and numerical abundance of Simien jackals likely has decreased during the Holocene. More recently, habitat loss and fragmentation have accelerated due to human population growth and agriculture.

Level of Genetic Variation

Restriction-site and sequence analyses of mtDNA showed that the population in the Bale Mountains National Park (Figure 4.1) had a single mitochondrial genotype, the most limited variability of any extant canid (Table 4.1). Subsequent analysis of 134 bp of mtDNA control-region sequence from two museum skins from an extinct northern population revealed two unique substitutions (Figure 4.2). Variability of microsatellite loci was also low in Simien jackals— only 46% of the mean heterozygosity and 38% of the mean allelic diversity typifying populations of other wolf-

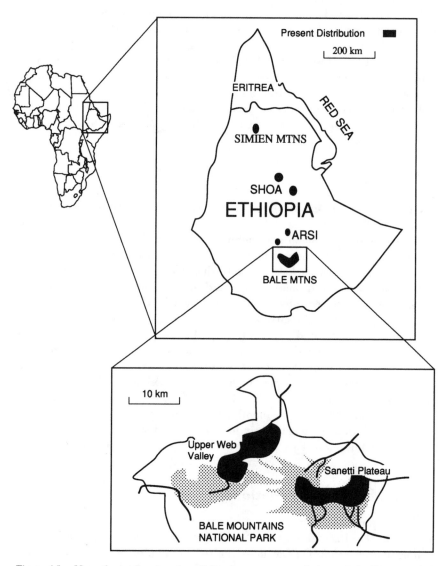

Figure 4.1. Map of sampling locations indicating extant populations of the Simien jackal (Gottelli et al., 1994).

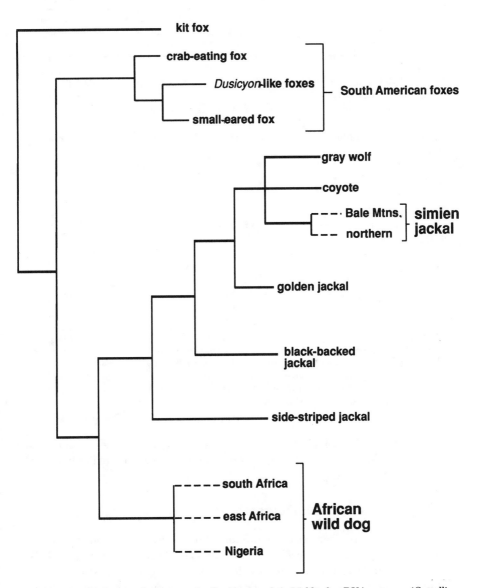

Figure 4.2. Phylogeny of representative Canidae based on 2.1 kb of mtDNA sequence (Gottelli et al., 1994; Roy et al, 1994a). African wild dog and Simien jackal genotypes were distinguished based on 134 bp of mtDNA control-region sequence (Roy et al., 1994a).

Table 4.2. Estimates of Variation of 10 Microsatellite Loci in Populations of Wolf-Like Canids*

Population	Sample size per locus	Mean number of alleles per locus	Mean expected heterozygosity[1]
Gray wolf			
Non-hybridizing (five pops., Alaska to Quebec)	17.7 (2.8)[2]	4.5 (1.1)	0.62 (0.07)
Mexican gray wolf	37.9 (0.1)	2.7 (0.3)	0.42 (0.07)
Coyote			
six pops., Alaska to Maine	17.0 (3.0)	5.9 (0.7)	0.68 (0.04)
Red wolf			
captive colony	29.9 (1.0)	5.3 (0.8)	0.55 (0.07)
Golden jackal			
Kenya	16.4 (0.7)	4.8 (0.8)	0.52 (0.10)
Black-backed jackal			
Kenya	13.7 (1.7)	5.0 (0.4)	0.67 (0.06)
Simien jackal[3]			
Ethiopia	19.6 (0.4)	2.4 (0.3)	0.24 (0.06)
African wild dog			
East Africa (Mara-Serengeti)	24.9 (0.7)	3.3 (0.3)	0.54 (0.05)
South Africa (Kruger)	77.5 (0.6)	3.5 (0.3)	0.56 (0.05)
Botswana	27.6 (2.3)	3.6 (0.4)	0.59 (0.04)
Domestic dog			
32 breeds	26.0 (5.3)	6.4 (0.8)	0.73 (0.05)

[1] Heterozygosity predicted from allele frequencies assuming Hardy–Weinberg equilibrium.
[2] Standard errors in parentheses are averaged over all populations.
[3] Only nine microsatellite loci were surveyed in this species (Gottelli et al., 1994).
*Source: Roy et al., 1994b; Gottelli et al., 1994.

like canids (Table 4.2). Such low levels of heterozygosity are consistent with an effective population size at equilibrium of only a few hundred individuals (Gottelli et al., 1994).

Hybridization with Domestic Dogs

Although loss of variation and inbreeding in isolated populations are concerns for endangered species, a more pressing issue with Simien jackals may be hybridization with domestic dogs. In many areas, domestic dogs are abundant and only loosely associated with humans. Many jackals have unusual coat coloration and morphology, and field researchers suspected the occurrence of hybridization because of reports of matings with dogs. Wild populations of wolves and coyotes are known to hybridize (to be discussed), and given the close phylogenetic relationship of Simien jackals to gray wolves, and the close kinship of gray wolves to domestic dogs, it is perhaps not surprising that dogs and Simien jackals might successfully mate as well. However, a firmer documentation of dog–jackal hybridization awaited the application of molecular genetic markers.

Initial genetic analysis showed that suspected hybrid individuals carried mtDNA genotypes identical to those in "pure" Simien jackals (Gottelli et al., 1994), consistent with field reports that if interspecific matings were taking place, they probably involved male domestic dogs with female jackals (such that a jackal mtDNA genotype would be expected in hybrids). Subsequent analysis of nine microsatellite loci (Gottelli et al., 1994) revealed that all animals thought to be hybrids had alleles not generally found in typical jackals, but present in domestic dogs (Figure 4.3). The presence of several dog marker alleles in phenotypically abnormal jackals adds support to the suspicion that hybridization has occurred, and provides an added justification for governmental control. Dogs not only hybridize with Simien jackals and compete with them for food, but also are reservoirs of canine diseases.

Conservation Relevance

In summary, molecular results currently available for Simien jackals suggest noticeable genetic effects of long-term population reduction compounded by recent habitat fragmentation. A great concern is the vulnerability of the few remaining populations to diseases (such as rabies, which is already thought to have eliminated about half of the Bale Mountain population) or to other stochastic demographic effects, as well as to possible effects of increased inbreeding and losses of genetic variation in small populations. Furthermore, there is a danger that loss of unique Simien jackal characteristics may occur via introgressive "swamping" from dogs. Hybridization with dogs is documented to have altered the genetic characteristics of the one population monitored thus far. No records exist of Simien jackals being bred in captivity (Ginsberg and Macdonald, 1990). Perhaps some of the remaining genetic diversity should be preserved in a captive breeding program, and such possibilities currently are under discussion.

AFRICAN WILD DOG

Lycaon pictus once ranged over most of Africa south of the Sahara Desert, inhabiting areas of dry woodland and savannah. However, due to habitat disturbance and disease, many populations have vanished or are severely reduced in number. The present populations are highly fragmented and total

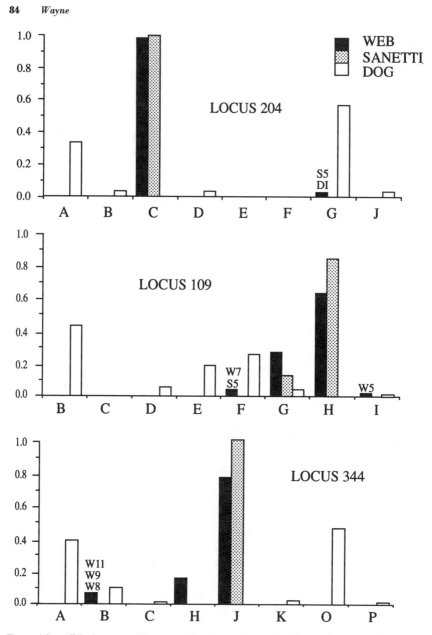

Figure 4.3. Allele frequency histograms for three microsatellite loci in domestic dogs, and in Simien jackals from the Sanetti plateau (where there are no dogs) and the Web Valley (where dogs are abundant). Consecutive letters indicate alleles that differ by a single two-bp repeat. Alleles found both in dogs, and in Simien jackals suspected to be hybrids (individuals S5, D1, W5, W7, W8, W9, and W11), are indicated above histogram bars.

no more than several thousand individuals (Ginsberg and Macdonald, 1990; Fanshawe et al., 1991). Importantly, the western populations are nearing extinction, yet are not represented in zoos (only south African wild dogs are bred in captivity). Severe losses also have occurred in east Africa. Populations in southern Africa currently are stabilized in protected areas (Fanshawe et al., 1991).

The pressing conservation genetic questions for the African wild dog involve the magnitude of genetic variability within and the degree of genetic subdivision among populations. Genetic variability was empirically examined using two approaches. First, levels of mtDNA variation were compared in extant versus extinct populations (the latter represented by tissues in museum collections [Roy et al., 1994a]). In present-day African wild dogs, the overall level of mtDNA polymorphism proved similar to that in North American gray wolves, but much less than that in coyotes (Table 4.1). Comparison with museum specimens showed that in the largest and best-protected population (Kruger National Park, South Africa), only one or a few genotypes had been lost in the past 100 years (Roy et al., 1994a). Genetic variability also was surveyed at 10 microsatellite loci (Table 4.2). Levels of polymorphism and heterozygosity in both south African and east African wild dogs proved similar to those of other outbred species, again suggesting that recent reductions in population size have not dramatically reduced genetic variation in this species.

To address the question of population subdivision, initially 104 African wild dogs from east Africa (Kenya and Tanzania) and south Africa (Zimbabwe and the Republic of South Africa) were examined for mtDNA RFLPs and cytochrome *b* sequences (Girman et al., 1993). Results showed that southern and eastern populations fell into two distinct monophyletic groups (Figure 4.2), which from molecular-clock considerations could have been separated for perhaps 500,000 years. The barrier to dispersal may have been the Rift Valley Lake system.

To determine if there were other hidden genetic units within the species, DNA was extracted from museum skins, and 130–250 bp of the hypervariable mtDNA control region was sequenced (Roy et al., 1994a). The one assayed museum skin from Nigeria displayed a highly distinct genotype, with sequence divergence at least as great as that between east and south African populations (Figure 4.2). Hence overall, at least two and perhaps three distinct population units may be indicated. Such subdivision was unexpected, given that dispersal in this species can occur over several hundred kilometers (Fuller et al., 1992), and studies of other highly mobile wolf-like canids in North America have found little evidence of regional differentiation (to be discussed). Perhaps wild dogs are more habitat-specific than North American wolf-like canids, and their dispersal more limited by the distribution of savannah woodland habitats (Girman et al., 1993). In any event, the genetic results suggest that zoos should consider establishing and maintaining African wild dogs from separate geographic regions to best preserve and represent the genetic diversity of the species.

GRAY WOLF

Canis lupus is a model species to address conservation genetic issues because wolf populations have been challenged by a variety of habitat alterations and demographic threats. For example, although the gray wolf is abundant and distributed continuously across large portions of its circumpolar distribution, in some areas, populations are small and highly isolated in habitats dramatically altered by human activities. Many populations have suffered severe mortality through predator-control activities, and some are just now beginning to expand and recolonize areas where they had been extirpated. Consequently, the gray wolf provides a model system, with internal undisturbed controls, to study the population genetic consequences of human disturbance.

Variability in an Isolated Population

In about 1950, a pair of gray wolves is thought to have crossed an ice bridge from the Canadian mainland to a moderate-sized island situated in northern Lake Superior. Thereafter, this isolated and well-monitored Isle Royale population increased to over 50 wolves by 1980, but then dramatically declined to a dozen or fewer individuals by 1990 (Peterson and Page, 1988; Peterson, unpublished data). For several years no new litters were born. Disease and changes in food abundance were first suggested as causes for the decline, but both became increasingly improbable as explanations. For example, the moose population (a main food source for wolves) has expanded dramatically in the past few years and carcass availability is high; and serological surveys and direct observations did not find evidence of high frequencies of canine disease (Mech et al., 1986; Peterson and Page, 1988).

To assess the genetic composition of Isle Royale gray wolves as it existed in the late 1980s, variability was monitored in mtDNA, allozymes, and genetic fingerprints as registered in assays of hypervariable nuclear loci. The population displayed a single mtDNA genotype, and only half the allozyme heterozygosity observed in an adjacent mainland population. Furthermore, results of multilocus DNA fingerprint assays suggested that the Isle Royale wolves were related about as closely as were known full siblings or parent–offspring pairs in captivity (Wayne et al., 1991a).

Thus, one possibility for the recent lack of reproductive success of the Isle Royale wolves is inbreeding depression, a phenomenon that indeed has been documented in captive wolf populations (Laikre and Ryman, 1991). Alternatively, prior to the population crash, it is probable that successfully mating wolves generally were born from different breeding pairs or were from widely spaced litters of the same parents, whereas after the population decline, remaining breeding-age wolves more likely were littermates or from closely spaced litters. If individuals have instinctual proclivities to avoid incestuous mating with co-reared animals, the post-crash population may have failed to recover because of behavioral difficulties in the mating process (Wayne et al., 1991a).

Pack Structure within Regions

Generally, a wolf pack consists of a mated pair, immediate offspring, and adult helper offspring (Mech, 1970, 1987). In areas with sufficient habitat to support many wolves, packs tend to set up territories with well-defined boundaries, and interpack aggression can be intense (Mech, 1970, 1977; Van Ballenberghe et al., 1975). These observations suggest that within packs, the dominant pair should have close genetic relationships to other adults and juveniles, and that dispersal between packs sharing a boundary might be uncommon. To test these and other predictions about pack structure, DNA fingerprinting methods were employed to assay genetic relationships among wolves from the same and neighboring packs in each of three geographic areas (Figure 4.4)

As based on levels of band-sharing in multilocus DNA fingerprints, a somewhat surprising finding was that many close genetic connections appeared to exist between packs within a region (Lehman et al., 1992). In Minnesota, for example, several interpack genetic-similarity values were as large as those between known siblings or parent–offspring pairs (Figure 4.5). In Denali National Park (Alaska), five such interpack connections were registered, but none was observed among gray wolf packs in the Inuvik region (Northwest Territories, Canada).

One possibility is that regional wolf populations exhibit inherently different levels of genetic relatedness in DNA fingerprint assays due to differences in demographic history (Packer et al., 1991). Another possibility is that the low values of genetic relatedness among gray wolves from different packs near Inuvik may reflect the rapid turnover of wolf packs there. Inuvik wolves are controlled by wildlife officials through hunting. In such managed areas, several packs may be removed at once, thus creating increased opportunity for the invasion of young, dispersing wolf pairs from outside the pack system. These invading wolves are unlikely to have close kinship to individ-

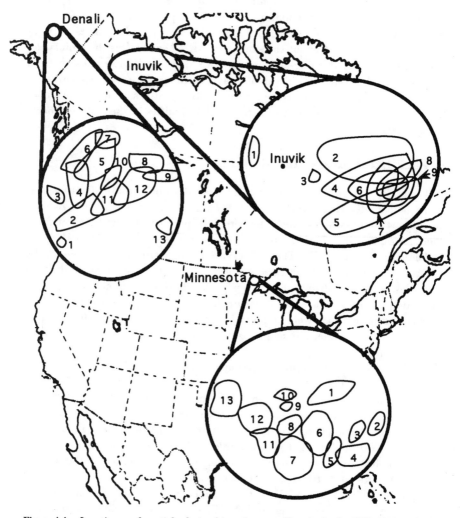

Figure 4.4.   Locations and spatial relationships of gray wolf packs in the 1989–1990 study by Lehman et al. (1992). For purposes of comparison with Figure 4.5, the following are the names of pack clusters in Minnesota: (1) Emerald; (2) Kawishiwi; (3) Isabella Lake; (4) Sawbill; (5) Pike; (6) Little Gabbro; (7) Nip Creek; (8) Birch; (9) Ely; (10) Winton; (11) Perch Lake; (12) Bear Island; and (13) Tower.

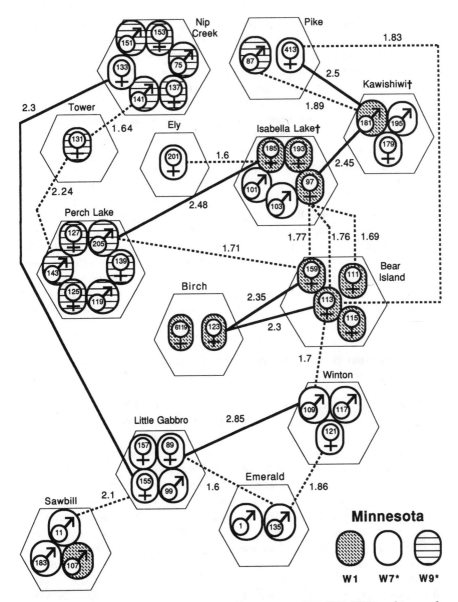

Figure 4.5. Pack membership (hexagons), mtDNA genotypes (W1, W7*, W9*), and interpack genetic connections (solid and dashed lines between packs, as suggested by high DNA fingerprint band similarities [numbers along lines; see Lehman et al., 1992]) for gray wolves in Minnesota. Solid lines indicate extremely high fingerprint similarities among wolves from different packs (suggestive of parent–offspring or full-sibling relationships); dotted lines indicate a possible but more distant genetic relationship.

uals from preexisting packs. In contrast, wolves in Minnesota are highly protected. In these undisturbed, more saturated pack systems, the occasional loss of individual pack members due to natural mortality may be counteracted by recruitment from nearby packs, thus leading to closer kinship ties between them. An intriguing hypothesis that we are now testing with additional samples is that close genetic ties between wolf packs (as observed in Minnesota) may be a more common occurrence in areas where wolf populations are abundant and highly protected.

If experience and long-term residency are important to the persistence of wolves (as is true, for example, in elephant societies; Estes, 1991), then newly established packs may have much lower survivorship and reproductive success than long-established packs. For example, the integration of young individuals into long-established packs might provide a means for transferring knowledge of an area. Consequently, predator-control measures, of dubious value to the preservation of game (Ginsberg and Macdonald, 1990), may be disturbing the genetic ties that lend cohesion and persistence to pack systems, resulting in an accelerated decline of wolf populations. Clearly, the behavioral and genetic effects of predator-control programs need to be studied carefully and their results incorporated into management plans.

Population Subdivision in North America and Europe

To compare the degree of macrogeographic genetic subdivision in western European gray wolves, where most populations are isolated and small, to those in North America, where populations are larger and more continuously distributed, mtDNA genotypes were assayed in 350 wolves from the two continents (Wayne et al., 1992). On a regional level, North American gray wolves proved not to be dramatically structured, as evidenced by the fact that population variability was high and several of the common mtDNA genotypes were widely distributed (Table 4.3). Nonetheless, mtDNA genotypic frequencies sometimes differed significantly between regions. For example, genotype W3 was common in Alaska and Northwest Territories but absent from populations in eastern Canada. Conversely, W1 was absent in Alaskan wolves but common in eastern Canada. Perhaps the regional differences reflect prior isolation in different Pleistocene refugia. Mexican wolves (currently extinct in the wild, but represented in captive populations initiated by small numbers of founders) appeared to be the most unique form of gray wolf in North America, displaying a single mtDNA genotype found nowhere else (Table 4.3). This genotype was also the most divergent of any observed on the continent (Wayne et al., 1992).

Table 4.3. Frequencies and Sample Size (in Parentheses) of mtDNA Genotypes (W1–W18) Observed in Gray Wolves from Localities in the Old and New World[1]

Old World	W3	W5	W6	W15	W16	W17	W18
Portugal	–	–	–	–	–	1.00 (3)	–
Sweden	1.00 (9)	–	–	–	–	–	–
Estonia	0.67 (2)	–	–	–	–	–	0.33 (1)
Italy	–	–	–	1.00 (14)	–	–	–
Israel	–	–	–	–	1.00 (2)	–	–
Iran	–	1.00 (2)	–	–	–	–	–
China	–	–	1.00 (2)	–	–	–	–

New World	W1	W2	W3	W4	W14
Alaska, Denali	–	–	0.73 (16)	0.27 (6)	–
Alaska, Anaktuvik	–	–	1.00 (14)	–	–
Alaska, Nome	–	–	1.00 (8)	–	–
Alaska, Kenai	–	–	0.17 (1)	0.83 (5)	–
Yukon Territory	–	–	1.00 (10)	–	–
Northwest Terr., Inuvik	0.75 (40)	–	0.23 (12)	0.02 (1)	–
Northwest Terr., Yellowknife	0.88 (7)	–	0.12 (1)	–	–
Brit. Columbia, Vancouver Is.	–	–	1.00 (15)	–	–
Alberta	0.25 (1)	–	0.75 (3)	–	–
Montana	0.20 (1)	–	0.60 (3)	0.20 (1)	–
Manitoba	–	1.00 (4)	–	–	–
Minnesota, northeast	0.95 (19)	–	–	0.05 (1)	–
Minnesota, north	1.00 (11)	–	–	–	–
Ontario, west	1.00 (8)	–	–	–	–
Ontario, central	0.67 (8)	–	–	0.33 (4)	–
Mexico	–	–	–	–	1.00 (4)

Coyote clade genotypes	W7*	W8*	W9*	W10*	W11*	W12*	W13*
Manitoba	1.00 (4)	–	–	–	–	–	–
Minnesota, northeast	0.52 (24)	–	0.48 (22)	–	–	–	–
Minnesota, north	0.22 (2)	–	0.78 (7)	–	–	–	–
Michigan, Isle Royale	–	1.00 (7)	–	–	–	–	–
Ontario, west	0.53 (8)	–	0.47 (7)	–	–	–	–
Ontario, central	0.39 (5)	0.08 (1)	0.46 (6)	–	–	–	0.08 (1)
Ontario, east	–	–	–	0.33 (1)	0.33 (1)	0.33 (1)	–
Quebec	–	–	–	0.07 (1)	0.07 (1)	0.43 (6)	0.43 (1)

[1] Genotypes W7–W13, highlighted by asterisks, were closely related to genotypes normally characteristic of coyotes.

Assays of microsatellite loci confirmed that genetic differentiation among regional populations in North America was low to moderate (Roy et al., 1994b). Statistical analysis of genotypic frequencies suggested that populations might be exchanging individuals at a rate on the order of perhaps a few individuals per generation, enough to overcome dramatic genetic differentiation by random drift (Slatkin, 1987). Moreover, genetic variability within populations was uniformly high, except for the Mexican wolf, which had about 50% of the heterozygosity typifying outbred gray wolf populations (Table 4.2).

In contrast, despite small sample sizes, greater mtDNA subdivision was evident in the Old World populations of gray wolves (Wayne et al., 1992); with one exception, each locality had a single, unique genotype (Table 4.3). Although the results suggest that contemporary wolf populations in Europe are genetically subdivided more so than their North American counterparts (a finding consistent with their smaller population sizes and greater degree of spatial isolation), the North American pattern might well reflect the ancestral condition in western Europe prior to human-mediated habitat fragmentation.

These findings on macrogeographic population genetic structure in gray wolves carry several conservation ramifications. First, because the endangered Mexican gray wolf is genetically recognizable, phenotypically distinct, and historically isolated from other gray wolves in North America (Nowak, 1979), efforts to breed pure Mexican wolves in captivity (for possible reintroduction into the wild) are necessary to preserve the unique history of this population. Second, because other wolf populations in North America are not strongly differentiated genetically, reintroductions into areas such as Yellowstone National Park need not consider only the nearest extant populations as source material. Although introduction of gray wolves from populations in which hybridization with coyotes has occurred (to be discussed) is perhaps not advisable, wolves from Alberta, British Columbia, and the Northwest Territories of Canada (for example) are genetically similar, and provide a relatively abundant source stock for contemplated translocations. Finally, it is ironic that the largest proportion of the captive breeding space for large canids in North America is currently occupied by "generic" gray wolves (Grisham et al., 1994). Perhaps the replacement of some captive gray wolves with endangered and distinctive taxa (such as the Simien jackal, African wild dog, or Mexican wolf) would make better use of the limited resources of zoos (Ginsberg and Macdonald, 1990).

Hybridization with the Coyote

Recent demographic changes in populations of gray wolves and coyotes seem closely associated. The gray wolf once ranged throughout most of

North America and parts of Mexico, but over the past few hundred years this species and its wilderness habitats have been systematically eliminated from the United States and Mexico (Mech, 1970; Nowak, 1979; Carbyn, 1987). Presently, stable populations remain only in northern Minnesota and Montana (Ginsberg and Macdonald, 1990). In eastern and central Canada, wolves are restricted to forested fragments where large game still exists and sufficient space remains for pack establishment. More or less concomitant with the decline in the gray wolf has been an increase in numbers and range of the coyote, which has expanded its distribution in the central United States and invaded the northern, eastern, and western United States and much of Canada (Nowak, 1979; Voigt and Berg, 1987). For example, coyotes arrived in Minnesota about 100 years ago and moved into areas of eastern Canada and Maine within the last 50 years (Nowak, 1979; Moore and Parker, 1992). A potential contributing factor is that predator-control programs directed at wild canids have a dramatic effect on wolves but may thereby also promote coyote population growth (Connolly, 1978). Another factor is that coyotes are habitat generalists and can live in urban areas, deserts, and grasslands, as well as forested wilderness (Bekoff and Wells, 1986; Voigt and Berg, 1987). Particularly within disturbed areas, coyotes can now be extremely abundant and greatly outnumber wolves.

One possible consequence of this pattern of abundance and distribution is that young dispersing wolves leaving saturated habitats of protected parkland might often encounter coyotes in smaller forest patches and agricultural areas, and perhaps occasionally mate with them. Indeed, mtDNA studies of North American gray wolves and coyotes provide evidence for such hybridization (Lehman et al., 1991). In Minnesota and eastern Canada, several mtDNA genotypes present in gray wolves were indistinguishable or very similar to otherwise distinctive genotypes normally found only in coyote populations (Figure 4.6). To the west and north, all wolves had mtDNA genotypes belonging to the phylogenetically distinct "gray wolf" clade, and none of the coyotes sampled had gray wolf–like genotypes. These findings were interpreted as evidence that transfers of mtDNA genotypes from coyotes to gray wolves had occurred through introgressive hybridization with the coyote (Lehman et al., 1991). This result is also consistent with the suspicion that young, dispersing gray wolf males mate with coyote females, rather than the reverse cross which seems less likely because smaller male coyotes could not dominate interspecific encounters.

The mtDNA results suggest that some interspecific hybridization has occurred between gray wolves and coyotes, but they do not provide a direct measure of this frequency nor the effect on the nuclear genetic composition of the hybridizing populations. Moreover, although mtDNA data indicate

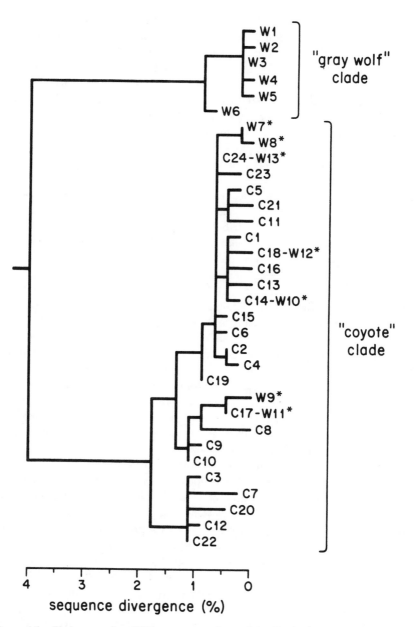

Figure 4.6. Phylogeny of mtDNA genotypes observed in North American gray wolves (W1–W13) and select coyotes (C1–C24) based on restriction-site data (Lehman et al., 1991). Genotypes with asterisks indicate gray wolf genotypes identical or very similar to those of coyotes (Table 4.3).

that female wolves and male coyotes rarely produce offspring who backcross to coyotes, the backcrossing of female coyote X male wolf offspring into the coyote population would not be detected because the resulting individuals would have coyote mtDNA genotypes. To measure the extent of introgression of nuclear genes, variation was next studied at 10 microsatellite loci in three sympatric or near-sympatric populations of gray wolves and coyotes that had not hybridized (based on mtDNA analysis), and two populations from Minnesota and eastern Canada that by the same evidence apparently had done so (Lehman et al., 1991; Roy et al., 1994b). Areas where only wolves (northern Quebec, Northwest Territories) or coyotes (California) are found also were included in the genetic analyses (Figure 4.7).

Based on an analysis of microsatellite allele frequencies that involved estimating differentiation in paired samples of sympatric or near-sympatric coyotes and gray wolves (Roy et al., 1994b), these two species in southern Quebec and Minnesota proved to be genetically more similar to each other than did sympatric populations of the two species elsewhere. In contrast, gray wolves of northern Quebec, from an area not yet colonized by coyotes, showed much less similarity to coyotes than did nearby gray wolf populations that are sympatric with coyotes and hybridize with them (Figure 4.8). These findings again appear consistent with an introgression scenario, and the data could be explained by perhaps two to three successful hybridization events per generation (Roy et al., 1994b). In these same assays, coyote populations in areas of suspected hybridization were genetically similar to those in areas without wolves, thus supporting the hypothesis that introgression is unidirectional, with the offspring of wolf–coyote matings successfully backcrossing only to gray wolves.

In summary, habitat succession and fragmentation caused by human activities may have shifted the distributions and numerical balance between gray wolves and coyotes such that these species now hybridize in some areas (but see Lehman et al., 1992 and Wayne et al., 1992 for other examples of contact without hybridization). Hybridization appears to have been mostly unidirectional with respect to gender and species, such that coyote mitochondrial and nuclear genotypes appear to have introgressed into some gray wolf populations but not vice versa. Interspecific hybridization may have had a significant phenotypic effect on wolf populations as well, because "wolves" of intermediate size and morphology have been described from the hybrid region (Kolenosky and Standfield, 1975; Hilton, 1978; Nowak, 1979; Schmitz and Kolenosky, 1985; Thurber and Peterson, 1991; Wayne, 1992; Lariviere and Crete, 1993). The observation that a wolf–coyote hybrid zone in eastern Canada has developed to a substantial size over a short period of 90 years or less (Figure 4.7) also raises the possibility that hybridization between some wolf-like canids, if extended over greater periods of time,

Figure 4.7. Sampling localities of gray wolves and coyotes. Locality names in italics indicate areas where sympatric or near-sympatric wolf and coyote populations were sampled; plain type indicates localities in Northwest Territories and Quebec, Canada, where only gray wolves are found, and in California where only coyotes are found. The north and west boundary of the hybrid zone between gray wolves and coyotes, as suggested by mtDNA evidence (Lehman et al., 1991), is indicated by a dashed line.

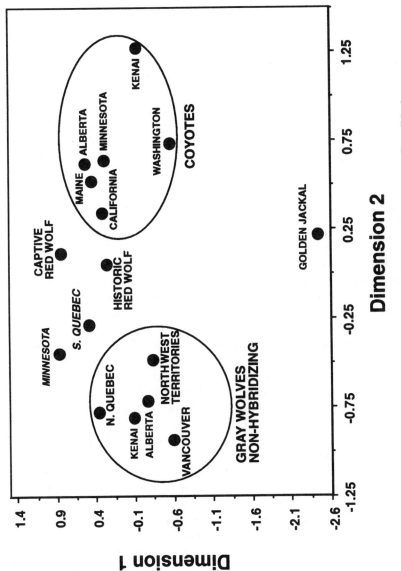

Figure 4.8. Similarities among populations of wolf-like canids in microsatellite allele frequencies, as summarized in the two dimensions of a multidimensional scaling axis (Roy et al., 1994b).

might totally obscure the genetic distinction of hybridizing forms. One possible example may involve the red wolf, as described next.

RED WOLF

The historic range of *Canis rufus* was the southcentral United States, where it overlapped with the gray wolf at its northern boundary, the Mexican gray wolf at its southern boundary, and the coyote throughout its central range. The red wolf declined precipitously in numbers early in this century, and went extinct in the wild about 1975. Generally intermediate in size between gray wolves and coyotes, the red wolf survives now primarily in a single captive breeding colony, from which reintroductions to nature are being attempted.

Genetic Evidence of Hybrid Origins

Genetic assays of red wolves have focused on three populations from different time periods (Wayne and Jenks, 1991; Roy et al., 1994a, 1994b). The first sample was obtained from the captive breeding colony of over 170 individuals, which was founded from 14 "pure" red wolves (culled from a much larger sample of individuals) caught in eastern Texas during 1974–1976. (Although this area was thought to contain the last pure representatives of red wolves, many of the other individuals captured were suspected of being coyote–red wolf hybrids.) The second sample consisted of preserved tissues from 77 canids captured from this same Texas population, and thereby provides a reference for diversity in the population ancestral to the captive population. (In this sample, 11% of individuals had been classified as red wolves, 31% as hybrids, and 58% as coyotes.) The third sample consisted of preserved tissues from six red wolves that died circa 1920, taken from several states. At that time, red wolves were still abundant and not thought to have hybridized extensively with coyotes (Nowak, 1979).

The first suspicion that red wolves might carry genes derived from hybridization with other wild canids came from studies of mtDNA restriction sites and sequences. Because the divergence of the red wolf had been thought to predate that of coyotes and gray wolves (Nowak, 1979, 1992), mtDNA genotypes were expected to delineate a distinctive red wolf clade. However, all mtDNA genotypes observed in the red wolf proved to be identical or very similar to those of gray wolves and coyotes (Wayne and Jenks, 1991). One possible explanation (not the only one— see Ferrell et al., 1980; Dowling et al., 1992; Nowak, 1992) was that the red wolf originated from repeated hybridization events between gray wolves and coyotes in historic times

(Wayne and Jenks, 1991; Jenks and Wayne, 1992). Such speculation could not be confirmed from the mitochondrial studies alone, however, due to the small sample sizes and the maternal mode of mtDNA inheritance.

Subsequent genetic analyses involved 10 microsatellite loci in large population samples of red wolves, gray wolves and coyotes (Roy et al., 1994b). At these hypervariable nuclear loci, the red wolf samples possessed no unique alleles compared to coyotes; moreover, allele frequency similarities otherwise placed the red wolf closest to hybridizing populations of gray wolves in Canada (Figure 4.8). These data were interpreted to support a hybrid derivation of red wolves from matings between coyotes and gray wolves in historic times, but with the introgressive process probably at a more advanced state than was true for the coyote–gray wolf hybrid zone in eastern Canada (Roy et al., 1994a, 1994b).

Conservation Relevance

One immediate conservation implication of these results is that the red wolf reintroduction program needs to be concerned with the effects of hybridization on reintroduced populations. Currently, only a few red wolf individuals are being released. Reintroduced male red wolves have been observed to pair with coyotes, and a hybrid litter has been born (Mike Phillips, personal communication). Conceivably, hybridization might be reduced if several red wolf packs were introduced simultaneously, because they might collectively exclude coyotes from a large area. However, hybridization could still occur on the fringes and between young dispersing red wolves and coyotes. Alternatively, coyotes could be removed from the release sites prior to the reintroductions, thereby allowing a seed population of red wolves to be established that could potentially exclude coyotes.

A larger issue concerns the taxonomic units that merit focused conservation attention (Ryder, 1986; Avise, 1989, 1994; Dizon et al., 1992; Wayne, 1992). The red wolf may be a species composed of various genetic intermediates between two distinct species that hybridized in historic times due to habitat changes and predator control programs that changed the relative abundance of coyotes and gray wolves. Should considerable resources be used to preserve this taxonomic entity when many other genetically distinctive endangered species receive little support? Similarly, should hybrids between gray wolves and coyotes in eastern Canada be preserved because they are distinct forms as well?

There are no easy answers. With respect to captive breeding and reintroduction, the red wolf program has been extremely successful. However, a planned reintroduction of red wolves into the Smoky Mountain National Park in Tennessee neither preserves new habitat nor increases the level of protection of existing endangered habitat. Whether or not the considerable management focus on the red wolf has been fully justifiable in its own sake, perhaps the skills, resources, and lessons learned in the red wolf program can now be applied to other endangered canids such as the Simien jackal, African wild dog, or Mexican gray wolf.

KIT FOX AND RED FOX

The analyses described previously of nuclear and mitochondrial loci generally support the conclusion that historical and contemporary gene flow among most populations of wolf-like canids has been sufficiently high to stymie strong genetic differentiation by random drift. However, small canids such as foxes may be much more restricted in dispersal ability and less able to traverse topographic barriers. Moreover, due to shorter dispersal distances, small canids may show more pronounced patterns of genetic differentiation with distance (Slatkin, 1993). Consequently, small canids may contain more "hidden" genetic units worthy of conservation than is true for their larger cousins.

Taxonomy and Genetics

The small arid-land foxes of North America are habitat specialists and relatively poor dispersers. In California, for example, the kit fox of the San Joaquin Valley, whose range is circumscribed by the coastal mountain range to the west and the Sierra-Nevada mountain range to the east, is considered a distinct subspecies (*Vulpes macrotis mutica*) and is protected by the U.S. Endangered Species Act (Hall, 1981; O'Farrell, 1987; Scott-Brown et al., 1987). Traditionally, populations to the east of the Rocky Mountains are collectively referred to as swift foxes (*V. velox*), and those to the west as kit foxes (*V. macrotis*) (Figure 4.9). However, the two forms hybridize in northcentral Texas and are recognized as conspecific by some authors (Packard and Bowers, 1970; Rohwer and Kilgore, 1973; Nowak and Paradiso, 1983; O'Farrell, 1987; Dragoo et al., 1990).

Have the lower dispersal abilities and increased habitat specificities of kit and swift foxes led to enhanced population genetic subdivision? Results of mtDNA analyses suggest that they have, and in a manner proportional to the distance between populations and the severity of the geographic barriers separating them (Mercure et al., 1993). First, a survey of 75 foxes from the northern and southern San Joaquin Valley identified three genotypes that together defined a significant monophyletic group (Figure 4.10). Second, as summarized in Figure 4.10, on a broader geographic scale, a major genetic

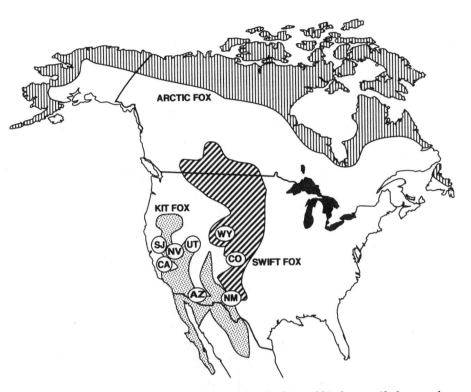

Figure 4.9. Present geographic range and sampling localities of kit foxes, swift foxes, and arctic foxes (Ginsberg and Macdonald, 1990). Localities are coded as follows: SJ, San Joaquin Valley; NV, Nevada; CA, Mohave, California; UT, Utah; AZ, central and southeast Arizona; NM, New Mexico; CO, Colorado; WY, Wyoming. The New Mexico (NM) locality is in the kit-swift fox hybrid zone.

subdivision within the kit–swift fox complex proved to distinguish populations from the east versus the west side of the Rocky Mountains, in a pattern consistent with the current taxonomic distinction between *V. macrotis* and *V. velox*. Indeed, the magnitude of the genetic distinction is nearly as great as the level of mtDNA divergence between this complex and the arctic fox, which typically is classified in a separate genus, *Alopex* (Figure 4.10). The phylogeographic pattern in the kit–swift fox complex contrasts with that observed in coyotes and gray wolves, in which conspecific populations on opposite sides of the Sierra Nevada and Rocky Mountain ranges showed little or no detectable mtDNA differentiation (Lehman and Wayne, 1991). Furthermore, within each of the two major kit–swift fox mtDNA clades, genetic distances among populations tended to be correlated with geographic distance (Mercure et al., 1993).

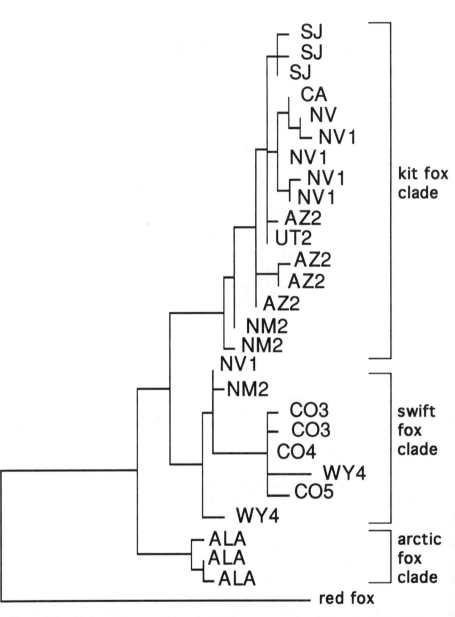

Figure 4.10. Phylogenetic tree of kit and swift fox genotypes based on analysis of mtDNA restriction site data (Mercure et al., 1993). Genotype codes indicate individual ID numbers and localities as in Figure 4.9. Outgroup species are as follows: ALA, *Alopex lagopus* (arctic fox), and *Vulpes vulpes* (red fox).

Conservation Relevance

These results carry both specific and general conservation implications. First, the San Joaquin kit fox is indeed genetically recognizable, although its mtDNA genotypes are closely related to genotypes found in other western states. This degree of distinction suggests a limited history of isolation and provides some support for special preservation efforts. Second, the scale of phylogeographic differentiation appears to differ between at least some large and small canid species, a finding which carries both taxonomic and conservation ramifications. For example, in early classification schemes, about 19 subspecies of the North American coyote were designated (see Nowak, 1979; Hall, 1981; Voigt and Berg, 1987), but the implied level of population subdivision may merely reflect a poor taxonomy, given the molecular genetic findings. However, in the arid-land foxes, existing taxonomies to a considerable degree predicted genetic patterns. In general, subspecific taxonomic distinctions should probably be viewed as more strongly suspect in highly mobile as opposed to less mobile canid species, unless there is independent historical geographic or other evidence for significant barriers to dispersal. With respect to the design of reintroduction programs, source stocks for large wolf-like canids can no doubt be drawn from larger geographic areas than may generally be true for the more highly subdivided fox-like canids, although the use of specific genetic information will remain critical in particular instances. For example, given the mtDNA findings (Figure 4.10), the recent use of foxes from Colorado and South Dakota (rather than New Mexico or Texas) as a source for a reintroduction into the Canadian province of Saskatchewan appears to have been appropriate (Carbyn, 1987).

CHANNEL ISLAND FOX

Some endangered species probably exist as metapopulations composed of a series of quasi-isolated populations that occasionally go extinct and are recolonized by a limited subset of individuals from neighboring areas (Hanski, 1989). Genetic characteristics of such populations reflect factors such as founder composition, selection, and historical changes in population size. Island populations provide a good natural model of this process, and in cases where historical information is also available, permit tests of genetic predictions based on colonization patterns and population sizes.

Historical Demography and Genetics

A good example involves *Urocyon littoralis*, an endangered species found only on the six Channel Islands off the coast of southern California (Gilbert et al., 1990; Wayne et al., 1991b). The island fox is an insular dwarf, about two thirds the size of its mainland progenitor, the gray fox *U. cinereoargenteus* (Collins, 1991; Wayne et al., 1991b). The general history of colonization and isolation of Channel Island foxes can be reconstructed through analysis of fossils and the geologic record (see Table 4.4 and Figure 4.11; Wayne et al., 1991b). About 16,000 years ago, the three current northern islands (which at that time were connected to one another) were colonized from the mainland. As sea level rose, 9,500 to 11,500 years ago, the northern islands were separated. About 4,000 years ago, foxes first arrived in the southern Channel Islands, probably via transport by Native Americans. This succession of events, combined with inferences about population size based on

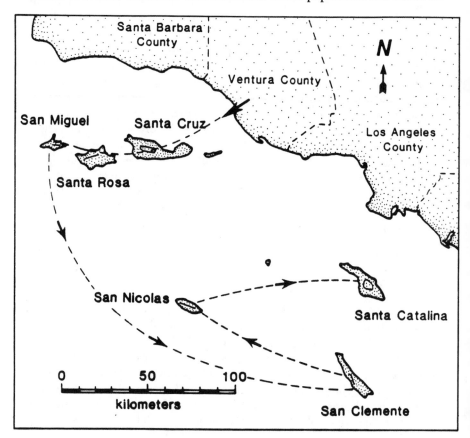

Figure 4.11. Relative size and location of six southern California Channel Islands where foxes are found. Solid lines on each island indicate transects along which fox samples were obtained. Dashed lines across water indicate hypothesized colonization routes based on fossil evidence and DNA fingerprint similarities (Gilbert et al., 1990; Wayne et al., 1991b).

Table 4.4. Physical Characteristics of the California Channel Islands and Empirically Estimated Levels of Genetic Variation in Their Fox Inhabitants

Island	Area (sq. km)	Distance to nearest island (km)	N_e[1]	Duration of isolation (yrs)	Alleles[2]/Heterozygosity[3]	
					Allozymes (20 loci)	Microsatellites (14 loci)
San Miguel	37	3	163	9,500	1.1/0.01	1.8/0.20
Santa Rosa	217	3	955	9,500	1.2/0.06	2.1/0.29
Santa Cruz	249	6	984	11,500	1.1/0.04	2.4/0.31
San Nicolas	58	80	247	2,200	1.0/0.00	1.0/0.00
Santa Catalina	194	34	979	800–3,800	1.0/0.00	2.6/0.36
San Clemente	145	34	551	3,400–4,300	1.1/0.01	2.4/0.29
Gray fox (mainland)	–	–	–	–	1.4/0.10	6.1/0.69

[1] Effective population size, estimated as half the census population size (see Reed et al., 1986; Wayne et al., 1991b).
[2] Mean number of alleles per locus.
[3] Predicted from allele frequencies assuming Hardy–Weinberg equilibrium.

censuses and island areas, allow predictions about relative levels of genetic variation in the various populations. For example, all else being equal, small islands such as San Miguel and San Nicolas, with only a few hundred foxes each, should have less variation than the larger islands such as Santa Rosa, with several thousand foxes. Furthermore, islands colonized last, such as San Clemente and Santa Catalina, should have lower levels of genetic variation due to sequential loss through colonization processes.

In general, these predictions were borne out in results of a variety of molecular genetic assays (Gilbert et al., 1990; Wayne et al., 1991b). The San Nicolas and San Miguel populations displayed the lowest levels of genetic variability overall, and indeed the San Nicolas population was absolutely invariant in multilocus DNA fingerprints (Gilbert et al., 1990), as well as completely monomorphic at 14 microsatellite loci that were highly variable in mainland foxes (Table 4.4). Such high levels of monomorphism have been described elsewhere only in inbred mice strains and in presumably inbred eusocial naked mole rats (Jeffreys et al., 1985; Reeve et al., 1990). Allozyme loci also showed the anticipated pattern, with, for example, the last-colonized islands of San Clemente and Santa Catalina having almost no genetic variation despite large present-day population sizes. On the other hand, patterns of within-island genetic variability did not always correspond precisely with probable colonization histories. Thus, in assays of microsatellite allele frequencies, these latter two populations showed relatively high genetic variation (Table 4.4; Gilbert et al., 1990). Perhaps the much higher mutation rate at particular microsatellite loci (10^{-4} to 10^{-5} per gamete per generation; Bruford and Wayne, 1993) compared to that of allozyme genes (about 10^{-7} mutations per gamete per generation; Nei, 1987), has facilitated a faster post-colonization recovery of variation at the microsatellite loci assayed.

Levels of allozyme heterozygosity appear higher than expected on the northern islands of Santa Rosa and Santa Cruz, given their small population sizes relative to the mainland (Table 4.4; Wayne et al., 1991b). Based on relative population sizes alone, even if all three northern islands were pooled, at equilibrium they would be expected to exhibit only about 15% of the heterozygosity in the mainland sample, but instead the reported values for Santa Rosa and Santa Cruz are about 50% as high as the mainland (Table 4.4). Because mutations are unlikely to have contributed much to the variability of allozyme loci over the period of 16,000 years, balancing selection may be implicated. In an infinite allele model for populations at drift–mutation equilibrium,

$$H = 4 N_e \mu / (4 N_e \mu + 1)$$

where H is the expected heterozygosity and μ is the mutation rate to neutral alleles (Crow and Kimura, 1970). Therefore, the mutation rate implied by the level of heterozygosity on Santa Rosa Island (0.055, see Table 4.4) is 3.2×10^{-4}, a value three orders of magnitude greater than the rate of 10^{-7} often quoted for allozyme loci (Nei, 1987). In contrast, the neutral mutation rate implied by levels of heterozygosity at microsatellite loci in the Santa Rosa foxes, more appropriately estimated using a stepwise mutation model

$$H = 1 - 1 / (1 + 8 N_e \mu)^{0.5}$$

(Ohta and Kimura, 1973; Valdes et al., 1993), is 1.3×10^{-4}, a value within the range of mutation rates (10^{-3} to 10^{-5}) typically given for microsatellite loci (Bruford and Wayne, 1993). This result raises the possibility that mechanisms to maintain allozyme heterozygosity, such as heterozygous fitness advantage or other forms of balancing selection, might be at work in the northern island populations.

Another aspect of the molecular analysis involved comparisons among the island populations (Wayne et al., 1991b). For example, based on DNA fingerprint assays, southern and northern islands formed distinct genetic groups, with additional substructure within regions also evident (Figure 4.12). In several cases, island samples were diagnosable by other genetic characters as well. For example, within the southern group of islands, the small population on San Nicolas also possessed a unique mtDNA genotype and a unique microsatellite allele.

Genetic Similarity

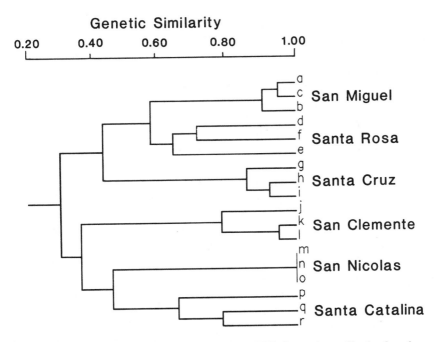

Figure 4.12. Phenogram based on overall similarity in DNA fingerprint profiles for three foxes (lettered) from each of six California Channel Islands (Gilbert et al., 1990).

Conservation Relevance

One conservation-relevant implication of these results is that small populations can persist for long time periods, even in the near absence of genetic variability. The population on San Nicolas is the most dramatic example, in which the molecular results suggest severe loss of genetic variation through founder effects and/or genetic drift, but the fossil record indicates the continued presence of foxes there for several thousand years. Although the San Nicolas population may have an unusually high frequency of skeletal abnormalities (Collins, 1982), it has rapidly recovered from a recent population decline (Kovach and Dow, 1985). Furthermore, an insulation from disease factors cannot entirely account for the persistence of this population, because an epidemiological survey indicated that all Channel Island populations have been exposed to several canine diseases (Garcelon et al., 1992).

On the other hand, the persistence of the San Nicolas population represents an isolated case, perhaps not representative of most situations. Populations with lower levels of genetic variation may well, on average, tend to go extinct with higher probability (for either genetic or demographic reasons). Nonetheless, the apparent success of the San Nicolas foxes implies that extremely inbred populations, such as might exist in captivity for many endangered species, at least potentially have the ability to persist and thrive in natural settings (e.g., O'Brien et al., 1985; O'Brien and Evermann, 1988; Roelke et al., 1993).

A second conservation implication is that genetic surveys of endangered species should carefully consider not only the effects of population history on levels of genetic variation, but also how selection and mutation rates may have influenced outcomes. For example, allozyme variability was absent in the Santa Catalina population, despite its large size and the presence of exceptionally high heterozygosity at mini- and microsatellite loci (Gilbert et al., 1990; Wayne et al., 1992). The allozyme results may be attributable to effects of a recent colonization of the island, coupled with a low mutation rate at allozyme loci, whereas the variability at mini- and microsatellite loci may simply have recovered due to a higher mutation rate, thereby blurring the genetic record of the colonization event. Conversely, allozyme variation was much higher than expected on some islands that were colonized early but have small populations (e.g., Santa Cruz and Santa Rosa). In these cases, mini- and microsatellite loci may provide a clearer picture of population history and variation in neutral, rapidly evolving loci, whereas levels of variation in allozymes may be at least partly indicative of balancing selection rather than demographic history alone (Allendorf and Leary, 1986).

The future preservation of Channel Island foxes is uncertain. In conservation plans, all island populations have been treated as a single taxonomic unit with a combined population of about 8,000 individuals (California Code of Regulations, 1992). However, particular island populations, by virtue of their isolation and small size, may be significantly more vulnerable than an equivalently sized mainland population. As has been shown, they are also more likely to be genetically differentiated from one another. In light of the unique genetic nature of the San Nicolas population, we recommend that these foxes receive additional attention and protection, perhaps as a distinctive subspecies. This may be administratively feasible, because San Nicolas (and Santa Clemente) are U.S. naval facilities. The northern islands are managed by the National Park Service and the Nature Conservancy. Continued protection of these island fox populations from human disturbance, and from diseases carried by domestic dogs, should be encouraged. Captive study and breeding of island foxes is not currently being done, a rather surprising circumstance considering that this species is such a unique and "charismatic" California canid, and a symbol of the distinctive southern California Channel Islands.

DARWIN'S FOX

Taxonomy and Genetics

On Chiloé Island, off the west coast of Chile, Charles Darwin observed and was the first to describe a small endemic fox, *Dusicyon fulvipes*. Darwin's fox has the smallest geographic range of any living canid (Osgood, 1943; Cabrera, 1958), and the unique island temperate rainforest it inhabits is not duplicated elsewhere. There are perhaps fewer than 500 foxes currently in existence (none in zoos). Darwin's fox is distinctive in having a small body size, short legs, and abbreviated muzzle (Osgood, 1943). Related foxes that are widespread on mainland Chile, and from which Darwin's fox presumably arose, are the South American gray fox *Dusicyon griseus*, and the culpeo fox *D. culpaeus*. The former is about 50% larger than Darwin's fox, and is often assumed to be conspecific with it (Honacki et al., 1982; Wozencraft, 1993). The latter tends to be larger still, but size variation within both the gray fox and the culpeo fox is so extreme that these species are difficult to distinguish in some areas (Fuentes and Jaksic, 1979).

Conventional wisdom has been that Darwin's fox was recently isolated from mainland stock because the channel separating Chiloé from the continent is only about five km wide, and the island was likely connected to South America when sea levels were lower during the last glaciation (ca. 13,000 years before present; Yahnke, 1994). However, recent reports of Darwin's fox on the mainland in central Chile (Medel et al., 1990), in a national park 350 km from the coast that is also inhabited by the mainland gray fox, refocused questions concerning the age and history of this species.

To address this issue genetically, blood samples were obtained from two Darwin's foxes on the mainland and one from Chiloé Island. Using universal primers, 354 base pairs from the hypervariable mtDNA control region were sequenced, and compared to those of the South American gray fox, culpeo fox, and other mainland fox species (Kocher et al., 1989; Yahnke et al., 1995). Results confirmed the close genetic relationship of mainland and island populations of Darwin's fox, and suggested a greater antiquity of the species than might have been predicted (Figure 4.13). Darwin's fox appears to have diverged early in the radiation of Chilean foxes, and is at least as divergent from the gray fox and culpeo fox as the latter two are from one another. Perhaps Darwin's fox is a relict form, having evolved from the first immigrant foxes to Chile after the land bridge formed between North and South America about 2–3 million years ago (Marshall et al., 1979; Webb, 1985).

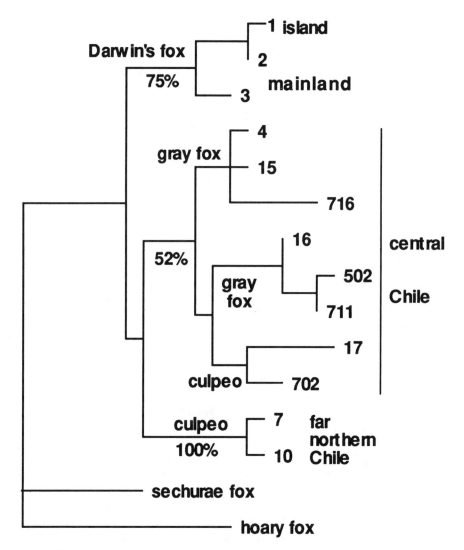

Figure 4.13. Phylogenetic tree of Chilean foxes based on analysis of 354 bp of mtDNA control-region sequence (Yahnke et al., 1995). The Sechurae fox (*Dusicyon sechurae*) is a northern outgroup species from Peru, and the hoary fox (*Lycalopex vetulus*) is an eastern outgroup species from Brazil. Bootstrap values (%) indicate relative support for various clades.

Conservation Relevance

Although the genetic evidence is as yet limited, Darwin's fox appears to by quite distinct phylogenetically. It also has a morphology unlike any living fox and occupies a restricted and unique temperate rainforest habitat. Yet, little census information exists for the species (Medel and Jaksic, 1988), nor is Darwin's fox being studied in captivity. Moreover, additional mainland populations may yet be discovered. Currently, about 70% of the captive breeding space for small canids is occupied by arctic and fennec foxes (Grisham et al., 1994). The former has a circumpolar distribution and is not endangered anywhere; the latter is a common fox of the Sahara and Middle East, and is also not endangered. Captive breeding and observation of Darwin's fox might help ensure that knowledge and resources are gained to preserve this species before it is reduced to desperately low numbers.

CONCLUSIONS AND PROSPECTS

Molecular studies of the Canidae have illustrated several uses of genetic techniques to better understand and preserve endangered species. The application of molecular methods to taxonomic questions can uncover previously unsuspected but distinctive genetic units (such as Darwin's fox), or diminish the presumed distinctiveness of previously recognized taxonomic forms (such as the red wolf). Speculations about the specific and subspecific taxonomies of endangered (and non-endangered) forms need to be enlightened by an understanding of the population subdivisions and gene flow patterns resulting ultimately from dispersal abilities, ecologies, and phylogeographic histories of populations. Molecular genetic approaches have proved well suited to addressing these issues. In general, small and relatively immobile canids, such as foxes, appear to be composed of a greater number of distinctive genetic units than their larger and more mobile relatives, such as wolves and coyotes, and such findings need to be taken into account in taxonomic revisions and conservation plans. As exemplified by studies of the Channel Island foxes, different perspectives afforded by alternative molecular genetic assays may also provide a more comprehensive view of the factors influencing genetic variation in endangered taxa. Finally, the genetic effects of hybridization may be extreme in some canids, with ramifications extending to a variety of taxonomic and conservation issues.

The application of molecular genetic techniques is expensive and labor intensive. Ironically, despite the relevance of genetic information to many conservation issues and the great potential for true gains in biological understanding, few governmental or private organizations have thus far funded molecular genetic studies of endangered species. Many genetic problems remain to be investigated. Perhaps the preliminary research on canids summarized in this chapter may help to illustrate the desirability of support for other important conservation genetic projects as well.

SUMMARY

1.   Levels of genetic variability within canid populations presumably reflect population size, the degree of population isolation, and the mutation rate of the surveyed loci. Widely abundant species (e.g., the coyote, gray wolf, and kit fox) tend to have high levels of genetic variation. Even the endangered African wild dog, which was once abundant but has declined dramatically during the past hundred years, still retains moderate levels of variability at mitochondrial and microsatellite loci. Conversely, the most endangered canid, the Simien jackal, has the lowest level of molecular genetic variation of any surveyed canid. Moreover, island populations of gray wolves and foxes, and isolated populations of European wolves, have reduced levels of variability generally consistent with smaller population sizes and extreme isolation.

2.   The degree of genetic subdivision within canid species reflects differences in the relative mobilities and historical demographies of species and the extent of topographic and habitat barriers that separate populations. North American gray wolves have higher levels of subdivision than coyotes (perhaps reflecting the differentiation of gray wolves into two well-separated Ice Age refugia), although both species can disperse over long distances. Highly fragmented populations of European gray wolves, east and south African populations of wild dogs, and populations of the diminutive kit and swift foxes on either side of the Rocky Mountains, are well differentiated. To understand the genetic units for conservation and captive breeding, biologists need to survey species for genetic differences and consider the effect of historical demography and dispersal abilities on the degree of subdivision expected within species.

3.   Because of the high mobility of many canid species, interspecific hybridization may lead to wide areas of genetic introgression, and thereby threaten the genetic integrity of endangered taxa. Hybrid zones have developed recently between gray wolves and coyotes in Minnesota and eastern Canada, perhaps reflecting habitat changes and predator-control measures favoring the immigration of coyotes and the explosive growth of their populations. An older and more extensive hybrid zone may have developed in the southcentral United States between coyotes, red wolves, and gray

wolves. In Africa, hybridization with domestic dogs threatens to obscure the unique genetic characteristics of the largest remaining population of the endangered Simien jackal.

4. Molecular systematic analyses of canid species have uncovered or bolstered the evidence for new distinct endangered taxa, such as Darwin's fox and the Mexican wolf. Such analyses have also revealed unexpected phylogenetic connections between Simien jackals and wolf-like canids, a finding consistent with the ability of domestic dogs and Simien jackals to hybridize. In other cases, molecular genetic data have cast doubt on previously recognized taxa. For example, mitochondrial and microsatellite assays suggest that the endangered red wolf may be composed of intergrades between gray wolves and coyotes.

REFERENCES

ALLENDORF, F.W. and R.F. LEARY. 1986. Heterozygosity and fitness in natural populations of animals. In *Conservation Biology: The Science of Scarcity and Diversity*, ed. M.E. Soulé, pp. 57–76. Sinauer, Sunderland, MA.

AMOS, B., C. SCHLOTTERER, and D. TAUTZ. 1993. Social structure of pilot whales revealed by analytical DNA profiling. *Science* 260:670–672.

AVISE, J.C. 1989. A role for molecular genetics in the recognition and conservation of endangered species. *Trends Ecol. Evol.* 4:279–281.

AVISE, J.C. 1991. Ten unorthodox perspectives on evolution prompted by comparative genetic findings on mitochondrial DNA. *Annu. Rev. Genet.* 25:45–69.

AVISE, J.C. 1992. Molecular population structure and the biogeographic history of a regional fauna: a case history with lessons for conservation biology. *Oikos* 63:62–76.

AVISE, J.C. 1994. *Molecular Markers, Natural History and Evolution.* Chapman & Hall, New York.

AVISE, J.C., J. ARNOLD, R.M. BALL, E. BERMINGHAM, T. LAMB, J.E. NEIGEL, C.A. REEB, and N.C. SAUNDERS. 1987. Intraspecific phylogeography: the mitochondrial DNA bridge between population genetics and systematics. *Annu. Rev. Ecol. Syst.* 18:489–522.

AVISE, J.C., J.E. NEIGEL, and J. ARNOLD. 1984. Demographic influences on mitochondrial DNA lineage survivorship in animal populations. *J. Molec. Evol.* 20:99–105.

AVISE, J.C. and W.S. NELSON. 1989. Molecular genetic relationships of the extinct dusky seaside sparrow. *Science* 243:646–648.

BEKOFF, M. and M.C. WELLS. 1986. Social ecology and behavior of coyotes. *Adv. Study Behav.* 16:251–338.

BROWN, W.M. 1986. The mitochondrial genome of animals. In *Molecular Evolutionary Biology*, ed. R. MacIntrye, pp. 95–128. Cornell University Press, Ithaca, NY.

BRUFORD, M.W. and T. BURKE. 1991. Hypervariable DNA markers and their applications in the chicken. In *DNA Fingerprinting: Approaches and Applications*, eds. T. Burke, G. Dolf, A.J. Jeffries and R. Wolff, pp. 230–242. Birkhäuser Verlag, Basel, Switzerland.

BRUFORD, M.W. and R.K. WAYNE. 1993. Microsatellites and their application to population genetic studies. *Curr. Biol.* 3:939–943.

CABRERA, A. 1958. Catálogo de los mamíferos de América del Sur. *Rev. del Mus. Argent. de Cienc. Nat. "Bernadino Rivadavia"* 4:1–307.

CALIFORNIA CODE OF REGULATIONS. 1992. In *California Fish and Game Commission, Depatment Fish and Game*, pp. 279–280. Section 670.5.(b)(6)(F).

CARBYN, L.N. 1987. Gray wolf and red wolf. In *Wild Furbearer Management and Conservation in North America*, eds. M. Novak, J.A. Baker, M.E. Obbard and B. Malloch, pp. 358–377. Ministry of Natural Resources, Toronto, Canada.

COLLINS, P.W. 1982. Origin and differentiation of the island fox: a study on the evolution of insular populations. MA thesis, University of California, Santa Barbara.

COLLINS, P.W. 1991. Interaction between island foxes (*Urocyon littoralis*) and Indians on the islands off the coast of southern California. I. Morphologic and archaeological evidence of human assisted dispersal. *J. Ethnobiol.* 11:51–81.

CONNOLLY, G.E. 1978. Predator control and coyote populations: a review of simulation models. In *Coyotes: Biology, Behavior, and Management*, ed. M. Bekoff, pp. 327–345. Academic Press, New York.

CROW, J.F. and M. KIMURA. 1970. *An Introduction to Population Genetics Theory*. Harper & Row, New York.

DIZON, A.E., C. LOCKYER, W.F. PERRIN, D.P. DEMASTER, and J. SISSON. 1992. Rethinking the stock concept: A phylogenetic approach. *Conserv. Biol.* 6:24–36.

DOWLING, T.E., W.L. MINCKLEY, M.E. DOUGLAS, P.C. MARSH, and B.D. DEMERAIS. 1992. Response to Wayne, Nowak, and Phillips and Henry: use of molecular characters in conservation biology. *Conserv. Biol.* 6:600–603.

DRAGOO, J.W., J.R. CHOATE, T.L. YATES, and T.P. O'FARRELL. 1990. Evolutionary and taxonomic relationships among North American arid land foxes. *J. Mamm.* 71:318–332.

ESTES, R. 1991. *The Behavior Guide to African Mammals: Including Hoofed Mammals, Carnivores, Primates*. University of California Press, Berkeley.

FANSHAWE, J.H., L.H. FRAME, and J.R. GINSBERG. 1991. The Wild dog! Africa's vanishing carnivore. *Oryx* 25:137–146.

FERRELL, R.E., D.C. MORIZOT, J. HORN, and C.J. CARLEY. 1980. Biochemical markers in a species endangered by introgression: the red wolf. *Biochem. Genet.* 18:39–49.

FISHER, R.A., W. PUTT, and E. HACKEL. 1976. An investigation of the products of 53 gene loci in three species of wild Canidae: *Canis lupus, Canis latrans,* and *Canis familiaris. Biochem. Genet.* 14:963–974.

FUENTES, E.R. and F.M. JAKSIC. 1979. Latitudinal size variation of Chilean foxes: tests of alternative hypotheses. *Ecology* 60:43–47.

FULLER, T.K., M.G.L. MILLS, M. BORNER, K. LAURENSON, S. LELO, R. BURROWS, and P.W. KAT. 1992. Long distance dispersal by African wild dogs in East and South Africa. *Afr. J. Zool.* 106:535–537.

GARCELON, D.K., R.K. WAYNE, and B.J. GONZALES. 1992. A serologic survey of the Island Fox (*Urocyon littoralis*) on the Channel Islands, California. *J. Wildl. Dis.* 28:223–229.

GEFFEN, E., A. MERCURE, D.J. GIRMAN, D.W. MACDONALD, and R.K. WAYNE. 1992. Phylogenetic relationships of the fox-like canids: mitochondrial DNA restriction fragment, site and cytochrome *b* sequence analyses. *J. Zool.* 228:27–39.

GILBERT, D.A., N. LEHMAN, S.J. O'BRIEN, and R.K. WAYNE. 1990. Genetic fingerprinting reflects population differentiation in the Channel Island fox. *Nature* 344:764–767.

GINSBERG, J.R. and D.W. MACDONALD. 1990. *Foxes, Wolves, Jackals and Dogs. An Action Plan for the Conservation of Canids.* International Union for Conservation of Nature and Natural Resources, Gland, Switzerland.

GIRMAN, D.J., P.W. KAT, M.G.L. MILLS, J.R. GINSBERG, M. BORNER, V. WILSON, J.H. FANSHAWE, C. FITZGIBBON, L.M. LAU, and R.K. WAYNE. 1993. Molecular genetic and morphological analyses of the African wild dog (*Lycaon pictus*). *J. Hered.* 84:450–459.

GOTTELLI, D. and C. SILLERO-ZUBIRI. 1992. The Ethiopian wolf— an endangered endemic canid. *Oryx* 26:205–214.

GOTTELLI, D., C. SILLERO-ZUBIRI, G.D. APPLEBAUM, M.S. ROY, D.J. GIRMAN, J. GARCIA-MORENO, E.A. OSTRANDER, and R.K. WAYNE. 1994. Molecular genetics of the most endangered canid: the Ethiopian wolf, *Canis simensis*. *Molec. Ecol.* 3:301–312.

GRISHAM, J., A. WEST, O. BYERS, and U. SEAL. 1994. *Canid, Hyena, and Aaardwolf— Conservation and Management Plan (CAMP)*. Oklahoma City Zoological Park, Oklahoma City, OK.

HAGELBERG, E., I.C. GRAY, and A.J. JEFFREYS. 1991. Identification of the skeletal remains of a murder victim by DNA analysis. *Nature* 352:427–429.

HALL, E.R. 1981. *The Mammals of North America*. Wiley, New York.

HANSKI, I. 1989. Metapopulation dynamics: Does it help to have more of the same? *Trends Ecol. Evol.* 4:113–114.

HILTON, H.H. 1978. Systematics and ecology of the eastern coyote. In *Coyotes: Biology, Behavior, and Management*, ed. M. Bekoff, pp. 209–228. Academic Press, New York.

HONACKI, J.H., K.E. KINMAN, and J.W. KOEPPL. 1982. *Mammal Species of the World. A Taxonomic and Geographic Reference*. Allen Press, Lawrence, KS.

HOSS, M., M. KOHN, S. PÄÄBO, F. KNAUER, and W. SCHRODER. 1992. Excremental analysis by PCR. *Nature* 359:199.

JEFFREYS, A.J., V. WILSON, and S.L. THEIN. 1985. Hypervariable minisatellite regions in human DNA. *Nature* 327:139–144.

JENKS, S.M. and R.K. WAYNE. 1992. Problems and policy for species threatened by hybridization: the red wolf as a case study. In *Wildlife 2001: Populations*, eds. D.R. McCullough and R.H. Barrett, pp. 237–251. Elsevier, London.

KENNEDY, P.K., M.L. KENNEDY, P.L. CLARKSON, and I.S. LIEPINS. 1991. Genetic variability in natural populations of the gray wolf, *Canis lupus*. *Can. J. Zool.* 69:1183–1188.

KINGDON, J. 1990. *Island Africa*. Collins Press, London.

KOCHER, T.D., W.K. THOMAS, A. MEYER, S.V. EDWARDS, S. PÄÄBO, F.X. VILLABLANCA, and A.C. WILSON. 1989. Dynamics of mitochondrial DNA evolution in animals: amplification and sequencing with conserved primers. *Proc. Natl. Acad. Sci. USA* 86:6196–6200.

KOLENOSKY, G.B. and R.O. STANDFIELD. 1975. Morphological and ecological variation among gray wolves (*Canis lupus*) of Ontario, Canada. In *The Wild Canids*, ed. M.W. Fox, pp. 62–72. Van Nostrand Reinhold, New York.

KOVACH, S.D. and R.J. DOW. 1985. *Island Fox Research on San Nicolas Island*. Annual Report, Department of the Navy, Pacific Missile Test Center, San Diego, CA.

LAIKRE, L. and N. RYMAN. 1991. Inbreeding depression in a captive wolf (*Canis lupus*) population. *Conserv. Biol.* 5:33–40.

LARIVIERE, S. and M. CRETE. 1993. The size of eastern coyotes (*Canis latrans*)— a comment. *J. Mamm.* 74:1072–1074.

LEHMAN, N., P. CLARKSON, L.D. MECH, T.J. MEIER, and R.K. WAYNE. 1992. The use of DNA fingerprinting and mitochondrial DNA to study genetic relationships within and among wolf packs. *Behav. Ecol. Sociobiol.* 30:83–94.

LEHMAN, N., A. EISENHAWER, K. HANSEN, L.D. MECH, R.O. PETERSON, P.J.P. GOGAN, and R.K. WAYNE. 1991. Introgression of coyote mitochondrial DNA into sympatric North American gray wolf populations. *Evolution* 45:104–119.

LEHMAN, N. and R.K. WAYNE. 1991. Analysis of coyote mitochondrial DNA genotype frequencies: estimation of effective number of alleles. *Genetics* 128:405–416.

MARSHALL, L.G., R.F. BUTLER, R.E. DRAKE, G.H. CURTIS, and R.H. TEDFORD. 1979. Calibration of the great American interchange. *Science* 204:272–279.

MECH, L.D. 1970. *The Wolf: The Ecology and Behavior of an Endangered Species*. University of Minnesota Press, Minneapolis.

MECH, L.D. 1977. Productivity, mortality, and population trends of wolves in north-eastern Minnesota. *J. Mamm.* 58:559–574.

MECH, L.D. 1987. Age, season, distance, direction, and social aspects of wolf dispersal from a Minnesota pack. In *Mammalian Dispersal Patterns*, eds. B.D. Chepko-Sade and Z.T. Halpin, pp. 55–74. University of Chicago Press, Chicago, IL.

MECH, L.D., S.M. GOYAL, and C.N. BOTA. 1986. Canine parvovirus infection in wolves (*Canis lupus*) from Minnesota. *J. Wildl. Dis.* 22:104–106.

MEDEL, R.G. and F.M. JAKSIC. 1988. Ecología de los canidos sudamericanos: Una revision. *Rev. Chiena Hist. Nat.* 61:67–79.

MEDEL, R.G., J.E. JIMENEZ, and F.M. JAKSIC. 1990. Discovery of a continental population of the rare Darwin's fox, *Dusicyon fulvipes* (Martin, 1837) in Chile. *Biol. Conserv.* 51:71–77.

MERCURE, A., K. RALLS, K.P. KOEPFLI, and R.K. WAYNE. 1993. Genetic subdivisions among small canids: mitochondrial DNA differentiation of swift, kit, and Arctic foxes. *Evolution* 47:1313–1328.

MOORE, G.C. and G.R. PARKER. 1992. Colonization by the eastern coyote (*Canis latrans*). In *Ecology and Management of the Eastern Coyotes*, ed. A.H. Boer, pp. 23–39. University of New Brunswick Press, Frederickton, Canada.

NEI, M. 1987. *Molecular Evolutionary Genetics*. Columbia University Press, New York.

NOWAK, R.M. 1979. North American Quaternary *Canis. Monogr. Mus. Nat. Hist.* University of Kansas, Lawrence.

NOWAK, R.M. 1992. The red wolf is not a hybrid. *Conserv. Biol.* 6:593–594.

NOWAK, R.M. and J.L. PARADISO. 1983. *Walker's Mammals of the World* (4th ed.). Johns Hopkins University Press, Baltimore, MD.

O'BRIEN, S.J. and J.F. EVERMANN. 1988. Interactive influence of infectious disease and genetic diversity of natural populations. *Trends Ecol. Evol.* 3:254–259.

O'BRIEN, S.J., M.E. ROELKE, L. MARKER, A. NEWMAN, C.A. WINKLER, D. MELTZER, L. COLLY, J.F. EVERMANN, M. BUSH, and D.E. WILDT. 1985. Genetic basis for species vulnerability in the cheetah. *Science* 227:1428–1434.

O'FARRELL, T.P. 1987. Kit Fox. In *Wild Furbearer Management and Conservation in North America*, eds. M. Novak, J.A. Baker, M.E. Obbard, and B. Malloch, pp. 423–431. Ministry of Natural Resources, Toronto, Canada.

OHTA, T. and M. KIMURA. 1973. The model of mutation appropriate to estimate the number of electrophoretically detectable alleles in a genetic population. *Genet. Res.* 22:201–204.

OSGOOD, W.H. 1943. The mammals of Chile. *Field Mus. Nat. Hist. Zool. Ser.* 30:1–268.

PACKARD, R.L. and J.H. BOWERS. 1970. Distributional notes on some foxes from western Texas and eastern New Mexico. *Southwest. Nat.* 14: 450–451.

PACKER, C., D.A. GILBERT, A.E. PUSEY, and S.J. O'BRIEN. 1991. A molecular genetic analysis of kinship and cooperation in African lions. *Nature* 351:562–565.

PAMILO, P. and M. NEI. 1988. Relationships between gene trees and species trees. *Molec. Biol. Evol.* 5:568–583.

PETERSON, R.O. and R.E. PAGE. 1988. The rise and fall of Isle Royale wolves, 1975–1986. *J. Mamm.* 69:89–99.

QUELLER, D.C., J.E. STRASSMAN, and C.R. HUGHES. 1993. Microsatellites and kinship. *Trends Ecol. Evol.* 8:285–288.

REED, J.M., P.D. DOERR, and J.R. WALTERS. 1986. Determining minimum population sizes for birds and mammals. *Wildl. Soc. Bull.* 14:255–261.

REEVE, H.K., D.F. WESTNEAT, W.A. NOON, P.W. SHERMAN, and C.F. AQUADRO. 1990. DNA fingerprinting reveals high levels of inbreeding in colonies of the eusocial naked mole rat. *Proc. Natl. Acad. Sci. USA* 87:2496–2499.

ROELKE, M.E., J.S. MARTENSON, and S.J. O'BRIEN. 1993. The consequences of demographic reduction and genetic depletion in the endangered Florida panther. *Curr. Biol.* 3:340–350.

ROHWER, S.A. and D.L. KILGORE, Jr. 1973. Interbreeding in the arid land foxes, *Vulpes velox* and *V. macrotis. Syst. Zool.* 22:157–165.

ROY, M.S., E. GEFFEN, D. SMITH, E. OSTRANDER, and R.K. WAYNE. 1994b. Patterns of differentiation and hybridization in North American wolf-like canids revealed by analysis of microsatellite loci. *Molec. Biol. Evol.* 11:553–570.

ROY, M.S., D.J. GIRMAN, and R.K. WAYNE. 1994a. The use of museum specimens to reconstruct the genetic variability and relationships of extinct populations. *Experientia* 50:551–557.

RYDER, O.A. 1986. Species conservation and systematics: the dilemma of subspecies. *Trends Ecol. Evol.* 1:9–10.

SCHMITZ, O.J. and G.B. KOLENOSKY. 1985. Wolves and coyotes in Ontario: morphological relationships and origins. *Can. J. Zool.* 63:1130–1137.

SCOTT-BROWN, J. M., S. HERRERO, and J. REYNOLDS. 1987. Swift fox. In *Wild Furbearer Management and Conservation in North America*, eds. M. Novak, J.A. Baker, M.E. Obbard, and B. Malloch, pp. 433–441. Ministry of Natural Resources, Toronto, Canada.

SHELDON, J.W. 1992. *Wild Dogs— A Natural History of the Nondomestic Canidae.* Academic Press, New York.

SLATKIN, M. 1987. Gene flow and the geographic structure of natural populations. *Science* 236:787–792.

SLATKIN, M. 1993. Isolation by distance in equilibrium and non-equilibrium populations. *Evolution* 47:264–279.

SLATKIN, M. and W.P. MADDISON. 1989. A cladistic measure of gene flow inferred from the phylogeny of alleles. *Genetics* 123:603–613.

STAINS, H.J. 1975. Distribution and taxonomy of the Canidae. In *The Wild Canids*, ed. M.W. Fox, pp. 3–26. Van Nostrand Reinhold, New York.

TAYLOR, A.C., W.B. SHERWIN, and R.K. WAYNE. 1994. Genetic variation of microsatellite loci in a bottlenecked species: the northern hairy-nosed wombat (*Lasiorhinus krefftii*). *Molec. Ecol.* 3:277–290.

THURBER, J.M. and R.O. PETERSON. 1991. Changes in body size associated with range expansion in the coyote (*Canis latrans*). *J. Mamm.* 72:750–755.

VALDES, A.M., M. SLATKIN, and N.B. FRIEMER. 1993. Allele frequencies at microsatellite loci: the stepwise mutation model revisited. *Genetics* 133:737–749.

VAN BALLENBERGHE, V., A.W. ERICKSON, and D. BYMAN. 1975. Ecology of the timber wolf in northeastern Minnesota. *Wildl. Monogr.* 43:1–43.

VOIGT, D.R. and W.E. BERG. 1987. Coyote. In *Wild Furbearer Management and Conservation in North America*, eds. M. Novak, J.A. Baker, M.E. Obbard and B. Malloch, pp. 345–357. Ministry of Natural Resources, Toronto, Canada.

WAYNE, R.K. 1992. On the use of molecular genetic characters to investigate species status. *Conserv. Biol.* 6:590–592.

WAYNE, R.K., S. GEORGE, D. GILBERT, P. COLLINS, S. KOVACH, D. GIRMAN, and N. LEHMAN. 1991b. A morphologic and genetic study of the island fox, *Urocyon littoralis. Evolution* 45:1849–1868.

WAYNE, R.K., D.A. GILBERT, A. EISENHAWER, N. LEHMAN, K. HANSEN, D. GIRMAN, R.O. PETERSON, L.D. MECH, P.J.P. GOGAN, U.S. SEAL, and R.J. KRUMENAKER. 1991a. Conservation genetics of the endangered Isle Royale gray wolf. *Conserv. Biol.* 5:41–51.

WAYNE, R.K. and S.M. JENKS. 1991. Mitochondrial DNA analysis supports extensive hybridization of the endangered red wolf (*Canis rufus*). *Nature* 351:565–568.

WAYNE, R.K., N. LEHMAN, M.W. ALLARD, and R.L. HONEYCUTT. 1992. Mitochondrial DNA variability of the gray wolf— genetic consequences of population decline and habitat fragmentation. *Conserv. Biol.* 6:559–569.

WAYNE, R.K., A. MEYER, N. LEHMAN, B. VAN VALKENBURGH, P.W. KAT, T.K. FULLER, D. GIRMAN, and S.J. O'BRIEN. 1990. Large sequence divergence among mitochondrial DNA genotypes within populations of East African black-backed jackals. *Proc. Natl. Acad. Sci. USA* 87:1772–1776.

WAYNE, R.K. and S.J. O'BRIEN. 1987. Allozyme divergence within the Canidae. *Syst. Zool.* 36:339–355.

WEBB, S.D. 1985. Late Cenozoic mammal dispersals between the Americas. In *The Great American Biotic Interchange*, eds. F.G. Stehli and S.D. Webb, pp. 357–382. Plenum, New York.

WORLD CONSERVATION UNION. 1993. *1994 IUCN Red List of Theatened Animals.* World Conservation Union, Gland, Switzerland.

WOZENCRAFT, W.C. 1993. Order Carnivora. In *Mammal Species of the World: A Taxonomic and Geographic Reference*, eds. D.E. Wilson and D.M. Reeder, pp. 279–348. Smithsonian Institution Press, Washington, DC.

YALDEN, D.W. 1983. The extent of high ground in Ethiopia compared to the rest of Africa. *Ethiop. J. Sci.* 6:35–38.

YAHNKE, C.J. 1994. Systematic relationships of Darwin's fox and the extinct Falkland Island wolf inferred from mtDNA sequence. MA thesis, Northern Illinois University, De Kalb.

YAHNKE, C.J., W.E. JOHNSON, E. GEFFEN, D. SMITH, F. HERTEL, M.S. ROY, C.F. BONACIC, T.F. FULLER, B. VAN VALKENBURGH, and R.K. WAYNE. 1995. Darwin's Fox: a distinct endangered species in a vanishing habitat. *Conserv. Biol., in press.*

5

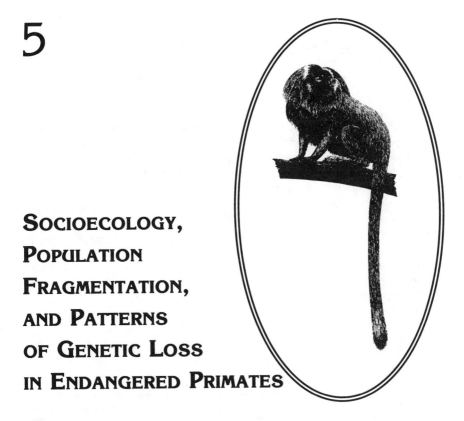

Socioecology, Population Fragmentation, and Patterns of Genetic Loss in Endangered Primates

Theresa R. Pope

INTRODUCTION

Primates comprise one of the most taxonomically diverse mammalian orders, with 14 families, 60 genera, and approximately 300 species. More than 55% of extant primate species are included in the IUCN Red List of Threatened Animals (Box 5.1). Although hunting and live capture for animal trade have had considerable impact on some species in some areas (Mittermeier, 1987, 1991), habitat destruction remains the single most important threat to nonhuman primate populations. More than 90% of primate species occur only in tropical forests, which are currently disappearing at a rate of nearly 17 million hectares per year (World Resource Institute, 1992). By 1980, the world's tropical forests were reduced to about half of their original expanse. Despite considerable public attention and conservation effort over the following decade, the rate of tropical deforestation doubled between 1980 and 1989 (Myers, 1989).

Box 5.1. Taxonomic and Conservation Status of Primates

The taxonomy of the order Primates is in a state of revision, particularly at the species/subspecies level. Thus, the number of species listed below can be considered approximate only. Depending on the taxonomic and listing source, be- tween 55% and 70% of primate species are considered threatened or endangered. Six families and 27 of the genera below are comprised entirely of species classi- fied as threatened or endangered by the World Conservation Union (1993).

	Families	Genera	Species	IUCN Red List (subsp. or sp.)
Lorisiformes				
African and Asian prosimians	1	5	14[1,2]	2
Lemuriformes				
Madagascar prosimians	5	14	46[3]	42
Tarsiiformes				
Tarsiers	1	1	4[1]	3
Platyrrhini				
New World monkeys	3	16	123[4,5]	59
Catarrhini				
Old World monkeys, apes	4	24	119[1]	75

[1] Lee et al., 1988; [2] Eudey, 1987; [3] Harcourt and Thronback, 1990; [4] Rylands et al., 1993; [5] Emmons, 1990

As tropical forests disappear, so do the primates that live in them. The consequences of deforestation for primates involve reduction of species–wide population size and fragmentation into local populations (typically contain- ing a few to hundreds of individuals) between which there is little or no opportunity for migration. Increased rates of allelic loss and a decrease in individual heterozygosity are expected in small populations due to genetic drift, and to the accumulation of inbreeding effects. Loss of species–wide genetic diversity is compounded by the increased susceptibility of remnant populations to demographic and environmental perturbations that can lead to extinction (Goodman, 1987; Shaffer, 1987).

Whereas some primate populations occur within parks or preserves and enjoy at least symbolic protection, the majority occupy public lands subject to human encroachment and to commercial exploitation such as logging and mining (Oates, 1986; Eudey, 1987; Myers, 1987). Protected status for pri- mates is often conferred on a forested area only after it has been heavily exploited, leaving a severely eroded resource base. The overburdened gov- ernments of developing tropical countries are faced with a human population explosion that will lead to a size doubling within the next 35 years (World Resource Institute, 1992). Adding to this problem, and to growing national debts, are already existing surpluses of landless people in cultures that remain reliant upon subsistence agriculture. Immediate economic gains from export of raw materials (such as timber), and immense population pressure, leave little incentive for governmental or public support for large-scale protection of remaining forests. Projections based on satellite imagery studies indicate that 90% of remaining tropical forest will be lost in less than 33 years if current deforestation rates persist (Terborgh, 1992).

Given this scenario, the question becomes one of triage: Which forest areas should be given priority in conservation efforts? What intensive management measures are necessary to maintain their primate populations? The concept of designing preserves based on the minimum viable population sizes of primates is not realistic. The sizes and locations of preserves are more likely to be determined by political and social considerations than by biological needs. Many remnant populations of threatened primates are already well below what is considered minimally viable based on existing models, and many more are likely to become so. Although captive propagation has been promoted as a means of preserving a reservoir of genetic variation for reintroduction to nature, zoos are far more likely to become museums than arks. Attempts to reintroduce captive–born primates into the wild have generally met with complete failure (Griffeth et al., 1989; Pearl, 1992). Translocations of wild animals under appropriate conditions have been more successful, but are still fraught with difficulties (Griffeth et al., 1989). Nonetheless, translocation may be the only conservation option for many primates, particularly when forests are due for imminent destruction, or are too small to maintain many individuals.

A variety of behavioral, ecological, and demographic factors must be considered in prioritization of conservation areas and in procedures of intensive management. Here, the focus will be on the interaction of these variables with population genetic structures within primate social systems. Most previous investigations of this relationship have involved theoretical treatments estimating effective population size (N_e) and projecting concomitant loss of genetic diversity (e.g., Reed et al., 1986; Lacy, 1987; Lande and Barrowclough, 1987). The concern here will be with evaluating assumptions and predictions of these models based on empirical patterns of genetic diversity within populations of threatened primates. How much genetic diversity remains? What does it say about how genetic diversity is lost? How do estimates of variability in small primate populations compare to those in larger populations of closely related, non–threatened species with similar socioecology?

The manner in which genetic variability is distributed among fragmented populations is primarily dependent on four factors: (1) the geographic distribution of genetic diversity prior to fragmentation; (2) the size and number of fragmented populations; (3) the length of time these populations have been isolated relative to species generation time; and (4) species socioecology (mating system, dispersal pattern, and the distribution in space of genetic lineages as a function of resource allocation). Comparative socioecology is especially important in primates because of the tremendous variation in social systems among species.

SOCIOECOLOGY AND POPULATION GENETIC STRUCTURE

Nearly all primate species are geographically subdivided into discrete or quasi-discrete populations, each of which may be further subdivided into social groups of varying size and composition. Gender-specific differences often exist in rates and patterns of intergroup movement, with a variety of population genetic consequences. Many primate species are organized into matrilineally-oriented social groups, with males accounting for most intergroup transfers. Examples include most of the Old World monkeys (Cercopithecidae), except for the baboon *Papio hamadryas* and some members of the subfamily Colobinae. In some cases, males dispersing into a given social group are genetically related. These transfers can occur simultaneously, as in the red howler monkey *Alouatta seniculus* (Pope, 1990) and some macaques (e.g., Drickamer and Vessey, 1973; Kawanaka, 1973), or sequentially when maturing males transfer preferentially into social groups already containing male relatives, as in the vervet monkey *Cercopithecus aethiops* (Cheney and Seyfarth, 1983). Conversely, in some primate species most intergroup movement is by females. For example, in the endangered muriqui (*Brachyteles arachnoides*) and the threatened Uganda red colobus (*Procolobus rufomitratus tephrosceles*), males remain in their natal social group and females disperse. Most South American monkeys for which good field data exist display either female-biased dispersal or dispersal by both sexes (exceptions include *Cebus* and *Saimiri*). In those species in which both sexes transfer between groups (e.g., the mantled howler monkey, *Alouatta palliata* and the endangered mountain gorilla, *Gorilla gorilla berengei*), social groups are usually composed of unrelated females and unrelated male(s), and are not oriented along kin lines of either sex.

The following sections highlight some population genetic ramifications anticipated for species that are hierarchically subdivided, and which may display matrifocal or patrifocal lineage structure.

Effective Population Size

Primates, like most vertebrates, live in structured populations that vary temporally in size. Tremendous variation in reproductive success among individuals, among families, and between the sexes, can effectively reduce the adult gametic pool sampled at each generation, such that loss of heterozygosity and polymorphism occurs at a faster rate than would be predicted based on census adult numbers. Effective population size (N_e) can be interpreted as the number of animals which, if the behavior were like an ideal population, would produce the same rate of loss of heterozygosity as that exhibited by an actual population of interest (Wright, 1931). In vertebrates, N_e is usually much smaller than actual population size (N). As the ratio N_e/N decreases for any given N, the expectation is for neutral allelic diversity to be lost more rapidly.

Differences between the sexes in survivorship affect the breeding sex ratio, and the extent to which mortality patterns differ by gender can in turn depend upon the social system. Both dispersal and attempted transfer are costly for males in terms of survival, because dispersers lose protection from predators and forfeit enhanced access to resources that social groups provide. Intrasexual competition among males, which nearly always involves aggression, also makes transfers into foreign social groups difficult (Pusey and Packer, 1987; Crockett and Pope, 1988). A successful male must then defend his position from challengers. Consequently, in male-transfer species, fewer males than females survive into and through adulthood (Dittus, 1977; Otis et al., 1981; Dunbar, 1987; Cheney et al., 1988; Struhsaker and Pope, 1991). On the other hand, female-biased emigration in species with patrilineal social systems appears less likely to result in strong gender-biased mortality. Female emigration is usually followed almost immediately by immigration into a new social unit, and with the possible exception of the common chimpanzee, *Pan troglodytes* (Pusey, 1980), is usually met with little or no resistance (Harcourt, 1979; Struhsaker and Leland, 1979; Strier, 1991).

Variance in reproductive success often differs between the sexes in primates, and this can further reduce N_e. In polygynous species, males usually experience higher variance in reproductive success than females, but the degree to which this holds can vary greatly depending on other aspects of the social system (reviewed in Clutton-Brock, 1988). Polygynous mating systems in primates include single-male harems, multimale/multifemale groups that range in size from less than 10 to more than 100 individuals, complex fusion–fission communities, solitary territoriality in which multiple female home ranges are incorporated into that of a single male, and multi-level societies that combine several of these features. Temporal–spatial distribution of females, synchrony of estrus, and female choice all influence the degree to which males can reproductively monopolize females. The effects of these variables on male competitive strategies and copulation success have been examined in a variety of primate species and reviewed extensively (e.g., Smuts, 1987).

Although variance in male lifetime reproductive success is one of the most important determinants of N_e in polygynous mammals (Harris and Allendorf, 1989), this parameter remains difficult to measure in long-lived species like primates. One salient outcome of the studies described earlier is that longevity itself contributes substantially to male reproductive success. In single-male harem mating systems such as those exhibited by many African guenons and South American howlers, males aggressively evicted from the group in which they hold breeding tenure are often able to take over another group of females (Struhsaker, 1988; Pope, 1990). Priority of access to estrous females in species characterized by multimale groups, such as baboons and macaques, is frequently (but not always) determined by rank in a dominance hierarchy (Silk, 1987). In some species such as the Japanese and rhesus macaque (*Macaca fuscata* and *M. mulatta*), rank increases with age and time lived in the group (Drickamer and Vessey, 1973; Sugiyama, 1976), whereas in the savannah baboon (*Papio cynocephalus*) the opposite pattern occurs (Smuts, 1985). In general, cumulative lifelong number of surviving offspring may tend to increase as a function of number of years spent resident in the group (Smuts, 1985; Altmann et al., 1988; Cheney et al., 1988).

Estimates of the variance in male reproductive success and breeding sex ratio are often based on slice-in-time data, which can lead to erroneous conclusions. Two caveats apply: (1) Variance in lifetime reproductive success is probably lower than the reproductive variance exhibited among males over the course of one, or even several successive breeding seasons; and (2) Male reproductive tenure or dominance rank is usually much shorter than generation time (by any definition). Both factors result in more males contributing gametes to the next generation than would be estimated from the ratio of breeding males to breeding females in any given year.

Because female reproductive rate is physiologically limited by gestation and lactation, variance in female reproduction is generally expected to be less than that exhibited by males. Nonetheless, competition among females for limiting resources can result in considerable variability in reproductive success. In primate species that live in large social groups with pronounced dominance hierarchies, high-ranking animals often have greater access to limiting foods or water (e.g., Dittus, 1977; Post et al., 1980; Whitten, 1983). Differences between high- and low-ranking females in fecundity, infant survivorship, or age at first reproduction have been demonstrated in numerous populations of baboons, macaques, and vervets (e.g., Drickamer, 1974; Dittus, 1979; Silk et al., 1981; Busse, 1982; Gouzoules et al., 1982; Whitten, 1983; Altmann et al., 1988). Even in species without overt dominance hierarchies, other forms of female competition can result in highly differential reproductive success. For example, up to 72% of female red howlers emigrate from their natal social group, after which they are aggressively prevented from transferring into other groups and have little chance of successfully reproducing (Pope, 1992; Crockett and Pope, 1993). Reproduction is often physiologically suppressed in all but one female in a social group in most Callitrichidae (marmosets and tamarins; Abbot, 1984; French et al., 1984; Terborgh and Goldizen, 1985; Garber et al., 1993), and possibly ruffed lemurs (*Varecia variegata*; White et al., 1992).

The example in Box 5.2, which compares two sympatric primates with different social systems, illustrates how some of the processes described above may contribute to observed differences in N_e/N ratios. In addition to the variables incorporated into this example, differences in reproductive success among social groups can also substantially reduce the proportion of population genetic variability represented in subsequent generations (see Robinson, 1988). The take-home message is that a variety of socioecological factors can influence N_e.

Lineage Effects and New Group Formation

The apportionment of allelic diversity among social groups, which can be estimated from molecular data such as that provided by allozymes, reflects both the amount of genetic diversity present at the time of population isolation, and the subsequent pattern of its loss. However, interpretation of such data (particularly from a single point in time) can be difficult, in part because of the usual lack of historical information about patterns of genetic diversity, and in part because of the expected influences of a variety of socioecological factors on group structure.

Gender-Biased Dispersal. As already mentioned, genetic lineages within primate social groups often tend to be either matrifocal or patrifocal. The organization of individuals into kin-oriented social groups can result in a highly non-random distribution of consanguineous alleles. Kin orientation is usually limited to the gender that is highly philopatric, remaining for life in the social group to which it was born. Breeding individuals of the dispersing sex, who may sometimes move among groups multiple times during their reproductive life span, are usually (but not always) unrelated to all group members except their own offspring.

Consider, for example, the genetic consequences of a common primate social system in which females remain in their natal social groups for life and males disperse to other groups before breeding. Coancestry among fe-

**Box 5.2. Comparison of N_e/N Ratios for the Red-tailed Guenon
and Red Colobus Populations of Kibale**

The Kibale National Park in western
Uganda is a 560 km² habitat island surrounded by extensive agricultural development. The two most abundant monkeys in Kibale are the Uganda red
colobus (*Procolobus rufomitratus tephrosceles*) and the red-tailed guenon (*Cercopithecus ascanius*), which differ markedly in ecology and social system. *P. r.
tephrosceles* is a large folivore that lives

in large ($N = 50$) multimale, patrilineal
social groups from which females disperse at adolescence. *C. ascanius* is a
small omnivore that lives in single–male,
matrilineal social groups of 30–35 individuals from which males disperse. The
demographic patterns caused by these
differences result in different proportions
of the parent gamete pool being represented in a subsequent generation.

A. Variability in lifetime reproductive success (LRS) for males versus females
(Struhsaker and Pope, 1991):

	Red-tailed guenon		Red colobus	
	Males	Females	Males	Females
Generation time	9.6 years		9.3 years	
Survival to adulthood (%)	19	60	34	47
Mean LRS ($\bar{k}$)	3.67	1.37	2.29	1.71
Variance LRS (δ^2)	18.13	1.58	8.03	1.78

B. Calculation of N_e/N when sex ratio and variance in progeny number differs
between the sexes (based on Lande and Barrowclough, 1987):

$$N_e = 4N_{ef}N_{em}/(N_{ef} + N_{em}),$$

where N_{ef} and N_{em} are the effective numbers of females and males, respectively, and

$$N_{ef} = (N_f\bar{k}_f - 1)/[\bar{k}_f - 1 + (\delta_f^2/\bar{k}_f)],$$
$$N_{em} = (N_m\bar{k}_m - 1)/[\bar{k}_m - 1 + (\delta_m^2/\bar{k}_m)],$$

$\bar{k}$ is the mean LRS of an individual, δ^2 is the variance of $\bar{k}$, and N_f and N_m are
the actual numbers of females and males in the population. Using the values for
these variables listed in A, and actual numbers of males and females based on
census densities for the entire reserve, the following estimates of N_e/N are obtained:

	N	N_e	N_e/N
Red colobus	40,520	14,160	0.35
Red-tailed guenon	17,710	3,196	0.18

males in a group will be higher than the average gene correlation among random individuals within a population of social groups. Female gametes within a group comprise a reduced, nonrandom fraction of the variability present among the population of gametes as a whole, resulting in genetic variance between social groups, or lineages. Thus, when a male originating from one lineage moves to another and breeds, heterozygosity in the first generation progeny will be in excess of Hardy–Weinberg expectations, and the number of excess heterozygotes is directly proportional to the genetic variance between groups (Chesser, 1991; Pope, 1992).

To quantitatively address the population genetic consequences of this form of sex-biased dispersal, Chesser (1991) calculated fixation indices and concomitant loss of gene diversity over 50 generations for populations containing different numbers and sizes of social lineages, and different numbers of breeding males per lineage. Some of the major qualitative findings were as follows (Figure 5.1):

1. All fixation indices (F_{LS}, F_{IS}, F_{IL}) came to an asymptote within a few generations and thereafter remained constant despite continued loss of genetic diversity within subpopulations and lineages [$(1 - \alpha)$ and $(1 - \theta)$, respectively];

2. F_{IL} values were consistently negative, indicating per-generation excesses of heterozygosity within lineages;

3. Smaller numbers of breeding males per lineage resulted in higher values of F_{LS}, indicating greater genetic differentiation among lineages, and more negative values of F_{IL} (numbers of females per lineage had the same effect, but to a lesser degree); and

4. F_{IS} values remained near zero.

Another finding was that the overall rate of loss of population genetic variability was inversely proportional to the number of social lineages (groups) in the population. For example, populations B and C in Figure 5.1 have identical total population size, breeding sex ratio, and variance in reproductive success, but population B contains only half as many groups. Cumulative loss of genetic variance $(1 - \alpha)$ occurred faster in population B as a consequence of indirect inbreeding. When fewer social lineages are available, a male is more likely to encounter and breed with female relatives. Summarized in Box 5.3 are empirical results exemplifying how the distribution of allozymic variation among social groups compares in three primate species with differing social systems.

In a primate social system in which males disperse and females are philopatric, increasing the rate at which males transfer between social groups is equivalent to increasing the number of breeding males per female generation. Contributions by more fathers reduces the variance between

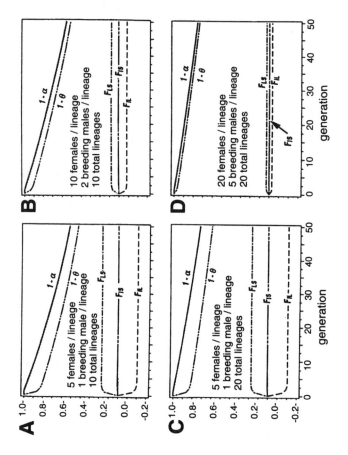

Figure 5.1. The influence of mating system on the distribution and rate of loss of population genetic diversity (from Chesser, 1991). In all four mating scenarios, breeding males contribute equally to offspring born within groups, females breed within their natal social groups, and males disperse randomly between groups before breeding. F_{IL} is the correlation of genes within individuals relative to correlation of genes within lineages (social groups); F_{LS} is the mean proportion of the genetic variance in the subpopulation that is found in lineages; F_{IS} is the correlation of genes within individuals relative to subpopulation allele frequencies; $(1 - \alpha)$ is the proportion of original genetic variance remaining in the subpopulation; and $(1 - \theta)$ is the proportion of original genetic variance remaining within lineages.

Box 5.3. Comparative Lineage Effects in Macaques and Red Howlers

Rhesus macaques (*Macaca mulatta*) and long–tailed macaques (*M. fascicularis*) both live in large, multimale, multi-female social groups in which females breed in their natal group and males disperse. Red howler monkeys (*Alouatta seniculus*) live in small social groups with one breeding male and several breeding females. Although many females emigrate from their natal group, they are rarely successful at transferring into others; instead, females in established groups are recruited from within, and are usually closely related. As in macaques, males disperse from the natal group and attempt to transfer to other social groups. In all three species, females first give birth at about five years of age.

Genetic markers were used to examine paternity in a population of *M. fascicularis* in Sumatra (de Ruiter et al., 1994) and a population of *A. seniculus* in Venezuela (Pope, 1990). In *M. fascicularis*, the α–male fathered 52–92% of the offspring conceived in a group during

his tenure, which lasted on average approximately two years. The β–male (often the displaced α–male) fathered 8–33% of a group's offspring during his tenure, which lasted on average one year. Remaining males fathered 2–9% of the offspring each. Thus, in any five-year period, group females ($N = 9$–10) gave birth to offspring fathered by at least two males (scenario B in Figure 5.1). In two–male groups of *A. seniculus*, the α–male fathered all or most of the offspring conceived during his tenure (i.e., in no case could the male genetically identified as fathering the majority of group offspring be excluded from fathering all). Male breeding tenure in single- and double-male groups combined lasted on average seven years. In most groups, only one male per female–generation time fathered offspring (scenario C in Figure 5.1). Partitioning of genetic variance among social groups in these two populations, and a population of *M. mulatta* in Pakistan, was as follows:

	No. Groups	Variable Loci	F_{ST}^1 ($\cong F_{LS}$)	F_{IS}^1 ($\cong F_{IL}$)	F_{IT}^1 ($\cong F_{IS}$)	Reference
M. mulatta	5	5	0.04	−0.08	−0.05	Melnick et al., 1984
M. fascicularis	6	11	0.04	0.02	0.06	de Jong et al., 1994
A. seniculus	14	9	0.22	−0.16	−0.10	Pope, 1992

[1] F–statistics calculated using Wright (1965, 1978) and Nei (1977) for *M. mulatta* and *A. seniculus*, and using Weir and Cockerham (1984) for *M. fascicularis;* see text and legend to Figure 5.1 for explanations.

uniting gametes within social groups by decreasing coancestry of females within lineages, and by increasing the proportion of total population genetic diversity represented in male gametes. However, the effects of increasing the rate of male transfer are counteracted when immigrant males are related. Then, the reduction in coancestry between females within lineages is reduced over what would be the case if males were unrelated, as is the proportion of population genetic diversity represented in male gametes. Even if all immigrant male kin were equally likely to breed (which is seldom true), genetic divergence between groups will be higher than expected based on numbers of migrants. Migration of kin groups between small local populations can actually increase the genetic variance between them (Fix, 1978).

In primate species such as mantled howlers, in which both sexes transfer between groups, social groups are comprised of unrelated females and unrelated males. All else being equal, genetic differentiation between social groups is expected to be lower in such species than in those with gender-limited dispersal.

Formation of New Social Groups.   Primate social groups are rarely all the same age (e.g., Pope, 1992). In most non–monogamous species, new groups form through lineal fission along matrilines or patrilines. This process increases mean genetic relatedness within group lineages, and also results in greater genetic variance between new groups than the population average. This fission phenomenon has been demonstrated for a variety of social systems, including Yanomama villages along patrilines (Smouse et al., 1981), and rhesus macaques (*Macaca mulatta*; Olivier et al., 1981) and Barbary macaques (*M. sylvanus*; Scheffrahn et al., 1993) along matrilines. A high rate of new group formation should, thus, be accompanied by greater mean genetic differentiation between social lineages than in a steady–state population (but see Melnick and Kidd, 1983). Among the factors that can augment new group formation are population growth, and increased intragroup competition resulting from an eroded resource base (e.g., Skorupa, 1986).

In some species, new group formation occurs when solitary, dispersing individuals of both sexes meet, form social bonds, and defend a territory. In red howlers, founding females are seldom related genetically, but, because the daughters of only one (reproductively dominant) female typically are successful at remaining in their natal group to breed, a single matriline ultimately is produced (Crockett and Pope, 1993). Only established groups form lineages, such that mean genetic variance between social groups is higher in a steady-state population than in one exhibiting a high rate of new group formation (Pope, 1992). This pattern is opposite of that expected in a species in which new group formation takes place via lineal fission.

Inbreeding.   The degree to which primates "avoid" breeding with relatives is a much-debated topic among behavioral ecologists (review in Moore, 1993), who are primarily concerned with *why* animals disperse. Is inbreeding avoidance a cause or a consequence of dispersal? This question is empirically difficult, but evidence from field studies suggests that the answer might be variable among species (Lemos de Sá, 1988; Pope, 1990). The question is also important from the perspective of conservation genetics. If inbreeding avoidance is merely a by-product of a process driven by gender differences in the distribution of limiting resources, then demographic changes in population structure (caused, for example, by habitat degradation) might influence the level of inbreeding more so than if the phenomenon were based strictly on avoidance of matings with close kin *per se*. For example, in some species, mean male breeding tenure in a group is shorter than female generation time, and there is little likelihood of males breeding with daughters. However, male tenure is usually limited by male competition, such that at sufficiently low population density, a male might maintain breeding tenure until well after his daughters become sexually mature. Thus, increased inbreeding driven by demographic changes could take place unless males explicitly avoid incestuous matings.

Genetic Structure Among Populations

Although many studies have examined partitioning of genetic variance within primate species, most have not considered social groups and geographic localities as separate levels of hierarchical subdivision (e.g., Kawamoto et al., 1984; Nozawa et al., 1991). Consequently, reported values of genetic differentiation tend to incorporate both variance between troops and between localities, obscuring relative contributions of the two.

A few studies have examined social and demic structure separately, and have found genetic differences among social units often comparable to or higher than those between populations. For example, Dracopoli et al. (1983) sampled 30 social groups of vervet monkeys from four separate localities in Kenya. Genetic differences "between groups within localities" and "between localities" were similar ($G_{CS} = 0.08$ and $G_{ST} = 0.09$, respectively). Melnick et al. (1984, 1986) found genetic differences between social groups of rhesus macaque in Pakistan ranging from $F_{ST} = 0.03 - 0.06$, whereas differentiation between populations in different countries was $G_{ST} = 0.08$. For red howler monkeys in Venezuela, F_{ST} values between social groups were 0.22 in one population and 0.14 in another, whereas the two populations differed little genetically ($F_{ST} = 0.02$; Pope, 1992). The latter is also the only primate study in which long–term observational data on dispersal have been compared with population genetic structure. The between–population F_{ST} estimated from the observed migration rate was 0.02, in excellent agreement with the genetic-based inference.

Genetic differentiation was much higher among island populations of the long–tailed macaque, *Macaca fascicularis*, than between social groups on the same island (Kawamoto et al., 1984). This outcome is probably more representative of what is to be expected for many threatened primate species, in which gene flow between fragmented populations has become impossible. Alleles lost due to genetic drift and founder effect can differ from one population fragment to the next. As will be seen in the following case studies, this pattern is indeed typical of primate species living in highly fragmented forests, and has many implications for their conservation.

CASE STUDIES

The Atlantic Forest of Brazil

The Brazilian coastal forest is a unique ecosystem, distinct from that of the Amazon Basin. Highly diverse biologically, it contains nearly 7% of the world's species, a large proportion of which is found nowhere else. Fifty-one species of mammals and 160 species of birds are endemic, as are 53% of trees, 64% of palms, and 74% of bromeliads (Quintela, 1990). Unfortunately, the Atlantic coast is also the industrial and agricultural hub of Brazil, containing its three largest cities and some 80 million people (World Resource Institute, 1992). As a consequence, more than 95% of the Atlantic forest has been destroyed (Fonseca, 1985), with the remaining 1% to 5% scattered in a mosaic of forest fragments. Here live 21 species of primates, 17 of which are endemic. Thirteen are listed by the IUCN as endangered, and three as vulnerable. Among those on the verge of extinction are the muriqui (*Brachyteles arachnoides*) and the golden lion tamarin (*Leontopithecus rosalia*). The two species are sympatric, such that the time frame over which population fragmentation occurred was identical. Nonetheless, different patterns of allelic diversity loss are displayed, largely as a consequence of species' differences in social structure, reproductive physiology, and generation length.

Leontopithecus. This genus, plus *Callithrix* (marmosets) and *Saguinus* (tamarins), comprise the New World Callitrichidae. *Leontopithecus* includes four highly endangered species of lion tamarins, all of which are endemic to the Atlantic forest (Figure 5.2): the golden lion tamarin (*L. rosalia*), golden-headed (*L. chrysomelas*), golden–rumped (*L. chrysopygus*), and the recently discovered black–faced lion tamarin (*L. caissara*, which is possibly a clinal variant of *L. chrysopygus*). The first three species were formerly considered subspecies of *L. rosalia*, but differences in dental and cranial morphology have led most recent taxonomists to recognize them as separate species (Rylands et al., 1993). *L. rosalia* has been the most publicized and, with the possible exception of *L. caissara*, probably has the smallest extant range. Less than 400 individuals of this species remain in the wild.

Genetic variation in three species of *Leontopithecus* was examined by

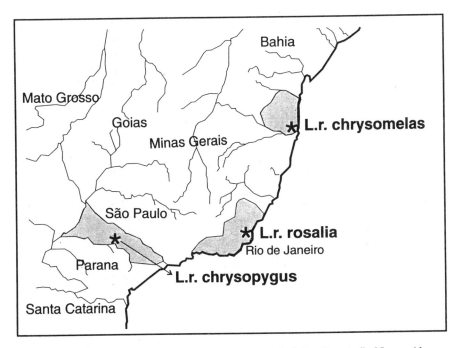

Figure 5.2. Former geographic distributions (shaded areas) of three "species" of *Leontopithecus* in the Atlantic coastal forest of Brazil (based on Rylands et al., 1993). Stars represent extant populations as follows: (1) *L.(r.) rosalia* (Poco das Antes Biological Reserve, where approximately 75% of the 400 individuals remaining in the wild are located); (2) *L.(r.) chrysopygus* (Morro de Diablo State Park, the largest of the two localities where this species is known to survive, the other being in the Caetetus State Reserve at the western edge of their former range); (3) *L.(r.) chrysomelas* (one of multiple isolated localities where this species remains).

Forman et al. (1986), who surveyed 47 allozyme loci in a total of 171 captive and wild animals (Table 5.1). Almost no intraspecific genetic variation was detected, yielding the lowest allelic diversity estimates yet reported for any primate. The near-absence of allozyme variation characterized both natural and captive populations of *L. chrysopygus* and *L. rosalia* (a species that has been bred extensively in zoos, and may have a larger population in captivity than in the wild). Recent attempts to uncover genomic variation in *L. rosalia* using multilocus minisatellite DNA probes have also yielded low variability estimates for most of the probes employed (W.F. Gergits, personal communication).

Low genetic variability in lion tamarins is perhaps not surprising, given the extremely small size and isolated nature of these populations. However, recent surveys of other callitrichid species suggest that low levels of genetic

Table 5.1. Allozymic Diversity (47 loci) in Three Species of Lion Tamarins Endemic to the Atlantic Forest of Brazil*

Sample	N^1	"Rare Alleles" Loci[2]	P^1	H^1
Leontopithecus rosalia				
Brazil, wild	73	2/2	0.03	0.01
U.S., captive	67	1/2	0.04	0.01
L. chrysomelas				
Brazil, captive	8	0/1	0.03	0.01
L. chrysopygus				
Brazil, wild	7	0/0	0	0
Brazil, captive	16	1/1	0.03	0.03

[1] N, sample size per locale; P, percent loci polymorphic/100; H, mean heterozygosity.

[2] Numerator, number of variable loci at which the frequency of the common allele was greater than 0.90; denominator, total number of variable loci.

* *Source:* Forman et al., 1986. All three species are listed as Endangered by the IUCN. The collecting locality for the wild *L. rosalia* sample was the 5.2 km[2] Poco das Antes Biological Reserve, the only remaining stronghold of this species in nature. This sample consisted of 63 free–ranging individuals and 10 wild–born animals and first–generation offspring that are now housed at the Rio de Janeiro Primate Center. The seven wild *L. chrysopygus* were collected from both localities in which the species is known to remain (see text).

variation may be the typical condition for this taxonomic family (Table 5.2). For example, allozyme studies of natural populations of seven species of *Saguinus* and *Callithrix* uncovered few polymorphic loci ($P = 0.05$–0.25) and consistently yielded low estimates of heterozygosity ($H = 0.01$–0.06). Some of the taxa assayed (e.g., *C. jacchus* and *S.m. niger*) are abundant and have large geographic distributions, but the ranges of three *Callithrix* species (*humeralifer*, *emiliae*, and *geoffroyi*) are small and more typical of the geographic extent of most callitrichids (Figure 5.3). As in *L. rosalia*, DNA fingerprints of wild-trapped families of *C. jacchus* also displayed low variability (Dixson et al., 1992). Band sharing among group members was so extensive that paternity exclusion was not possible.

Class I genes of the major histocompatibility complex (MHC), which encode glycoproteins that play an important role in the immunological response, also display low variability in callitrichid species. These loci are normally hypervariable in other organisms, presumably because allelic polymorphisms are maintained through balancing selection (Hughes and Nei, 1988). Watkins et al. (1991) examined MHC class I genes from five callitrichid species (*S. oedipus*, *S. geoffroyi*, *S. fuscicollis*, *C. jacchus*, and *L. rosalia*), all of which proved to exhibit little nucleotide sequence variation. The only other mammals thus far reported to show comparably low levels of MHC class I polymorphism are the eusocial and highly inbred naked mole rat (Nizetic et al., 1988), the Syrian hamster (Darden and Streilein, 1984), and the cheetah (which is believed to have gone through several recent

Table 5.2. Allozymic Diversity (20 loci) in Eight Species of Tamarins (*Saguinus*) and Marmosets (*Callithrix*) in Brazil. *C. geoffroyi* is listed as vulnerable by the IUCN; the other species are not threatened

Species	No. Collecting Localities	N^1	"Rare Alleles" Loci[1]	P^1	H^1	Reference
Saguinus						
S. *fuscicollis*	2	138	0/3	0.15	0.04	Melo et al., 1992
S. *midas midas*	3	13	1/5	0.25	0.06	Melo et al., 1992
S. *midas niger*	?	123	2/4	0.21	0.02	Melo et al., 1992
Callithrix						
C. *humeralifer*	1	14	2/4	0.20	0.04	Meireles et al., 1992
C. *emiliae* (pop. 1)	1	63	4/5	0.25	0.03	Meireles et al., 1992
(pop. 2)	1	22	0/1	0.05	0.01	Meireles et al., 1992
C. *jacchus*	1	35	1/2	0.10	0.02	Meireles et al., 1992
C. *penicillata*	1	10	0/4	0.20	0.05	Meireles et al., 1992
C. *geoffroyi*	1	6	0/1	0.05	0.03	Meireles et al., 1992

[1] See footnote to Table 5.1.

population bottlenecks; O'Brien et al., 1985; see Chapter 3). Furthermore, extensive sharing of MHC class I alleles was observed between some of the callitrichid species. The MHC class I genes of S. *oedipus* and S. *fuscicollis* were indistinguishable from each other, and one of the two genic forms in all assayed S. *geoffroyi* was also shared by S. *oedipus* (Watkins et al., 1991).

Low levels of genetic variability are consistent with unique aspects of callitrichid social structure and breeding physiology. These are the only monkeys that regularly bear fraternal twins, as opposed to single young. Although the twins are dizygotic, they are connected *in utero* by a vascular network through which blood and bone marrow elements are exchanged, resulting in the establishment of a stable, haemopoietic bone marrow chimera. Twins are born with bone marrow elements and leukocytes shared by their co–twin, such that DNA fingerprints based on DNA derived from leukocytes of twins yield identical banding patterns (Dixson et al., 1988). Watkins et al. (1991) presented evidence that bone marrow chimerism may have been an important selective pressure in minimizing the evolution of genetic diversity at MHC class I loci in callitrichids. Because of the immune system's role in rejection of foreign tissues, twins with highly divergent MHC class I genes could experience problems exchanging bone marrow (see Fleischhauer et al., 1990). Similarities at MHC loci might also contribute to explanations as to why some callitrichid species successfully hybridize in the laboratory.

In terms of social organization, callitrichids are the only monkeys that frequently exhibit a mating system of cooperative polyandry, wherein two

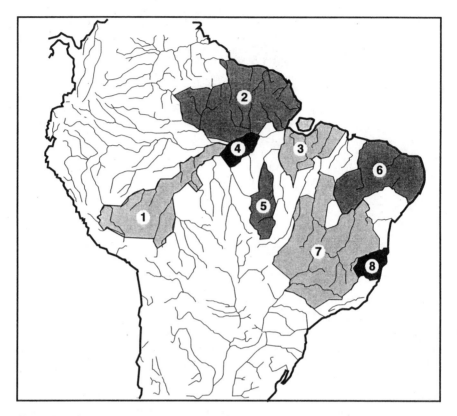

Figure 5.3. Geographic distributions (based on Rylands et al., 1993) of eight callitrichid species for which population genetic data are described in Table 5.2: (1) *Saguinus fuscicollis weddelli;* (2) *S. midas midas;* (3) *S.m. niger;* (4) *Callithrix humeralifer;* (5) *C. emiliae;* (6) *C. jacchus;* (7) *C. penicillata;* (8) *C. geoffroyi.*

males, both of whom copulate with a single female, cooperate in rearing her young. Males carry the infants and, in insectivorous species, feed them insects once the young begin eating solid foods (Goldizen, 1987a). Siblings from previous years also assist in infant feeding and carrying. Species (including those of *Leontopithecus*) for which long–term data exist may display polyandrous, monogamous, and multimale/multifemale social groups within the same natural population (Ferrari, 1988; Goldizen and Terborgh, 1989; Baker et al., 1993; Garber et al., 1993; reviewed in Ferrari and Ferrari, 1989). Reproduction is usually (but not always) suppressed in all but one breeding female (Ferrari and Ferrari, 1989). The breeding system exhibited by a social group at a given time appears to be primarily dependent upon two factors:

1. *Number of non–reproductive helpers available to raise offspring.* Goldizen (1987b) found that in saddle–back tamarins (*S. fuscicollis*), monogamous pairs without offspring helpers never attempted reproduction during 33 group years of observation. She suggested that costs of lactation and of carrying infants (which at birth can be 14–24% of the mother's body weight) were simply too great for lone pairs to raise twin offspring.

2. *Number of breeding opportunities available in the surrounding area.* When only one breeding female at a time is tolerated within a social group, successful reproduction by an individual of either sex requires that he or she must either form a new group, or wait for a breeding position to open in an existing group.

Consequently, both sexes tend to disperse from their natal social group. Like most callitrichids, *Leontopithecus* are territorial. In a long-term study of *L. rosalia* at the Poco de Antes Reserve, Baker et al. (1993) found that virtually all available habitat was "owned" by the 45 resident social groups, such that breeding opportunities for young animals were extremely limited. This was also found to be the case in studies of *S. fuscicollis* (Goldizen and Terborgh, 1989) and *S. emperator* (Garber et al., 1993). All three studies concluded that breeding positions were infrequently vacant, and that the majority of dispersing males and females moved into an immediately adjacent social group.

The basic consequences of this type of breeding system for genetic variability in callitrichid populations can be summarized as follows:

1. N_e *can be greatly reduced.* Because of polyandry and female reproductive suppression within a social group, the number of breeding males per generation can be much higher than the number of breeding females. Variance in reproductive success can be extremely high for both sexes, particularly in high-density populations in saturated habitats where sexually mature females accumulate in groups as they await breeding opportunities. Both of these factors will reduce the ratio of N_e/N.

2. *Inbreeding may be frequent.* Individuals of both sexes are highly philopatric, either migrating to immediately adjacent groups or sometimes breeding in the natal group. Movement patterns such as these in a species with overlapping generations are likely to result in some level of inbreeding. If migration events occur every four years (the maximum time to sexual maturity) in a callitrichid species with a minimum reproductive life span of 10 years (Garber et al., 1993), then inbreeding with close relatives (e.g., full siblings) is likely unless it is specifically avoided behaviorally. Furthermore, if genetic divergence at MHC loci is detrimental to fetal development, then matings between relatives might have been selected for during early stages of callitrichid evolution.

3. *Generation time effects may reduce heterozygosity.* Callitrichids have a shorter generation time than most anthropoid primates, reaching sexual maturity in captivity at 1–2 years of age compared to 4–5 years in most Cebidae (the other family of New World monkeys). All else being equal, this too will accelerate the loss of heterozygosity by drift or inbreeding, especially in small populations.

In summary, the low levels of genetic variation exhibited by extant populations of *Leontopithecus* were probably typical of these species prior to their fragmentation into small, isolated forest patches. These species may in fact be adapted to a homozygous genetic background, and be less likely to suffer from the deleterious effects of inbreeding than species with more heteroselected genomes. Increased susceptibility to disease, for example, has been cited as a possible deleterious consequence of loss of genetic variation at MHC loci due to inbreeding and population bottlenecks (O'Brien et al., 1985; Hughes, 1991). This would be much less of a difficulty for a callitrichid that is already adapted to low levels of allelic diversity at these loci. As shown next, the situation may be quite different for another endangered primate that lives in the Atlantic forest of Brazil: the muriqui.

Brachyteles.   This genus is monotypic and contains the largest New World primate, the muriqui, *B. arachnoides.* Only 650–750 individuals are known to remain. The current distribution is limited to 15–18 isolated forest remnants (Figure 5.4), 11 of which contain fewer than 25 animals (Mittermeier et al., 1987; Martuscelli et al., 1994) (smaller than an average natural social group). Unlike *Leontopithecus,* muriqui fare poorly under captive conditions. The world captive population consists of less than 15 animals at the Rio de Janeiro Primate Center.

Virtually all of the populations containing less than 25 animals occupy small, isolated patches of forest in which there is little or no possibility of population expansion (Mittermeier et al., 1987; Nishimura et al., 1988). This fact, combined with a female interbirth interval of nearly two years (Strier, 1991), seriously limits the potential for *in situ* recovery. The population at Fazenda Esmeralda, for example, consists of a single social group living in a 0.44 km² patch of forest surrounded by cattle pasture. This group has been studied extensively since 1983 (Fonseca, 1985; Lemos de Sá, 1988). Although reproduction rate among adult females has remained comparable to that reported in other muriqui populations (cf. Strier, 1991), births have not offset losses by deaths, disappearances of subadult females, and lack of immigrants. Group size declined from 18 in 1986 to 11 in 1991. The largest remaining tracts of forest are in São Paulo state in the south, where the majority of surviving muriqui probably resides. The possibility of translocat-

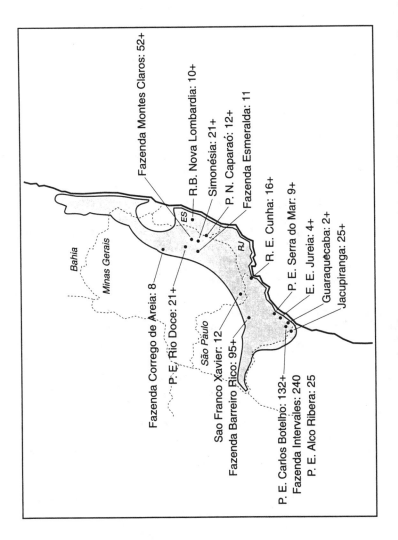

Figure 5.4. Geographic distribution of *Brachyteles arachnoides* in the Atlantic coastal forest of Brazil (based on Martuscelli et al., 1994, and Mittermeier et al., 1987). The encircled area describes the historic geographic distribution. The remaining few hundred individuals are distributed among 15 isolated forest remnants (represented by dots, followed by locality name and the number of individuals known to survive at each site).

ing demographically doomed northern populations to these areas is complicated by the fact that the northern and southern forms of *Brachyteles* differ
both morphologically and behaviorally (Lemos de Sá et al., 1993). This has
generated considerable debate as to whether the two forms should be considered different subspecies (Vieira, 1955; Kinzey, 1982; Torres de Assumpção, 1983).

The historic range of the muriqui (Figure 5.4) is generally coincident with
that of *Leontopithecus* (Figure 5.2), and both taxa have been subject to the
same rate and pattern of habitat fragmentation. Nonetheless, genetic variability in muriqui populations has proved dramatically different from that in
the lion tamarins. In 1990, a pilot study was aimed at collecting baseline
data on variability in morphology, genetics, and disease exposure in remnant
populations of muriqui. Ten of the 12 remaining animals in the Fazenda
Esmeralda (FE) population in the north were captured, as were two animals
from Fazenda Barriero Rico (FBR) in the south. Genetic variation was
examined electrophoretically at 32 allozyme loci, and using one multilocus
DNA probe. Although sample sizes were small, considerable genetic variability was detected (Table 5.3). Indeed, levels of polymorphism and heterozygosity are among the highest recorded for any primate species (Forman
et al., 1986; see Tables 5.1–5.5).

Although no records indicate precisely how long the FE animals have
been isolated, a census conducted throughout the muriqui species range in
1970 indicated a total population size at that time of at least 3,000 animals
(Aguirre, 1971). By 1987, the population had been reduced to its current
status (Mittermeier et al., 1987; a 70–75% reduction in species population
size in less than 20 years, or probably less than one muriqui life span).
Current levels of gene diversity present in both remnant populations suggest
that high genetic variability was characteristic of this species throughout its

**Table 5.3. Allozymic Diversity (32 loci) in Two Populations of the Muriqui, a Highly Endangered
Species Endemic to the Atlantic Forest of Brazil.** Fewer than 800 animals are known
to remain, distributed among 15 isolated forest remnants, two of which are represented below. Fazenda Esmeralda contains 12 animals and Fazenda Barriero Rico
contains about 95 individuals (see Figure 5.4)

Population	Race	N^1	"Mean No. Alleles" per Locus	P^1	H^1	Hardy–Weinberg Expected H
Fazenda Esmeralda	Northern	9	1.4	0.28	0.11	0.12
Fazenda Barriero Rico	Southern	2	1.1	0.12	0.11	0.08

Between–population genetic variance (F_{ST}): 0.41
Between–population genetic distance (Nei, 1972): 0.14

[1] See footnote to Table 5.1.

former range. Population decline has been rapid relative to generation length, such that much of the variation present in the historic population is still retained.

Several aspects of muriqui social structure and ecology probably contribute to the high levels of genetic diversity apparently characteristic of this species. Individuals live in large social groups (mean $N = 25$), and adult sex ratios within groups are near unity (Strier, 1991). Females in estrus mate with multiple males in the group, among which there is no apparent priority for access (Strier, 1986; Lemos de Sá, 1988). These factors tend to minimize differences between males and females in reproductive success, thereby increasing N_e/N. Also, muriqui occur in a diverse array of habitats, ranging from high-montane rainforest to severely degraded secondary forest with dense understory of bamboo and lianas. The species is characterized as folivorous, but a broad assortment of fruits and flowers is included in the diet (Nishimura et al., 1988). Gut morphology, dentition, and breadth in diet and habitat are highly convergent with those of South American species of howler monkeys (Zingeser, 1973), which also exhibit high genetic diversity (to be discussed). High levels of genetic variation might conceivably be favored in such habitat generalists. For example, folivores must physiologically detoxify a wide array of plant secondary compounds.

Should the small northern populations of muriqui be translocated to the larger reserves in the south? Although the molecular results are tentative due to small numbers of individuals sampled, they indicate that the northern and southern forms are rather strongly divergent (Table 5.3). These genetic findings can be interpreted to give added weight to distinctions between the two forms as recognized previously in morphology and behavior (Milton, 1985; Strier, 1986; Lemos de Sá et al., 1993). Whether or not they are officially recognized taxonomically, the northern and southern forms represent unique gene pools that have been evolving independently for some time. Careful translocation of subadult females between forest patches with social groups might offer the best management hope for the rapidly disappearing northern race.

Central America: The Howler Bottlenecks

Central America currently has the second highest rate of deforestation of any tropical region, exceeded only by that of West Africa (World Resource Institute, 1992). Tropical forests that once extended from southern Mexico to the tip of Panama remained relatively intact until 1950 (Collins, 1990), but over the next 40 years, more than 50% of the existing forest was destroyed. The river of forest that once ran the length of the Central American Isthmus has been reduced to a series of unconnected puddles that are shrinking at a rate of 1.4 million hectares per year (World Resource Institute, 1992).

Seven species of primates live in the forests of Central America, of which four are endemic and four are listed by the IUCN as vulnerable or endangered. One of the threatened endemics is the Mexican black howler, *Alouatta pigra*, which occurs in the Yucatan peninsula of Mexico, Belize, and Guatemala. The mantled howler, *A. palliata*, ranges from southern Mexico through Panama and east of the Andes to northern Peru. In the South American part of its range, it is listed by the IUCN as a separate, endangered subspecies, *A. palliata equatorialis* (the Central American subspecies is not formally listed as threatened, but given current rates of habitat destruction, should probably become so within the next five years). The geographic ranges of both species have been highly fragmented in the last 30–40 years.

Whereas two South American howler species that were examined genetically (*A. belzebul* and *A. seniculus*) both exhibit among the highest levels of allozymic diversity found in any primate, the paucity of genetic variation in the Central American species (Table 5.4) rivals that of *Leontopithecus* (Table 5.1). Malmgren and Brush (1978) examined 20 allozyme loci in 170 *A. palliata* from 17 social groups in Costa Rica. They found polymorphisms at only two loci ($P = 0.10$), and one of these was suspect due to posttranslational changes associated with pregnancy. Mean individual heterozygosity was $H = 0.01$. Examination of allozyme loci in a population of *A. palliata* in Panama yielded similar results (one locus varied, with $H = 0.01$; Thorington, personal communication), but went unpublished due to the "negative" nature of the findings. James (1992) surveyed 36 allozyme loci in 10 social groups of *A. pigra* ($N = 47$) from Belize. Only two loci varied ($P = 0.05$; $H = 0.02$). These results contrast sharply with allozyme studies of the two South American howler species (Table 5.4), neither of which is considered threatened.

All four howler monkey species are similar ecologically. Central America displays a mosaic of climatic conditions due to the combined effects of its location between two major ocean systems and the mountain ranges that bisect its length. Seasonal rainfall and temperature vary dramatically from the east to the west side of the Isthmus. Both of the Central American howlers occupy a range of forest types comparable to that described above for *Brachyteles* and *A. seniculus*, and presumably encounter similar ecological challenges (Glander, 1977). The social systems of *A. pigra* and *A. belzebul* are not well known, but group size and composition in both species are similar to those described for *A. seniculus* groups, which function as small single-male harems (Pope, 1990). *A. palliata* exhibits the most aberrant social structure, with relatively large groups that are usually multimale, although the adult sex ratio within groups is comparable to that found in

Table 5.4. Allozymic Diversity in South American and Central American Howler Monkeys (*Alouatta*). *A. pigra* is listed as vulnerable in the IUCN Red List; both Central American species are on the U.S. List of Endangered and Threatened Taxa. The South American species are not threatened

Sample	No. Collecting Localities	N^1	No. Loci	"Rare Alleles" Loci[1]	P^1	H^1	Reference
South America							
A. belzebul (Brazil)							
	3	35	13	4/7	0.54	0.07	Schneider et al., 1991
	1	26	13	1/5	0.38	0.08	Schneider et al., 1991
	3	18	13	1/4	0.31	0.04	Schneider et al., 1991
	Various	971	13	5/8	0.61	0.07	Schneider et al., 1991
A. seniculus (Venezuela)							
	1	137	29	1/10	0.34	0.10	Pope, 1992
A. seniculus (Brazil)							
	1	16	26	4/9	0.35	0.10	Lima et al., 1990
Central America							
A. pigra (Belize)							
	1	47	36	0/2	0.05	0.02	James, 1992
A. palliata (Costa Rica)							
	1	170	20	?	0.10	0.01	Malmgren and Brush, 1978

[1] See footnote to Table 5.1.

the other species (Crockett and Eisenberg, 1987). Differences between the dispersal patterns of *A. seniculus* and *A. palliata* (discussed earlier) suggest less variance in reproductive success for *A. palliata* females, which would result in a proportionally larger ratio of N_e/N. Based on this alone, one would predict a slower loss of alleles from a population of *A. palliata* than from a same–sized *A. seniculus* population.

The most conspicuous difference between the Central and South American howlers is the nature of their historic distributions. Multiple population crashes attributed to yellow fever epidemics (Collias and Southwick, 1952; Baldwin, 1976) and hurricanes (Hartshorn, 1984) have been documented in Central American howlers since the turn of the century. James (1992) concluded that low genetic variation in an *A. pigra* population resulted from four severe population bottlenecks— three due to hurricanes and one to yellow fever— that had occurred since 1930. A crash in 1949 of the Panama population of *A. palliata* was attributed to yellow fever (Collias and Southwick, 1952). Between 1951 and the time the population was sampled in the mid–1970s, population size grew from 239 to more than 1,500 animals (Milton, 1982). Prior to 1950, the forests between Mexico and Colombia formed a continuous corridor along which disease epidemics like yellow fever could spread throughout a large portion of the geographic range of each species. By contrast, the geographic distributions of the South American howlers are much larger and far less contiguous. The range of *A. seniculus* extends from east of the Andes throughout the Guianas, Surinam, Venezuela, Colombia, Peru, Bolivia, and Brazil north of the Amazon, and the range of *A. belzebul* extends throughout the south Amazon Basin from east of the Rio Purus to the coast. Disease epidemics and weather catastrophes are likely to affect only a small percentage of either species' range.

It is possible that mortality factors such as these reduced Central American howler populations to local pockets of survivors in much the same way that deforestation is doing now. Although this would reduce allelic diversity within local populations, the alleles that were lost or reduced in frequency would not be expected to be the same in every locality. Thus, considerable polymorphism could remain within the total species' range. Whether the alleles that are fixed in one howler population are the same as those present in others, however, remains to be tested.

Asia: The Island Macaques

Members of the genus *Macaca* have radiated extensively throughout most of Asia. Many of the 18+ species are confined to single islands or archipelagoes, whereas others have enormous distributions that span many countries. Six species and four subspecies are listed by the IUCN as vulnerable or endangered. Six of these are confined to islands, and two others are mainland species: the Barbary macaque (*M. sylvanus*), confined to isolated pockets of forest in Algeria and Morocco in northern Africa; and the lion-tailed macaque (*M. silenus*), confined to wet evergreen forests in western India. The lion-tailed macaque is the most critically endangered of all the *Macaca*, with fewer than 1,000 individuals distributed among 32 forest remnants (Ali, 1985). Two additional species, *M. arctoides* of mainland Southeast Asia and *M. thibetiana* of China, are suspected of being endangered but insufficient information is available to determine their exact status (Eudey, 1987). Most macaques live in large multimale, multifemale social groups between which males migrate. Details of their mating system were described earlier in this Chapter.

All macaque species have experienced range fragmentation due to deforestation, with effects on population numbers ranging from moderate to severe. Two Barbary macaque populations have been sampled genetically, each containing between 1,300 and 2,000 animals living in forest patches of several hundred square kilometers (Scheffrahn et al., 1993). Both showed moderate to high levels of allozyme variation (Table 5.5). Barbary macaques are considered vulnerable to extinction because there are only a few such forest remnants remaining within their former geographic distribution. Thus, although threatened as a species, it appears that local populations retain considerable genetic variation. Macaques have been genetically monitored more extensively than most other primates. Characteristically, their populations exhibit relatively high genetic variability, with individual heterozygosities as high as $H = 0.12$ (Table 5.5).

A notable exception involves the threatened Japanese macaque, *M. fuscata*, for which Nozawa et al. (1982) reported mean heterozygosity values usually less than 0.02 (based on $N = 1,646$ individuals; see Table 5.5). Japanese macaques are distributed throughout Japan south of Hokkaido, but most populations are small. Approximately 20% of *M. fuscata* are confined to isolated suburban parks, which typically contain only two or three social groups maintained by human provisioning. Other small populations live in more mountainous areas severely fractionated by clear–cutting (Eudey, 1987). Loss of alleles through drift and the accumulation of inbreeding is hardly surprising in such populations. Genetic differentiation between groups within a locality ($F_{GL} = 0.01$) proved negligible (Nozawa et al., 1982), as might be expected when males are exchanged among a few social groups generation after generation. Note, however, that overall polymorphism (P_T) for all populations combined was 0.66. Although within–population gene diversity was extremely low, differentiation of allelic contents across populations produced substantial genetic diversity in the species overall. This effect was not a consequence of divergence between islands alone; seven of the nine most differentiated populations were located on the main island of Honshu ($F_{RT} = 0.32$; Nozawa et al., 1991).

A similar pattern occurs in the long–tailed macaque *M. fascicularis*, which occurs throughout the Indonesian archipelago, Borneo, the Philippines, and a large part of mainland Southeast Asia. This species is found primarily in riparian habitats along waterways and in mangrove swamps. Although parts of its range have been heavily deforested, others such as Borneo and some Indonesian islands retain large tracts of forested land. Consequently, the degree of population fragmentation in this species is highly variable. Kawamoto et al. (1984) examined genetic variation within and between populations of *M. fascicularis* ($N = 456$ individuals) on five islands in Indonesia. Highest within-population genetic diversities occurred on Sumatra (Table 5.5), the largest and most forested of the islands surveyed (Collins, 1990). Although polymorphism on the other islands was comparable to that observed for *M. fuscata*, usually ranging from $P = 0.03 - 0.15$, heterozygosity estimates were generally higher and more variable ($H = 0.01 - 0.08$). As in *M. fuscata*, total polymorphism for all islands combined ($P_T = 0.52$) was much higher than in local populations. Genetic variation among populations was substantial, with the "among-localities-within-islands" component ranging from $F_{LR} = 0.10 - 0.30$ for the five islands, and "between-island" $F_{RT} = 0.41$ (Kawamoto et al., 1984).

Table 5.5. Allozymic Diversity within Populations of Macaques. *Macaca fuscata* and *M. sylvanus* are listed as vulnerable in the IUCN Red List (World Conservation Union, 1993), and *M. fuscata yakui* and *M. silenus* are considered endangered. See text for details regarding species distributions and degrees of population fragmentation.

No. Loci	Sample	P^1 (and P_T)	H^1 (and H_T)	Reference
1. Non-Threatened Species				
Macaca mulatta: rhesus macaque				
35	Dunga Gali, Pakistan	0.14	0.05	Melnick et al., 1984
21	Pakistan[2]	0.29	0.05	Melnick et al., 1986[3]
30	India[2]	0.27	0.08	Melnick et al., 1986[3]
30	Thailand[2]	0.40	0.09	Melnick et al., 1986[3]
30	China[2]	0.40	0.08	Melnick et al., 1986[3]
M. fascicularis: long-tailed macaque				
23[4]	Sumatra (6 troops, 1 locale)	0.48[4]	—	de Ruiter et al., 1994
33	Sumatra (4 troops, 3 locales)	0.12–0.30 (0.42)	0.04–0.08 (0.06)	Kawamoto et al., 1984
33	Java (6 troops, 3 locales)	0.03–0.15 (0.15)	0.01–0.05 (0.04)	Kawamoto et al., 1984
33	Bali (8 troops, 5 locales)	0.06–0.15 (0.27)	0.01–0.04 (0.02)	Kawamoto et al., 1984
33	Lombok (4 troops, 3 locales)	0.09–0.12 (0.12)	0.03–0.04 (0.03)	Kawamoto et al., 1984
33	Sumbawa (7 troops, 5 locales)	0.09–0.12 (0.12)	0.04–0.06 (0.04)	Kawamoto et al., 1984
33	Indonesia (5 islands above)	0.12–0.42 (0.52)	0.02–0.06 (0.04)	Kawamoto et al., 1984
31	Angaur Is., Micronesia	0.26	0.10	Kawamoto et al., 1988
32	Mauritius Is.[5]	0.16	0.06	Kondo et al., 1993
31	Thailand (12 troops)	0.45	0.05–0.12 (0.09)	Kawamoto et al., 1989
M. sinica: toque macaque				
32	Sri Lanka (20 troops, 13 locales)	0.12–0.28 (0.34)	0.04–0.09 (0.08)	Shotake et al., 1991

Table 5.5. (*cont.*)

No. Loci	Sample	P^1 (and P_T)	H^1 (and H_T)	Reference
2. Threatened Species				
M. fuscata: Japanese macaque				
32	Honshu Is. (19 troops, 9 locales)	0.03–0.22 (0.50)	0.00–0.03 (0.01)	Nozawa et al., 1982
32	Awajishima Is. (2 troops, 1 locale)	0.03 (0.03)	0.00–0.01 (0.01)	Nozawa et al., 1982
32	Shodishima Is. (2 troops, 1 locale)	0.03–0.06 (0.06)	0.01–0.02 (0.02)	Nozawa et al., 1982
32	Kashima Is. (1 troop)	0.03	0.02	Nozawa et al., 1982
32	Shikoku Is. (2 troops, 2 locales)	0.03–0.06 (0.06)	0.00–0.01 (0.01)	Nozawa et al., 1982
32	Kyushu Is. (6 troops, 4 locales)	0.03–0.16 (0.25)	0.00–0.02 (0.01)	Nozawa et al., 1982
32	Koshima Is. (1 troop)	0.09	0.01	Nozawa et al., 1982
32	Japan (38 locales, 8 islands)	0.03–0.28 (0.66)	0.00–0.06 (0.02)	Nozawa et al., 1991
M. fuscata yakui: Yakushima macaque				
32	Yaku Is. (2 locales)	0.03	0.00–0.01	Nozawa et al., 1991
M. sylvanus: Barbary macaque				
23	Algeria (2 locales, 1 region)	0.17–0.22 (0.22)	0.05–0.08	Scheffrahn et al., 1993
M. silenus: lion-tailed macaque				
29	India (2 regions)	0.12–0.21	—	Fooden and Lanyon, 1989

[1] See footnote to Table 5.1; here, the ranges presented are values across social groups, whereas P_T and H_T values are for an entire island treated as a single locality.

[2] Samples from laboratory stocks in which origin was known only by country.

[3] Analyses based on earlier publications by various authors.

[4] de Ruiter et al. (1994) state that 17 of 29 loci showed polymorphism, but six of these did not give reproducible results, and hence were not included in the calculation of polymorphism.

[5] Introduced in the early 17th century.

147

Because macaque males typically leave their natal social group but females remain for life, even more among-population differentiation might be expected for maternally transmitted mitochondrial (mt) DNA genes than is true for typical autosomal loci. In a summary of the literature involving conspecific population structures of four macaque species (*M. fascicularis*, *M. mulatta*, *M. fuscata*, and *M. sinica*) as registered by molecular genetic markers, Melnick and Hoelzer (1992) discovered just such an outcome. The among-population component of population differentiation was much greater for mtDNA.

IMPLICATIONS FOR CONSERVATION

Population subdivision has been promoted as an effective means for conserving overall allelic diversity in captive populations (e.g., Allendorf, 1986; Fuerst and Maruyama, 1986; Lacy, 1987). For example, Lacy used a series of computer simulations to show that subdivided populations collectively retain genetic diversity better than do panmictic populations of the same size. Although within-subpopulation variability was lost rapidly, more total genetic diversity was preserved through the operation of genetic drift in solidifying between-population differences. It is ironic that the greatest threat to the survival of primate species worldwide, habitat fragmentation, may also be the means by which allelic diversity is being preserved in the face of shrinking species' ranges. Unfortunately, we have had little control over the size and number of population subunits. Nobody, for example, would advocate population fragmentation on the scale seen in *Brachyteles*, where loss of demographic viability has become of paramount concern. Heterozygosity has already been almost completely lost in many threatened species (e.g., *Macaca fuscata*, *Alouatta pigra*), in which the most critical problem is continued shrinkage and loss of existing subpopulations. Because allelic diversity is "stored" as between–population differences, extinction of remnant populations results in irretrievable loss of alleles from the species gene pool. Whereas some heterozygosity can be reinstated through translocation of individuals between extant populations, lost alleles are essentially gone forever.

The fitness effects of heterozygosity loss should vary depending on the genetic history of a species. Species whose local populations are adapted to homozygous genetic backgrounds (such as appears to be the case in many Callitrichidae) should suffer fewer deleterious consequences than those typically exhibiting high levels of local gene diversity. Even heterozygous populations that experience a bottleneck can become adapted to local inbreeding after an initial period of inbreeding depression during which deleterious recessive alleles are culled (Templeton, 1987; but the process can be more complicated than the simple elimination of harmful recessives— see Brewer et al., 1990). There are numerous examples of populations that have thrived following severe bottlenecks, including some Japanese macaques (Itani, 1975) and the *A. pigra* population described earlier.

How serious a threat to local population viability is inbreeding depression? Although such depression is widely assumed to occur in small natural populations, detailed studies of how the phenomenon may alter demographic variables in nature have not been undertaken. Inbreeding depression does not necessarily result in negative population growth. A typical female monkey that lives for 20 years and begins reproduction at five years of age is biologically capable of producing 10–15 offspring during her lifetime. Only two of these offspring need survive and reproduce to maintain stationary population size. In the often-cited study by Ralls and Ballou (1982), mean infant mortality in highly inbred captive primates was 47.9%, compared to 29.8% in noninbred colonies of the same species. Most female monkeys and apes resume cycling soon after the death of an infant, whereas a female that nurses an infant to weaning age may not come again into estrus until one or more years after her infant is born. In the common chimpanzee, for example, mean interbirth interval is 5–6.5 years, but if the first offspring dies during this period the female will conceive again within 4–8 months after the death (Richard, 1985). Thus, a high infant-mortality rate can more than double the annual birthrate. This will tend to counteract both increased infant mortality and a reduction in conception rate due to decreased fertility, even in a worst–case scenario in which a formerly outbred population is suddenly subjected to regular inbreeding between close relatives. In natural populations, it is far more likely that rates of inbreeding increase gradually as populations shrink in size and probabilities of immigration into a social group containing relatives increase. Eradication of deleterious recessive alleles due to mild inbreeding over several generations can be sufficient to eliminate subsequent inbreeding depression (Templeton, 1987).

Forest fragmentation on the scale described earlier for the Atlantic forest of Brazil is typical throughout west and east Africa, Madagascar, and large parts of Asia and Central America (Collins, 1990). Nearly 80% of the world's primate species live in these forests, and all exhibit moderate to extreme population subdivision. For example, less than 15% remains of the original rainforests in Africa west of the Sanaga River in Cameroon (Sayer et al., 1992), where 16 primate species and five subspecies are endemic (14 of which are listed by the IUCN as endangered or vulnerable). Hunting pressure within the scattered forest fragments is so intense that many local primate populations have been nearly exterminated (Oates, 1994). Forest primates endemic to east Africa are equally threatened. Less than 10% of the forests of Uganda, Rwanda, and Tanzania remain (Sayer et al., 1992). Endangered primates in this area include the mountain gorilla (*Gorilla gorilla berengei*), which is now restricted to two distant forest remnants, and the Gordon's red colobus (*Procolobus gordonorum*) and the Sanje mangabey (*Cercocebus galeritus sanjei*), which are limited to scattered forest patches in the Uzungwa Mountains of Tanzania (Lee et al., 1988). Similarly dire conditions for primates exist in countries such as El Salvador, Guatemala, Vietnam, Sri Lanka, Bangladesh, the Philippines, Java, Bali, Madagascar, and indeed, throughout most of the world.

Local population sizes in many primate species are already well below levels necessary to stem rapid heterozygosity loss, as demonstrated in several of the case studies described earlier. Nonetheless, small populations that remain *demographically* viable can thrive even after losing substantial genetic variability, and can also be important reservoirs for a species' composite genetic diversity. Furthermore, management programs designed to conserve only one or a few larger populations include no guarantee that higher species-wide genetic variability will be retained, and could have other undesired consequences. Knowledge of a species' mating system and social organization are critical to evaluating alternative management schemes, even with respect to the focused issue of genetic variability maintenance. As demonstrated earlier, some types of dispersal patterns and breeding systems augment genetic differentiation among social groups more than others, which affects loss of polymorphism in small populations. Generally, the more social groups (or lineages) present in a population of a given N_e, the longer the time frame over which allelic diversity is maintained.

The intent here is not to advocate the conservation of small primate populations at the expense of large ones, but rather to promote this as a complementary strategy. The importance of conserving both large (where possible) and small populations should be appreciated. Unfortunately, the protection of large areas of forest is not an option within what remains of the geographic ranges of many critically threatened primates. This tends to make these regions less attractive to the large national and international aid agencies that provide the majority of funding for tropical conservation. A common perception is that small parks provide less biodiversity return per unit of conservation effort. To the extent that this may be true, the conservation of biodiversity and the preservation of endangered species may require different sets of management priorities and guidelines.

SUMMARY

1. More than 55% of extant primate species are rare and currently listed as endangered or vulnerable to extinction. Habitat destruction and fragmen-

tation due to tropical deforestation have been the major causes of the worldwide decline in primate populations, and remain the most important threats to their continued existence. Most threatened primates occupy small, isolated forest remnants between which there is little or no opportunity for migration. Genetic variation within and among these small populations depends upon the distribution of genetic diversity prior to fragmentation, the size and number of subpopulations, the time frame over which subdivision occurred, and species socioecology.

2. Low levels of genetic diversity are characteristic of some primate taxa. Many of the Callitrichidae exhibit exceedingly low variability at both allozyme and MHC loci, possibly as a consequence of a unique reproductive physiology and social system. In these monkeys, low polymorphism and heterozygosity are exhibited by both large and small populations, and are not necessarily indicative of genetic drift or inbreeding. These species appear to be adapted to a homozygous genetic background, and should be less likely to suffer from deleterious effects of inbreeding than species with more heteroselected genomes. Low genetic diversity is also characteristic of both endangered and non-endangered Central American howler monkeys, but this appears to be a consequence of repeated population bottlenecks caused by yellow fever epidemics and hurricanes.

3. Some primate taxonomic groups, such as macaques, typically display high within-population genetic variation, but may exhibit heterozygosity loss when populations become small and isolated. The alleles lost often differ among populations, such that overall genetic diversity within a species remains high and is displayed largely as a between-population variance. The highly endangered muriqui of Brazil still retains both high polymorphism and heterozygosity within remnant populations, but the extreme fragmentation of this species' range has occurred very recently relative to generation time.

4. Loss of genetic diversity within population fragments is influenced by the proportion of the adult gamete pool sampled each generation, and by lineage effects. Primate social systems are characterized by specific patterns of variation in reproductive success that are determined by the direction of kin orientation within social groups, the sex bias and frequency of dispersal between social groups, and the mating pattern within groups. These variables can be used to estimate the proportion of the population gene pool that is passed on each generation. The smaller this proportion, the faster gene diversity is lost relative to overall population size. The number of genetic lineages within a typical social group, and social group size, are also species' characteristics. These factors influence not only the rate of allelic loss, but also the amount of genetic diversity at the time a population becomes isolated.

5. In most primate species, dispersing individuals tend to transfer into immediately adjacent social groups, such that the opportunity to breed with close and distant relatives arises frequently. The degree to which inbreeding is behaviorally "avoided" influences the rate at which inbreeding effects accumulate in small populations containing few social groups. In some primate species, dispersal rates are mediated by population density and intrasexual competition; in such species, changes in population structure caused by habitat degradation can result in higher rates of close inbreeding. Natal emigration rate in other species does not appear to be influenced by demographic variables, and is less affected by environmental perturbations.

6. Although close inbreeding could potentially increase infant mortality, this effect may be counteracted demographically by higher birthrates. The prolonged period of maternal care characteristic of monkeys and apes is terminated by death of the infant, after which most females quickly resume cycling. This can reduce the interbirth interval by more than 50%. Natural primate populations that have undergone bottlenecks, such as Central American howler species and many Japanese macaques, have all exhibited positive population growth rates despite an almost complete lack of genetic diversity as assessed by allozymic techniques.

7. Setting aside large areas of protected forest is no longer an option within what remains of the geographic ranges of many endangered primates. The most critical problem for these species is continued shrinkage and loss of existing forest fragments, within which local population sizes are already well below that required to prevent rapid decay of heterozygosity. Case studies indicate that substantial heterozygosity may already have been lost throughout the geographic ranges of many of these species, but that composite species' diversity can be maintained as between–population genetic variance. Thus, continued loss of forest remnants results in an irretrievable loss of species' genetic diversity.

REFERENCES

ABBOTT, D.H. 1984. Behavioural and physiological suppression of fertility in subordinate marmoset monkeys. *Amer. J. Primatol.* 6:169–186.
AGUIRRE, A.C. 1971. *O mono Brachyteles arachnoides (E. Geoffroy). Situação atual da espécie no Brasil.* Academia Brasileira de Ciências, Rio de Janeiro.
ALI, R. 1985. An overview of the status and distribution of the lion–tailed macaque. In *The Lion–Tailed Macaque: Status and Conservation*, ed. P.G. Heltne, pp. 13–26. Alan R. Liss, New York.
ALLENDORF, F.W. 1986. Genetic drift and the loss of alleles versus heterozygosity. *Zoo Biol.* 5:181–190.

ALTMANN, J., G. HAUSFATER, and S.A. ALTMANN. 1988. Determinants of reproductive success in savannah baboons, *Papio cynocephalus*. In *Reproductive Success: Studies of Individual Variation in Contrasting Breeding Systems*, ed. T.H. Clutton-Brock, pp. 403–418. University of Chicago Press, Chicago, IL.

BAKER, A.J., J.M. DIETZ, and D.G. KLEIMAN. 1993. Behavioural evidence for monopolization of paternity in multi–male groups of golden lion tamarins. *Anim. Behav.* 46:1091–1103.

BALDWIN, L.A. 1976. Vocalizations of howler monkeys (*Alouatta palliata*) in southwestern Panama. *Fol. Primatol.* 26:81–108.

BREWER, B.A., R.C. LACY, M.L. FOSTER, and G. ALAKS. 1990. Inbreeding depression in insular and central populations of *Peromyscus* mice. *J. Hered.* 81:257–266.

BUSSE, C.D. 1982. Social dominance and offspring mortality among female chacma baboons (abstract). *Int. J. Primatol.* 3:267.

CHENEY, D.L. and R. SEYFARTH. 1983. Nonrandom dispersal in free–ranging vervet monkeys: social and genetic consequences. *Amer. Nat.* 122:392–412.

CHENEY, D.L., R.M. SEYFARTH, S.J. ANDELMAN, and P.C. LEE. 1988. Reproductive success in vervet monkeys. In *Reproductive Success: Studies of Individual Variation in Contrasting Breeding Systems*, ed. T.H. Clutton–Brock, pp. 384–402. University of Chicago Press, Chicago, IL.

CHESSER, R.K. 1991. Gene diversity and female philopatry. *Genetics* 127:437–447.

CLUTTON-BROCK, T.H. (ed.). 1988. *Reproductive Success: Studies of Individual Variation in Contrasting Breeding Systems*. University of Chicago Press, Chicago, IL.

COLLIAS, N. and C. SOUTHWICK. 1952. A field study of population density and social organization in howling monkeys. *Proc. Am. Philos. Soc.* 96:143–156.

COLLINS, M. 1990. *The Last Rain Forests: A World Conservation Atlas*. Oxford University Press, New York.

CROCKETT, C.M. and J.F. EISENBERG. 1987. Howlers: variations in group size and demography. In *Primate Societies*, eds. B.B. Smuts, D.D. Cheney, R.M. Seyfarth, R.W. Wrangham, and T.T. Struhsaker, pp. 54–68. University of Chicago Press, Chicago, IL.

CROCKETT, C.M. and T.R. POPE. 1988. Inferring patterns of aggression from red howler monkey injuries. *Amer. J. Primatol.* 15:289–308.

CROCKETT, C.M. and T.R. POPE. 1993. Consequences of sex differences in dispersal for juvenile red howler monkeys. In *Juvenile Primates: Life History, Development, and Behavior*, eds. M.E. Periera and L.A. Fairbanks, pp. 104–118. Oxford University Press, New York.

DARDEN, A.G. and J.W. STREILEIN. 1984. Syrian hamsters express two monomorphic class I major histocompatibility complex molecules. *Immunogenetics* 20:603–622.

DE JONG, G., J.R. DE RUITER, and R. HARING. 1994. Genetic structure of a population with social structure and migration. In *Conservation Genetics*, eds. V. Loeschcke, J. Tomiuk, and S.J. Jain, pp. 147–164. Birkhäuser Verlag, Basel, Switzerland.

DE RUITER, R.N., F. HARING, G. DE JONG, W. SCHEFFRAHN, and J. VAN HOOF. 1994. The influence of social organization on genetic variability and relatedness within stable social groups: A computer simulation study based on paternity testing and population genetic analysis of long–tailed macaques (*Macaca fascicularis*). In *Behavior and Genes in Natural Populations of Long-tailed Macaques* (Macaca fascicularis), published PhD dissertation, J. de Ruiter, pp. 109–130. Universiteit Utrecht, Netherlands.

DITTUS, W. 1977. The social regulation of population density and age–sex distribution in the toque monkey. *Behaviour* 63:281–322.

DITTUS, W. 1979. The evolution of behavior regulating density and age-specific sex ratios in a primate population. *Behaviour* 69:265–302.

DIXSON, A.F., M.G. ANZENBERGER, M.A.O. MONTIERO DA CRUZ, I. PATEL, and A.J. JEFFREYS. 1992. DNA fingerprinting of free–ranging groups of common marmosets (*Callithrix jacchus*) in NE Brazil. In *Paternity in Primates: Genetic Tests and Theories*, eds. R.D. Martin, A.F. Dixson, and E.J. Wickings, pp. 192–202. Karger, Basel, Switzerland.

DIXSON, A.F., N. HASTIE, I. PATEL, and A.J. JEFFRIES. 1988. DNA "fingerprinting" of captive family groups of common marmosets (*Callithrix jacchus*). *Fol. Primatol.* 51:52–55.

DRACOPOLI, N.C., F.L. BRETT, T.R. TURNER, and C.J. JOLLY. 1983. Patterns of genetic variability in the serum proteins in the Kenyan vervet monkey (*Cercopithecus aethiops*). *Amer. J. Phys. Anthropol.* 61:39–49.

DRICKAMER, L.C. 1974. A ten–year summary of reproductive data for free-ranging *Macaca mulatta*. *Fol. Primatol.* 21:61–80.

DRICKAMER, L.C. and S. VESSEY. 1973. Group changing in free–ranging male rhesus monkeys. *Primates* 14:359–368.

DUNBAR, R.I.M. 1987. Demography and reproduction. In *Primate Societies*, eds. B.B. Smuts, D.D. Cheney, R.M. Seyfarth, R.W. Wrangham, and T.T. Struhsaker, pp. 240–249. University of Chicago Press, Chicago, IL.

EMMONS, L.H. 1990. *Neotropical Rainforest Mammals: A Field Guide*. University of Chicago Press, Chicago, IL.

EUDEY, A.A. 1987. *IUCN/SSC Primate Specialist Group Action Plan for Asian Primate Conservation: 1987–91*. IUCN, Gland, Switzerland.

FERRARI, S.F. 1988. The behavior and ecology of the buffy-headed marmoset, *Callithrix flaviceps* (O. Thomas, 1903). PhD dissertation, University College, London, UK.

FERRARI, S.F. and M.A.L. FERRARI. 1989. A re–evaluation of the social organization of the Callitrichidae, with references to the ecological differences between genera. *Fol. Primatol.* 52:132–147.

FIX, A.G. 1978. The role of kin-structured migration in genetic microdifferentiation. *Ann. Hum. Genet.* 41:329–339.

FLEISCHHAUER, K., N.A. KERNAN, R.J. O'REILLY, B. DUPONT, and S.Y. YANG. 1990. Bone marrow–allograft rejection by T–lymphocytes recognizing a single amino acid difference in HLA–B44. *New Eng. J. Med.* 323:1818–1822.

FONSECA, G.A. 1985. Observations on the ecology of the muriqui (*Brachyteles arachnoides* E. Geoffroy 1806). *Prim. Conserv.* 5:48–52.

FOODEN, J. and S.M. LANYON. 1989. Blood–protein allele frequencies and phylogenetic relationships in *Macaca*: a review. *Amer. J. Primatol.* 17:209–241.

FORMAN, L., D.G. KLEIMAN, R.M. BUSH, J.M. DIETZ, J.D. BALLOU, L.G. PHILLIPS, A.F. COIMBRA–FILHO, and S.J. O'BRIEN. 1986. Genetic variation within and among lion tamarins. *Amer. J. Phys. Anthropol.* 7:1–11.

FRENCH, J.A., D.H. ABBOTT, and C.T. SNOWDON. 1984. The effect of social environment on estrogen excretion, scent marking and sociosexual behavior in tamarins (*Saguinus oedipus*). *Amer. J. Primatol.* 6:155–168.

FUERST, P.A. and T. MARUYAMA. 1986. Considerations on the conservation of alleles and of genic heterozygosity in small managed populations. *Zoo Biol.* 5:171–180.

GARBER, P.A., F. ENCARNACION, L. MOYA, and J.D. PRUETZ. 1993. Demographic and reproductive patterns in moustached tamarin monkeys (*Saguinus mystax*): implications for reconstructing Platyrrhine mating systems. *Amer. J. Primatol.* 29:235–254.

GLANDER, K.E. 1977. Poison in a monkey's garden of eden. *Nat. Hist.* 86:34–41.

GLANDER, K.E. 1992. Dispersal patterns in Costa Rican mantled howling monkeys. *Int. J. Primatol.* 13:415–436.

GOLDIZEN, A.W. 1987a. Tamarins and marmosets: communal care of offspring. In *Primate Societies*, eds. B.B. Smuts, D.D. Cheney, R.M. Seyfarth, R.W. Wrangham, and T.T. Struhsaker, pp. 34–43. University of Chicago Press, Chicago, IL.

GOLDIZEN, A.W. 1987b. Facultative polyandry and the role of infant-carrying in wild saddleback tamarins (*Saguinus fuscicollis*). *Behav. Ecol. Sociobiol.* 20:99–109.

GOLDIZEN, A.W. and J. TERBORGH. 1989. Demography and dispersal patterns of a tamarin population: possible causes of delayed breeding. *Amer. Nat.* 134:208–224.

GOODMAN, D. 1987. The demography of chance extinction. In *Viable Populations for Conservation*, ed. M.E. Soulé, pp. 11–34. Cambridge University Press, Cambridge, UK.

GOUZOULES, H., S. GOUZOULES, and L. FEDIGAN. 1982. Behavioural dominance and reproductive success in female Japanese monkeys (*M. fuscata*). *Anim. Behav.* 30:1138–1151.

GRIFFETH, B., J.M. SCOTT, J.W. CARPENTER, and C. REED. 1989. Translocation as a species conservation tool: Status and strategy. *Science* 245:477–480.

HARCOURT, A.H. 1979. Social relationships among female mountain gorillas. *Anim. Behav.* 27:251–264.

HARCOURT, C. and J. THORNBACK. 1990. *Lemurs of Madagascar and the Comoros: The IUCN Red Data Book.* IUCN, Gland, Switzerland.

HARRIS, R.B. and F.W. ALLENDORF. 1989. Genetically effective population size of large mammals: an assessment of estimators. *Conserv. Biol.* 3:181–191.

HARTSHORN, G.S. 1984. *Belize Country Environmental Profile: A Field Study.* Nicolait, Belize City.

HUGHES, A.L. 1991. MHC polymorphism and the design of captive breeding programs. *Conserv. Biol.* 5:249–251.

HUGHES, A.L. and M. NEI. 1988. Pattern of nucleotide substitution at major histocompatibility complex class I loci reveals overdominant selection. *Nature* 335:167–170.

ITANI, J. 1975. Twenty years with Mt. Takasaki monkeys. In *Primate Utilization and Conservation*, eds. G. Bermant and D.G. Lindberg, pp. 101–125. Wiley, New York.

JAMES, R.A. 1992. Genetic variation in Belizian black howler monkeys (*Alouatta pigra*). PhD dissertation, Rutgers University, New Brunswick, NJ.

KAWAMOTO, Y., T.M. ISCHAK, and J. SUPRIATNA. 1984. Genetic variations within and between troops of the crab–eating macaque (*Macaca fascicularis*) on Sumatra, Java, Bali, Lombok and Sumbawa, Indonesia. *Primates* 25:131–159.

KAWAMOTO, Y., T. ISHIDA, J. SUZUKI, O. TAKENAKA, and P. VARAVUDHI. 1989. A preliminary report on the genetic variations of crab–eating macaques in Thailand. *Kyoto University Overseas Research Report Studies on Asian Non–human Primates* 7:94–103.

KAWAMOTO, Y., K. NOZAWA, K. MATSUBAYASHI, and S. GOTOH. 1988. A population-genetic study of crab–eating macaques (*Macaca fascicularis*) on the island of Anguar, Palau, Micronesia. *Fol. Primatol.* 51:169–181.

KAWANAKA, K. 1973. Intertroop relations among Japanese monkeys. *Primates* 14:113–159.

KINZEY, W.G. 1982. Distribution of some neotropical primates and the model of Pleistocene forest refugia. In *Biological Diversification in the Tropics*, ed. G.T. Prance, pp. 455–482. Columbia University Press, New York.

KONDO, M., Y. KAWAMOTO, K. NOZAWA, K. MATSUBAYASHI, T. WATANABE, O. GRIFFITHS, and M.A. STANLEY. 1993. Population genetics of crab–eating macaques (*Macaca fascicularis*) on the island of Mauritius. *Amer. J. Primatol.* 29:167–182.

LACY. R.C. 1987. Loss of genetic diversity from managed populations: interacting effects of drift, mutation, immigration, selection, and population subdivision. *Conserv. Biol.* 1:143–158.

LANDE, R. and G.F. BARROWCLOUGH. 1987. Effective population size, genetic variation, and their use in population management. In *Viable Populations for Conservation*, ed. M.E. Soulé, pp. 87–123. Cambridge University Press, Cambridge, UK.

LEE, P.C., J. THORNBACK, and E.L. BENNETT. 1988. *Threatened Primates of Africa. The IUCN Red Data Book.* IUCN, Gland, Switzerland.

LEMOS DE SÁ, R.M. 1988. Situação de uma População de Mono–carvoeiros, *Brachyteles arachnoides*, em Fragmento de Mata Atlântica (M.G.), e Implicações para sua Conservação. MS thesis, Universidade de Brasília.

LEMOS DE SÁ, R.M., T.R. POPE, T.T. STRUHSAKER, and K.E. GLANDER. 1993. Sexual dimorphism in canine length of woolly spider monkeys (*Brachyteles arachnoides*, E. Geoffroy 1806). *Int. J. Primatol.* 14:755–762.

LIMA, M.M.C., M.I.C. SAMPAIO, M.P.C. SCHNEIDER, W. SCHEFFRAHN, H. SCHNEIDER, and F.M. SALZANO. 1990. Chromosome and protein variation in red howler monkeys. *Rev. Brasil. Genet.* 13:789–802.

MALMGREN, L.A. and A.H. BRUSH. 1978. Isozymes and plasma proteins in eight groups of golden mantled howling monkeys (*Alouatta palliata*). In *Recent Advances in Primatology*, eds. D.J. Chivers and K.A. Joysey, vol. 3, pp. 283–285. Academic Press, New York.

MARTUSCELLI, P., L.M. PETRONI, and F. OLMOS. 1994. Fourteen new localities for the muriqui *Brachyteles arachnoides*. *Neotrop. Prim.* 2:12–15.

MEIRELES, C.M.M., M.I.C. SAMPAIO, H. SCHNEIDER, and M.P.C. SCHNEIDER. 1992. Protein variation, taxonomy and differentiation in five species of marmosets. *Primates* 33:227–238.

MELNICK, D.J. and G.A. HOELZER. 1992. Differences in male and female macaque dispersal lead to contrasting distributions of nuclear and mitochondrial DNA variation. *Int. J. Primatol.* 13:379–393.

MELNICK, D.J., C.J. JOLLY, and K.K. KIDD. 1984. The genetics of a wild population of rhesus macaques (*Macaca mulatta*). I. Genetic variability within and between social groups. *Amer. J. Phys. Anthropol.* 63:341–360.

MELNICK, D.J., C.J. JOLLY, and K.K. KIDD. 1986. The genetics of a wild population of rhesus macaques (*Macaca mulatta*). II. The Dunga Gali population in a species–wide perspective. *Amer. J. Phys. Anthropol.* 71:129–140.

MELNICK D.J. and K.K. KIDD. 1983. The genetic consequences of social group fission in a wild population of rhesus monkeys (*Macaca mulatta*). *Behav. Ecol. Sociobiol.* 12:229–236.

MELNICK, D.J., M.C. PEARL, and A.F. RICHARD. 1984. Male migration and inbreeding avoidance in wild rhesus monkeys. *Amer. J. Primatol.* 7:229–243.

MELO, A.C.A., M.I.C. SAMPAIO, M.P.C. SCHNEIDER, and H. SCHNEIDER. 1992. Biochemical diversity and genetic distance in two species of the genus *Saguinus*. *Primates* 33:217–225.

MILTON, K. 1982. Dietary quality and demographic regulation in a howler monkey population. In *The Ecology of a Tropical Forest: Seasonal Rhythms and Long–Term Changes*, eds. E.G. Leigh, Jr., A.S. Rand, and D.M. Windsor, pp. 535–550. Smithsonian Institution Press, Washington, DC.

MILTON, K. 1985. Mating patterns of woolly spider monkeys, *Brachyteles arachnoides*: implications for female choice. *Behav. Ecol. Sociobiol.* 17:53–59.

MITTERMEIER, R.A. 1987. Effects of hunting on rain forest primates. In *Primate Conservation in the Tropical Rain Forest*, eds. C.W. Marsh and R.A. Mittermeier, pp. 109–146. Alan R. Liss, New York.

MITTERMEIER, R.A. 1991. Hunting and its effect on wild primate populations in Suriname. In *Neotropical Wildlife Use and Conservation*, eds. J.G. Robinson and K.H. Redford, pp. 93–107. University of Chicago Press, Chicago, IL.

MITTERMEIER, R.A., C.M.C. VALLE, I.B. SANTOS, C.A.M. PINTO, K.B. STRIER, A.L. YOUNG, E.M. VEADO, I.O. CONSTABLE, S.G. PACCAGNELLA, and R.M. LEMOS DE SÁ. 1987. Current distribution of the muriqui in the Atlantic forest region of Brazil. *Prim. Conserv.* 8:143–149.

MOORE, J. 1993. Inbreeding and outbreeding in primates: what's wrong with the dispersing sex? In *The Natural History of Inbreeding and Outbreeding: Theoretical and Empirical Perspectives*, ed. M.W. Thornhill, pp. 392–426. University of Chicago Press, Chicago, IL.

MYERS, N. 1987. Trends in destruction of rain forests. In *Primate Conservation in the Tropical Rain Forest*, eds. C.W. Marsh and R.A. Mittermeier, pp. 3–22. Alan R. Liss, New York.

MYERS, N. 1989. *Deforestation Rates in Tropical Forests and their Climatic Implications.* A Friends of the Earth Report. London.

NEI, M. 1972. Genetic distance between populations. *Amer. Nat.* 106:283–292.

NEI, M. 1977. F–statistics and analysis of gene diversity in subdivided populations. *Ann. Hum. Genet.* 41:225–233.

NISHIMURA, A., G.A.B. FONSECA, R.A. MITTERMEIER, A.L. YOUNG, K.B. STRIER, and C.M.C. VALLE. 1988. The muriqui, genus *Brachyteles*. In *Ecology and Behavior of Neotropical Primates*, eds. R.A. Mittermeier, A.B. Rylands, A. Coimbra–Filho, and G.A.B. Fonseca, vol. 2, pp. 577–610. World Wildlife Fund, Washington, DC.

NIZETIC, D., M. STEVANOVIC, B. SOLDATOVIC, I. SAVIC, and R. CRKVENJAKOV. 1988. Limited polymorphism of both classes of MHC genes in four different species of the Balkan mole rat. *Immunogenetics* 28:91–98.

NOZAWA, K., T. SHOTAKE, Y. KAWAMOTO, and Y. TANABE. 1982. Population genetics of Japanese monkeys: II. Blood protein polymorphisms and population structure. *Primates* 23:252–271.

NOZAWA, K., T. SHOTAKE, M. MINEZAWA, Y. KAWAMOTO, K. HAYASAKA, S. KAWAMOTO, and S. ITO. 1991. Population genetics of Japanese monkeys: III. Ancestry and differentiation of local populations. *Primates* 32:411–435.

OATES, J.F. 1986. *IUCN/SSC Primate Specialist Group Action Plan for African Primate Conservation: 1986–90.* IUCN, Gland, Switzerland.

OATES, J.F. 1994. Africa's primates in 1992: Conservation issues and options. *Amer. J. Primatol.* 34:61–71.

O'BRIEN, S.J., M.E. ROELKE, L. MARKER, A. NEWMAN, C.A. WINDLER, D. MELTZER, L. COLLY, J.F. EVERMAN, M. BUSH, and D.E. WILDT. 1985. Genetic basis for species vulnerability in the cheetah. *Science* 227:1428–1434.

OLIVIER, T.J., C. OBER, J. BUETTNER–JANUSCH, and D.S. SADE. 1981. Genetic differentiation among matrilines in social groups of rhesus monkeys. *Behav. Ecol. Sociobiol.* 8:279–285.

OTIS, J.S., J.W. FROELICH, and R.W. THORINGTON. 1981. Seasonal and age–related differential mortality by sex in the mantled howler monkey, *Alouatta palliata. Int. J. Primatol.* 2:197–205.

PEARL, M. 1992. Conservation of Asian primates: Aspects of genetics and behavioral ecology that predict vulnerability. In *Conservation Biology*, eds. P.L. Fiedler and S.K. Jain, pp. 297–320. Chapman & Hall, New York.

POPE, T.R. 1990. The reproductive consequences of male cooperation in the red howler monkey: paternity exclusion in multi–male and single–male troops using genetic markers. *Behav. Ecol. Sociobiol.* 27:439–446.

POPE, T.R. 1992. The influence of dispersal patterns and mating system on genetic differentiation within and between populations of the red howler monkey (*Alouatta seniculus*). *Evolution* 46:1112–1128.

POST, D.G., G. HAUSFATER, and S.A. McCUSKEY. 1980. Feeding behavior of yellow baboons (*Papio cynocephalus*): relationship to age, gender, and dominance rank. *Fol. Primatol.* 34:170–195.

PUSEY, A.E. 1980. Inbreeding avoidance in chimpanzees. *Anim. Behav.* 28:543–582.

PUSEY, A.E. and C. PACKER. 1987. Dispersal and philopatry. In *Primate Societies*, eds. B.B. Smuts, D.D. Cheney, R.M. Seyfarth, R.W. Wrangham, and T.T. Struhsaker, pp. 250–266. University of Chicago Press, Chicago, IL.

QUINTELA, C.E. 1990. An SOS for Brazil's beleaguered Atlantic Forest. *Nat. Cons. Mag.* 40(2):14–19.

RALLS, K. and J. BALLOU. 1982. Effects of inbreeding on infant mortality in captive primates. *Int. J. Primatol.* 3:491–505.

REED, J.M., P.D. DOER, and J.R. WALTERS. 1986. Determining minimum population sizes for birds and mammals. *Wildl. Soc. Bull.* 14:255–261.

RICHARD, A.F. 1985. *Primates in Nature.* W.H. Freeman & Company, New York.

ROBINSON, J.G. 1988. Group size in wedge–capped capuchin monkeys *Cebus olivaceus* and the reproductive success of males and females. *Behav. Ecol. Sociobiol.* 23:187–197.

RYLANDS, A.B., A.F. COIMBRA–FILHO, and R.A. MITTERMEIER. 1993. Systematics, geographic distribution, and some notes on the conservation status of the Callitrichidae. In *Marmosets and Tamarins: Systematics, Behaviour, and Ecology,* ed. A.B. Rylands, pp. 11–77. Oxford University Press, New York.

SAYER, J.A., C.S. HARCOURT, and N.M. COLLINS. 1992. *The Conservation Atlas of Tropical Forests: Africa.* IUCN and Macmillan, Cambridge, UK.

SCHEFFRAHN, W., N. MENARD, D. VALLET, and B. GACI. 1993. Ecology, demography, and population genetics of Barbary macaques in Algeria. *Primates* 43:381–394.

SCHNEIDER, H., M.I.C. SAMPAIO, M.P.C. SCHNEIDER, J.M. AYRES, C.M.L. BARROSO, A.R. HAMEL, B.T.F. SILVA, and F.M. SALZANO. 1991. Coat color and biochemical variation in Amazonian wild populations of *Alouatta belzebul. Amer. J. Phys. Anthropol.* 85:85–93.

SHAFFER, M. 1987. Minimum viable populations: coping with uncertainty. In *Viable Populations for Conservation,* ed. M.E. Soulé, pp. 69–86. Cambridge University Press, Cambridge, UK.

SHOTAKE, T., K. NOZAWA, and C. SANTIAPILAI. 1991. Genetic variability within and between the troops of the toque macaque, *Macaca sinica,* in Sri Lanka. *Primates* 32: 283–299.

SILK, J.B. 1987. Social behavior in evolutionary perspective. In *Primate Societies,* eds. B.B. Smuts, D.D. Cheney, R.M. Seyfarth, R.W. Wrangham, and T.T. Struhsaker, pp. 318–329. University of Chicago Press, Chicago, IL.

SILK, J.B., C.B. CLARK-WHEATLEY, P.S. RODMAN, and A. SAMUELS. 1981. Differential reproductive success and facultative adjustment of sex ratios among captive female bonnet macaques (*Macaca radiata*). *Anim. Behav.* 29:1106–1120.

SKORUPA, J.P. 1986. Responses of rainforest primates to selective logging in Kibale forest, Uganda: a summary report. In *Primates: The Road to Self–Sustaining Populations,* ed. K. Benirschke, pp. 57–70. Springer–Verlag, New York.

SMOUSE, P.E., V.J. VITZTHUM, and J.V. NEEL. 1981. The impact of random and lineal fission on the genetic divergence of small human groups: a case study among the Yanomama. *Genetics* 98:179–197.

SMUTS, B.B. 1985. *Sex and Friendship in Baboons.* Aldine, New York.

SMUTS, B.B. 1987. Sexual competition and mate choice. In *Primate Societies,* eds. B.B. Smuts, D.D. Cheney, R.M. Seyfarth, R.W. Wrangham, and T.T. Struhsaker, pp. 385–489. University of Chicago Press, Chicago, IL.

STRIER, K.B. 1986. The behavior and ecology of the woolly spider monkey, or muriqui (*Brachyteles arachnoides* E. Geoffroy 1806). PhD dissertation, Harvard University, Cambridge, MA.

STRIER, K.B. 1991. Demography and conservation of an endangered primate, *Brachyteles arachnoides. Conserv. Biol.* 5:214–218.

STRUHSAKER, T.T. 1988. Male tenure, multi–male influxes, and reproductive success in redtail monkeys (*Cercopithecus ascanius*). In *A Primate Radiation,* eds. A. Gautier–Hion, F. Bouliere, J. Gautier, and J. Kingdon, pp. 340–363. Cambridge University Press, Cambridge, UK.

STRUHSAKER, T.T. and L. LELAND. 1979. Socioecology of five sympatric monkey species in the Kibale Forest, Uganda. In *Advances in the Study of Behavior,* eds. J.S. Rosenblatt, R.A. Hinde, C. Beer, and M.C. Busnel, vol. 9, pp. 159–228. Academic Press, New York.

STRUHSAKER, T.T. and T.R. POPE. 1991. Mating system and reproductive success: a comparison of two African forest monkeys (*Colobus badius* and *Cercopithecus ascanius*). *Behaviour* 117:182–205.

SUGIYAMA, Y. 1976. Life history of male Japanese monkeys. In *Advances in the Study of Behavior*, eds. J.S. Rosenblatt, R.A. Hinde, E. Shaw, and C. Beer, vol. 7. Academic Press, New York.

TEMPLETON, A.R. 1987. Inferences on natural population structure from genetic studies on captive mammalian populations. In *Mammalian Dispersal Patterns: The Effects of Social Structure on Population Genetics*, eds. B.D. Chepko–Sade and Z.T. Halpin, pp. 257–272. University of Chicago Press, Chicago, IL.

TERBORGH, J. 1992. *Diversity and the Tropical Rainforest*. Scientific American Library, New York.

TERBORGH, J. and A.W. GOLDIZEN. 1985. On the mating system of the cooperatively breeding saddle–backed tamarin (*Saguinus fuscicollis*). *Behav. Ecol. Sociobiol.* 16:293–299.

TORRES DE ASSUMPÇÃO, C. 1983. Ecological and behavioral information on *Brachyteles arachnoides*. *Primates* 24:584–593.

VIEIRA, C.O.C. 1955. Lista remissiva dos mamíferos do Brasil. *Arg. Zool.* São Paulo 8:341–474.

WATKINS, D.I., T.L. GARBER, Z.W. CHEN, A.L. HUGHES, and N.L. LEVINS. 1991. Evolution of New World primate MHC class I genes. In *Molecular Evolution of the Major Histocompatibility Complex*, eds. J. Klein and D. Klein, pp. 177–191. Springer–Verlag, Heidelberg, Germany.

WEIR, B.S. and C.C. COCKERHAM. 1984. Estimating F–statistics for the analysis of population structure. *Evolution* 38:1358–1370.

WHITE, F.J., A.S. BURTON, S. BUCHOLZ, and K.E. GLANDER. 1992. Social organization of free–ranging ruffed lemurs, *Varecia variegata variegata*: mother–adult daughter relationship. *Amer. J. Primatol.* 28:281–287.

WHITTEN, P.L. 1983. Diet and dominance among female vervet monkeys (*Cercopithecus aethiops*). *Amer. J. Primatol.* 5:139–159.

WORLD CONSERVATION UNION. 1993. *1994 IUCN Red List of Threatened Animals*. World Conservation Union, Gland, Switzerland.

WORLD RESOURCES INSTITUTE. 1992. *World Resources: A Guide to the Global Environment 1992–93*. Oxford University Press, New York.

WRIGHT, S. 1931. Evolution in Mendelian populations. *Genetics* 16:97–159.

WRIGHT, S. 1965. The interpretation of population structure by F–statistics with special regard to systems of mating. *Evolution* 19:395–420.

WRIGHT, S. 1978. *Evolution and the Genetics of Populations, Vol. 4: Variability Within and Among Natural Populations*. University of Chicago Press, Chicago, IL.

ZINGESER, M.R. 1973. Dentition of *Brachyteles arachnoides* with reference to Alouattine and Atelinine affinities. *Fol. Primatol.* 20:351–390.

6

AVIAN
CONSERVATION
GENETICS

Susan M. Haig
and John C. Avise

INTRODUCTION

Habitat loss and fragmentation via human activities represent the most
serious threats to global biodiversity (Wilson, 1992). Often, the impact of
these forces on avian populations and species can be better understood by
consideration of several genetic factors: the level of genetic diversity within
a population and its possible relevance to population viability; the degree of
genetic fragmentation of geographic populations within a species and the
associated magnitudes and patterns of gene flow; the genetic distinctiveness
of a taxon under consideration; and the extent of that taxon's hybridization
and introgression with related forms. In recent years, advances in molecular
genetics, statistical genetics, and population viability analyses have vastly
improved our ability to assess these factors.

Addressing these issues for bird populations in nature can be difficult. For
example, nearly all avian species in North America are protected (appropriately so) by state and federal laws (Box 6.1), but the procedures for
obtaining scientific permits for genetic work with even the most common of
species are cumbersome. A related logistic difficulty is that many avian taxa
are geographically widespread and/or migratory, and thus come under
multiple governmental jurisdictions. Once avian materials are available for
genetic analysis, a second challenge has been to uncover sufficient polymor-

Box 6.1. Taxonomic and Conservation Status

Nearly 10,000 extant species of birds (roughly 20% of all vertebrate species) are recognized worldwide (Monroe and Sibley, 1993). For better or worse, avian taxa are well represented in taxonomic lists of compromised species. In the 1993 list of endangered and threatened taxa (U.S. Endangered Species Act), nearly 250 avian taxa are recorded; in the 1994 IUCN Red List, 970 avian taxa appear. These values constitute nearly one third of all vertebrates represented in these re-spective lists. Most birds are legally protected from physical harm in North America, Japan, and Russia by the Migratory Bird Treaty Act of 1917. Endangered birds are further protected under CITES (Convention on International Trade of Endangered Species) and the Wild Bird Conservation Act of 1992. In addition, many countries afford legal protection to resident and migratory avian taxa.

phisms for robust estimation of paternity, gene flow, and other relevant genetic parameters. In protein-electrophoretic assays, birds generally have displayed relatively low to moderate levels of within-population variation, and limited between-population differences (Avise and Aquadro, 1982; Barrowclough, 1983). In recent years, however, direct assays of DNA (including mitochondrial [mt] DNA, and the hypervariable regions underlying "DNA fingerprints") have added a wealth of variable molecular markers for avian studies, leading to qualitatively new perspectives about such topics as mating systems, population structure, and systematics (Avise, 1996).

Conversely, rationales for conservation-oriented genetic studies of birds can be unusually compelling. Many species are easily observed and studied in nature, so genetic studies can be coupled with field observations on ecology and natural history. Like the miners' canaries, birds can be key indicators of ecosystem health, and conservation of their populations may provide useful umbrella functions in protecting other less-understood taxa. Indeed, for this reason, birds have figured prominently in the broader conservation literature. For example, Rachel Carson's dramatic account of the decline of bird populations in her 1962 classic *Silent Spring* (credited by many as a genesis of the environmental movement) alerted the world to the dangers of pesticide misuse. John Terborgh's (1989) thought-provoking *Where Have All the Birds Gone?* further highlighted the relationship between increasing human numbers and the decline in avian diversity. Because birds tend to elicit public attention and empathy, many avian conservation efforts can also serve a broader role in promoting environmental education and awareness.

Unlike several other chapters in this book, in which multifaceted genetic approaches have been focused on one or a few narrowly circumscribed taxa or geographic regions, comprehensive genetic analyses have not as yet been directed toward any particular avian assemblage (with the possible exception of the red-cockaded woodpecker, *Picoides borealis*, to be discussed). Rather, a series of smaller but informative case histories have appeared in which idiosyncratic genetic issues of conservation relevance have been examined across a miscellany of endangered and threatened taxa. Thus, this chapter consists of a series of vignettes, the goal being to illustrate the diversity of questions in avian conservation addressed through the use of genetic techniques. This chapter also serves to point out the surprising paucity of avian conservation-genetic analyses, and the need for further studies of a more comprehensive nature.

SPECIES IDENTIFICATION
AND EVOLUTIONARY DISTINCTIVENESS

Determination of whether a group of organisms constitutes a "species," "subspecies," or "population" always has been a central topic of debate among evolutionary biologists (e.g., Darwin, 1859; Mayr, 1940; Templeton, 1989). Several competing or overlapping species concepts and definitions exist (see Cracraft, 1983; Avise and Ball, 1990; Avise, 1994), such that considerable scope remains for discussion about species-level and other taxonomic boundaries in birds (e.g., Krajewski, 1994). Delimitation of avian "subspecies" is especially problematic (Ryder, 1986; O'Brien and Mayr, 1991; Ball and Avise, 1992). The U.S. Endangered Species Act currently protects all endangered "species, subspecies, or distinct populations." Nonetheless, from a pragmatic conservation perspective, designation of subspecies and species (and the perceived phylogenetic relationships of these forms to other taxa) can exert considerable influence on how limited resources are allocated by management agencies for recovery of threatened or endangered forms. Thus, several molecular genetic studies have been conducted towards the goal of clarifying genetic relationships among avian taxa in a conservation context.

One early and noteworthy case in which taxonomic designation played an important role in recognition and management of an endangered population involved *Ammodramus maritimus nigrescens*, a coastal sparrow from eastcentral Florida. The dusky seaside sparrow had been described in the 1800s as a distinct species, but later was demoted to the status of one (among about nine) recognized subspecies of seaside sparrows that inhabit coastal marshes along the Gulf of Mexico and the Atlantic seaboard. The dusky population declined precipitously beginning in the 1960s, and the last dusky individual died in captivity in 1987. A retrospective genetic study of the seaside sparrow complex based on mtDNA failed to detect any significant differences between the dusky and other Atlantic coast subspecies, but did suggest a striking distinction between Atlantic coast and Gulf coast birds (Avise and Nelson, 1989). This pattern, when interpreted in conjunction with similar phylogeographic patterns reported in a number of other coastal-restricted vertebrate and invertebrate species (Avise, 1992; see Chapter 14), suggests that long-term biogeographic separations are responsible for the genetic distinctiveness of Atlantic and Gulf populations. In this case, a taxonomy for the seaside sparrow complex (which had been the basis for conservation strategies) probably was an inadequate if not misleading reflection of evolutionary relationships within the assemblage. This case history is elaborated in Chapter 14.

Studies on Hawaiian honeycreepers (Box 6.2) illustrate how assessment of genetic variability and evolutionary relationships among several still-extant taxa can provide a useful backdrop for conservation efforts. Relationships among these species initially were assessed in DNA–DNA hybridization studies, which suggested that the ancestral form was the first passerine to inhabit the islands (Sibley and Ahlquist, 1982). Results from subsequent allozyme surveys were interpreted to indicate that population

---

Box 6.2. Hawaii's Avifauna

The tragic decline of Hawaii's avifauna by hunting, loss of habitat, and introduction of non-native species and avian diseases (reviewed in Warner, 1968; Ralph and van Riper, 1985; Scott and Kepler, 1985; James and Olson, 1991; Ehrlich et al., 1992) provides a sad backdrop for genetic studies in a conservation arena. During the tenure of the Polynesians, which began about 400 AD and peaked with perhaps 200 thousand inhabitants in the mid-1600s, more than 40 bird species were extirpated. Following the arrival of Captain Cook and the first Western culture in 1778, another approximately 20 species were lost. Only about one fourth of the native avian taxa still survives, and many of these species soon may be extinct. More than

20 endemic Hawaiian birds are federally listed as threatened or endangered.

One devastated taxonomic group is the finch subfamily Drepanidinae, the Hawaiian honeycreepers (which have been the subject of several genetic analyses; see text). These beautiful and fascinating birds had radiated to fill numerous niches on the Hawaiian Islands, evolving a remarkable array of distinctive morphologies that covered a broad spectrum of passerine bird adaptations. More than 25 species were described historically and an additional 14+ species have been identified through fossils (James and Olson, 1991). At least 10 historic species have gone extinct, and most of the remaining species (ca. 18) survive today only as small remnant populations.

bottlenecks may have accompanied or followed the founding events, because seven of nine assayed species displayed exceptionally low protein-electrophoretic variation (mean heterozygosity, 0.015; Johnson et al., 1989). Recently, mtDNA analyses of the 'Amakihi species complex (genus *Hemignathus*) indicated that mean inter-island sequence divergence ($d = 0.037$) is an order of magnitude higher than intra-island variation ($d = 0.003$; Tarr and Fleischer, 1993). Closer examination of intra- and inter-island variation, and comparisons against results of other avian taxa similarly assayed, led Tarr and Fleischer to recommend that the subspecies *H. virens stejnegeri* and *H.v. chloris* be afforded the status of separate species, a suggestion generally consistent with the allozymic distinctiveness of *H.v. stejnegeri* (Johnson et al., 1989).

Evolutionary relationships among the cranes (Box 6.3), many of which are endangered or threatened with extinction, have long been of interest. Recent molecular analyses based on DNA-DNA hybridization resulted in the phylogenetic estimate summarized in Figure 6.1 (Krajewski, 1989, 1990). In all methods of data analysis, the crowned crane (*Balearica pavonina*) proved to be the evolutionary outlier, followed by the Siberian crane (*Grus leucogeranus*). Furthermore, the three species of Australasian *Grus* formed a clade, as did five predominantly Palearctic *Grus* species (Figure 6.1). Krajewski (1994) then considered phylogenetic measures of biodiversity

---

Box 6.3. Captive Breeding and Cross-Fostering in Cranes

Of the 14 species of cranes (Gruidae) worldwide, nearly all are listed as threatened or endangered under the Convention on International Trade in Endangered Species, and seven are placed on the 1994 *Red List of Threatened Animals* (World Conservation Union, 1993). Populations of several species have become so small that captive breeding and/or cross-fostering programs have been used either to boost wild populations, or to maintain species in captivity.

Cross-fostering refers to the conservation practice whereby eggs (or young) from one species (usually rare) are placed in the nests of related species for rearing by foster parents. For indeterminate egg layers, this procedure allows production of much larger numbers of potentially rearable young than would otherwise be the case. This management strategy has been applied successfully to endangered whooping cranes (*Grus americana*), whose young can be cross-fostered by the more common sandhill crane (*G. canadensis*). Love and Deininger (1992) recently developed a species-specific molecular probe (involving a repetitive sequence element) for the whooping crane. This probe should prove useful in cross-fostering programs, where pure whooping crane young must be differentiated from hybrids.

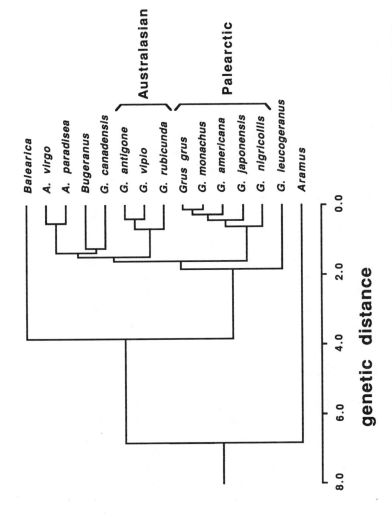

Figure 6.1. Evolutionary relationships among cranes as estimated from DNA–DNA hybridization (after Krajewski, 1989). Shown is a cluster phenogram (which proved similar in structure to those from other distance-based algorithms). A subsequent study based on direct nucleotide sequencing of the mitochondrial cytochrome *b* gene also supported major features of this phylogeny (Krajewski and Fetzner, 1994).

which weight species according to their contribution to the overall genetic diversity of a clade. By this criterion, the greatest single contributor to family diversity was the crowned crane, and the smallest contributor was the common crane (*G. grus*). Within the genus *Grus*, the sandhill crane (*G. canadensis*) also contributed disproportionately to composite evolutionary diversity.

HYBRIDIZATION

Species designations can be complicated further by hybridization and introgression phenomena, which appear to be widespread in birds. Many avian species separated for millions of years are known to retain the anatomical and physiological capacity for hybrid production (Prager and Wilson, 1975), and over 10% of recognized avian species are estimated to have engaged in hybridization producing viable offspring (Grant and Grant, 1992). Such hybridization can have important ecological and evolutionary consequences (Hewitt, 1989).

Ducks and geese are well known for their propensity to hybridize (Grant and Grant, 1992). Cases in point with conservation ramifications involve the common and widely introduced mallard duck (*Anas platyrhynchos*), and several close relatives with which the mallard is suspected of hybridizing. Two of these latter taxa, the endangered Hawaiian duck (*A. wyvilliana*) and the New Zealand grey duck (*A. superciliosa*), are local or regional endemics. Allozyme assays suggest that there has been extensive introgressive hybridization between mallards and Hawaiian ducks on the island of Oahu, resulting in the near disappearance of Hawaiian duck alleles (Browne et al., 1993). Similarly, genetic assays of mtDNA by Rhymer et al. (1994) were interpreted to support earlier morphological indications (Gillespie, 1985) that mallards have hybridized extensively with New Zealand grey ducks, perhaps to the extent that few (< 5%) pure grey ducks remain on the South Island. These studies should raise consciousness about the potential for genetic swamping and loss of localized endemic taxa through introgressive hybridization with more common and widespread forms.

On the other hand, for closely related taxa possessing "heterospecific" alleles, it often may be difficult to completely disentangle whether these alleles have resulted from introgressive hybridization (now or in the past) versus an alternative possibility that they represent shared retentions of ancestral polymorphisms by weakly differentiated descendents. The black duck (*A. rubripes*) of eastern North America is another close relative of the mallard which has declined dramatically in this century. Concomitant with the black duck's decline has been an expansion of mallard populations into their range, leading to suspicions that introgressive hybridization with the mallard is partly responsible. However, in allozyme assays, the two species failed to exhibit the kinds of fixed allelic differences that would be most useful in documenting ongoing hybridization (Ankney et al., 1986; Patton and Avise, 1986). In mtDNA assays, black ducks represented a subset of the lineage diversity present in mallards (Avise et al., 1990). These genetic results can be interpreted as reflecting a paraphyletic relationship of the mallard to the black duck (such that the black duck is a recent evolutionary derivative of a more broadly distributed mallard–black duck ancestor), and by themselves do not require (or eliminate) the possibility of extensive post-speciational introgression.

POPULATION VARIABILITY AND FRAGMENTATION

Continued habitat loss has escalated the need to understand possible relationships between the loss or redistribution of genetic variability in fractionated gene pools and population sustainability. At the very least, estimates of molecular diversity within and among populations provide empirical access to the genetic structures of species, which in turn can illuminate intraspecific evolutionary processes, such as the histories of populations and their levels of genetic and demographic connectedness. In a conservation arena, such information might be useful, for example, in identifying appropriate demographic conservation units, or perhaps in assessing donor–recipient populations in transplantation or reintroduction programs.

Population Variability in Endangered Species

Traditionally, intraspecific molecular studies have involved protein electrophoresis, and more than a dozen such surveys now are available for endangered avian species (Table 6.1). Evans (1987) reviewed the broader allozyme literature for birds, and estimated mean population heterozygosity for non-endangered avian species at $H = 0.065$ (79 species). Comparison of Evans' data against Table 6.1 indicates that many endangered avian species (11 of 13) have lower than average heterozygosities, but also that this is not invariably true. For example, both the endangered wood stork (*Mycteria americana*) and red-cockaded woodpecker display unusually high population heterozygosities (0.093 and 0.078, respectively). In recent years, several studies have estimated variability within avian populations using levels of band-sharing in multilocus DNA fingerprints. In general, these assays have

Table 6.1. Genetic Variation Within and Among Endangered Avian Populations, Based on Allozyme Data Unless Otherwise Noted

Species	N	Number pops.	Polymorphic loci	Mean heterozygosity	F_{ST}[1]	Source
Wood stork (*Mycteria americana*)	396	15	9/20	0.093	0.02	Stangel et al., 1990
Trumpeter swan (*Cygnus buccinator*)	229	3	5/19	0.010	–	Berrett and Vyse, 1982
Hawaiian duck (*Anas wyvilliana*)	20	2	5/20	0.035	–	Browne et al., 1993
Laysan duck (*Anas laysanensis*)	7	1	1/20	0.014	–	Browne et al., 1993
Blue duck (*Hymenolaimus malacorhynchos*)	58	5	2/24	0.002	–	Triggs et al., 1992
Lesser prairie chicken (*Tympanuchus pallidicinctus*)	13	1	0/27	0.000	–	Gutiérrez et. al., 1983
Guam rail (*Rallus owstoni*)	102	1	4/26	0.030	–	Haig and Ballou, 1995
Piping plover (*Charadrius melodus*)	122	5	4/36	0.016	0.02	Haig and Oring, 1988b
Kakapo (*Strigops habroptilus*)	13	3	5/27	0.019–0.044	–	Triggs et. al., 1989
Spotted owl (*Strix occidentalis*)	107	7	1/23	0.000–0.022	0.55	Barrowclough and Gutiérrez, 1990
Micronesian kingfisher (*Halcyon cinnamomina*)	48	1	0/29	0.000	–	Haig and Ballou, 1995
Red-cockaded woodpecker (*Picoides borealis*)	442	26	7/16	0.078	0.14	Stangel et al., 1992
(*Picoides borealis*)[2]	101	14	–	–	0.19	Haig et al., 1994a
Laysan finch (*Telespiza cantans*)	138	4	5/33	0.011–0.031	0.05	Fleischer et al., 1991

[1] Standardized interpopulational variance in allele frequency (see Chapter 9).

[2] Data from RAPD markers.

also indicated lower mean genetic diversities (higher DNA similarities) within populations of endangered as compared to non-endangered taxa (Figure 6.2), but here, too, exceptions exist. For example, one of the most endangered avian species in the world is the Waldrapp ibis *(Geronticus eremita)*, for which a recent molecular genetic survey (Signer et al., 1994) of a captive colony founded by six individuals revealed high genetic diversity in DNA fingerprint profiles (mean ca. 50% band sharing).

Interpretation of such findings in a conservation context remains problematic. For example, even if the low genetic variabilities commonly observed in endangered species reflect population bottleneck effects and associated inbreeding in small populations, they may well be the outcomes rather than the causes of the reductions in population size. In general, causal links between population "health" and genetic variability as measured in molecular assays have not as yet been firmly established (or critically examined) for avian species. Furthermore, empirical estimates of genetic variability may vary depending on the genetic techniques employed. For example, Micronesian kingfishers *(Halcyon cinnamomina cinnamomina)* and New Zealand blue ducks *(Hymenolaimus malacorhynchos)* have no measureable allozyme diversity, but moderate variability in DNA fingerprints (Triggs et al., 1992; Haig and Ballou, 1995). Regardless of the adaptive relevance (or lack thereof) of genetic variation as detected in current assays, molecular markers *can* provide conservation-relevant applications, such as in the assessment of population structure and parentage (to be discussed).

Population Structure in Non-migratory Species

Some of the most detailed work in avian conservation genetics has been conducted on the endangered red-cockaded woodpecker (Figure 6.3). This non-migratory, cavity-nesting species was one of the first avian taxa to be listed under the U.S. Endangered Species Act. Its decline was brought about by severe habitat fragmentation and loss of old-growth pine forests that the species requires for nest cavities. Current conservation efforts include translocation of red-cockaded woodpeckers off private land and from one federal landholding to another, to establish new colonies or to relocate old ones. Care should be taken to insure that only appropriate populations are mixed. Toward that end, molecular markers from the nuclear genome have been used to measure genetic variability within populations and genetic distances between them.

In an extensive allozyme survey, Stangel et al. (1992) found only a weak correlation between the magnitude of allozyme heterozygosity and population size, relatively large between-population differences in allelic frequencies

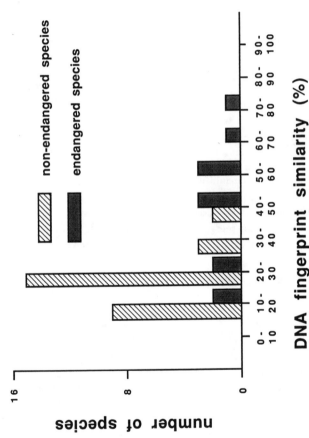

Figure 6.2. Frequency distribution summarizing mean DNA similarities within numerous endangered and non-endangered avian species, as registered in DNA fingerprint assays of conspecific individuals (which in most cases were not known to be close relatives). Compiled from the following references: Burke and Bruford, 1987; Burke et al., 1989; Birkhead et al., 1990; Gibbs et al., 1990; Gyllensten et al., 1990; Meng et al., 1990; Morton et al., 1990; Rabenold et al., 1990; Westneat, 1990, 1993; Jones et al., 1991; Lifjeld et al., 1991, 1993; Longmire et al., 1991, 1992; Ashworth and Parkin, 1992; Brock and White, 1992; Hanotte et al., 1992; Oring et al., 1992; Triggs et al., 1992; Wetton et al., 1992; Yamagishi et al., 1992; Choudhury et al., 1993; Haig et al., 1993a, 1994b, 1994c, 1995; Pinxten et al., 1993; Fleischer et al., 1994; Rave et al., 1994.

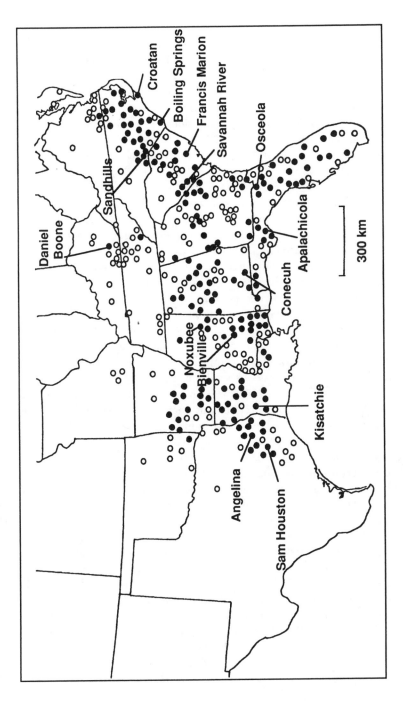

Figure 6.3. Current (closed circles) and historic (open circles) distribution by counties of the red-cockaded woodpecker in the southeastern United States. The named locations indicate sampling sites in the Haig et al. (1994a, 1994b) genetic studies.

171

($F_{ST} = 0.14$; Table 6.1), and a modest increase of genetic distance with geographic distance. A later study using random amplified polymorphic DNA (RAPD) markers (Haig et al., 1994a) indicated even stronger population differentiation ($F_{ST} = 0.19$), and again a general tendency for closer genetic similarity between geographically proximate populations (Figure 6.4). These studies were interpreted to suggest that translocations should take place primarily among birds from nearby populations (provided that ecological and demographic concerns also were satisfied). Special caution was urged in moving birds great distances from south to north, because other evidence suggests that southern birds may be at a size/thermoregulatory disadvantage at higher latitudes (Haig et al., 1994a).

Another non-migratory, endangered species whose populations are threatened by loss of old-growth forest habitat is the spotted owl, *Strix occidentalis*. Barrowclough and Gutiérrez (1990) investigated genetic variability at 23 loci in 107 individuals representing the three named subspecies (*S.o. caurina* from Oregon and northern California, *S.o. occidentalis* from central and southern California, and *S.o. lucida* from New Mexico; Table 6.1). No genetic variation was observed among the six populations of *S.o. caurina*, but a major allele frequency difference between the Pacific Coast populations and those in New Mexico ($F_{ST} = 0.55$) was interpreted to indicate that these disjunct populations have been long isolated and may represent two species. Further studies of geographic variation based on DNA-level assays are currently in progress. In addition, recent behavioral and morphological evidence for natural hybridization between the northern spotted owl and the barred owl (*Strix varia*) in the Pacific northwest indicates the need for genetic analyses of the extent of hybridization and its conservation ramifications (Hamer et al., 1994).

Avian species that inhabit islands can be at great risk of extinction due to limited dispersal options, founder effects, introduced predators, and demographic fluctuations associated with small population sizes. Evidence of this vulnerability is striking: 93% of avian species that went extinct between 1600 and 1980 were island endemics (King, 1980), as were 73% of birds that vanished in North America and U.S. territories between 1492 and 1987 (Williams and Nowak, 1987). In New Zealand, the blue duck has been the subject of genetic analyses designed to assess population genetic structure within an island endemic. This species has declined severely due to agricultural development and loss of its riverine habitat (isolated mountain streams) on both the North and South Islands. Unusually high similarities in DNA fingerprint profiles within and among several nearby populations were suggestive of within-population inbreeding, and also indicated closer genetic relationships between adjacent than between distant populations (Triggs et al., 1992). Results supported field reports of consanguineous

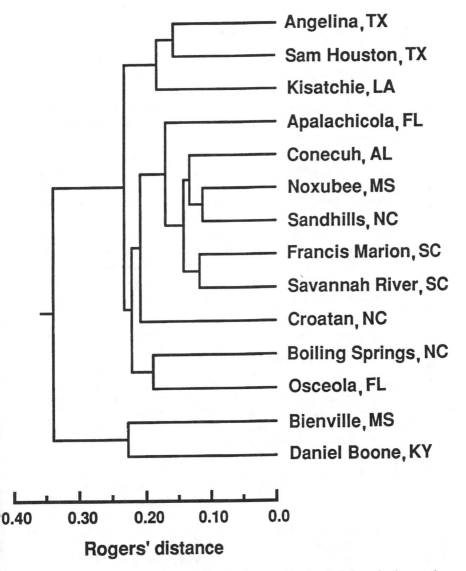

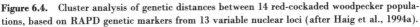

Rogers' distance

Figure 6.4. Cluster analysis of genetic distances between 14 red-cockaded woodpecker populations, based on RAPD genetic markers from 13 variable nuclear loci (after Haig et al., 1994a).

matings, and also appeared consistent with the natural history of the species, which includes low dispersal of individuals from natal sites.

Molecular Markers in Migratory Birds

Migratory species can be at risk from problems encountered at breeding, migratory, or wintering sites. Recent concern over loss of habitat in the tropics as well as North America has resulted in a large-scale effort (Box 6.4) to identify and rectify factors causing the decline in neotropical migrant birds (Finch and Stangel, 1993; Hagan and Johnston, 1993). Despite these concerns, it remains difficult to design recovery strategies until factors can be pinpointed that most impact specific populations during their annual cycle. Color marking or banding of individuals has provided much data on individual movements and migration routes (e.g., Haig and Oring, 1988a; Haig and Plissner, 1993), but normally only a small fraction of marked birds is recovered. Molecular identification of population-specific markers offers a new alternative for large-scale population monitoring.

In migratory shorebirds, groundwork recently has been laid for some of the first genetic studies of this nature. In the dunlin (*Calidris alpina*), DNA sequence analyses of hypervariable mtDNA regions from 73 individuals yielded 35 different mtDNA genotypes, many of which appeared population-specific (Wenink et al., 1993). Phylogenetic analyses of these genotypes

---

Box 6.4. Decline of Neotropical Migrants

Over the past quarter century or more, results from the Breeding Bird Survey and other sources indicate that the numbers of many neotropical migrant songbirds have declined through all or major portions of their respective ranges (Robbins et al., 1989; Askins et al., 1990). Significant loss of habitat (on breeding, migration, and wintering sites) presumably has contributed to the decline. As a result, a major conservation effort coordinated through Partners in Flight has been launched in the United States (and soon in Canada). Partners in Flight, established by the North American Fish and Wildlife Foundation, involves working groups that review, standardize, and recommend monitoring methods, as well as promote interagency and interinvestigator cooperation (Finch and Stangel, 1993). As part of the effort, a program has been established to monitor avian predation and survivorship at mistnetting sites throughout the U.S. (DeSante et al., 1993). Further protocols have been established to relate productivity to habitat conditions in a program called the Breeding Biology Research and Monitoring Database (Martin and Geupel, 1993). The large-scale, cooperative approach taken by Partners in Flight is perhaps the first of its kind, and may provide a useful model for coordinated conservation efforts for other taxonomic groups.

revealed five distinct groups of birds (Alaska, west coast of North America, Gulf of Mexico, western Europe, and the Taymyr Peninsula in Russia) that probably trace to different geographic origins in the circumpolar breeding range of the species. Such genetic markers in principle should allow identification of the geographic source of individuals at any time during the life cycle, and allow for quantification of gene flow among breeding sites. On the other hand, similar studies of the ruddy turnstone (*Arenaria interpres*) revealed little population differentiation, perhaps due to recent range expansion from a localized ancestral source population that had undergone a bottleneck in numbers (Wenink et al., 1994).

Longmire et al. (1991) applied multilocus DNA fingerprint assays to questions of migrational behavior in the Arctic subspecies of the peregrine falcon (*Falco peregrinus tundrius*), individuals of which appeared to possess a unique feature in DNA fingerprint profiles. Results from samples taken along traditional migration routes and at a wintering site in Peru led to the conclusion that many birds migrating along the eastern and gulf coasts of the United States were probably wintering elsewhere in Central or South America.

Although such genetic efforts are still in their infancy for birds, the hope is that knowledge of where specific breeding populations migrate and winter eventually may also help to identify the environmental factors affecting their survival. To cite but one example where such information is needed, the winter distributions of particular breeding populations of the endangered piping plover (*Charadrius melodus*) remain unknown, yet might provide important clues as to why this species is in sharp decline (Haig and Plissner, 1993).

GENETIC RELATIONSHIPS WITHIN POPULATIONS

For management of small populations, it is often important to understand levels of relatedness among individuals, yet the molecular genetic means to accomplish this for organisms of unknown pedigree are only now being developed. For example, various DNA fingerprinting approaches recently have been examined as tools for assessing parentage, extended kinship, social systems, population structure, and effective population size (N_e).

Parentage

In an example that illustrates how knowledge of parentage might be useful in small-population conservation plans, Longmire et al. (1992) used

DNA fingerprinting to assess paternity within a small captive flock of the endangered whooping crane (*Grus americana*). To maximize production of fertile eggs, several females had been inseminated with semen from multiple males, but this procedure also made paternity uncertain. However, the *a posteriori* molecular analyses permitted recovery of information on genetic paternity. This knowledge should prove useful in the design of future matings to maximize genetic diversity or avoid inbreeding within the population. A similar example involves the endangered Hawaiian goose or nene (*Branta sandvicensis*), in which assessment of relatedness by DNA fingerprints in a captive colony permitted verification and completion of pedigree records (Rave et al., 1994).

A similar example from nature involved assessment of the mating system in red-cockaded woodpeckers. This species lives in groups in which several helpers may assist breeding pairs in raising offspring (Walters, 1990). One critical question has been whether male helpers (often sons of the breeding pair) achieve successful fertilizations. Assays based on DNA fingerprinting were used to evaluate parentage in a small South Carolina population (36 birds sampled; Haig et al., 1993a) and in a larger North Carolina population (224 birds sampled; Haig et al., 1994b). In neither study was there evidence for multiple paternity of broods. Thus, management programs based on considerations of effective population size (which is influenced by the sex ratio among breeders) can proceed under the assumption of a monogamous mating system (without significant helper reproduction) for this species.

For several animal groups, growing evidence from molecular genetic markers suggests that conventional behavioral observations of copulation frequency, position in a dominance hierarchy, or other presumed fitness correlates are often inadequate if not poor predictors of true biological parentage (Avise, 1994). An avian case in point involves the New Zealand pukeko or purple swamphen (*Porphyrio porphyrio*), which sometimes breeds in communal groups of as many as 20 individuals. Genetic fingerprint studies of populations, conducted in conjunction with behavioral observations, revealed no consistent relationship between social dominance, frequency of copulations, and parentage among males. Results led Lambert et al. (1994) to conclude that evolutionary biologists must be more cautious in assigning functional explanations to observed behaviors. Similar cautions could be directed to captive animal breeders or to other biologists employing behavioral surrogates to assess biological parentage in a conservation context.

Extended Relatedness

In the design of recovery strategies for small populations, it would often be desirable to know genetic relatedness (r) among all individuals, but this task has proved to be far more challenging than that of assessing parentage (Ashworth and Parkin, 1992; Geyer and Thompson, 1992; Geyer et al., 1993; Haig et al., 1993a, 1994a, 1994b, 1994c, 1995). One suggested approach is to use multilocus DNA fingerprinting to empirically estimate genetic relationships among individuals when at least a partial pedigree has been established from other evidence. For example, in two field-monitored populations of red-cockaded woodpeckers in North Carolina and South Carolina, genetic similarity values (proportions of DNA bands shared) proved to be correlated significantly with relatedness (r) calculated from putative pedigrees (Figure 6.5). After such calibrations between r and DNA similarity have been established for populations of known pedigree, this knowledge might then be used to extrapolate from additional DNA fingerprints to estimate r among individuals of unknown relationship. Even then, however, caution is urged, because empirical experience with red-cockaded woodpeckers (Haig et al., 1993a, 1994b), as well as with lions (Gilbert et al., 1991; Packer et al., 1991),

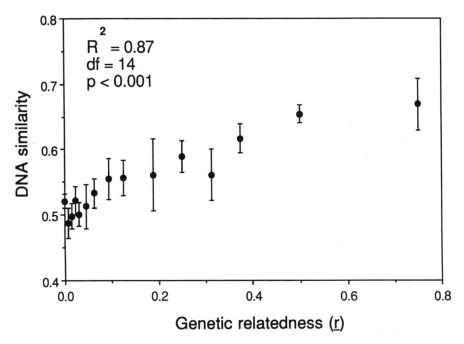

Figure 6.5. Relationship between mean genetic similarity in DNA fingerprint assays, and probable relatedness values calculated from a putative pedigree, for red-cockaded woodpeckers (224 birds sampled) in a North Carolina population (after Haig et al., 1994b). The least-squares linear regression was highly significant. Similar analyses of a South Carolina population (36 birds sampled) also yielded a significant relationship between DNA band sharing and r ($R^2 = 0.59$; Haig et al., 1993a).

indicates that calibrations between DNA band sharing can differ among conspecific populations. Furthermore, distributions of DNA similarity values for adjacent classes of *r* usually overlap (Figure 6.5) such that only qualitative guidelines to general levels of relatedness can be anticipated from this approach (see also Lynch, 1988).

A further matter is how information on genetic relatedness might then be utilized in *population viability analyses* (Box 6.5), or in managed breeding programs. For captive populations, a number of genetic management stra-

---

Box 6.5. Population Viability Analysis (PVA)

A *viable population*, defined as a population capable of self-maintenance without continuing manipulation or intervention by humans, is the desired goal in most restoration programs for threatened or endangered species. Formal population viability analyses are now conducted routinely for threatened and endangered species. In a PVA, a desired survival probability usually is specified for a stated period of time, and management guidelines are developed to achieve this objective. For example, new recovery goals for the threatened piping plover were derived from a recent study, which estimated that about 3,300 breeding pairs would be needed to insure survival of the Northern Great Plains population for the next 100 years, with 95% probability (Ryan et al., 1993; Haig et al., 1994d). In principle, numerous genetic, demographic, and habitat considerations should be taken into account in deriving such guidelines. In practice, many uncertainties and untested assumptions greatly complicate matters. However, the PVA approach is a continually evolving process that has added sophistication to endangered species recovery programs.

The recovery goal for the endangered red-cockaded woodpecker has been stated in terms of population viability.

Before the species is delisted it must reach an effective population size of 500 animals (250 clans) in each of 15 populations across the species' range and occurring in representative habitat (U.S. Fish and Wildlife Service, 1985). On a local level, population viability analysis of a small South Carolina population indicated a precarious future for this deme (with a projected mean time of 41.5 years to extinction, and 72% probability of extinction within 200 years), unless new birds were brought in from other areas (Haig et al., 1993a, 1993b). To guide the translocation plan, a PVA was used to model the genetic and demographic effects of adding birds of varying sex ratios. Surprisingly, the analysis suggested that the successful annual introduction of only three foreign females and two males for 10 years would more than double the projected mean time to extinction, and decrease the likelihood of extinction within 200 years to 4%. Alternative models involving annual translocations of 50 males and 50 females for 25 years did not significantly increase population viability over the three-female–two-male scenario. Donor populations currently are being chosen on the basis of the RAPD analyses described in the text (Haig et al., 1994a).

Table 6.2. Breeding Strategies Considered in the Choice of Founders for a New Population of Guam Rails on the Island of Rota. After Haig et al., 1990

"Gene drop" pedigree analyses were used to evaluate the indicated strategies for choice of captive breeding pairs to found the Rota population (see Table 6.3 and text).

1. RANDOMLY choose individuals from captive populations to produce birds for Rota.
2. Choose the most FECUND captive breeders to produce birds for Rota.
3. Choose captive individuals with highest ALLOZYME DIVERSITY to produce birds for Rota.
4. Choose captive individuals whose chicks will represent EQUAL FOUNDER CONTRIBUTION on Rota.
5. Choose captive individuals whose chicks will represent MAXIMUM ALLELIC DIVERSITY on Rota.
6. Choose captive individuals whose chicks will represent MAXIMUM FOUNDER GENOME EQUIVALENTS on Rota.

---

tegies have been evaluated (Table 6.2), most of which operate under the general assumption that maximizing intrapopulational genetic diversity is desirable (Geyer et al., 1989; Lacy, 1989). Some empirical examples follow to show how DNA fingerprinting and other molecular methods to estimate relatedness have been utilized in a conservation context.

Island Endemics

The Palila (*Loxioides bailleui*) is an endangered Hawaiian honeycreeper (Box 6.2) that has undergone extensive range contraction in recent years. The species now survives only on Mauna Kea, a dormant volcano on the big island of Hawaii. The severe population decline and low hatchability of eggs led Fleischer et al. (1994) to evaluate genetic variability and structure using DNA fingerprint assays in the two remaining populations. They found no evidence for extrapair fertilizations, low levels of intrapopulation DNA similarity (ca. 0.26; Figure 6.2), and relatively high DNA similarity between the two sites. The authors concluded that the populations were not inbred, that they had been in recent genetic contact, and that reciprocal translocations probably would be appropriate, at least from a genetic point of view.

Another Hawaiian endemic that has suffered a severe population decline is the nene. Recent assays of mtDNA restriction sites and sequences revealed that this morphologically distinct goose is allied most closely to the Canada goose (*Branta canadensis*), and that its ancestors probably arrived in Hawaii from North America about 900 thousand years ago (Quinn et al., 1991). Beginning in the late 1800s, the species declined (due to habitat loss, predation from introduced species, and hunting) from about 25,000 birds to fewer that 17 birds in the late 1940s, when a captive breeding program was inaugurated. Some nene have since been released to the wild, whereas others are held in captivity at several locations. Recently, genetic diversity in captive populations of nene was assessed by DNA fingerprinting (Rave et al., 1994). Mean genetic similarities between individuals were high (0.67–0.88) relative to reports for other birds (Figure 6.2), suggesting a recent history of inbreeding and/or the effects of population bottlenecks. Plans are to utilize such genetic findings to further assess the status of wild populations and to make recommendations for breeding plans in captive colonies.

The island of Guam currently is experiencing a crisis in avian extinction (Box 6.6). Two endangered taxa endemic to this small island, the Guam rail (*Rallus owstoni*) and Micronesian kingfisher (*Halcyon cinnamomina cinnamomina*), have been the subject of recent molecular genetic studies. Both species were brought into captivity in 1983 and 1984, and subsequently went extinct in the wild (Haig and Ballou, 1995). Unfortunately, no information was available about genetic relationships among the 21 Guam rails or the 29 Micronesian kingfishers that are considered founders of the captive populations. The Guam rail breeding program has been successful thus far, with the population expanding to 173 individuals; however, Micronesian kingfishers remain on the brink of extinction (Box 6.6).

Box 6.6. Guam's Avifauna

The western Pacific island of Guam, a 480-square-kilometer member of the Northern Marianas chain, was home to about 18 native avian species prior to 1940. Among these, two species and three subspecies were endemic (Enbring and Pratt, 1985). During World War II, inadvertent introduction of the brown tree snake (*Boiga irregularis*) resulted in the loss of 7 of 11 of Guam's native forest birds (Savidge, 1987). The remaining taxa were emergency-listed as endangered in 1984 (U.S. Fish and Wildlife Service, 1984a, 1984b). The decline was so rapid that only the Guam rail (an endemic species) and Micronesian kingfisher (an endemic subspecies) were rescued and brought into captivity (see text). The brown tree snake remains abundant on Guam, but the nearby island of Rota is snake-free and has been selected as the site for reintroduction of Guam rails to nature. Unfortunately, the status of Micronesian kingfishers remains desperate. They have not reproduced well in captivity, and the presence of congeners on nearby islands makes the success of introductions doubtful. Thus, current management efforts include improved husbandry techniques for the captive population (based in part on recommendations from DNA analyses), and attempts to "snake-proof" certain trees on Guam. The hope is that Micronesian kingfishers could be released on Guam even if only a few natural sites were protected.

To better manage matings within these captive populations (and to reestablish wild populations), genetic analyses were carried out to determine genetic diversity within each species and relatedness among founders. Results from both allozymic analyses (Table 6.1) and DNA fingerprinting indicated relatively low levels of genetic diversity within the two species, although no significant loss of genetic variation could be documented subsequent to the establishment of the captive populations (Haig et al., 1994c, 1995; Haig and Ballou, 1995). Significant correlations ($R^2 = 0.95$ for Guam rails, 0.90 for Micronesian kingfishers) were demonstrated between known levels of interindividual r within the captive-breeding pedigree and DNA similarity as reflected in DNA fingerprints. Unfortunately, overlap in levels of DNA similarity for various categories of organismal relationship compromised attempts to assign exact relatedness to specific pairs of founders on the basis of DNA evidence alone. Nonetheless, cluster analyses of DNA similarities among founders helped to reclassify two groups of Guam rails (three individuals each) as probable first-degree relatives (siblings or parent–offspring), and similarly prompted reclassification of three pairs of Micronesian kingfisher founders as close relatives (Figure 6.6). Breeding strategies have now been changed for both species so that these close relatives are not paired.

Establishment of genetic relationships among founders prompted further efforts to evaluate the six genetic management options described in Table 6.2 (Haig et al., 1990). Management plans for the captive populations were needed for both Guam rails and Micronesian kingfishers. Furthermore, Guam rails currently are being introduced to the nearby island of Rota, and genetic factors were considered for choosing which birds to introduce. The options were evaluated after establishing a pedigree for Guam rails and then using a pedigree analysis technique called a *gene drop*. In this technique, two unique alleles assigned to each founder are "dropped" through the pedigree using Monte Carlo simulations. The end result is an estimate of the number of founder alleles, the number of founders from which at least one allele has survived ("founder contribution"), and heterozygosity in the current (or Rota) population after, for example, 10,000 simulations. For Guam rails, the greatest levels of genetic diversity (e.g., heterozygosity, allelic diversity, founder contribution, etc.) were found using a strategy of "maximizing founder genome equivalents" (Table 6.3; Lacy, 1989), that is, managing the captive pedigree so that in the current population (and the population to be introduced to Rota), founder contribution is equalized as the number of retained founder alleles is maximized. Guam rails and Micronesian kingfishers (and several other species/populations) now are managed with this goal in mind.

In conclusion, relatively few genetic studies focused explicitly on issues of avian conservation have as yet appeared. Clearly, many conservation

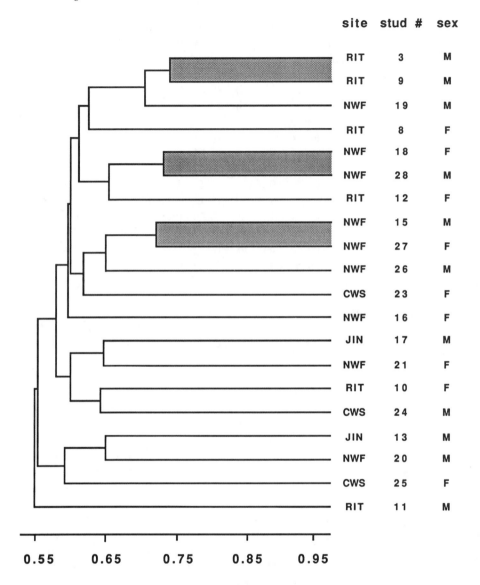

DNA fingerprint similarity

Figure 6.6.   Identification of close relatives among founders of the Micronesian kingfisher captive population (Haig et al., 1995). Shown is a cluster phenogram based on DNA similarity values estimated from multilocus DNA fingerprint gels. Stud indicates founder studbook numbers and letters refer to capture sites in northern Guam. Birds in shaded clusters (stud nos. 3 and 9, 18 and 28, and 15 and 27) were interpreted to be particularly close relatives.

Table 6.3. Consequences of Options Evaluated via "Gene Drop" Pedigree Analysis for Reintroduction of Guam Rails to the Island of Rota. After Haig et al., 1990; see also Table 6.2 and text

Status and option	Heterozygosity retained [1]	Allelic diversity	Founders represented	Founder genome equivalent	Pairs needed [2]
Founders (initial population)	1.00	42.0	21	21.0	–
Current population	0.98	31.5	16	10.5	–
Management options considered (see Table 6.2)					
1. Randomly choose individuals	0.95	24.1	13	9.4	23
2. Choose most fecund breeders	0.98	20.5	12	8.3	8
3. Choose for high allozyme diversity	1.00	18.9	12	7.1	13
4. Equalize founder contribution	0.98	27.2	15	13.4	8
5. Maximize unique "gene drop" alleles	1.00	29.3	15	13.7	23
6. Maximize founder genome equivalents	1.00	29.2	15	14.4	16

[1] With the exception of "founders" and "current population" data, results represent values that would be present in the new Rota population if the indicated option were adopted.
[2] Indicates the number of captive breeding pairs that would be needed to achieve the option being evaluated.

problems exist, and molecular technologies are available for providing various types of genetic information of potential relevance to management goals. Although detailed molecular studies cannot be carried out on every threatened or endangered avian population, in propitious situations they can provide an excellent return of genetic information from modest investment.

SUMMARY

1. Birds tend to be highly visible and sensitive indicators of environmental disturbance, and perhaps for these reasons are strongly represented on lists of threatened, endangered, and recently extirpated species. Molecular appraisals of rare taxa have allowed examination of the genetic consequences of the decline and fragmentation of avian populations.

2. Avian taxa have figured prominently in discussions of species concepts and taxonomic boundaries, and these issues often come into pragmatic focus in conservation plans for endangered forms. In several instances, molecular genetic techniques have been brought to bear on questions concerning the taxonomic and evolutionary distinctiveness of threatened populations and species that are under consideration for limited management resources.

3. Hybridization and introgression are relatively common phenomena in many avian groups, such as waterfowl, and can confound both taxonomic assignments and conservation strategies. Genetic evidence suggests that some rare and endemic ducks have been "swamped" with introgressed alleles from more common relatives. However, the appearance of "foreign" alleles in a species is not by itself definitive proof of current or past introgressive hybridization. An alternative possibility is the parallel retention of ancestral polymorphisms by weakly differentiated descendant taxa.

4. Many, but not all, endangered avian taxa have relatively low levels of within-population genetic variation as measured by protein and DNA assays, presumably because of population bottlenecks and/or small effective population sizes. However, causal links between molecular variation and a population's genetic "health" remain to be firmly established for birds.

5. Molecular analyses of geographic populations in several endangered avian species have revealed how genetic variation is distributed within and among sites. In the case of the red-cockaded woodpecker, this information has led to recommendations about which populations should be utilized in ongoing translocation programs. For some migratory species, genetic markers specific to particular breeding populations have been identified and provide a basis for attempts to identify natal sites of birds collected at any time during their life cycle. Such information could help to determine where particular populations migrate and winter, and thereby help to identify environmental factors contributing to annual mortality.

6. Captive populations of some endangered species (such as the Guam rail and Micronesian kingfisher) are maintained under breeding schemes designed to enhance genetic variation, for example, by equalizing genetic contributions from the original founders. Toward this end, DNA fingerprinting has been employed to assess genetic relationships among the founders taken from nature, and to further examine parentage and kinship within the captive pedigree. First-order relationships (parent–offspring or full siblings) often can be determined, but identification of precise relatedness among more distant relatives remains problematic.

REFERENCES

ANKNEY, C.D., D.G. DENNIS, L.N. WISHARD, and J.E. SEEB. 1986. Low genic variation between black ducks and mallards. *Auk* 103:701–709.
ASHWORTH, D. and D.T. PARKIN. 1992. Captive breeding: can genetic fingerprinting help? *Symp. Zool. Soc. Lond.* 64:135–149.
ASKINS, R.A., J.F. LYNCH, and R. GREENBERG. 1990. Population declines in neotropical birds in eastern North America. *Curr. Ornithol.* 7:1–57.

AVISE, J.C. 1992. Molecular population structure and the biogeographic history of a regional fauna: a case history with lessons for conservation biology. *Oikos* 63:62–76.

AVISE, J.C. 1994. *Molecular Markers, Natural History and Evolution.* Chapman & Hall, New York.

AVISE, J.C. 1996. Three fundamental contributions of molecular genetics to avian ecology and evolution. *Proc. XXI Int. Ornithol. Congr.,* (in press).

AVISE, J.C., C.D. ANKNEY, and W.S. NELSON. 1990. Mitochondrial gene trees and the evolutionary relationship of mallard and black ducks. *Evolution* 44:1109–1119.

AVISE, J.C. and C.F. AQUADRO. 1982. A comparative summary of genetic distances in the vertebrates: patterns and correlations. *Evol. Biol.* 15:151–185.

AVISE, J.C. and R.M. BALL, Jr. 1990. Principles of genealogical concordance in species concepts and biological taxonomy. *Oxford Surv. Evol. Biol.* 7:45–67.

AVISE, J.C. and W.S. NELSON. 1989. Molecular genetic relationships of the extinct dusky seaside sparrow. *Science* 243:646–648.

BALL, R.M., Jr. and J.C. AVISE. 1992. Mitochondrial DNA phylogeographic differentiation among avian populations and the evolutionary significance of subspecies. *Auk* 109:626–636.

BARRETT, V.A. and E.R. VYSE. 1982. Comparative genetics of three trumpeter swan populations. *Auk* 99:103–108.

BARROWCLOUGH, G.F. 1983. Biochemical studies of microevolutionary processes. In: *Perspectives in Ornithology,* eds. A.H. Brush and G.A. Clark, pp. 223–261. Cambridge University Press, New York.

BARROWCLOUGH, G.F. and R.J. GUTIÉRREZ. 1990. Genetic variation and differentiation in the spotted owl (*Strix occidentalis*). *Auk* 107:737–744.

BIRKHEAD, T.R., T. BURKE, R. ZANN, F.M. HUNTER, and A.P. KRUPNA. 1990. Extrapair paternity and intraspecific brood parasitism in wild zebra finches *Taeniopygia guttata,* revealed by DNA fingerprinting. *Behav. Ecol. Sociobiol.* 27:315–324.

BROCK, M.K. and B.N. WHITE. 1992. Application of DNA fingerprinting to the recovery program of the endangered Puerto Rican parrot. *Proc. Natl. Acad. Sci. USA* 89:1121–1125.

BROWNE, R.A., C.R. GRIFFIN, P.R. CHANG, M. HUBLEY, and A.E. MARTIN. 1993. Genetic divergence among populations of the Hawaiian duck, Laysan Duck, and mallard. *Auk* 110:49–56.

BURKE, T. and M.W. BRUFORD. 1987. DNA fingerprinting in birds. *Nature* 327:149–152.

BURKE, T., N.B. DAVIES, M.W. BRUFORD, and B.J. HATCHWELL. 1989. Parental care and mating behavior of polyandrous dunnocks *Prunella modularis* related to paternity by DNA fingerprinting. *Nature* 338:249–251.

CARSON, R. 1962. *Silent Spring.* Houghton Mifflin, Boston, MA.

CHOUDHURY, S., C.S. JONES, J.M. BLACK, and J. PROP. 1993. Adoption of young and intraspecific nest parasitism in barnacle geese. *Condor* 95:860–868.

CRACRAFT, J. 1983. Species concepts and speciation analysis. In *Current Ornithology,* ed. R.F. Johnston, pp. 159–187. Plenum Press, New York.

DARWIN, C. 1859. *On the Origin of Species.* Murray, London.

DeSANTE, D.F., O.E. WILLIAMS, and K.M. BURTON. 1993. The Monitoring Avian Productivity and Survivorship (MAPS) program: overview and progress. In *Status and Management of Neotropical Migratory Birds,* eds. D.M. Finch and P.W. Stangel, pp. 208–222. USDA Forest Service General Technical Report RM-229, Fort Collins, CO.

EHRLICH, P.R., D.S. DOBKIN, and D. WHEYE. 1992. *Birds in Jeopardy.* Stanford University Press, Stanford, CA.

ENBRING, J. and H.D. PRATT. 1985. Endangered birds in Micronesia: their history, status, and future prospects. *Bird Conserv.* 2:71–106.

EVANS, P.G.H. 1987. Electrophoretic variability of gene products. In *Avian Genetics: A Population and Ecological Approach*, eds. F. Cooke and P.A. Buckley, pp. 105–162. Academic Press, New York.

FINCH, D.M. and P.W. STANGEL (eds.). 1993. *Status and Management of Neotropical Migratory Birds*. USDA Forest Service General Technical Report RM-229, Fort Collins, CO.

FLEISCHER, R.C., S. CONANT, and M.P. MORIN. 1991. Genetic variation in native and translocated populations of the Laysan Finch (*Telespiza cantans*). *Heredity* 66:125–130.

FLEISCHER, R.C., C.L. TARR, and T.K. PRATT. 1994. Genetic structure and mating system in the palila, an endangered Hawaiian honeycreeper, as assessed by DNA fingerprinting. *Molec. Ecol.* 3:383–392.

GEYER, C.J., O.A. RYDER, L.G. CHEMNICK, and E.A. THOMPSON. 1993. Analysis of relatedness in California condors from DNA fingerprints. *Molec. Biol. Evol.* 10:571–589.

GEYER, C.J. and E.A. THOMPSON. 1992. Constrained Monte Carlo maximum likelihood for dependent data. *J. Royal Stat. Soc.* B54:657–699.

GEYER, C.J., E.A. THOMPSON, and O.A. RYDER. 1989. Gene survival in the Asian wild horse (*Equus przewalskii*): II. Gene survival in the whole population, in subgroups, and through history. *Zoo Biol.* 8:313–329.

GIBBS, H.L., P.J. WEATHERHEAD, P.T. BOAG, B.N. WHITE, L.M. TABAK, and D.J. HOYSAK. 1990. Realized reproductive success of polygynous red-winged blackbirds revealed by DNA markers. *Science* 250:1394–1397.

GILBERT, D.A., C. PACKER, A.E. PUSEY, J.C. STEPHENS, and S.J. O'BRIEN. 1991. Analytical DNA fingerprinting in lions: parentage, genetic diversity, and kinship. *J. Hered.* 82:378–386.

GILLESPIE, G.D. 1985. Hybridization, introgression, and morphometric differentiation between mallard (*Anas platyrhynchos*) and grey ducks (*Anas superciliosa*) in Otago, New Zealand. *Auk* 102:459–469.

GRANT, P.R. and B.R. GRANT. 1992. Hybridization of bird species. *Science* 256:193–197.

GUTIÉRREZ, R.J., R.M. ZINK, and S.Y. YANG. 1983. Genic variation, systematic, and biogeographic relationships of some galliform birds. *Auk* 100:33–47.

GYLLENSTEN, U.B., S. JACOBSSON, and H. TEMRIN. 1990. No evidence for illegitimate young in monogamous and polygynous warblers. *Nature* 343:168–170.

HAGAN, J.M. III and D.W. JOHNSTON (eds.). 1993. *Ecology and Conservation of Neotropical Migrant Landbirds*. Smithsonian Institution Press, Washington, DC.

HAIG, S.M. and J.D. BALLOU. 1995. Genetic diversity in two avian species formerly endemic to Guam. *Auk* 112: in press.

HAIG, S.M., J.D. BALLOU, and N.J. CASNA. 1994c. Identification of kin structure among Guam rail founders: a comparison of pedigrees and DNA profiles. *Molec. Ecol.* 4:109–119.

HAIG, S.M., J.D. BALLOU, and N.J. CASNA. 1995. Genetic identification of kin in Micronesian Kingfishers. *J. Hered.* 86: in press.

HAIG, S.M., J.D. BALLOU, and S.R. DERRICKSON. 1990. Management options for preserving genetic diversity: reintroduction of Guam rails to the wild. *Conserv. Biol.* 4:290–300.

HAIG, S.M., J.R. BELTHOFF, and D.H. ALLEN. 1993a. Examination of population structure in red-cockaded woodpeckers using DNA profiles. *Evolution* 47:185–194.

HAIG, S.M., J.R. BELTHOFF, and D.H. ALLEN. 1993b. Population viability analysis for a small population of red-cockaded woodpeckers and an evaluation of population enhancement strategies. *Conserv. Biol.* 7:289–301.

HAIG, S.M., T. EUBANKS, R. LOCK, L. PFANNMULLER, E. PIKE, M. RYAN, and J. SIDLE. 1994d. *Recovery Plan for Piping Plovers Breeding on the Great Lakes and Northern Great Plains*. U.S. Fish and Wildlife Service, Twin Cities, MN.

HAIG, S.M. and L.W. ORING. 1988a. Distribution and dispersal in the piping plover. *Auk* 105:630–638.

HAIG, S.M. and L.W. ORING. 1988b. Genetic differentiation of piping plovers across North America. *Auk* 105:260–267.

HAIG, S.M. and J.H. PLISSNER. 1993. Distribution and abundance of piping plovers: results and implications of the 1991 International Census. *Condor* 95:145–156.

HAIG, S.M., J.M. RHYMER, and D.G. HECKEL. 1994a. Population differentiation in randomly amplified polymorphic DNA of red-cockaded woodpeckers *Picoides borealis*. *Molec. Ecol.* 3: 581–595.

HAIG, S.M., J.R. WALTERS, and J.H. PLISSNER. 1994b. Genetic evidence for monogamy in the red-cockaded woodpecker, a cooperative breeder. *Behav. Ecol. Sociobiol.* 23:295–303.

HAMER, T.E., E.D. FORSMAN, A.D. FUCHS, and M.L. WALTERS. 1994. Hybridization between barred and spotted owls. *Auk* 111:487–492.

HANOTTE, O., M.W. BRUFORD, and T. BURKE. 1992. Multilocus DNA fingerprints in gallinaceous birds: general approaches and problems. *Heredity* 68:481–494.

HEWITT, G.M. 1989. The subdivision of species by hybrid zones. In *Speciation and Its Consequences*, eds. D. Otte and J.A. Endler, pp. 85–110. Sinauer, Sunderland, MA.

JAMES, H.F. and S.L. OLSON. 1991. Descriptions of 32 new species of birds from the Hawaiian Islands: Part II. Passeriformes. *Ornith. Monog.* 46. American Ornithologists Union, Washington, DC.

JOHNSON, N.K., J.A. MARTEN, and C.J. RALPH. 1989. Genetic evidence for the origin and relationships of Hawaiian honeycreepers. *Condor* 91:379–396.

JONES, C.S., C.M. LESSELLS, and J.R. KREBS. 1991. Helpers-at-the-nest in European bee-eaters (*Meriops apaiaster*): a genetic analysis. In *DNA Fingerprinting: Approaches and Applications*, eds. T. Burke, G. Dolf, A.J. Jeffreys, and R. Wolff, pp. 169–192. Birkhäuser Verlag, Basel, Switzerland.

KING, W.B. 1980. Ecological basis of extinction in birds. *Acta XVII Cong. Int. Ornith.* 17:905–911.

KRAJEWSKI, C. 1989. Phylogenetic relationships among cranes (Gruiformes: Gruidae) based on DNA hybridization. *Auk* 106:603–618.

KRAJEWSKI, C. 1990. Relative rates of single-copy DNA evolution in cranes. *Molec. Biol. Evol.* 7:65–73.

KRAJEWSKI, C. 1994. Phylogenetic measures of biodiversity: a comparison and critique. *Biol. Conserv.* 69:33–39.

KRAJEWSKI, C. and J.W. FETZNER, Jr. 1994. Phylogeny of cranes (Gruiformes: Gruidae) based on cytochrome-*B* DNA sequences. *Auk* 111:351–365.

LACY, R.C. 1989. Analysis of founder representation in pedigrees: founder equivalents and founder genome equivalents. *Zoo Biol.* 8:111–123.

LAMBERT, D.M., C.D. MILLAR, K. JACK, S. ANDERSON, and J.L. CRAIG. 1994. Single- and multilocus DNA fingerprinting of communally breeding pukeko: Do copulations or dominance ensure reproductive success? *Proc. Natl. Acad. Sci. USA* 91:9641–9645.

LIFJELD, J.T., P.O. DUNN, R.J. ROBERTSON, and P.T. BOAG. 1993. Extra-pair paternity in monogamous tree swallows. *Anim. Behav.* 45:213–229.

LIFJELD, J.T., T. SLAGSVOLD, and H.M. LAMPE. 1991. Low frequency of extra pair paternity in pied flycatchers revealed by DNA fingerprinting. *Behav. Ecol. Sociobiol.* 29:95–101.

LONGMIRE, J.L., R.E. AMBROSE, N.C. BROWN, T.J. CADE, T.L. MAECHTLE, S.W. SEEGAR, F.P. WARD, and C.M. WHITE. 1991. Use of sex-linked minisatellite fragments to investigate genetic differentiation and migration of North American populations of the peregrine falcon (*Falco peregrinus*). In *DNA Fingerprinting: Approaches and Applications*, eds. T. Burke, G. Dolf, A.J. Jeffreys and R. Wolff, pp. 217–229. Birkhäuser Verlag, Basel, Switzerland.

LONGMIRE, J.L., G.F. GEE, C.L. HARDEKOPF, and G.M. MARK. 1992. Establishing paternity in whooping cranes (*Grus americana*) by DNA analysis. *Auk* 109:522–529.

LOVE, J. and P. DEININGER. 1992. Characterizations and phylogenetic significance of a repetitive DNA sequence from whooping cranes (*Grus americana*). *Auk* 109:73–79.

LYNCH, M. 1988. Estimation of relatedness by DNA fingerprinting. *Molec. Biol. Evol.* 5:584–589.

MARTIN, T.E. and G.R. GEUPEL. 1993. Nest-monitoring plots: methods for locating nests and monitoring success. *J. Field Ornithol.* 64:507–519.

MAYR, E. 1940. Speciation phenomena in birds. *Amer. Nat.* 74:249–278.

MENG, A., R.E. CARTER, and D.T. PARKIN. 1990. The variability of DNA fingerprints in three species of swan. *Heredity* 64:73–80.

MONROE, B.L., Jr. and C.G. SIBLEY. 1993. *A World Checklist of Birds.* Yale University Press, New Haven, CT.

MORTON, E.S., L. FORMAN, and M. BRAUN. 1990. Extrapair fertilizations and the evolution of colonial breeding in purple martins. *Auk* 107:275–283.

O'BRIEN, S.J. and E. MAYR. 1991. Bureaucratic mischief: recognizing endangered species and subspecies. *Science* 251:1187–1188.

ORING, L.W., R.C. FLEISCHER, J.M. REED, and K.E. MARSDEN. 1992. Cuckoldry through stored sperm in the sequentially polyandrous spotted sandpiper. *Nature* 359:631–633.

PACKER, C., D.A. GILBERT, A.E. PUSEY, and S.J. O'BRIEN. 1991. A molecular genetic analysis of kinship and cooperation in African lions. *Nature* 351:562–564.

PATTON, J.C. and J.C. AVISE. 1986. Evolutionary genetics of birds. IV. Rates of protein divergence in waterfowl (Anatidae). *Genetica* 68:129–143.

PINXTEN, R., O. HANOTTE, M. EENS, R.F. VERHEYEN, A.A. DHONDT, and T. BURKE. 1993. Extra-pair paternity and intraspecific brood parasitism in the European starling (*Sturnus vulgaris*): evidence from DNA fingerprinting. *Anim. Behav.* 45:795–809.

PRAGER, E.M. and A.C. WILSON. 1975. Slow evolutionary loss of the potential for interspecific hybridization in birds: a manifestation of slow regulatory evolution. *Proc. Natl. Acad. Sci. USA* 72:200–204.

QUINN, T.W., G.F. SHIELDS, and A.C. WILSON. 1991. Affinities of the Hawaiian goose based on two types of mitochondrial DNA data. *Auk* 108:585–593.

RABENOLD, P.P., K.N. RABENOLD, W.H. PIPER, J. HAYDOCK, and S.W. ZACK. 1990. Shared paternity revealed by genetic analysis in cooperatively breeding tropical wrens. *Nature* 348:538–540.

RALPH, C.J. and C. VAN RIPER III. 1985. Historical and current factors affecting Hawaiian native birds. *Bird Conserv.* 2:7–42.

RAVE, E.H., R.C. FLEISCHER, F. DUVALL, and J.M. BLACK. 1994. Genetic analyses through DNA fingerprinting of captive populations of Hawaiian geese. *Conserv. Biol.* 8:744–751.

RHYMER, J.M., M.J. WILLIAMS, and M.J. BRAUN. 1994. Mitochondrial analysis of gene flow between New Zealand mallards (*Anas platyrhynchos*) and grey ducks (*A. superciliosa*). *Auk* 111: in press.

ROBBINS, C.S., J.R. SAUER, R. GREENBERG, and S. DROEGE. 1989. Population declines in North American birds that migrate to the Neotropics. *Proc. Natl. Acad. Sci. USA* 86:7658–7662.

RYAN, M.R., B.G. ROOT, and P.M. MAYER. 1993. Status of the piping plover on the Great Plains of North America. *Conserv. Biol.* 7:581–585.

RYDER, O.A. 1986. Species conservation and systematics: the dilemma of subspecies. *Trends Ecol. Evol.* 1:9–10.

SAVIDGE, J.A. 1987. Extinction of an island forest avifauna by an introduced snake. *Ecology* 68:660–668.

SCOTT, J.M. and C.B. KEPLER. 1985. Distribution and abundance of Hawaiian native birds: a status report. *Bird Conserv.* 2:43–70.

SIBLEY, C.G. and J.E. AHLQUIST. 1982. Relationships of the Hawaiian honeycreepers (Drepaninini) as indicated by DNA–DNA hybridization. *Auk* 99:130–140.

SIGNER, E.N., C.R. SCHMIDT, and A.J. JEFFRIES. 1994. DNA variability and parentage testing in captive Waldrapp ibises. *Molec. Ecol.* 3:291–300.

STANGEL, P.W., M.R. LENNARTZ, and M.H. SMITH. 1992. Genetic variation and population structure of red-cockaded woodpeckers. *Conserv. Biol.* 6:283–292.

STANGEL, P.W., J.A. RODGERS, Jr., and A.L. BRYAN. 1990. Genetic variation and population structure of the Florida wood stork. *Auk* 107:614–619.

TARR, C.L. and R.C. FLEISCHER. 1993. Mitochondrial–DNA variation and evolutionary relationships in the Amakihi complex. *Auk* 110:825–831.

TEMPLETON, A.R. 1989. The meaning of species and speciation: a genetic perspective. In *Speciation and Its Consequences*, eds. D. Otte and J.A. Endler, pp. 3-27. Sinauer, Sunderland, MA.

TERBORGH, J. 1989. *Where Have All the Birds Gone?* Princeton University Press, Princeton, NJ.

TRIGGS, S.J., R.G. POWLESLAND, and C.H. DAUGHERTY. 1989. Genetic variation and conservation of Kakapo (*Strigops habroptilus*: Psittaciformes). *Conserv. Biol.* 3:92–96.

TRIGGS, S.J., M.J. WILLIAMS, S.J. MARSHALL, and G.K. CHAMBERS. 1992. Genetic structure of blue duck (*Hymenolaimus malacorhynchos*) populations revealed by DNA fingerprinting. *Auk* 109:80–89.

U.S. FISH AND WILDLIFE SERVICE. 1984a. Determination of endangered status for the Guam rail. 50 CFR Part 17. *Fed. Reg.* 49(71):14354–14356.

U.S. FISH AND WILDLIFE SERVICE. 1984b. Determination of endangered status for seven birds and two bats of Guam and the Northern Mariana Islands. 50 CFR Part 17. *Fed. Reg.* 49(167):33881–33885.

U.S. FISH AND WILDLIFE SERVICE. 1985. *Endangered Species Recovery Plan for the Red-Cockaded Woodpecker.* U.S. Fish and Wildlife Service, Atlanta, GA.

WALTERS, J.R. 1990. Red-cockaded woodpeckers: a "primitive" cooperative breeder. In: *Cooperative Breeding in Birds: Long-term Studies of Ecology and Behavior*, eds. P.B. Stacey and W.D. Koenig, pp. 69-101. Cambridge University Press, Cambridge, UK.

WARNER, R.E. 1968. The role of introduced diseases in the extinction of the endemic Hawaiian avifauna. *Condor* 70:101–120.

WENINK, P.W., A.J. BAKER, and M.G.J. TILANUS. 1993. Hypervariable-control-region sequences reveal global population structuring in a long-distance migrant shorebird, the dunlin (*Calidris alpina*). *Proc. Natl. Acad. Sci. USA* 90:94–98.

WENINK, P.W., A.J. BAKER, and M.G.J. TILANUS. 1994. Mitochondrial control region sequences in two shorebird species, the turnstone and the dunlin, and their utility in population genetic studies. *Molec. Biol. Evol.* 11:22–31.

WESTNEAT, D.F. 1990. Genetic parentage in the Indigo bunting: a study using DNA fingerprinting. *Behav. Ecol. Sociobiol.* 27:67–76.

WESTNEAT, D.F. 1993. Polygyny and extrapair fertilizations in eastern red-winged blackbirds (*Agelaius phoeniceus*). *Behav. Ecol.* 4:49–60.

WETTON, J.H., D.T. PARKIN, and R.E. CARTER. 1992. Use of genetic markers for parentage analysis in *Passer domesticus* (house sparrows). *Heredity* 69:243–254.

WILLIAMS, J.D. and R.M. NOWAK. 1987. Vanishing species in our own backyard: extinct fish and wildlife of the U.S. and Canada. In *The Last Extinction*, eds. L. Kaufman and K. Mallory, pp. 107–139. MIT Press, Cambridge, MA.

WILSON, E.O. 1992. *The Diversity of Life.* Belknap Press, Cambridge, MA.

WORLD CONSERVATION UNION. 1993. *1994 IUCN Red List of Threatened Animals.* World Conservation Union, Gland, Switzerland.

YAMAGISHI, S., I. NISHIUMI, and C. SHIMODA. 1992. Extrapair fertilization in monogamous bull-headed shrikes revealed by DNA fingerprinting. *Auk* 109:711–721.

7

CONSERVATION GENETICS
OF MARINE TURTLES

Brian W. Bowen and John C. Avise

INTRODUCTION

The first known turtles adapted for a marine existence appear as fossils
dating to 150 million years ago (Pritchard, 1979). Today, seven extant
species of marine turtles are recognized. These organisms provide model
subjects for genetic analyses in a conservation context. First, all species are
listed as threatened or endangered (Box 7.1), and they are quintessential
subjects of public conservation awareness and concern. Second, turtles tend
to be morphologically conservative, and as a result numerous phylogenetic
questions of relevance to management programs persist, especially at lower
taxonomic levels. Third, the long generation lengths, oceanic habits, and
protracted migrations of marine turtles make these animals difficult to
observe directly, thus concealing several life-history components that are
relevant to conservation strategies. Indeed, turtle biologists take pride in
developing ingenious methods to uncover otherwise inaccessible aspects of
sea turtle biology. For example, reproductive histories of females are deduced from laparoscopic examination of oviducts (Limpus and Reed, 1985);
migratory pathways are chronicled from tag recoveries and from the composition of epibiota attached to turtle carapaces (Caine, 1986; Eckert and

---

Box 7.1. Taxonomic and Conservation Status of Marine Turtles

The current taxonomy for extant marine turtle species (superfamily Chelonioidea) is as follows:

A. Family Dermochelyidae
   1. *Dermochelys coriacea* (leatherback)
B. Family Cheloniidae
   2. *Chelonia mydas* (green; a related form, *Chelonia agassizi* [black turtle], is of questionable species status; see text)
   3. *Caretta caretta* (loggerhead)

4. *Eretmochelys imbricata* (hawksbill)
5. *Lepidochelys kempi* (Kemp's ridley)
6. *Lepidochelys olivacea* (olive ridley)
7. *Natator depressus* (flatback)

All species regularly occurring in U.S. waters (numbers 1–5 above) are listed as threatened or endangered under the U.S. Endangered Species Act of 1973. All seven species are listed as threatened by the World Conservation Union (1993).

Eckert, 1988); and age and growth rates are deduced from patterns of bone deposition (Zug et al., 1986; Klinger and Musick, 1995). The genetic findings described in this chapter extend this tradition of scientific sleuthing into a molecular realm.

For reasons that will become apparent, assays of maternally inherited mitochondrial (mt) DNA have figured prominently in genetic studies of marine turtles. One unexpected generality to emerge from these and other molecular assays is that as a group, the Testudines (marine, freshwater, and terrestrial turtles) appear to be "conservative" in terms of molecular variation, typically exhibiting low levels of within-species diversity and low rates of molecular evolution relative to many other organismal groups. For example, allozyme surveys across global populations of the green turtle led Bonhomme et al. (1987) to conclude that the pace of protein evolution in *Chelonia mydas* is intrinsically low (but see Smith et al., 1977). Restriction-site and nucleotide sequence analyses at intraspecific and interspecific levels led Avise et al. (1992) and Bowen et al. (1993b) to conclude that mtDNA evolution in turtles proceeds at a several-fold lower rate than the "conventional" vertebrate pace (Brown et al., 1979; Wilson et al., 1985). Single-copy nuclear DNAs also show a pattern of low nucleotide diversity (Karl et al., 1992), and sequence regions flanking hypervariable microsatellite loci show extreme conservatism across turtle species (FitzSimmons et al., 1995). Chromosomal characters also are known to be conservative in the Testudines: in several taxa, chromosomal banding patterns are nearly unchanged over 200 million years (Bickham, 1981). Hybridization between relatively ancient (10–50 + million-year-old) marine turtle lineages (Wood et al., 1983; Conceicão et al., 1990; Karl et al., 1995) indicates that reproductive barriers evolve slowly in marine turtles as well. Hypotheses about the evolutionary mechanisms responsible for these patterns have been forwarded (Avise et al., 1992; Martin et al., 1992; Rand, 1994), but these need not concern us here. Notwithstanding the conservative nature of molecular and organismal evolution in marine turtles, numerous genetic polymorphisms have been uncovered. The serviceability of these molecular markers in the context of conservation biology is the focus of this chapter.

NATURAL HISTORY BACKGROUND

Variation within and among species exists in particular phases of marine turtle life history, but a generalized life cycle includes the following elements (Figure 7.1; Ernst and Barbour, 1989). Every few years, adult females migrate to a nesting locale that may be hundreds or thousands of kilometers from resident foraging grounds. The female ascends a sandy beach, excavates a nest, and lays 60–150 eggs. A female typically lays several clutches during the nesting season, and then returns to coastal or pelagic foraging areas. About 50–70 days after the nesting event, hatchlings emerge from the sand (typically *en masse*) and enter the ocean environment, which they will inhabit for the remainder of their lives, excepting the terrestrial forays of nesting females. Relatively little is known about the movements or habits of juvenile turtles, but the young of most species may remain in the pelagic realm for several years before recruiting onto coastal foraging grounds (Pritchard, 1976a; Carr, 1987; but see Walker and Parmenter, 1990). Males spend virtually their entire lives at sea. In at least some species, mating is believed to take place in courting areas adjacent to the nesting colony (Eckert and Eckert, 1988; Limpus et al., 1992), but it is possible that matings also take place in feeding areas or along migrational routes.

Gender is known to be influenced by the temperature of incubation, with lower temperatures producing a high proportion of males (Morreale et al., 1982; Mrosovsky et al., 1984). Marine turtles have exceptionally long generation lengths, and estimates of age at sexual maturity usually exceed 20 years (e.g., Limpus and Walter, 1980; Frazer and Ladner, 1986; Klinger and Musick, 1995).

Hatchlings appear to be nearly omnivorous (as indicated by the behavior of captive specimens) but the diets of adults are more specialized and differ among species— leatherback turtles feed on jellyfish, hawksbill turtles are spongivorous, green turtles are herbivorous, and ridley, loggerhead, and flatback turtles are mostly carnivorous. Feeding areas vary accordingly. Leatherbacks occur in the pelagic zone, hawksbills are associated with reefs and rocky areas, and green turtles are found primarily in shallow sea grass pastures along continental margins. The leatherback turtle is partially ho-

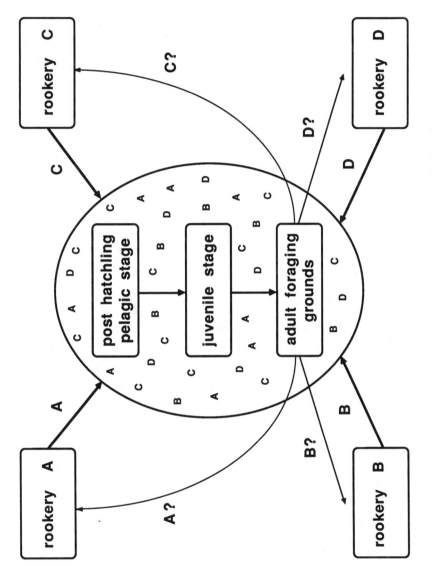

Figure 7.1. Schematic outline of the generalized marine turtle life cycle.

meothermic (Frair et al., 1972; Greer et al., 1973) and can forage in cold temperate waters as far north as the Gulf of Alaska. Loggerheads and Kemp's ridley turtles can feed in warm temperate zones, but greens, olive ridleys, and hawksbill turtles are almost exclusively tropical. The flatback turtle is confined to the coastal waters of northern Australia and New Guinea, but other species (with the exception of the Kemp's ridley— to be discussed) occur at suitable latitudes in the Atlantic, Indian, and Pacific Oceans.

GENETIC STRUCTURE OF ROOKERY POPULATIONS AND NATAL HOMING

Green Turtle

Females of this species (Box 7.2) typically migrate long distances between foraging and nesting sites. For example, females that nest on Ascension Island on the mid-Atlantic ridge utilize feeding grounds along the Brazilian coast, some 2,000 km away (Carr, 1975). Nesting is strongly colonial, with females often using specific beaches while adjacent habitat remains unvisited (Carr and Carr, 1972).

Over the past four decades, biologists have tagged tens of thousands of nesting females and used tag recoveries to deduce reproductive behavior and the migratory pathways of adults (Figure 7.2). The results demonstrate

---

Box 7.2. Conservation Concerns for the Green Turtle

When Christopher Columbus navigated the western Atlantic, green turtles were so abundant that they routinely collided with his boats, apparently causing consternation among the crew. Over the ensuing five centuries, numbers declined by about 99%, from an estimated 50 million adults to perhaps a few hundred thousand green turtles throughout the Caribbean today. By the end of the 17th century, an enormous rookery at Grand Cayman Island was extirpated by overharvesting, and many other rookeries throughout the world suffered a similar fate (Parsons, 1962; King, 1982).

The green turtle (named for the emerald tint of its muscle tissue) is prized for its meat, eggs, and for a gelatinous tissue (calipee) used in soup. The greatest population insults from human harvesting occur on nesting beaches, where green turtles are especially vulnerable (Frazier, 1975, 1979). However, excessive slaughter on feeding grounds can have significant population impacts as well. Other problems stemming from human activities involve degradation of nesting beaches and feeding habitats.

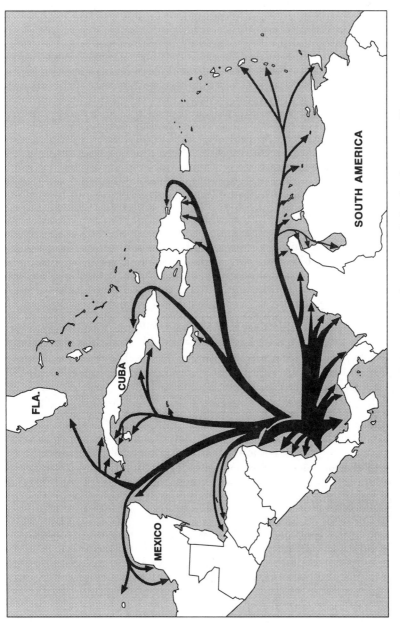

Figure 7.2. Migratory routes and foraging areas for green turtle females that nest at Tortuguero, Costa Rica, as revealed by recoveries of physical tags applied at the rookery (after Carr and Carr, 1972).

195

unequivocally that nesting females return with high fidelity to the same rookery in successive reproductive migrations (Carr and Ogren, 1960). Indeed, only two instances are known in which a green turtle switched nesting sites: One female tagged at the Aves Island rookery (200 km west of Dominica in the Caribbean) later was recovered nesting at Mona Island, Puerto Rico, 560 km distant (Kontos et al., 1988); and one female tagged on Tromelin Island in the Indian Ocean subsequently was observed nesting on Europa Island, over 2,000 km away (LeGall and Hughes, 1987).

The strong site fidelity of nesting females prompted Carr (1967) and others to propose that mature females return to nest on their natal beach. However, Hendrickson (1958; see also Owens et al., 1982) proposed an alternative scenario that also is consistent with nest-site fidelity by adult females. Under this "social facilitation" hypothesis, first-time breeders follow experienced females to a rookery site, and after a successful experience, fix on that site for future nestings. Another theoretical possibility (henceforth the "chance-encounter" hypothesis) is that first-time breeders "randomly" encounter a suitable beach (suitability might include the presence of nesting turtles), but again fix on that site for subsequent nesting. In principle, these hypotheses could be tested directly by recoveries of nesting adults tagged as young, but no physical tag has been developed that when applied to a 40-g hatchling persists for several decades in the marine environment during the maturation of a 150-kg adult (Carr, 1986). However, the "natal homing" hypothesis generates a testable prediction about population genetic structure that is not shared by the social facilitation or chance encounter models (Figure 7.3): Under strict natal homing, rookeries should be well differentiated with respect to female lineages.

To critically address these competing models of female migratory behavior, mtDNA haplotypes (as defined by restriction-site analyses) were assayed for individuals representing more than 225 nests at 15 major green turtle rookeries in the Atlantic, Indian, and Pacific Oceans (Figure 7.4). Genotypic distributions are summarized in Table 7.1.

A robust test of the natal homing versus social facilitation hypotheses requires nesting populations which are known to overlap on foraging grounds. Two well documented examples (based on tagging data) are known among genetically sampled green turtle colonies. The first involves the Ascension Island and Surinam nesting populations, which overlap extensively in foraging areas along the Brazilian coast (Figure 7.5). Among 50 assayed nests, no mtDNA genotypes were shared between these colonies (Figure 7.5). The second case involves Caribbean nesting populations that share feeding pastures in Nicaragua and elsewhere in the region. As indicated in Table 7.1, the two largest rookeries (Aves Island, Venezuela and Tortuguero, Costa Rica) also are characterized by nearly fixed differences in mtDNA genotypes (Meylan et al.,

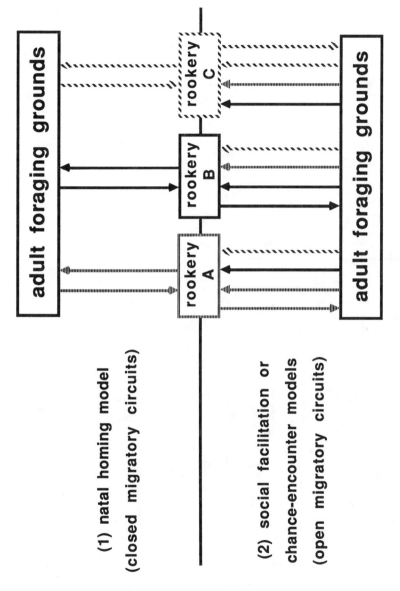

Figure 7.3. Schematic outline of alternative scenarios for the lifetime movements of marine turtle females. Closed, rookery-specific migratory circuits characterize the natal homing model, whereas open migratory circuits between rookeries are predicted under the social facilitation and chance-encounter models (see text).

197

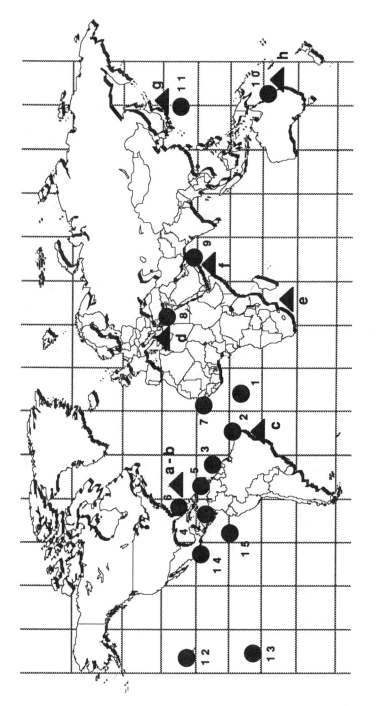

Figure 7.4. Rookery collection sites for green turtles (circles, numbered) and loggerheads (triangles, lettered) in mtDNA restriction-site surveys (Bowen et al., 1992, 1994b). Locales are as listed in Table 7.1.

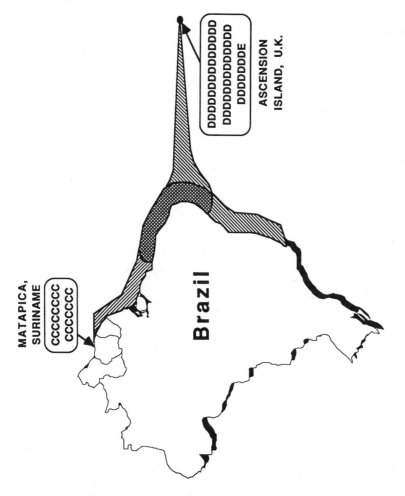

Figure 7.5. MtDNA genotypes observed in green turtles from Suriname ($n = 15$ nests) and Ascension Island ($n = 35$). Adult females that nest at these respective locales are known to overlap extensively in foraging ranges along the Brazilian coast, as indicated (Carr, 1975; Schulz, 1982; Mortimer and Carr, 1987).

Table 7.1. Numbers of Nests with Indicated mtDNA Genotypes Observed in Green Turtles and Loggerheads (from Restriction-site Surveys of Bowen et al., 1992, 1994b) [1]

1. Green turtle rookeries

mtDNA genotype

	A	B	C	D	E	F	G	H	I	J	K	L	M	N
Costa Rica (4)	15	–	–	–	–	–	–	–	–	–	–	–	–	–
Florida (6)	21	3	–	–	–	–	–	–	–	–	–	–	–	–
Venezuela (5)	1	–	7	–	–	–	–	–	–	–	–	–	–	–
Surinam (3)	–	–	–	15	–	–	–	–	–	–	–	–	–	–
Ascension (1)	–	–	–	34	1	–	–	–	–	–	–	–	–	–
Brazil (2)	–	–	–	15	–	1	–	–	–	–	–	–	–	–
Bissau (7)	–	–	–	–	–	–	13	–	–	–	–	–	–	–
Cyprus (8)	–	–	–	–	–	–	–	10	–	–	–	–	–	–
Oman (9)	–	–	–	–	–	–	–	–	15	–	–	–	–	–
Mexico (14)	–	–	–	–	–	–	–	–	7	–	–	–	–	–
Galapagos (15)	–	–	–	–	–	–	–	–	8	–	–	–	–	–
Hawaii (12)	–	–	–	–	–	–	–	–	–	6	16	–	–	–
Polynesia (13)	–	–	–	–	–	–	–	–	–	–	2	1	–	–
Australia (10)	–	–	–	–	–	–	–	–	–	–	–	15	–	–
Japan (11)	–	–	–	–	–	–	–	–	–	–	–	1	5	14

2. Loggerhead rookeries

mtDNA genotype

	A	B	C	D	E	F	G	H
USA, a (Georgia/South Carolina)	2	60	–	–	1	–	–	–
USA, b (Florida)	–	9	1	19	–	–	–	–
Brazil (c)	–	–	11	–	–	–	–	–
Greece (d)	–	–	–	21	–	–	–	–
South Africa (e)	–	–	–	15	–	–	–	–
Oman (f)	–	–	–	–	–	8	–	–
Australia (h)	–	–	–	–	–	–	14	–
Japan (g)	–	–	–	–	–	–	–	15

[1] Numbers and letters in parentheses refer to the locales mapped in Fig. 7.4. No mtDNA genotypes were shared between the species, notwithstanding the shared letter designations used in this table and chapter.

1990). These findings contradict the social facilitation scenario, which predicts extensive interchange of female lineages among rookeries.

This pattern of geographic structuring of maternal genotypes has also been reported in a survey of 10 Indo–Pacific rookeries based on mtDNA control-region sequences (Norman et al., 1994) and in reexaminations of Atlantic population structure with control-region sequences (Allard et al., 1994; Lahanas et al., 1994; Encalada, 1995).

Tests to distinguish between the natal homing and the chance-encounter hypotheses require comparisons among rookeries within an ocean basin, such that unescorted individuals are not geographically (physically) disbarred from encountering the relevant nesting beaches (whether or not foraging ranges overlap). Many such genetic comparisons are available within the Atlantic–Mediterranean and Indian–Pacific Oceans. Figure 7.6 summarizes estimates of interrookery gene flow of maternal lineages (expressed as Nm values) based on the observed mtDNA genotypic distributions (Table 7.1). Most such values are low (< 1.0). In general, because Nm values may be interpreted as estimates of the mean number of individuals exchanged between populations per generation (Slatkin, 1987), results indicate a strong propensity for philopatry to natal site by nesting females. In only a few cases were rookeries not cleanly distinguished by restriction fragment length polymorphism (RFLP) analyses of mtDNA (Table 7.1). Although these instances might reflect high maternal gene flow between particular rookery pairs, this explanation is contrary to the overall pattern of haplotype dis-

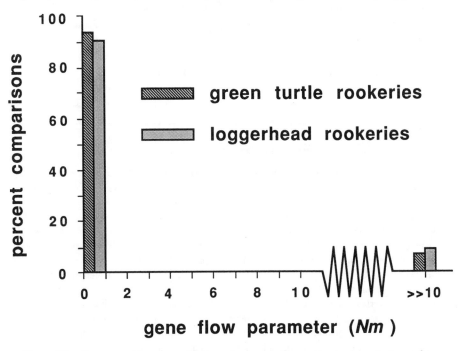

gene flow parameter (*Nm*)

Figure 7.6. Frequency histogram of interrookery gene flow estimates in pairwise rookery comparisons ($n = 43$ and $n = 21$ tests, respectively) between conspecific green turtle and loggerhead colonies *within* ocean basins, based on mtDNA (Bowen et al., 1992, 1994b). The Nm values were calculated by Slatkin's (1989) method (values in the mode to the right are arbitrary, merely indicating that the colonies were statistically indistinguishable in mtDNA genotypic frequencies). Similar estimates of matriarchal gene flow were obtained by the Takahata and Palumbi (1985) method.

tribution (Figure 7.6). More likely, the assays failed to detect genetic differences that do exist (see Allard et al., 1994; Encalada, 1995), and/or the colonies are tightly connected in a historical sense (to be discussed).

Loggerhead Turtle

Loggerhead hatchlings (Box 7.3) leave the nesting beach to occupy oceanic current systems (such as the North Atlantic Gyre), where they drift passively for a few years before recruiting to coastal neritic zones (Carr, 1986). Adult migrations between nesting and foraging areas range in length from tens to thousands of kilometers (Bjorndal et al., 1983; Limpus et al., 1992). Turtles from separate rookeries can overlap extensively on feeding areas (Meylan, 1982). Limpus et al. (1992) speculated that the resident foraging grounds of adults correspond to the first suitable habitat encountered upon completion of the juvenile life-history stages.

Tagging data indicate that the great majority (e.g., 98%) of adult females returns repeatedly to the same beach (Limpus, 1985), but a noticeable minority of individuals moves between nearby nesting locales both within and between nesting seasons. For example, Bjorndal et al. (1983) reviewed 25 cases of nest-site relocations in the southeastern United States, and

---

Box 7.3. Conservation Concerns for the Loggerhead Turtle

In most areas, loggerhead turtles have been spared from intensive human harvesting due to a lower palatability of *Caretta* meat. However, populations have been impacted substantially by fishing and shellfishing industries. For example, shrimping operations in the southeastern United States and Gulf of Mexico inadvertently drown many thousands of turtles each year (Ross, 1982; National Research Council, 1990). Such levels of mortality probably are not sustainable, and this has led to calls for wider adoption of fisheries procedures that reduce incidental turtle deaths (such as "turtle excluder devices" on trawl nets). Incidental capture in longline and driftnet fisheries is another source of mortality for juvenile loggerheads in the open ocean. For example, the longlines for swordfish in the Mediterranean capture an estimated 20,000 juvenile turtles per year, of which at least 20% perish (Groombridge, 1990; Aguilar et al., 1994). In the north and west Pacific, driftnet fisheries drown many thousands of turtles each year (Nishimura and Nakahigashi, 1990; Wetherall et al., 1993). Although a moratorium on use of driftnets in the north Pacific currently is in effect, a developing longline fishery (based primarily in Hawaii) soon may replace the driftnet fisheries as a major source of loggerhead turtle mortality.

Gyuris and Limpus (1988) reported 13 instances of female movement (79–105 km) between nesting beaches in Australia. LeBuff (1974) reported that one female tagged on a west Florida rookery was observed nesting on an east Florida rookery (550 km distant) four years later. It remains unclear from tagging data how much interrookery genetic exchange is mediated by such female movements, or whether long-distance dispersal events (e.g., across ocean basins) occur as well.

Rookeries of the southeastern United States are well suited for addressing these issues with genetic markers. This region hosts one of the largest reproductive aggregates of loggerheads in the world, with nesting unevenly and discontinuously distributed from Virginia to Texas. Southern Florida hosts about 90% of the nesting, and Georgia and South Carolina host most of the remaining 10% (Murphy and Hopkins-Murphy, 1989). Bowen et al. (1993a) assayed mtDNA genotypes from 92 loggerhead nests in these three states, and observed one genotype ("D") which, although present at frequencies of 67% and 64% in samples from east Florida ($n = 15$) and west Florida ($n = 14$), respectively, was absent from the Georgia ($n = 44$) and South Carolina ($n = 19$) collections. The significant genetic difference between the Florida and the Georgia–South Carolina samples indicates the existence of at least two nesting populations of loggerhead turtles in this region.

At face value, the pronounced mtDNA genetic structure among regional loggerhead populations appears inconsistent with the propensity for nest-site relocations as revealed in tag recoveries, because in theory the reciprocal exchange of even a few migrants per generation should be sufficient to maintain similar genotypic frequencies in populations at equilibrium between gene flow and genetic drift (Allendorf, 1983). One likely explanation for the enigma involves the geographic scale of observed nesting relocations, most of which involve proximate sites. For example, Richardson (1982) reported 20% crossover between adjacent nesting islands in Georgia, but only 2–4% crossover between more distant islands along the Georgia coast. Hence, the population genetic pattern may reflect a propensity for regional natal homing by loggerhead females, tempered by a behavioral plasticity in nest-site selection within that area. Natal homing (at least on a regional scale) also is indicated by genetic data from Indian–Pacific rookeries, where samples from all four assayed loggerhead colonies (Japan, Australia, South Africa, and Oman) exhibited fixed differences in mtDNA genotype (Bowen et al., 1994b; Table 7.1).

Tag-recapture studies indicate that the two genetically defined nesting populations in the southeastern United States tend to utilize different foraging areas (Richardson, 1982; Meylan et al., 1983). However, a genetic test of the social facilitation hypothesis requires the co-occurrence on foraging areas of turtles from separate rookeries. To address this issue, genetic comparisons of Mediterranean and western Atlantic rookeries proved appropriate. Loggerhead juveniles derived from west Atlantic nesting beaches probably inhabit the Mediterranean, as evidenced by the following observations: (1) More juvenile loggerheads occupy Mediterranean foraging grounds than can be generated by Mediterranean rookeries alone (Carr, 1987); (2) One turtle tagged in Texas subsequently was recovered in the Adriatic (Manzella et al., 1988); and (3) Movement of juveniles into the Mediterranean may be facilitated by the North Atlantic Gyre (which is known to carry juvenile west Atlantic loggerheads past the Azores). Groombridge (1990) elaborated on this scenario with the speculation that some juveniles from the west Atlantic become trapped in the Mediterranean and remain there to breed. Laurent et al. (1993) provided the first genetic data indicating that juveniles from the west Atlantic occupy feeding habitats in the Mediterranean: a mtDNA genotype observed in U.S. rookeries but absent in sampled Mediterranean rookeries was found in 22% of loggerhead specimens from feeding grounds in the western Mediterranean. Under these circumstances, a critical test of the natal homing versus social facilitation scenarios is possible. Based on restriction-site assays, Bowen et al. (1993a) showed that the largest Mediterranean rookery (Zakynthos, Greece) contained only one of the two mtDNA genotypes that dominate west Atlantic samples; estimates of cross-oceanic matriarchal exchange ranged from $Nm = 0.0–1.3$. These data are inconsistent with a major role for social facilitation in the nesting behavior of female loggerheads.

Hawksbill Turtle

Information on early life stages of hawksbill turtles (Box 7.4) is scarce, but neonates are assumed to occupy pelagic habitats, as do loggerhead and green turtles (Witzell, 1983). Hawksbill nesting populations are considered to be less colonial than other cheloniid turtles, but this may in part be an artifact of the steep decline in abundance in recent decades (Carr and Meylan, 1980). Early reports suggested that adults might be less migratory than other marine turtles, but more recent research has demonstrated that some females travel long distances between nesting seasons (Meylan, 1982). Tag recoveries indicate a high degree of adult nest-site fidelity (Diamond, 1976; Limpus et al., 1983). Indeed, one of the earliest reports of philopatry by a marine turtle involved a hawksbill turtle, tagged with a brass ring in 1794, which returned to the same nesting location in Sri Lanka for 32 years (Flowers, 1925).

Do female hawksbill turtles return to nest on their natal beach? Broderick et al. (1994) reported significant differences in mtDNA haplotype frequency

The shells of hawksbill turtles have been used for thousands of years to make "tortoiseshell" jewelry, and many populations have been depleted by demand for the highly prized shells. For example, possibly the largest rookery in the world, at Chiriqui, Panama, was extirpated in recent decades by harvesting of adult females (King, 1982). The international trade in Caribbean hawksbill was active until December 1992, with Cuba, Haiti, and Panama being the largest exporters, and Japan being the largest importer (Milliken and Tokunaga, 1987; Canin,

1989). International commerce was discontinued by Japan, but a thriving local trade persists in many areas, and efforts are being made to reopen the international market. Between 1970 and 1986, approximately 250 thousand Caribbean hawksbill turtles were harvested (Milliken and Tokunaga, 1987). As a result, less than 10,000 nesting females remain in the Caribbean region (Meylan, 1989), and the hawksbill turtle is second only to the Kemp's ridley (Box 7.6) in risk of extinction.

between nesting areas in northeastern and northwestern Australia. However, pairs of nesting beaches within these two regions were indistinguishable, consistent with tagging studies indicating that hawksbill turtles may move between adjacent nesting habitats within a particular region (Limpus et al., 1983). Bass (1994) documented significant mtDNA haplotype frequency shifts between five of six surveyed Caribbean rookeries that from further genetic considerations (Bowen et al., 1995b; see below) appear to overlap on feeding grounds. Taken together, these molecular data support a natal homing model for the reproductive migrations of hawksbill turtles.

Conservation Relevance

The mtDNA data for green turtles, loggerheads, and hawksbills are generally consistent with a natal homing model for female nesting behavior on a local or regional scale, and are inconsistent with extensive interrookery matrilineal gene flow (under either the social facilitation or chance-encounter models). The primary relevance of these findings to conservation programs is that *genetic* differentiation in mtDNA strongly implies *demographic* independence among rookeries, at least over ecological timescales relevant to species recovery plans (Avise, 1994, 1995). This low level of interrookery matrilineal exchange suggests that colonies overexploited by humans or extirpated by natural causes will not likely recover via natural recruitment of non-indigenous females over short ecological timeframes. This genetics-based inference is consistent with empirical experience— several rookeries extirpated by human activities over the past four centuries, including the huge green turtle rookery on Grand Cayman Island in the Caribbean, have yet to be recolonized (Parsons, 1962). The conservation implications are clear: Each remaining turtle rookery should be viewed for management purposes as an autonomous demographic entity.

This demographic autonomy of rookeries over ecological timescales holds regardless of the level of interrookery gene flow mediated by males and the mating system (i.e., if females mate with males from other rookeries). Two recent studies have addressed the issue of nuclear gene flow in these species. Gyuris and Limpus (1988) assayed allozyme allele frequencies among loggerhead rookeries in Queensland, Australia, and observed significant population genetic structure. Karl et al. (1992) assayed RFLP allele frequencies at several anonymous nuclear gene regions in green turtles and found significant genetic population structure within ocean basins. Nonetheless, the inferred Nm values were somewhat larger for these biparentally inherited markers than for mtDNA, implicating males as a vehicle for interrookery gene flow. However, even if nesting populations within an ocean basin were linked by frequent interrookery matings, a strong matrilineal population structure would imply a high degree of demographic independence among nesting populations, because each rookery's reproductive output is inextricably tied to the nesting success of indigenous females.

ROOKERY LIFE SPANS

Intraoceanic Genetic Patterns

Natal homing by female green turtles, loggerheads, and hawksbills cannot be perfect because rookeries for these species are scattered throughout available habitat in every ocean. How persistent are nesting beaches, and how old are individual colonies? Speculations about rookery age have varied by three or four orders of magnitude. Some present-day nesting colonies occur at sites suspected of being only a few thousand years old. For example, fossils dating to 1,100 years BP were reported from the active green turtle rookery at Raine Island, Australia (Limpus, 1987). The vertical margins of this coral cay almost certainly precluded nesting during lower sea levels associated with the Wisconsin glaciation (18,000–10,000 years BP), such that the colony must be between 1,000 and 10,000 years old. Climatic cooling associated with glacial advances no doubt contracted the northern and southern limits of turtle nesting range repeatedly during the Pleistocene, such that many present-day rookeries in higher latitudes (e.g., green turtle and loggerhead colonies in the southeastern United States and the Mediterranean) are probably less than 10,000 years old (Bowen et al., 1992; Encalada, 1995). At the other extreme, Carr and Coleman (1974) posited that the green turtle colony on Ascension Island has persisted for more than 70 million years.

This latter suggestion warrants elaboration. Ascension Island, situated just south of the equator on the mid-Atlantic ridge, is home to one of the most isolated green turtles rookeries in the world. Females that nest on Ascension utilize foraging pastures along the South American coast (Figure 7.5). Why do Ascension females undertake long-distance reproductive migrations that involve great exertional and navigational feats, when closer sites suitable for nesting are available (as judged by the presence of other green turtle rookeries on the coast of South America)? To account for this remarkable migratory behavior, Carr and Coleman (1974) hypothesized that the ancestors of Ascension green turtles colonized volcanic islands adjacent to South America in the late Cretaceous, soon after the opening of the equatorial Atlantic Ocean by continental drift. Over the ensuing 70 million years, these islands were gradually displaced from South America by seafloor spreading (at a rate of about two cm per year), and in response the turtles may have developed progressively longer migratory circuits. This innovative hypothesis hinges on several assumptions that were untested at the time: natal homing by females, the continuous availability of suitable nesting sites on proto-Ascension island(s), and an unbroken chain of ancestral–descendent populations on these islands.

The Carr–Coleman hypothesis was tested critically by Bowen et al. (1989) using mtDNA data. If the Ascension Island population evolved independently over tens of millions of years, its mitochondrial genomes should be nearly saturated with base-substitutional differences from those of other rookeries in the Atlantic. However, the magnitude of mtDNA sequence divergence between Ascension nesters and other Atlantic green turtles proved to be small; indeed, subsequent assays of a Brazilian rookery uncovered a mtDNA genotype indistinguishable in restriction-site assays (Table 7.1) and in control-region sequences (Encalada, 1995) from the most common Ascension genotype. Clearly, the colonization of Ascension Island (or at least extensive maternal lineage input into the population) has occurred in recent evolutionarily time.

Despite the significant differences in mtDNA genotypic frequency that characterize conspecific rookeries of assayed sea turtle species, the magnitudes of mtDNA sequence divergence typically remain small, particularly within ocean basins. For example, mtDNA haplotypes observed in green turtles from the Atlantic Ocean and Mediterranean differed by only zero to four mutational steps in restriction-site assays (Figure 7.7), and by zero to 12 nucleotide substitutions in control-region sequences (Encalada, 1995). These translate into estimates of sequence divergence ranging from

INDIAN-PACIFIC CLADE

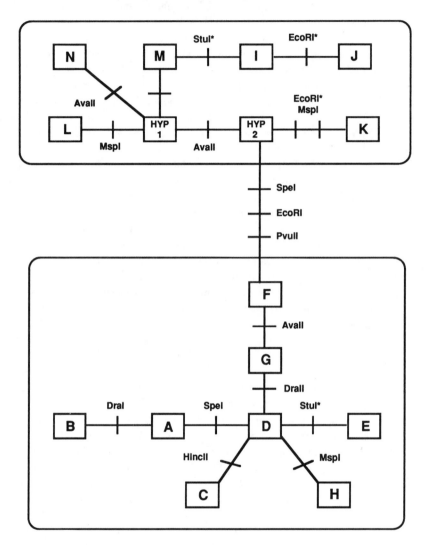

ATLANTIC-MEDITERRANEAN CLADE

Figure 7.7. Parsimony network summarizing interrelationships among mtDNA genotypes (Table 7.1) observed among green turtle rookeries worldwide (after Bowen et al., 1992). Restriction-site changes are indicated along branches, and stars indicate incidences of inferred homoplasy. HYP 1 and HYP 2 are hypothetical genotypes, not actually observed.

$p = 0.000–0.003$ for RFLPs (Bowen et al., 1992), and $p = 0.000–0.044$ for control-region sequences (Encalada, 1995). Based on tentative mtDNA clock calibrations for the Testudines formulated from RFLP data (0.2–0.4% sequence divergence between pairs of lineages per million years; Avise et al., 1992), these values suggest that most Atlantic–Mediterranean green turtles shared a common maternal ancestor within approximately the last one million years. Under "conventional" clock calibrations for vertebrate mtDNA, which are severalfold faster, the common ancestor dates approximately to within the last 100-200 thousand years. In any event, the mtDNA phylogeny for green turtles in the Atlantic–Mediterranean is evolutionarily "shallow." Similar conclusions apply to green turtles in the Indo–Pacific (Norman et al., 1994; Figure 7.7), to hawksbills in the Indo–Pacific (Broderick et al., 1994) and Caribbean (Bass, 1994), and to loggerhead turtles within the Atlantic and Pacific Oceans (Bowen et al., 1994b).

These small genetic distances between mtDNA lineages indicate that conspecific rookeries within an ocean basin do not constitute long-separated populational units. Bowen et al. (1989) suggested that appropriate nesting beaches must be ephemeral over evolutionary time, continually arising and disappearing with catastrophic events (such as hurricanes), and with changes in the physical and biotic environment (such as climate and sea level, presence of nest predators, competition for nesting space, or disease). Such environmental challenges inevitably displace or extinguish some rookeries, and this process must have been balanced by colonization events involving "mistakes" in female natal homing. Thus, a recurrent genetic restructuring of colonies is postulated to result from this process of extinction and colonization, the net effect being that most rookeries within an ocean remain evolutionarily close in terms of mitochondrial (and nuclear) DNA sequences. Notwithstanding a strong propensity for natal homing by females that produces striking geographic structure of maternal lineages at particular horizons in time, a process of intraoceanic "concerted" evolution has kept colonies tightly connected in a historical, genealogical sense.

Given the brief duration of marine turtle rookeries over geological time frames, absolute natal homing by females would be a sure formula for population extinction (Carr et al., 1978). Considering the ephemeral nature of rookeries, and the inevitable lapses in natal homing behavior required for rookery establishment, why then does any detectable genetic structure among colonies exist? Under neutrality theory, partitions in mtDNA gene genealogies are expected to become concordant with geographic partitions in population structure only after about $2N_{f(e)}$ generations of population isolation (Neigel and Avise, 1986; Pamilo and Nei, 1988), where $N_{f(e)}$ is the evolutionary effective population size of females. One consequence is that, all else being equal, longer generation length is expected to increase the sidereal time required for lineage-sorting processes to generate fixed mtDNA differences between rookeries. If we conservatively assume a generation length of 20 years for green and loggerhead turtles, and $N_{f(e)} = 1,000$ per rookery, then daughter colonies descended from an ancestral source population would require about 40,000 years on average to establish, through lineage-sorting processes, fixed mtDNA differences of the type commonly observed. These time intervals are probably considerably longer than the evolutionary life spans of most rookeries.

A resolution of this paradox is that values of $N_{f(e)}$ might be much lower than current rookery census sizes might otherwise suggest (see Avise et al., 1988). This would be the case, for example, if new nesting beaches were colonized by only one or a few gravid females. The low nucleotide diversities and small numbers of mtDNA genotypes observed in most surveyed rookeries can be construed as evidence in support of this possibility (Lahanas et al., 1994; Encalada, 1995). For example, in the global genetic survey of green turtles (Bowen et al., 1992), most rookeries exhibited either a single mtDNA haplotype, or only two haplotypes differing by a single restriction-site change (Table 7.1; Figure 7.7). Only the Japanese, Polynesian, and Venezuelan samples contained divergent mtDNA genotypes necessitating a hypothesis of multiple colonization events. Another possibility is that values of rookery $N_{f(e)}$ typically are low for reasons other than founder events, such as extreme fluctuations in population sizes of established colonies, or large variances among females in reproductive success. In either case, the sifting of mitochondrial lineages through relatively few females could induce detectable population genetic structure over short evolutionary timescales.

Conservation Relevance

In light of the ephemeral nature of nesting beaches, and the shallow evolutionary separations of conspecific turtle rookeries within ocean basins as revealed in genetic assays, it now appears unlikely that specific migratory routes are genetically programmed in marine turtles (this conclusion cannot be stated with finality, because extremely rapid evolution of route-specific migratory behavior has been demonstrated for some birds; Berthold, 1988). Although a predisposition to utilize environmental cues must surely have a genetic basis in marine turtles, the positional information essential for navigation to a particular locality probably is learned (imprinted) rather than inherited. The nature of these cues remains uncertain. Marine turtles possess a fine olfactory sense such that orientation by chemoreception has been proposed (Koch et al., 1969; Grassman et al., 1984), perhaps in concert with celestial, inertial, or other guidance systems (Carr et al., 1978; Owens et al., 1982). A geomagnetic navigational sense is strongly implicated for green and loggerhead turtles (Lohmann, 1992; Lohmann and Lohmann, 1994). Regardless of the orientation mechanisms, natal homing by imprinting permits a flexible response to altered nesting conditions, such that new migrational circuits could be established in a single generation. Imprinting on a new habitat requires no genetic modification, and is consistent with (if not necessary for) a successful colonization strategy.

One ramification of a learned rather than genetically ordained migratory circuit is that human-mediated translocation programs (designed to establish or augment foreign rookeries) are not necessarily doomed to failure because of site-specific genetic programming. Marine turtle biologists have long been intrigued by the prospect of relocating eggs or young from one beach to another, and a few such ventures (Carr, 1967) already have been undertaken (sometimes in conjunction with "headstart" programs; Box 7.5). On the other hand, the potential feasibility of a proposed translocation program does not alone justify this approach, and several additional considerations suggest caution (Frazer, 1992; Taubes, 1992). In general, undesirable consequences of geographic translocations include the following (see also Chapter 14): (1) the possibility of disease or parasite spread; (2) inevitable erosion of the overall genetic diversity within a species, much of which is known to be geographically partitioned; and (3) the irretrievable loss of historical genetic records of conspecific populations. Additional translocation concerns specific to marine turtles are outlined in Boxes 7.5 and 7.6.

Box 7.5. Headstarting

Headstarting describes the management program whereby turtle hatchlings are reared in captivity for at least several months before release, the intent being to reduce hatchling and post-hatchling mortality (which normally is high in nature). Despite decades of effort, no turtle of headstart origin has been recovered on a nesting beach (National Research Council, 1990). For this and other reasons, the approach is controversial. Detractors contend that headstart programs are: (1) expensive; (2) perhaps interfere with proper imprinting by juveniles to the nesting beach; (3) may produce young that are too naive to survive in the wild; and (4) seldom address the actual cause of wildlife depletion (Frazer, 1992; Bowen et al., 1994a). They also note that earlier headstart programs (before the temperature-dependent nature of sex determination in marine turtles was understood) sometimes were counterproductive by producing male-biased sex ratios (Shaver et al., 1988; Wibbels et al., 1989). Supporters of headstarting counter that: (1) survivors may have been missed or may yet nest at release sites; and (2) research and public-awareness components of headstarting programs generate substantial benefits (Caillouet, 1987).

Box 7.6. Conservation Concerns for the Kemp's Ridley Turtle

The Kemp's ridley is by far the most endangered of the marine turtles, with nesting confined to the western Gulf of Mexico, primarily in the state of Tamaulipas, Mexico. In the latter half of this century, numbers of nesters declined from tens of thousands in the 1940s to a few hundred in recent years (Woody, 1985), initially as a result of heavy harvesting of both eggs and nesting females during the 1950s. The egg trade was halted by Mexico in 1966 (Pritchard, 1969a), but a developing shrimp fishery then began to drown adults on coastal feeding grounds (Pritchard, 1976b; National Research Council, 1990). Turtle excluder devices (TEDs) now are required for shrimp trawls operating in U.S. territorial waters, but the level of compliance is uneven and TEDs are not required in Mexican waters.

In response to the decline of the Kemp's ridley, a cooperative headstart program (Box 7.5) between the United States and Mexico was initiated in 1978 (Caillouet, 1987). Over the ensuing 15 years, tens of thousands of headstarted hatchlings from the Tamaulipas rookery were released in Texas, in efforts to establish a new nesting colony. The program was not demonstrably successful with respect to this goal (National Research Council, 1990; Taubes, 1992), and for this and other reasons the program has been discontinued.

Another ramification of the observed intraocean matrilineal structure is that mtDNA markers may be used to assess the rookery source of turtles captured on feeding grounds or migratory corridors (Norman et al., 1994). For example, four of the largest loggerhead rookeries in the Indian–Pacific region appear to be distinguished by diagnostic restriction-site differences (Table 7.1), and this opens an opportunity to identify the origin of turtles in foraging areas far removed from the nesting beaches, as described in the next section.

GENETIC COMPOSITION OF POPULATIONS ON FEEDING GROUNDS AND MIGRATORY ROUTES

Whereas mtDNA markers have illuminated the population structure of nesting colonies, the genetic composition of feeding populations is poorly understood. Resident foraging grounds may be hundreds or thousands of kilometers from nesting habitat (Pritchard, 1976a; Mortimer and Carr, 1987), and these spatial scales can tend to obscure the migrational links between specific breeding populations and feeding aggregates (but see Limpus et al., 1992). Thus, questions of how nesting and foraging populations are related has represented a significant gap in the scientific understanding of marine turtle natural history. Links between nesting populations and juvenile assemblages have been especially problematic because juveniles are hard to track, tag, and capture in the marine environment. These issues relate directly to population demography (Avise, 1995), and also raise an important conservation concern. Thousands of sea turtles are drowned each year in commercial fisheries (National Research Council, 1990; Wetherall et al., 1993; Aguilar et al., 1994). Does such activity impact the few remaining large and resilient breeding populations, or does this mortality draw from smaller nesting colonies that may be dwindling toward extinction?

To determine the origin of feeding populations, marine turtle researchers have borrowed an approach from fishery management. In this method (previously applied, for example, to anadromous salmon; see Chapter 8), differences in allele frequencies among breeding stocks are used to estimate the relative contributions of spawning populations to a mixed-stock fishery. As discussed previously, rookery-specific mtDNA genotypes often characterize marine turtle populations, and these "private" markers can be especially useful for elucidating the relative contribution of candidate rookeries to turtle populations on feeding grounds or migratory corridors (Avise and Bowen, 1994; Broderick et al., 1994). Even when nesting populations share genetic markers but in different frequencies, rookery contributions to particular non-nesting locales can be estimated using statistical procedures such as maximum likelihood (Pella and Milner, 1987; Millar, 1987) as implemented in available computer programs (Masuda et al., 1991; Xu et al., 1994).

Juvenile Green Turtles

Based on mtDNA control-region sequences from the major nesting populations in the western Atlantic (Encalada, 1995), researchers assessed the rookery origins of juvenile green turtles ($n = 80$) in a major foraging area near Great Inagua, Bahamas (Lahanas et al., 1995). Maximum likelihood analyses indicate that juvenile green turtles recruit to this feeding area from nesting populations throughout the greater Caribbean region, in cohorts roughly proportional to the size of contributing rookeries. The authors interpreted this pattern as suggestive evidence of extensive mixing during the pelagic post-hatching stage.

Juvenile Loggerhead Turtles

Juvenile loggerheads occur from New England to the Caribbean, but are particularly abundant in bays and estuaries of the southeastern United States. Inadvertent mortality from commercial fishing and shipping may be considerable, and represents a major conservation concern. To assess the rookery origin of juveniles in the vicinity of Charleston, South Carolina, Sears et al. (1995) utilized RFLP markers observed in the Florida and Georgia–South Carolina nesting populations. Endemic mtDNA haplotypes from both nesting populations were observed among the juveniles in the entrance channel to Charleston Harbor, in frequencies indicating a disproportionate contribution by the northern rookeries. This finding is consistent with tag-return data demonstrating that adult females from the northern nesting assemblage tend to preferentially utilize feeding grounds along this Atlantic coastline (Bell and Richardson, 1978; Meylan et al., 1983).

As discussed earlier, several lines of evidence suggest that neonates from the west Atlantic rookeries recruit to juvenile feeding grounds in the Mediterranean. Laurent et al. (1993) used cytochrome *b* sequences from mtDNA to test this possibility. A haplotype endemic to west Atlantic nest samples was present in 22% of turtles ($n = 59$) captured in the Spanish longline fishery in the western Mediterranean (Figure 7.8; Box 7.3). Based on these genetic data, a maximum likelihood estimate of stock composition indicates that about 57% of the juveniles in the western Mediterranean are derived from west Atlantic nesting beaches (Bowen, 1995).

Juvenile loggerheads recently have been reported in large numbers off the coast of Baja California (Pitman, 1990), and were recently discovered in the north Pacific driftnet fishery (Wetherall et al., 1993). This distribution is remarkable because loggerhead nesting is unknown in the central and eastern Pacific (Ross, 1982; Frazier, 1985). The nearest nesting colonies for this species are in Japan and Australia, over 10,000 km from Baja California. Earlier researchers speculated that the North Pacific Current passively transports hatchlings from Japan to the feeding areas in the eastern Pacific, but this hypothesis is supported by only a single tag recovery (Uchida and Teruya, 1991). To test this hypothesis, and to document the migratory pathways of juveniles, Bowen et al. (1995a) used rookery-specific markers derived from a 350 bp fragment of the mtDNA control-region sequence. Results indicate that the Japanese nesting beaches contributed about 95% (57 of 60) of the juveniles sampled in the north Pacific driftnet fishery and on the Baja feeding grounds (Figure 7.9). The remaining individuals (three of 60) possessed a mtDNA genotype characteristic of the Australian nesting beaches. This suggests a small but detectable contribution of Southern-Hemisphere rookeries to the north Pacific population of juvenile loggerheads. In this case, genetic markers may document one of the longest migrations for any marine vertebrate.

Hawksbill Turtles

Broderick et al. (1994) used the mtDNA genotypic distributions observed in four Australian nesting populations to assess the origins of hawksbill

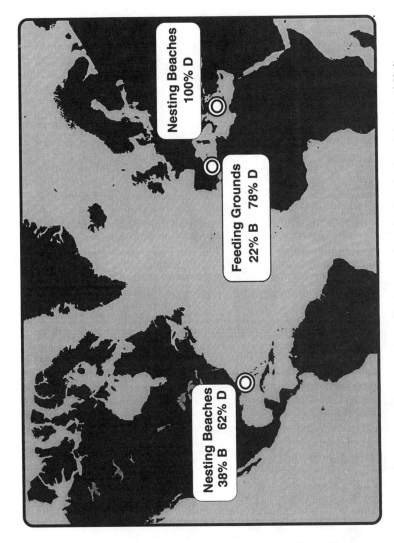

Figure 7.8. Distribution of mtDNA haplotypes in loggerhead turtles from Atlantic and Mediterranean nesting areas, and a Mediterranean feeding ground (Bowen et al., 1993a; Laurent et al., 1993). For simplicity, haplotypes that occur in less than 5% of nesting samples are not included. Maximum likelihood analysis indicates that approximately 57% of turtles in this western Mediterranean feeding area are derived from west Atlantic nesting beaches (Bowen, 1995).

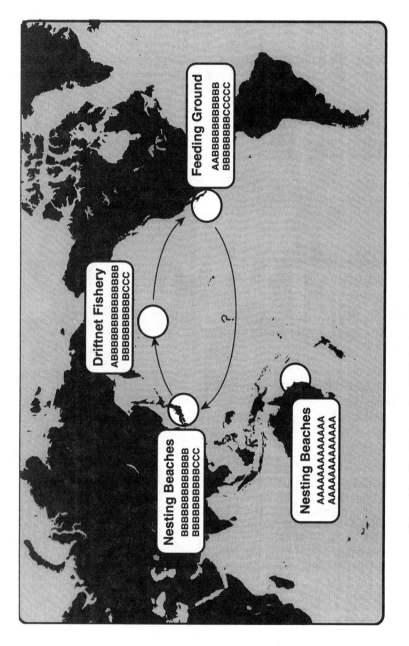

Figure 7.9. Distribution of mtDNA haplotypes in loggerhead turtles from Pacific nesting beaches, the north Pacific driftnet fishery, and feeding grounds adjacent to Baja California (Bowen et al., 1995a). Maximum likelihood analysis indicates that approximately 95% of driftnet mortalities and Baja feeding individuals are derived from Japanese nesting beaches.

turtles on two feeding areas in northern Australia. Both foraging areas were characterized by haplotype frequencies that differed significantly from those of neighboring rookeries (approximately 500 to 1,000 km distant). The authors concluded that substantial recruitment into these foraging areas must have occurred from more distant nesting sites.

Bowen et al. (1995b) used mtDNA control-region sequence data from seven west Atlantic nesting populations (Bass, 1994) to estimate the contribution of regional rookeries to a foraging area at Mona Island, Puerto Rico ($n = 41$). Results indicated that this feeding ground is not composed primarily of turtles from the adjacent nesting beach on Mona Island, but is occupied by turtles from nesting populations throughout the Caribbean region. Highly divergent haplotypes from a south Atlantic nesting population (Bahia, Brazil; Bass, 1994) were not detected in the Caribbean feeding population. From these data, Bowen et al. (1995b) concluded that hawksbill turtles recruit to feeding grounds on a scale greater than 100 km, but less than the 7,000 km that separate Mona Island from Bahia.

Conservation Relevance

The feeding grounds and migrational corridors of marine turtles are subject to human and other impacts, and inherited genetic markers allow researchers to identify the natal source of turtles in affected areas. Such determinations have legal implications. In the language of international conservation, *range states* are nations impacted by the consumption of natural resources at a remote location. The 1982 United Nations (U.N.) Convention on the Law of the High Seas recognizes that nations which host the developmental habitat for a species (anadromous salmon in the original case) hold exclusive fishing rights for these animals in the open ocean (Van Dyke, 1993); and the 1983 U.N. Convention on the Conservation of Migratory Species prohibits the taking of migratory endangered species on the high seas. Under these principles as applied to marine turtles, countries that host nesting and developmental habitats have some level of jurisdiction over fisheries impacting "their" turtles. Genetic markers may eventually provide a basis for international agreements defining acceptable mortality in coastal and oceanic fisheries.

Both tagging studies and genetic analyses indicate that overlap on regional feeding grounds may be a general feature of marine turtle population biology. Conservation programs for marine turtle foraging habitat will require a regional perspective that recognizes how several nesting populations can be jointly impacted by human activities (or other sources of mortality) during non-nesting phases of the life cycle (see Ottenwalder and Ross, 1992). This aspect of demographic interconnection among rookeries via mortality on shared foraging grounds or migration routes contrasts with the demographic autonomy of rookeries with respect to reproduction.

GLOBAL PHYLOGEOGRAPHY
AND INTRASPECIFIC SYSTEMATICS

Genetic Patterns

For marine organisms that produce passively transported pelagic young (and/or in which adults move extensively), genetic populations often are recognizable on huge spatial scales commensurate with oceanic basins and with the environmental boundaries that define species' distributions (Chapter 11). Marine turtles provide a counterpoint to this pattern. Whereas pelagic-stage greens, loggerheads, and hawksbill turtles may drift for years in oceanic currents, most mature individuals apparently return to reproduce on or near a natal beach. This migratory behavior contributes to the intraoceanic population structure of these species and counteracts the genetic consequences of dispersal at early life-history stages. Do physical barriers to dispersal also play a role in shaping the population genetic structures of marine turtles?

Because green turtles are adapted primarily to tropical conditions, populations in the Atlantic presumably are isolated by geographic barriers from those in the Pacific at the present time. When were these populations last connected? The global mtDNA survey of green turtle rookeries by Bowen et al. (1992) provides some clues. A striking feature of the mtDNA data proved to be the phylogenetic grouping of observed genotypes into two assemblages that correspond precisely to major ocean basins: (1) the Atlantic Ocean and Mediterranean Sea; and (2) the Indian and Pacific Oceans. This interoceanic distinction was evident in both parsimony (Figure 7.7) and distance (Figure 7.10) analyses. The estimate of net mtDNA sequence divergence (after correction for within-ocean variability) was $p = 0.006$, a value which translates to about 1.5–3.0 million years of population separation under the aforementioned clock calibrations for Testudines.

This phylogeographic pattern (Avise et al., 1987) is consistent with the geographic and climatic boundaries that currently define green turtle distributions. Populations from these two major ocean basins probably are isolated from one another by the cold temperate conditions around the southern tips of Africa and South America, whereas no obvious physical barriers to turtle movement exist among rookeries within the Atlantic–Mediterranean or within the Indian–Pacific basins. Thus, mtDNA data indicate that the geographic partitioning of the world's oceans by continental land-

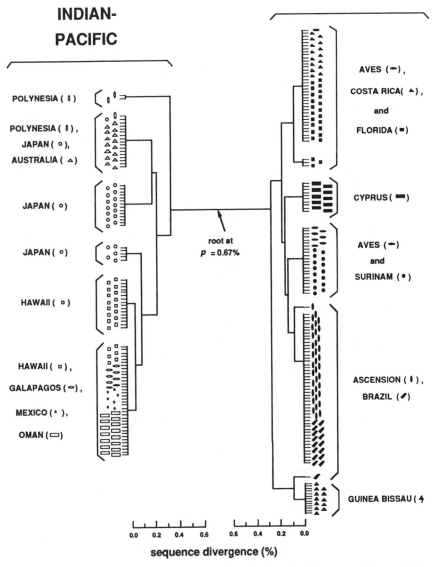

Figure 7.10.   Cluster phenogram summarizing relationships among 226 sampled nests of the green turtle (after Bowen et al., 1992). To conserve space, this is presented on mirror-image divergence axes centered around the root leading to two distinct clonal lineages confined to the two major ocean basins.

masses has been of overriding significance in shaping the global matriarchal phylogeny of *C. mydas*.

In global genetic surveys of the loggerhead turtle (Bowen et al., 1994b), a primary topological feature of the intraspecific mtDNA phylogeny was again a bifurcation: Two distinct mtDNA lineages (Figure 7.11) differed at a mean level of sequence divergence ($p = 0.009$) nearly identical to that noted previously for the green turtle. However, in this case, the phylogenetic partition does not correspond to geographic partitions, because both logger-head mtDNA lineages were observed in the Atlantic and Indian Oceans.

In contrast to tropical green turtles, loggerheads nest primarily in warm temperate regimes (e.g., an Indian Ocean rookery at Natal, South Africa is within 1,000 km of the south Atlantic). This temperate distribution may have facilitated transfers of matrilines around southern Africa in recent times. One possible scenario is as follows (Bowen et al., 1994b). During cooler periods of the Pleistocene, perhaps loggerhead populations were isolated by geography and climate into Atlantic and Indian–Pacific Ocean basins, leading to evolution of the two distinct mtDNA lineages. Subsequently, warmer temperatures associated with interglacial periods allowed an expansion of loggerhead habitat to higher latitudes, thereby opening a temperate corridor around southern Africa. At least some mtDNA lineage transfers must have been recent, as judged by the similarity of RFLP genotypes shared between the ocean basins. An alternative possibility— that both major mtDNA lineages were retained as global polymorphisms for much longer periods of time— cannot be excluded completely. However, recent interoceanic exchange of matrilines is strongly implicated by the close genetic affinity of mtDNA genotypes in separate oceans.

The temperate distribution of loggerhead turtles apparently has not led to an "antitropical" phylogeographic pattern. Thus, not only do the mtDNA lineage distributions suggest interoceanic gene flow when climate and geography permit, but they also indicate that this species can colonize readily across the equator.

Conservation Relevance

Good taxonomies, including those at the intraspecific level, are among the essential foundations of conservation practice (Avise, 1989; Daugherty et al., 1990). Although mtDNA analyses incorporate but a minute fraction of the phylogenetic legacy of a species, it is of interest to compare observed patterns of mtDNA against inferred patterns of organismal evolution based on morphology and biogeography.

Green turtle colonies show much variability in details of ethology and morphology (e.g., Carr, 1980; Pritchard, 1980; Mrosovsky, 1983). Nesting

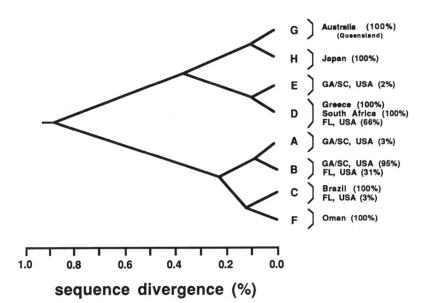

sequence divergence (%)

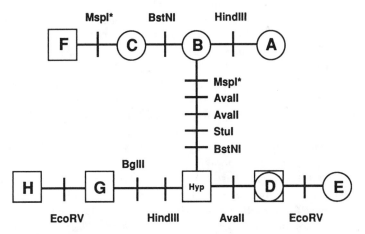

Figure 7.11. Phylogenetic relationships among mtDNA genotypes in loggerhead turtles as deduced from restriction-site surveys (Bowen et al., 1994b). *Above:* Cluster phenogram (genotypes labeled as in Table 7.1), with rookery sources (and frequencies) indicated. *Below:* maximum parsimony network (one of two equally parsimonious alternatives), showing inferred restriction-site changes along branches. Circles designate Atlantic–Mediterranean genotypes, squares denote Indian–Pacific genotypes, stars indicate an instance of presumed homoplasy, and HYP is a hypothetical genotype not observed.

seasons differ widely in timing and duration, even among rookeries in the same region. Some locales support both nesting and foraging aggregates, and non-migratory populations may exist at these sites. With respect to morphology, south Atlantic nesters are notably larger than Caribbean nesters, which are larger than those in the east Pacific and Mediterranean Sea. Green turtles also differ in features such as carapace shape and coloration (Frazier, 1971). When describing such differences between rookeries and regional populations, earlier researchers pondered the extent to which these might reflect evolutionary genetic divergence (Carr and Goodman, 1970). Opinions ranged the gamut from speculation that no evolutionary distinction exists between populations (even in separate ocean basins), to the possibility that nearly every rookery is a distinct evolutionary entity (Mrosovsky, 1983). Subspecific status has been proposed for numerous regional forms of the green turtle, including populations in the Caribbean (*C.m. viridis*), south Atlantic (*C.m. mydas*), Indo–west Pacific (*C.m. japonica*), Gulf of California (*C.m. carrinegra*), and east Pacific (*C.m. agassizi*). The latter is most consistently recognized taxonomically, and indeed often is accorded full species status as the "black turtle," *C. agassizi*. In reviewing the status of these various taxonomic units, Carr (1967) concluded that throughout the Pacific and Indian Oceans, *Chelonia* is a taxonomic shambles.

This taxonomic chaos contrasts with the relatively straightforward phylogeographic patterns revealed in mtDNA. Genetic structure of rookeries along matrilines certainly does exist within oceans, but appears evolutionarily shallow; and, a relatively deep genetic separation distinguishes Atlantic–Mediterranean from Indian–Pacific Ocean rookeries. Notably, this partition is consistent with recognized domains in marine zoogeography (Briggs, 1974). If evolutionary patterns in mtDNA alone were to be used as a taxonomic guide, then the *Chelonia* complex might reasonably be divided into two ocean-confined subspecies, with additional population-level differentiation appreciated within each ocean basin. However, it is inadvisable to rest taxonomic conclusions on a single line of evidence, or in this case on a single gene genealogy. Kamezaki and Matsui (1995) support subspecies status for the black turtle, based on morphometric analysis of skull characters from across the range of *C. mydas*. Clearly, additional lines of evidence (including those from nuclear genes; see Karl et al. 1992) should be gathered before final taxonomic conclusions are drawn for the *Chelonia* complex.

These data for green turtles illustrate a distinction between *management units* (MUs) and *evolutionarily significant units* (ESUs) as defined by Moritz (1994a, 1994b). Based on the mtDNA data, nesting populations within an ocean basin are effectively independent from one another demographically (albeit closely linked phylogenetically), and hence probably qualify as MUs. The Atlantic–Mediterranean and Indian–Pacific assemblages appear to be phylogenetically distinct at a much greater evolutionary depth (perhaps most of the Pleistocene), and hence may qualify as ESUs.

For the loggerhead turtle, subtle morphological differences prompted earlier recognition of Atlantic and Indian–Pacific subspecies (*C.c. caretta* and *C.c. gigas*, respectively), but recent reviews tend to reject these subspecific assignments on the grounds of overall morphological similarity between loggerhead turtles on a global scale (Pritchard and Trebbau, 1984; Dodd, 1988). The mitochondrial data are consistent with this latter interpretation in the sense that some closely related mtDNA genotypes are shared between oceans (Bowen et al., 1994b).

Another conservation-relevant byproduct of global phylogenetic surveys of marine turtles is the identification of genetic markers for forensic investigations and law enforcement. Genetic markers provided by mtDNA can reveal the source (species and region of origin) of marine turtle specimens or derived products. For example, Encalada et al. (1994) used mtDNA polymorphisms to determine the region of origin of a green turtle surrendered by a cargo vessel in the Port of San Francisco. The boat had traveled from west Africa through the Panama Canal, and up the west coast of Central America; mtDNA sequence information determined that the turtle had been captured in the southern or eastern Atlantic. Based on these data, personnel from the National Marine Fisheries Service were able to plan for repatriation of this animal. In general, forensic applications are particularly powerful when coupled with polymerase chain reaction (PCR) methods that can recover sequence information from processed or partially degraded material (see Chapter 2). Such approaches demonstrate the desirability of range-wide genetic surveys of endangered species that are subject to human exploitation and international trade.

SPECIES-LEVEL AND HIGHER SYSTEMATICS

Ridley Turtles

These species (Box 7.6) display a unique *arribada* behavior in which large numbers of females nest synchronously. Animals aggregate off the nesting beach for days or weeks, and then come ashore *en masse* to lay eggs. The resulting near-simultaneous hatching of millions of eggs (some 50–55 days later) may serve to reduce hatchling mortality by saturating terrestrial and aquatic predators. However, this ridley nesting behavior also makes rookeries highly susceptible to human exploitation.

Two species of ridley turtles are recognized: (1) the olive ridley, which occurs throughout the Indian, Pacific, and Atlantic Oceans (excepting the northwest Atlantic); and (2) the Kemp's ridley, which nests primarily at one site in the western Gulf of Mexico and feeds along the coastal margins of the north Atlantic (Box 7.6). According to Carr (1967), these distributions "make no sense at all under modern conditions of climate and geography." Furthermore, a close morphological similarity between ridley forms has raised questions regarding their phylogenetic distinctiveness. Garman (1880) first described the Kemp's ridley as a taxonomic entity; Loveridge and Williams (1957) applied the subspecies designations *Lepidochelys olivacea kempi* and *L.o. olivacea*; and Pritchard (1969b) endorsed specific status for the two forms, based on differences in shell configuration, skull morphology, and scute counts.

To further address systematic issues in the ridley complex, Bowen et al. (1991) compared mtDNA restriction-fragment profiles between the Kemp's ridley and olive ridleys from the Atlantic (Suriname) and Pacific (Costa Rica) coasts of the Americas. Results were consistent with current taxonomy. Olive ridleys from the two oceans proved to be closely related, whereas the Kemp's ridleys differed from the olive ridleys at a mean level of mtDNA sequence divergence ($p = 0.012$) greater than *any* genetic distances observed among conspecific populations of either green turtles or loggerheads around the world (Figure 7.12). As expected, the genetic distance between the Kemp's and olive ridleys was much lower than that between ridleys and loggerheads (mean $p = 0.04$; Figure 7.12). All of these conclusions were bolstered by parallel findings based on mitochondrial cytochrome *b* sequences (Bowen et al., 1993b).

Pritchard (1969b) presented a biogeographic scenario for the ridley complex that may explain the observed genetic patterns in mtDNA (see also Hughes, 1972; Hendrickson, 1980). Based on a synthesis of distributional and morphological information, Pritchard suggested that an ancestral ridley population was sundered by rise of the Isthmus of Panama into proto-Kemp's and proto-olive forms in the Atlantic and Pacific basins, respectively, and that only recently have olive ridleys colonized the Atlantic Ocean via the Cape of Good Hope. Based on one aforementioned mtDNA molecular clock calibration for Testudines (0.4% sequence divergence per million years), the observed Kemp's ridley–olive ridley genetic distance translates to about three million years of separation, a time span consistent with the known rise of the Panamanian Isthmus. Furthermore, genetic identity of Atlantic and Pacific samples of the olive ridley at more than 100 assayed mtDNA restriction sites suggests that the proposed secondary colonization of the Atlantic occurred within the last 100 thousand years. These findings support Pritchard's (1969b) biogeographic scenario, but analyses of additional olive ridley rookeries (in the Indian Ocean and elsewhere) will be necessary to resolve the global phylogeography of this species.

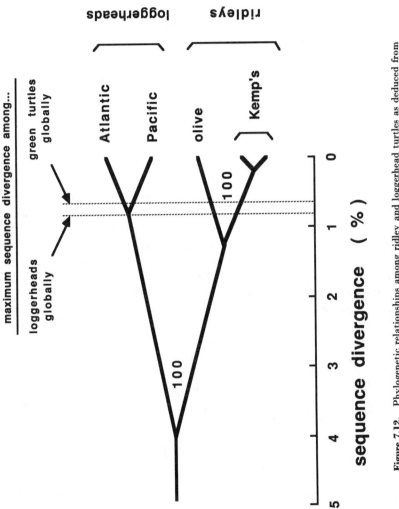

Figure 7.12. Phylogenetic relationships among ridley and loggerhead turtles as deduced from restriction-site comparisons of mtDNA (after Bowen et al., 1991). Shown is a cluster phenogram, with topology identical to that of a parsimony network whose bootstrap values are indicated.

Higher–Level Phylogenetic Relationships

Controversial areas also exist in the higher systematics of marine turtles. Among the most problematic questions are the following (see Bowen et al., 1993b):

1. Is the leatherback the sister taxon to the cheloniid turtles? (One alternative possibility is that turtles with marine adaptations arose more than once in evolution.)
2. Is the flatback turtle a close phylogenetic ally of the green (as a former placement within *Chelonia* might suggest), or is it related to loggerheads and ridleys?
3. Is the spongivorous hawksbill allied more closely to herbivorous green turtles, or to the carnivorous loggerheads?

These questions fall outside the appropriate window of resolution for conventional mtDNA restriction-site assays, but they have been addressed through direct sequencing of 503 bp fragments of the mitochondrial cytochrome *b* gene (Bowen et al., 1993b). A consensus phylogenetic summary of results is presented in Figure 7.13, from which were drawn the following provisional answers to the questions posed herein. Consistent with current classification (Box 7.1), the leatherback proved to be the most genetically divergent of the marine turtles (but whether it is the sister taxon to other marine species could not be decided conclusively from available evidence). The flatback turtle proved quite distinct from all other species, supporting recent removal from the genus *Chelonia* and resurrection of the genus *Natator* (Limpus et al., 1988). The flatback turtle appeared about equidistant genetically from the green turtles (tribe Chelonini; $p = 0.109$) and from the loggerhead–ridley complex (tribe Carettini; $p = 0.108$), a finding consistent with the proposal of a third tribe, Natatorini, within the family Cheloniidae (Zangerl et al., 1988). The hawksbill turtle was allied with the loggerhead–ridley complex rather than the green turtles, a finding that is supported by immunological assays (Frair, 1979). This grouping implies that the spongivorous dietary habit of the hawksbill evolved from a carnivorous rather than herbivorous ancestor (Bowen et al., 1993b).

Conservation Relevance

No arbitrary level of genetic divergence can define biological species boundaries, particularly among allopatric forms such as the Kemp's and olive ridleys. Furthermore, no realignment of taxonomy can be definitive with a single-character system such as mtDNA. Nonetheless, genetic data can be quite useful for elucidating the evolutionary relationships upon which

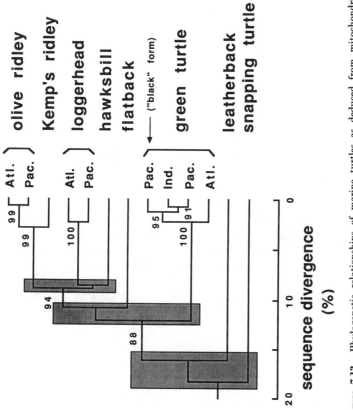

Figure 7.13. Phylogenetic relationships of marine turtles as deduced from mitochondrial cytochrome *b* sequences (after Bowen et al., 1993b). Shown is a composite summary (the basic framework of which is a cluster phenogram) that also shows putative clades (and associated bootstrap values) recognized in a parsimony analysis. Shaded boxes encompass nodes that were not consistently resolved (i.e., < 85% bootstrapping support under parsimony, or those that varied across data analysis methods). The non-marine snapping turtle (*Chelydra serpentina*) was included as a putative outgroup.

taxonomies and management programs ultimately rest, especially when interpreted in conjunction with biogeography, morphology, and other lines of evidence. Under the U.S. Endangered Species Act (and parallel international regulations), legal protection in principle can be extended to organisms that are distinct at the species, subspecies, or population levels. In practical terms, however, most (ca. 80%) of the entities added to the list for U.S. federal protection have been recognized as full species (Wilcove et al., 1993). Thus, the perceived taxonomic or evolutionary "uniqueness" of organisms is unquestionably an implicit factor in establishing conservation priorities.

The international program to protect the Kemp's ridley (Box 7.6) represents the largest conservation effort for any marine turtle, yet as described previously, taxonomic boundaries and evolutionary relationships in the ridley complex were controversial. Genetic data from mtDNA helped corroborate the suspected evolutionary distinctiveness of the Kemp's ridley, and thereby bolstered the rationale for special conservation efforts.

At higher taxonomic levels, phylogenetic uniqueness has been proposed as a basis for prioritizing taxa with regard to conservation value, under the rationale that distinct phylogenetic lineages make a disproportionate contribution to overall evolutionary diversity (Vane-Wright et al., 1991). If strictly applied, this philosophy would dictate, for example, that a higher conservation priority be extended to the leatherback turtle than to the Kemp's ridley, because the latter belongs to a recent and more speciose clade (Figure 7.13). On the other hand, to assign lower conservation priorities to extant species of recent origin could be counterproductive, especially if these recent lineages are a potential source of future biodiversity (Erwin, 1991). Clearly, other factors such as risk of extinction, biogeographic isolation, ecological role, evolutionary history, and economic importance must be considered in establishing conservation priorities.

SUMMARY

1.   Decade-long generation lengths, oceanic habits, and long-distance migrations have inhibited direct field studies of marine turtles, and a general morphological conservatism has obscured phylogenetic relationships within and among the seven extant species, all of which are listed as threatened or endangered. In recent years, molecular genetic studies have illuminated several aspects of marine turtle natural history and evolution that are of relevance to conservation programs.

2.   Based on striking differences in the frequency of mtDNA genotypes among conspecific nesting populations of green, loggerhead, and hawksbill

turtles, interrookery exchange of maternal lineages appears to be low. Results are consistent with natal homing by females, and inconsistent with high gene flow under models of social facilitation or chance rookery encounter. Genetic differentiation in maternal lineages implies a demographic independence of rookeries (over ecological timescales), such that for management purposes, each nesting population should be viewed as effectively autonomous with respect to female reproduction.

3. Low magnitudes of mtDNA sequence divergence among rookery-specific genotypes indicate that regional nesting colonies are tightly connected phylogenetically. This "concerted" evolution presumably is attributable at least in part to rookery extinctions and colonizations over short evolutionary timeframes. Results imply that the *particular* migratory routes employed by natally homing females probably result not from genetic programming, but rather from imprinting on locale-specific environmental cues. Thus, translocation programs for eggs and hatchlings, designed to establish or augment rookeries, are not inevitably doomed to failure on the basis of genetic programming. However, several other biological considerations diminish the rationale for translocation strategies in marine turtle management.

4. Observed differences in mtDNA genotype between rookeries provide a genetic basis for assigning individuals captured on feeding grounds or migrational routes to their natal site. Such assignments can carry legal and jurisdictional ramifications of conservation relevance. Furthermore, several genetic studies indicate that turtles from multiple rookeries can co-occur on particular feeding areas or migrational routes. Thus, with respect to mortality factors during non-nesting periods, regional rookeries can be demographically connected. An effective conservation program may require a regional perspective that recognizes how multiple rookeries can be jointly impacted by human activities (or other sources of mortality) in coastal and oceanic marine turtle habitats.

5. On a global scale, green turtle populations display a relatively deep phylogenetic partition in mtDNA that corresponds to a historical geographic separation between tropical waters in the Atlantic–Mediterranean versus the Indian–Pacific Oceans. Loggerhead turtles lack this phylogeographic pattern, presumably because of recent interoceanic gene flow around South Africa. Results are relevant to intraspecific taxonomies of these species, which remain controversial.

6. Genetic distinctiveness of the Kemp's ridley population from related olive ridleys has bolstered the rationale underlying special conservation efforts for this species. Genetic assays at higher taxonomic levels also have contributed to an understanding of phylogenetic relationships among marine

turtles, and have raised issues regarding phylogenetic uniqueness as a basis for conservation priority.

REFERENCES

AGUILAR, R., J. MAS, and X. PASTOR. 1994. Impact of Spanish swordfish longline fisheries on the loggerhead sea turtle *Caretta caretta* population in the western Mediterranean. In *Proceedings 12th Annual Symposium on Sea Turtle Biology and Conservation*, comps. T.H. Richardson and J.I. Richardson, pp. 91–96. NOAA Tech. Memo. NMFS-SEFC. National Technical Information Service, Springfield, VA.

ALLARD, M.W., M.M. MIYAMOTO, K.A. BJORNDAL, A.B. BOLTEN, and B.W. BOWEN. 1994. Support for natal homing in green turtles from mitochondrial DNA sequences. *Copeia* 1994:34–41.

ALLENDORF, F.W. 1983. Isolation, gene flow, and genetic differentiation among populations. In *Genetics and Conservation*, eds. C.M. Schonewald-Cox, S.M. Chambers, B. MacBryde, and L. Thomas, pp. 51–65. Benjamin/Cummings, London.

AVISE, J.C. 1989. A role for molecular genetics in the recognition and conservation of endangered species. *Trends Ecol. Evol.* 4:279–281.

AVISE, J.C. 1994. *Molecular Markers, Natural History and Evolution.* Chapman & Hall, New York.

AVISE, J.C. 1995. Mitochondrial DNA polymorphism and a connection between genetics and demography of relevance to conservation. *Conserv. Biol.* 9:686–690.

AVISE, J.C., J. ARNOLD, R.M. BALL, E. BERMINGHAM, T. LAMB, J.E. NEIGEL, C.A. REEB, and N.C. SAUNDERS. 1987. Intraspecific phylogeography: the mitochondrial DNA bridge between population genetics and systematics. *Annu. Rev. Ecol. Syst.* 18:489–522.

AVISE, J.C., R.M. BALL, Jr., and J. ARNOLD. 1988. Current versus historical population sizes in vertebrate species with high gene flow: a comparison based on mitochondrial DNA lineages and inbreeding theory for neutral mutations. *Molec. Biol. Evol.* 5:331–344.

AVISE, J.C. and B.W. BOWEN. 1994. Investigating sea turtle migrations using DNA markers. *Curr. Opin. Genet. Devel.* 4:882–886.

AVISE, J.C., B.W. BOWEN, T. LAMB, A.B. MEYLAN, and E. BERMINGHAM. 1992. Mitochondrial DNA evolution at a turtle's pace: evidence for low genetic variability and reduced microevolutionary rate in the Testudines. *Molec. Biol. Evol.* 9:457–473.

BASS, A.L. 1994. Conservation genetics of the hawksbill marine turtle, *Eretmochelys imbricata*, in the Caribbean and West Atlantic. MS thesis, Louisiana State University, Baton Rouge.

BELL, R. and J.I. RICHARDSON. 1978. An analysis of tag recoveries from loggerhead sea turtles (*Caretta caretta*) nesting on Little Cumberland Island, Georgia. *Flor. Mar. Res. Publ.* 3:1–66.

BERTHOLD, P. 1988. Evolutionary aspects of migratory behavior in European warblers. *J. Evol. Biol.* 1:195–209.

BICKHAM, J.W. 1981. Two-hundred-million-year-old chromosomes: deceleration of the rate of karyotypic evolution in turtles. *Science* 212:1291–1293.

BJORNDAL, K.A., A.B. MEYLAN, and B.J. TURNER. 1983. Sea turtle nesting at Melbourne Beach, Florida: I. Size, growth, and reproductive biology. *Biol. Conserv.* 26:65–77.

BONHOMME, F., S. SALVIDIO, A. LeBEAU, and G. PASTEUR. 1987. Comparison génétique des tortues vertes (*Chelonia mydas*) des Oceans Atlantique, Indien et Pacifique. *Genetica* 74:89–94.

BOWEN, B.W. 1995. Voyages of the ancient mariners: tracking marine turtles with genetic markers. *BioScience, in press.*

BOWEN, B.W., F.A. ABREU-GROBOIS, G.H. BALAZS, N. KAMEZAKI, C.J. LIMPUS, and R.J. FERL. 1995a. Trans-Pacific migrations of the loggerhead turtle (*Caretta caretta*) demonstrated with mitochondrial DNA markers. *Proc. Natl. Acad. Sci. USA* 92:3731–3734.

BOWEN, B.W., J.C. AVISE, J.I. RICHARDSON, A.B. MEYLAN, D. MARGARITOULIS, and S.R. HOPKINS-MURPHY. 1993a. Population structure of loggerhead turtles (*Caretta caretta*) in the northwestern Atlantic Ocean and Mediterranean Sea. *Conserv. Biol.* 7:834–844.

BOWEN, B.W., A.L. BASS, A. GARCIA-RODRIGUEZ, A. BOLTEN, C.E. DIEZ, R. VAN DAM, K.A. BJORNDAL, and R.J. FERL. 1995b. Origin of hawksbill turtles (*Eretmochelys imbricata*) in a Caribbean feeding area, as indicated by mtDNA sequence analysis. *Ecol. Appl., in press.*

BOWEN, B.W., T.A. CONANT, and S.R. HOPKINS-MURPHY. 1994a. Where are they now? The Kemp's ridley headstart project. *Conserv. Biol.* 8:853–856.

BOWEN, B.W., N. KAMEZAKI, C.J. LIMPUS, G.R. HUGHES, A.B. MEYLAN, and J.C. AVISE. 1994b. Global phylogeography of the loggerhead turtle (*Caretta caretta*) as indicated by mitochondrial DNA genotypes. *Evolution* 48:1820–1828.

BOWEN, B.W., A.B. MEYLAN, and J.C. AVISE. 1989. An odyssey of the green turtle: Ascension Island revisited. *Proc. Natl. Acad. Sci. USA* 86:573–576.

BOWEN, B.W., A.B. MEYLAN, and J.C. AVISE. 1991. Evolutionary distinctiveness of the endangered Kemp's ridley sea turtle. *Nature* 352:709–711.

BOWEN, B.W., A.B. MEYLAN, J. PERRAN ROSS, C.J. LIMPUS, G.H. BALAZS, and J.C. AVISE. 1992. Global population structure and natural history of the green turtle (*Chelonia mydas*) in terms of matriarchal phylogeny. *Evolution* 46:865–881.

BOWEN, B.W., W.S. NELSON, and J.C. AVISE. 1993b. A molecular phylogeny for marine turtles: trait mapping, rate assessment, and conservation relevance. *Proc. Natl. Acad. Sci. USA* 90:5574–5577.

BRIGGS, J.C. 1974. *Marine Zoogeography.* McGraw-Hill, New York.

BRODERICK, D., C. MORITZ, J.D. MILLER, M. GUINEA, R.I.T. PRINCE, and C.J. LIMPUS. 1994. Genetic studies of the hawksbill turtle *Eretmochelys imbricata*: evidence for multiple stocks in Australian waters. *Pacif. Conserv. Biol.* 1:123–131.

BROWN, W.M., M. GEORGE, Jr., and A.C. WILSON. 1979. Rapid evolution of animal mitochondrial DNA. *Proc. Natl. Acad. Sci. USA* 76:1967–1971.

CAILLOUET, C.W., Jr. 1987. *Reports on Efforts to Prevent Extinction of Kemp's Ridley Sea Turtles Through Head Starting.* NOAA Tech. Memo. NMFS-SEFC-188. National Technical Information Service, Springfield, VA.

CAINE, E.A. 1986. Carapace epibionts of nesting loggerhead sea turtles: Atlantic coast of U.S.A. *J. Exp. Mar. Biol. Ecol.* 95:15–26.

CANIN, J. 1989. International trade in sea turtle products. In *Proceedings 9th Annual Workshop on Sea Turtle Biology and Conservation*, comps. S.A. Eckert, K.L. Eckert, and T.H. Richardson, pp. 27–29. NOAA Tech. Memo. NMFS-SEFC-232. National Technical Information Service, Springfield, VA.

CARR, A. 1967. *So Excellent a Fishe: A Natural History of Sea Turtles.* Scribner, New York.

CARR, A. 1975. The Ascension Island green turtle colony. *Copeia* 1975:547–555.

CARR, A. 1980. Some problems of sea turtle ecology. *Amer. Zool.* 20:489–498.

CARR, A. 1986. Rips, FADS, and little loggerheads. *BioScience* 36(2):92–100.

CARR, A. 1987. New perspectives on the pelagic stage of sea turtle development. *Conserv. Biol.* 1:103–121.

CARR, A. and M.H. CARR. 1972. Site fixity in the Caribbean green turtle. *Ecology* 53:425–429.

CARR, A., M.H. CARR, and A.B. MEYLAN. 1978. The ecology and migrations of sea turtles: 7. The west Caribbean green turtle colony. *Bull. Amer. Mus. Nat. Hist.* 162:1–46.

CARR, A. and P.J. COLEMAN. 1974. Seafloor spreading theory and the odyssey of the green turtle from Brazil to Ascension Island, Central Atlantic. *Nature* 249:128–130.

CARR, A.F. and D. GOODMAN. 1970. Ecologic implications of size and growth in *Chelonia*. *Copeia* 1970:783–786.

CARR, A. and A. MEYLAN. 1980. Extinction or rescue for the hawksbill. *Oryx* 15:449–450.

CARR, A. and L. OGREN. 1960. The ecology and migrations of sea turtles, IV: the green turtle in the Caribbean Sea. *Bull. Amer. Mus. Nat. Hist.* 121:1–48.

CONCEIÇÃO, M.B., J.A. LEVY, L.F. MARINS, and M.A. MARCOVALDI. 1990. Electrophoretic characterization of a hybrid between *Eretmochelys imbricata* and *Caretta caretta* (Cheloniidae). *Comp. Biochem. Physiol.* 97B:275–278.

DAUGHERTY, C.H., A. CREE, J.M. HAY and M.B. THOMPSON. 1990. Neglected taxonomy and continuing extinctions of tuatara (*Sphenodon*). *Nature* 347:177–179.

DIAMOND, A. 1976. Breeding biology and conservation of hawksbill turtles, *Eretmochelys imbricata* L., on Cousin Island, Seychelles. *Biol. Conserv.* 9:199–215.

DODD, C.K., Jr. 1988. Synopsis of the biological data on the loggerhead sea turtle *Caretta caretta* (Linnaeus, 1758). *U.S. Fish and Wildl. Serv. Biol. Rept.* 88(14):1–110. National Technical Information Service, Springfield, VA.

ECKERT, K.L. and S.A. ECKERT. 1988. Pre-reproductive movements of leatherback sea turtles (*Dermochelys coriacea*) nesting in the Caribbean. *Copeia* 1988:400–406.

ENCALADA, S.E., S.A. ECKERT, and B.W. BOWEN. 1994. Forensic applications of mitochondrial DNA markers: origin of a confiscated green turtle. *Mar. Turt. Newsl.* 66:1–3.

ENCALADA, S.E. 1995. Phylogeography and conservation genetics of the green sea turtle (*Chelonia mydas*) in the Atlantic Ocean and Mediterranean Sea: a mtDNA control region sequence assessment. MS thesis, University of Florida, Gainesville.

ERNST, C. and R.W. BARBOUR. 1989. *Turtles of the World*. Smithsonian Institution Press, Washington, DC.

ERWIN, T. 1991. An evolutionary basis for conservation strategies. *Science* 253:750–752.

FITZSIMMONS, N.N., C. MORITZ, and S.S. MOORE. 1995. Conservation and dynamics of microsatellite loci over 300 million years of marine turtle evolution. *Molec. Biol. Evol.* 12:432–440.

FLOWERS, S. 1925. Contributions to our knowledge of the duration of life in vertebrate animals: 3. Reptiles. *Proc. Zool. Soc. Lond.* 37:911–981.

FRAIR, W. 1979. Taxonomic relations among sea turtles elucidated with serological tests. *Herpetologica* 35:239–244.

FRAIR, W., R.G. ACKMAN, and N. MROSOVSKY. 1972. Body temperature of *Dermochelys coriacea*: warm turtle from cold water. *Science* 177:791–793.

FRAZER, N.B. 1992. Sea turtle conservation biology and halfway technology. *Conserv. Biol.* 6:179–184.

FRAZER, N.B. and R.C. LADNER. 1986. A growth curve for green sea turtles, *Chelonia mydas*, in the U.S. Virgin Islands. *Copeia* 1986:798–802.

FRAZIER, J. 1971. Observations on sea turtles at Aldabra Atoll. *Phil. Trans. Roy. Soc. Lond.* B260:373–410.

FRAZIER, J. 1975. Marine turtles in the western Indian Ocean. *Oryx* 13:162–175.

FRAZIER, J. 1979. Marine turtle management in the Seychelles: a case-study. *Environ. Conserv.* 6:225–230.

FRAZIER, J. 1985. Misidentification of sea turtles in the East Pacific: *Caretta caretta* and *Lepidochelys olivacea*. *J. Herpetol.* 19:1–11.

GARMAN, S. 1880. On certain species of Chelonioidae. *Bull. Mus. Comp. Zool. Harvard* 6:123–126.

GRASSMAN, M.A., D.W. OWENS, J.P. MCVEY, and R. MÁRQUEZ. 1984. Olfactory-based orientation in artificially imprinted sea turtles. *Science* 224:83–84.

GREER, A.E., J.D. LAZELL, and R.M. WRIGHT. 1973. Anatomical evidence for counter-current heat exchanger in the leatherback turtle *Dermochelys coriacea*. *Nature* 144:181.

GROOMBRIDGE, B. 1990. Marine turtles in the Mediterranean; distribution, population status, conservation. *Report to the Council of Europe Environment Conservation and Management Division, Nature and Environment Series 48.* Strasbourg, France.

GYURIS, E. and C.J. LIMPUS. 1988. The loggerhead turtle *Caretta caretta* in Queensland: population breeding structure. *Aust. J. Wildl. Res.* 15:197–209.

HENDRICKSON, J.R. 1958. The green sea turtle, *Chelonia mydas* (Linn.) in Malaya and Sarawak. *Proc. Zool. Soc. Lond.* 130:455–535.

HENDRICKSON, J.R. 1980. Ecological strategies of sea turtles. *Amer. Zool.* 20:597–608.

HUGHES, G.R. 1972. The olive ridley sea-turtle (*Lepidochelys olivacea*) in South-east Africa. *Biol. Conserv.* 4:128–134.

KAMEZAKI, N. and M. MATSUI. 1995. Geographic variation in skull morphology of the green turtle, *Chelonia mydas*, with a taxonomic discussion. *J. Herpetol.* 29:51–60.

KARL, S.A., B.W. BOWEN, and J.C. AVISE. 1992. Global population structure and male-mediated gene flow in the green turtle (*Chelonia mydas*): RFLP analyses of anonymous nuclear DNA regions. *Genetics* 131:163–173.

KARL, S.A., B.W. BOWEN, and J.C. AVISE. 1995. Hybridization among the ancient mariners: characterization of marine turtle hybrids with molecular genetic assays. *J. Heredity* 86:262–268.

KING, F.W. 1982. Historical review of the decline of the green turtle and hawksbill. In *Biology and Conservation of Sea Turtles*, ed. K.A. Bjorndal, pp. 183–188. Smithsonian Institution Press, Washington, DC.

KLINGER, R.C. and J.A. MUSICK. 1995. Age and growth of loggerhead turtles (*Caretta caretta*) from Chesapeake Bay. *Copeia* 1995:204–209.

KOCH, A.L., A. CARR, and D.W. EHRENFELD. 1969. The problem of open-sea navigation: the migration of the green turtle to Ascension Island. *J. Theoret. Biol.* 22:163–179.

KONTOS, A., S. ECKERT, K. ECKERT, J.L. GOMEZ, R. LEE, and R. VAN DAM. 1988. Inter-island migration of nesting green turtle, *Chelonia mydas*. *Mar. Turt. Newsl.* 42:10–11.

LAHANAS, P.N., K.A. BJORNDAL, A.B. BOLTEN, S. ENCALADA, M.M. MIYAMOTO, and B.W. BOWEN. 1995. Genetic composition of a green turtle feeding ground population: evidence for multiple origins. (Submitted).

LAHANAS, P.N., M.M. MIYAMOTO, K.A. BJORNDAL, and A.B. BOLTEN. 1994. Molecular evolution and population genetics of Greater Caribbean green turtles (*Chelonia mydas*) as inferred from mitochondrial DNA control region sequences. *Genetica*. 94:57–67.

LAURENT, L., J. LESCURE, L. EXCOFFIER, B.W. BOWEN, M. DOMINGO, M. HADJICHRISTOPHOROU, L. KORNARAKI, and G. TRABUCHET. 1993. Genetic studies of relationships between Mediterranean and Atlantic populations of loggerhead turtle *Caretta caretta* (Cheloniidae). Implications for conservation. *Compt. Rend. Acad. Sci., Paris* 316:1233–1239.

LEBUFF, C.R., Jr. 1974. Unusual nesting relocation in the loggerhead turtle, *Caretta caretta*. *Herpetologica* 30:29–31.

LEGALL, J.-Y. and G.R. HUGHES. 1987. Migration de la tortue verte *Chelonia mydas* dans l'Ocean Indien Sud-Ouest observees a partir des marquages sur les sites de pontes Europa e Tromelin (1970–1985). *Amphib.–Reptil.* 8:277–282.

LIMPUS, C.J. 1985. A study of the loggerhead sea turtle, *Caretta caretta*, in Eastern Australia. PhD dissertation, University of Queensland, St. Lucia, Queensland.

LIMPUS, C.J. 1987. A turtle fossil on Raine Island, Great Barrier Reef. *Search* 18:254–256.

LIMPUS, C.J., E. GYURIS, and J.D. MILLER. 1988. Reassessment of the taxonomic status of the sea turtle genus *Natator* McCulloch, 1908, with redescription of the genus and species. *Trans. Royal Soc. S. Aust.* 112:1–9.

LIMPUS, C.J., J.D. MILLER, V. BAKER, and E. MCLACHLAND. 1983. The hawksbill turtle, *Eretmochelys imbricata* (L.), in north-eastern Australia: the Campbell Island rookery. *Austr. J. Wildl. Res.* 10:185–197.

LIMPUS, C.J., J.D. MILLER, C.J. PARMENTER, D. REIMER, N. MCLACHLAND, and R. WEBB. 1992. Migration of green (*Chelonia mydas*) and loggerhead (*Caretta caretta*) turtles to and from eastern Australian rookeries. *Wildl. Res.* 19:347–358.

LIMPUS, C.J. and P.C. REED. 1985. The green turtle *Chelonia mydas* in Queensland: population structure in a coral reef feeding ground. In *Biology of Australasian Frogs and Reptiles*, eds. G.C. GRIGG, R. SHINE, and H. EHMANN, pp. 342–351. Surrey Beatty Sydney, Australia.

LIMPUS, C.J. and D.G. WALTER. 1980. The growth of immature green turtles (*Chelonia mydas*) under natural conditions. *Herpetologica* 36:162–165.

LOHMANN, K. 1992. How sea turtles navigate. *Sci. Amer.* 266(1):100–106.

LOHMANN, K.J. and C.M.F. LOHMANN. 1994. Detection of magnetic inclination angle by sea turtles: a possible mechanism for determining latitude. *J. Exp. Biol.* 194:23–32.

LOVERIDGE, A. and E.E. WILLIAMS. 1957. Revision of the African tortoises and turtles of the suborder Cryptodira. *Bull. Mus. Comp. Zool. Harvard* 115:163–557.

MANZELLA, S.A., C.T. FONTAINE, and B. SCHROEDER. 1988. Loggerhead sea turtle travels from Padre Island, Texas to the mouth of the Adriatic Sea. *Mar. Turt. Newsl.* 42:7.

MARTIN, A.P., G.J.P. TAYLOR, and S.R. PALUMBI. 1992. Rates of mitochondrial DNA evolution in sharks are slow compared with mammals. *Nature* 357:153–155.

MASUDA, M., S. NELSON, and J. PELLA. 1991. *User Manual for GIRLSEM, GIRLSYM, and CONSQRT.* U.S.–Canada Salmon Program, National Marine Fisheries Service, Juneau, AK.

MEYLAN, A.B. 1982. Sea turtle migration— evidence from tag returns. In *Biology and Conservation of Sea Turtles*, ed. K.A. BJORNDAL, pp. 91–100. Smithsonian Institution Press, Washington, DC.

MEYLAN, A.B. 1989. Status report of the hawksbill turtle. In *Proceedings 2nd Western Atlantic Turtle Symposium*, eds. L. Ogren, F. Berry, K. Bjorndal, H. Kumpf, R. Mast, G. Medina, H. Reichart, and R. Witham, pp. 101–115. NOAA Technical Memorandum NMFS-SEFC-226. National Technical Information Service, Springfield, VA.

MEYLAN, A.B., K.A. BJORNDAL, and B.J. TURNER. 1983. Sea turtles nesting at Melbourne Beach, Florida: II. Post-nesting movement of *Caretta caretta*. *Biol. Conserv.* 26:79–90.

MEYLAN, A.B., B.W. BOWEN, and J.C. AVISE. 1990. A genetic test of the natal homing versus social facilitation models for green turtle migration. *Science* 248:724–727.

MILLAR, R.B. 1987. Maximum likelihood estimation of mixed stock fishery composition. *Can. J. Fish. Aquat. Sci.* 44:583–590.

MILLIKEN, T. and H. TOKUNAGA. 1987. *Japanese Sea Turtle Trade, 1970–1986.* Center for Marine Conservation, Washington, DC.

MORITZ, C. 1994a. Application of mitochondrial DNA analysis in conservation: a critical review. *Molec. Ecol.* 3:401–411.

MORITZ, C. 1994b. Defining 'evolutionary significant units' for conservation. *Trends Ecol. Evol.* 9:373–375.

MORREALE, S.J., G.J. RUIZ, J.R. SPOTILA, and E.A. STANDORA. 1982. Temperature-dependent sex determination: current practices threaten conservation of sea turtles. *Science* 216:1245–1247.

MORTIMER, J.A. and A. CARR. 1987. Reproduction and migrations of the Ascension Island green turtle (*Chelonia mydas*). *Copeia* 1987:103–113.

MROSOVSKY, N. 1983. *Conserving Sea Turtles.* British Herpetological Society, London.

MROSOVSKY, N., S.R. HOPKINS-MURPHY, and J.I. RICHARDSON. 1984. Sex ratio of sea turtles: seasonal changes. *Science* 225:739–741.

MURPHY, T.M. and S.R. HOPKINS-MURPHY. 1989. *Sea Turtle and Shrimp Fishing Interactions: A Summary and Critique of Relevant Information.* Center for Marine Conservation, Washington, DC.

NATIONAL RESEARCH COUNCIL. 1990. *Decline of the Sea Turtles.* National Academy Press, Washington, DC.

NEIGEL, J.E. and J.C. AVISE. 1986. Phylogenetic relationships of mitochondrial DNA under various demographic models of speciation. In *Evolutionary Processes and Theory*, eds. E. Nevo and S. Karlin, pp. 513–534. Academic Press, New York.

NISHIMURA, W. and S. NAKAHIGASHI. 1990. Incidental capture of sea turtles by Japanese research and training vessels: results of a questionnaire. *Mar. Turt. Newsl.* 51:1–4.

NORMAN, J.A., C. MORITZ, and C.J. LIMPUS. 1994. Mitochondrial DNA control region polymorphisms: genetic markers for ecological studies of marine turtles. *Molec. Ecol.* 3:363–373.

OTTENWALDER, J.A. and J.P. ROSS. 1992. The Cuban sea turtle fishery: description and needs for management. In *Proceedings 11th Annual Workshop on Sea Turtle Biology and Conservation*, comps. M. Salmon and J. Wyneken, pp. 90–92. NOAA Tech. Memo. NMFS-SEFC-302. National Technical Information Service, Springfield, VA.

OWENS, D.W., M.A. GRASSMAN, and J.R. HENDRICKSON. 1982. The imprinting hypothesis and sea turtle reproduction. *Herpetologica* 38:124–135.

PAMILO, L. and M. NEI. 1988. Relationships between gene trees and species trees. *Molec. Biol. Evol.* 5:568–583.

PARSONS, J. 1962. *The Green Turtle and Man.* University of Florida Press, Gainesville, FL.

PELLA, J.J. and G.B. MILNER. 1987. Use of genetic marks in stock composition analysis. In *Population Genetics and Fishery Management*, eds. N. Ryman and F. Utter, pp. 247–276. University of Washington Press, Seattle.

PITMAN, R. 1990. Pelagic distribution and biology of sea turtles in the eastern tropical Pacific. In *Proceedings 10th Annual Workshop on Sea Turtle Biology and Conservation*, comps. T.H. Richardson, J.I. Richardson, and M. Donnelly, pp. 143–148. NOAA Tech. Memo. NMFS-SEFC-278. National Technical Information Service, Springfield, VA.

PRITCHARD, P.C.H. 1969a. The survival status of ridley sea-turtles in American waters. *Biol. Conserv.* 2:13–17.

PRITCHARD, P.C.H. 1969b. Studies of the systematics and reproductive cycles of the genus *Lepidochelys.* PhD dissertation, University of Florida, Gainesville.

PRITCHARD, P.C.H. 1976a. Post-nesting movements of marine turtles (Cheloniidae and Dermochelyidae) tagged in the Guianas. *Copeia* 1976:749–754.

PRITCHARD, P.C.H. 1976b. Endangered species: Kemp's ridley turtle. *Flor. Natur.* 49(3):15–19.

PRITCHARD, P.C.H. 1979. *Encyclopedia of Turtles.* TFH Press, Neptune, NJ.

PRITCHARD, P.C.H. 1980. The conservation of sea turtles: practices and problems. *Amer. Zool.* 20:609–617.

PRITCHARD, P.C.H. and P. TREBBAU. 1984. The Turtles of Venezuela. *Contrib. Herpetol. 2, Society for the Study of Amphibians and Reptiles.* Fundacion de Internados Rurales, Caracas, Venezuela.

RAND, D.M. 1994. Thermal habit, metabolic rate and the evolution of mitochondrial DNA. *Trends Ecol. Evol.* 9:125–131.

RICHARDSON, J.I. 1982. A population model for adult female loggerhead sea turtles (*Caretta caretta*) nesting in Georgia. PhD dissertation, University of Georgia, Athens.

ROSS, J.P. 1982. Historical decline of loggerhead, ridley, and leatherback sea turtles. In *Biology and Conservation of Sea Turtles*, ed. K.A. Bjorndal, pp. 189–195. Smithsonian Institution Press, Washington, DC.

SCHULZ, J.P. 1982. Status of sea turtle populations nesting in Surinam with notes on sea turtles nesting in Guyana and French Guiana. In *Biology and Conservation of Sea Turtles*, ed. K.A. BJORNDAL, pp. 435–437. Smithsonian Institution Press, Washington, DC.

SEARS, C.J., B.W. BOWEN, R.W. CHAPMAN, S.B. GALLOWAY, S.R. HOPKINS-MURPHY, and C.M. WOODLEY. 1995. Demographic composition of the juvenile loggerhead sea turtle (*Caretta*

caretta) population of Charleston, South Carolina: evidence from mitochondrial DNA markers. *Mar. Biol., in press.*

SHAVER, D.J., D.W. OWENS, A.H. CHANEY, C.W. CAILLOUET, Jr., P. BURCHFIELD, and R. MÁRQUEZ. 1988. Styrofoam box and beach temperatures in relation to incubation and sex ratios of Kemp's ridley sea turtles. In *Proceedings 8th Annual Workshop on Sea Turtle Conservation Biology*, comp. B.A. Schroeder, pp. 103–108. NOAA Tech. Memo. NMFS-SEFC-214. National Technical Information Service, Springfield, VA.

SLATKIN, M. 1987. Gene flow and the geographic structure of natural populations. *Science* 236:787–792.

SLATKIN, M. 1989. Detecting small amounts of gene flow from the phylogeny of alleles. *Genetics* 121:609–612.

SMITH, M.H., H.O. HILLESTAD, M.N. MANLOVE, D.O. STRANEY and J.M. DEAN. 1977. Management implications of genetic variability in loggerhead and green sea turtles. *Int. Cong. Game Biol.* 13:302–312.

TAKAHATA, N. and S.R. PALUMBI. 1985. Extranuclear differentiation and gene flow in the finite island model. *Genetics* 109:441–457.

TAUBES, G. 1992. A dubious battle to save the Kemp's ridley sea turtle. *Science* 256:614–616.

UCHIDA, S. and H. TERUYA. 1991. Transpacific migration of a tagged loggerhead, *Caretta caretta*, and tag-return results of loggerheads released from Okinawa, Japan. In *International Symposium of Sea Turtles in Japan*, ed. I. Uchida, pp. 171–182. Hemiji City Aquarium, Hemiji City, Japan.

VAN DYKE, J.M. 1993. International governance and stewardship of the high seas and its resources. In *Freedom for the Seas in the 21st Century: Ocean Governance and Environmental Harmony*, eds. J.M. Van Dyke, D. Zaelke, and G. Hewison, pp. 13–20. Island Press, Washington, DC.

VANE-WRIGHT, R.l., C.J. HUMPHRIES, and P.H. WILLIAMS. 1991. What to protect— systematics and the agony of choice. *Biol. Conserv.* 55:235–254.

WALKER, T.A. and C.J. PARMENTER. 1990. Absence of a pelagic stage in the life cycle of the flatback turtle, *Natator depressa* (Garman). *J. Biogeogr.* 17:275–278.

WETHERALL, J.A., G.H. BALAZS, R.A. TOKUNAGA, and M.Y.Y. YONG. 1993. Bycatch of marine turtles in North Pacific high-seas driftnet fisheries and impacts on the stocks. In *INPFC Symposium on Biology, Distribution, and Stock Assessment of Species Caught in the High Seas Driftnet Fisheries in the North Pacific Ocean*, eds. J. Ito, W. Shaw, and R.L. Burgner, pp. 519–538. Int. North Pacific Fish. Comm., Vancouver, Canada.

WIBBELS, T.R., Y.A. MORRIS, D.W. OWENS, G.A. DIENBURG, J. NOELL, J.K. LEONG, R.E. KING, and R. MÁRQUEZ. 1989. Predicted sex ratios from the international Kemp's ridley sea turtle head-start research project. In *Proceedings 1st International Symposium on Kemp's Ridley Sea Turtle Biology, Conservation, and Management*, eds. C.W. Caillouet, Jr. and A.M. Landry, pp. 77–81. TAMU-SG-89-105. Sea Grant College Program, Texas A&M University, Galveston, TX.

WILCOVE, D.S., M. McMILLAN, and K.C. WINSTON. 1993. What exactly is an endangered species? An analysis of the U.S. endangered species list: 1985–1991. *Conserv. Biol.* 7:87–93.

WILSON, A.C., R.L. CANN, S.M. CARR, M. GEORGE, Jr., U.B. GYLLENSTEN, K.M. HELM-BY-CHOWSKI, R.G. HIGUCHI, S.R. PALUMBI, E.M. PRAGER, R.D. SAGE and M. STONEKING. 1985. Mitochondrial DNA and two perspectives on evolutionary genetics. *Biol. J. Linn. Soc.* 26:375–400.

WITZELL, W.N. 1983. Synopsis of biological data on the hawksbill turtle *Eretmochelys imbricata* (Linnaeus, 1766). *FAO Fisheries Synopsis No. 137.* Food and Agriculture Organization of the United Nations, Rome.

WOOD, J.R., F.E. WOOD, and K. CRITCHLEY, 1983. Hybridization of *Chelonia mydas* and *Eretmochelys imbricata. Copeia* 1983:839–842.

WOODY, J.B. 1985. Kemp's ridley continues decline. *Mar. Turt. Newsl.* 35:4–5.

WORLD CONSERVATION UNION. 1993. *1994 IUCN Red List of Threatened Animals.* World Conservation Union, Gland, Switzerland.

XU, S., C.J. KOBAK, and P.E. SMOUSE. 1994. Constrained least squares estimation of mixed stock composition from mtDNA haplotype frequency data. *Can. J. Fish. Aquat. Sci.* 51:417–425.

ZANGERL, R., L.P. HENDRICKSON, and J.R. HENDRICKSON. 1988. A redescription of the Australian flatback sea turtle, *Natator depressus. Bishop Mus. Bull. Zool.* 1:1–69.

ZUG, G.R., A.H. WYNN, and C. RUCKDESCHEL. 1986. Age determination of loggerhead sea turtles, *Caretta caretta*, by incremental growth marks in the skeleton. *Smithsonian Contrib. Zool.* 427:1–34.

8

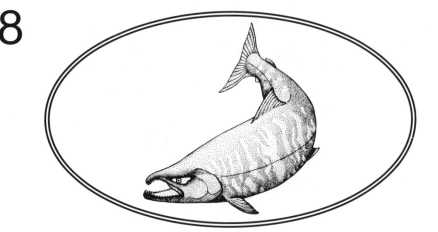

CONSERVATION AND GENETICS
OF SALMONID FISHES

Fred W. Allendorf and Robin S. Waples

INTRODUCTION

Fishes in the family Salmonidae (Figure 8.1) include approximately 75 extant species that are native throughout the Northern Hemisphere. Salmonids have been widely introduced into the Southern Hemisphere because of their ease of propagation and favor with anglers, and they now dominate cold freshwaters throughout the World. These fish have fascinated humans for thousands of years because of their value as food and their occurrence in large and spectacular spawning aggregations. They have played a central role in most cultures in North America, Asia, and Europe.

Nonetheless, many salmonid populations and species throughout these regions are also threatened today with extinction (Box 8.1). The demise of salmon and trout in the western United States has become a conservation crisis of enormous biological, economical, and political significance (Hedrick and Miller, 1994). Two recent papers have brought the magnitude of this crisis to national attention by summarizing the status of salmon in the western United States, excluding Alaska (Nehlsen et al., 1991; Williams et al., 1992). These reports concluded that over 100 major populations of salmon and anadromous trout have been extirpated, and that an additional

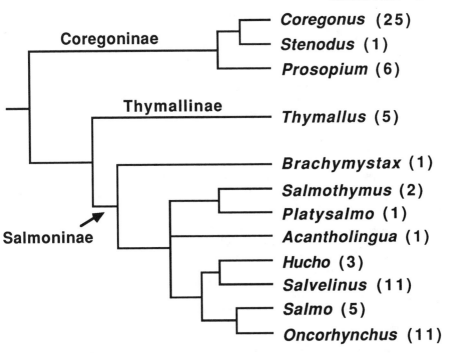

Figure 8.1. Phylogeny of salmonid genera (after Stearley and Smith, 1993). In parentheses are the approximate numbers of species in each genus (Nelson, 1994).

160 local populations of salmon and trout from California, Idaho, Oregon, and Washington are at high or moderate risk of extinction. Efforts to conserve and reestablish salmon confront the primary economic interests of this region: the hydroelectric power industry, agriculture, forestry, and commercial fishing.

In the mid-1800s, approximately 10–16 million anadromous salmon and trout returned to spawn each year in the Columbia River Basin; by the 1980s this number had dwindled to approximately 2.5 million (Northwest Power Planning Council, 1987). Furthermore, an estimated 80% of current returning fish result from hatchery production. Thus, return of naturally produced anadromous salmon and trout is less than 5% of historic numbers. According to Williams et al. (1992), approximately 75% of historical populations in the Columbia River Basin are either extinct, at immediate risk of extinction, or of special concern.

Many factors have contributed to this decline. Overfishing was probably the cause of the initial decline; commercial harvest of chinook salmon peaked in 1883 (Mullan, 1987). Large-scale irrigation projects began around 1900 and caused substantial reductions in spawning and rearing habitat in some

---

Box 8.1. Scientific and common names of the salmonid species mentioned in this chapter

In the family as a whole, 22 species appear in the 1994 IUCN Red List of Threatened Animals (World Conservation Union, 1993). Within North America, various populations of the species indicated by asterisk are listed for protection under the ESA.

Coregonus
    C. clupeaformis   Lake whitefish

Oncorhynchus
* *O. apache*   Apache trout
   O. chrysogaster   Mexican golden trout
* *O. clarki*   Cutthroat trout
* *O. gilae*   Gila trout
   O. gorbuscha   Pink salmon
   O. keta   Chum salmon
   O. kisutch   Coho salmon
   O. masou   Masu salmon
   O. mykiss   Rainbow trout, steelhead, golden trout, redband trout
* *O. nerka*   Sockeye salmon, kokanee
* *O. tshawytscha*   Chinook salmon

Salmo
   S. salar   Atlantic salmon
   S. trutta   Brown trout

Salvelinus
   S. alpinus   Arctic char
   S. confluentus   Bull trout
   S. fontinalis   Brook trout
   S. malma   Dolly Varden
   S. namaycush   Lake trout

Thymallus
   T. arcticus   Arctic grayling

---

areas (Northwest Power Planning Council, 1986). A major decline in overall numbers coincided with construction of hydroelectric dams in the 1930s. Approximately one third of natural spawning and rearing habitat has been lost by flooding or rendered inaccessible by dams, including most spawning habitat on the main-stem Columbia (Northwest Power Planning Council, 1987).

The threat of widespread application of the U.S. Endangered Species Act (ESA) to protect these fish has caused many people to question the principles upon which the Act was established, and indeed the value of the ESA itself. The recognition of local populations as endangered "species" has been especially controversial. Under the ESA, a *species* is defined to include "any distinct population segment of any species of vertebrate fish or wildlife which interbreeds when mature" [Public Law 95-632 (1978), 92 Stat. 3751]. The complex life history of anadromous salmon and trout, combined with their tendency to home to natal streams after several years in the ocean, has produced a complex assortment of many genetically and ecologically distinct local populations within each salmon species (Altukhov and Salmenkova, 1991; Waples, 1991b; Utter et al., 1993).

The primary objective of this chapter is to present an overview of genetic issues underlying the current crisis in salmonid conservation. The empirical focus is on western North American representatives of the genus *Oncorhynchus*, but the basic issues are general to salmonid taxa throughout the world. In addition, several aspects of salmonid conservation and genetics that are somewhat different from those associated with other species in this volume will be emphasized (e.g., inheritance patterns associated with ancestral polyploidy, artificial propagation, and human exploitation).

Phylogenetic History

The Salmonidae is a primitive teleost family whose relationship to other families is not well understood (Lauder and Liem, 1983; Behnke, 1992). The family includes three subfamilies (Figure 8.1; Stearley and Smith, 1993) that are widely distributed throughout the Northern Hemisphere: Coregoninae (whitefishes and ciscoes), Thymallinae (grayling), and Salmoninae (char, trout, and salmon). These species comprise a monophyletic clade descended from a single tetraploid ancestor that originated some 50–100 million years ago (Allendorf and Thorgaard, 1984; Behnke, 1992), although the salmonid fossil record is extremely scanty during this period (Norden, 1961; Cavender, 1970). The earliest fossil definitely attributed to Salmonidae is *Eosalmo driftwoodensis* from Eocene deposits (40–50 million years ago) in western Canada (Wilson, 1977).

Behnke (1992) has suggested that *Salmo* and *Oncorhynchus* diverged from other members of the subfamily Salmoninae 30–40 million years ago, and that these genera separated into an Atlantic Ocean group (*Salmo*) and a Pacific Ocean group (*Oncorhynchus*) about 15 million years ago. By five million years ago, *Oncorhynchus* had diverged into two separate lineages, one leading to Pacific salmon and the other to the trout of western North America (Figure 8.2). The taxonomy of western trout has long troubled ichthyologists because of extreme genetic, morphological, life-historical, and ecological variation within species (Jordan and Everman, 1896; Allendorf and Leary, 1988; Behnke, 1992).

Phylogenetic reconstruction of *Oncorhynchus* is further complicated because of historical interspecific hybridization and introgression (Smith, 1992; Utter and Allendorf, 1994). The most compelling evidence of this are strik-

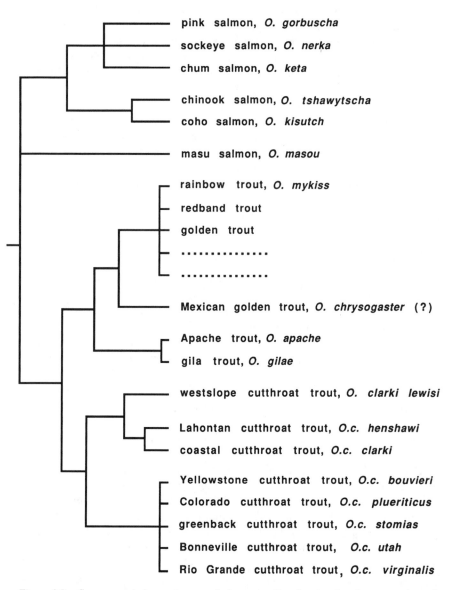

Figure 8.2. Consensus phylogenetic tree of the genus *Oncorhynchus* based on a review of morphological and molecular data (after Utter and Allendorf, 1994). Branch lengths are approximate to the relative divergences among taxa. The question mark reflects uncertainty about the supposed Mexican golden trout collection. The redband trout and golden trout are often considered conspecific with the rainbow trout.

ing discordancies in relationships based upon nuclear and mitochondrial genes for cutthroat trout subspecies and rainbow trout (Allendorf and Leary, 1988). In such cases, branching dichotomous trees, as in Figures 8.1 and 8.2, cannot accurately represent the phylogenetic relationships of all genes because of occasional genetic transfers between branches (reticulate evolution) mediated by interspecific hybridization.

Natural History

Salmonids have several different major life-history types (Box 8.2). Pacific salmon of the genus *Oncorhynchus* are primarily anadromous. That is, they reproduce in freshwater but spend part of their life in the ocean. This life history sometimes involves extensive migrations. For example, some chinook salmon migrate over 3,200 km upstream to spawn in headwaters of the Yukon River (Healey, 1991) and some sockeye salmon undertake annual feeding migrations in the ocean of over 3,700 km (Burgner, 1991).

Many salmonids remain in freshwater throughout their lives. However, some freshwater populations are migratory and undertake extensive feeding and spawning migrations. For example, bull trout that emerge in the North Fork of the Flathead River in Alberta, Canada, migrate over 150 km to reside in Flathead Lake in western Montana (Pratt, 1992). Other freshwater populations of bull trout spend their entire lives in small streams or lakes.

Box 8.2. Salmonid Life Cycle

The generalized salmonid life cycle begins with external fertilization of eggs in cold freshwater. In most species, the eggs are laid in a depression dug in the gravel, called a *redd*. The eggs hatch when the *alevins* still possess a yolk sac, but the small fish remain in the gravel. The fish become *fry* after the yolk sac is adsorbed and the alevins emerge from the gravel. The fry quickly develop a series of vertical bars on their sides (parr marks) and are then called *parr*; this stage may last a few months or even years. Many salmonids undergo extensive migrations from their natal streams. In some species, the fry emerge from the gravel and immediately migrate to a lake. In anadromous populations, parr transform into *smolts* that migrate to the ocean. The parr–smolt transformation involves profound changes in morphology, physiology, and behavior in preparation for change from a freshwater to saltwater habitat. Some freshwater salmonids migrate hundreds of kilometers to spend the majority of their lives in large lakes or rivers. As sexual maturity approaches, the maturing adult undergoes a return migration to spawn in its home stream. Some species are semelparous (i.e., all individuals die after spawning), whereas other species are iteroparous.

Most aspects of life history vary greatly at all taxonomic levels in salmonids. On a broad scale, life-history variability is largely concordant with phylogeny (Hutchings and Morris, 1985; Stearley, 1992). For example, truly anadromous forms are found only in Salmoninae (Hutchings and Morris, 1985). Nevertheless, there is also extraordinary intraspecific variability, especially within *Oncorhynchus*. Most *Oncorhynchus* species have both anadromous and nonanadromous forms.

The sockeye salmon is a case in point. Individuals of this species are typically anadromous, usually migrating to the ocean after one or more years in a freshwater lake. However, freshwater resident forms are also widespread. Kokanee are discrete populations of sockeye salmon that are fully adapted to freshwater existence and presumably have evolved many times from anadromous populations in recent geologic times (Foote et al., 1989). Some kokanee populations exist sympatrically with sockeye salmon populations (Wood and Foote, 1990; Burgner, 1991). Another freshwater form of sockeye salmon is referred to as residual sockeye (Burgner, 1991). Residual sockeye are mostly male progeny of anadromous parents that become sexually mature without migrating to the ocean.

Salmonids in other genera also show extensive life-history variability. Four forms of bull trout are distinguishable by life-history characteristics (Pratt, 1992): (1) resident bull trout that spend their entire lives in small streams; (2) anadromous bull trout that occur in some coastal river systems (Haas and McPhail, 1991); (3) adfluvial bull trout that mature in lakes and spawn in tributaries, where young reside for 1–3 years (Fraley and Shepard, 1989); and (4) fluvial bull trout that have a similar life history except they move between main rivers and tributaries. Individuals of these latter three forms can make extensive spawning migrations, usually do not attain sexual maturity until age five or six, and can reach a size exceeding 10 kg (Fraley and Shepard, 1989).

Complex Life History

Stearns (1976) has defined a life-history *strategy* as a set of coadapted reproductive traits resulting from selection in a particular environment. Some anadromous salmonids have spectacularly complex life histories. For example, sockeye salmon from the upper Fraser River emerge from the gravel in freshwater at an elevation of 1,000 m, migrate to a nursery lake, and generally spend two years growing in the lake. They then undergo smoltification and migrate some 1,000 km downstream to the ocean where they remain for 2–3 years. In the ocean, sockeye salmon undergo long feeding migrations. Sexually mature adults return to the mouth of the Fraser River, retrace their journey of 1,000 km upstream to their natal stream, and spawn.

The life history of sockeye salmon (Figure 8.3) generally includes three distinct habitats (nursery lake, ocean, and spawning stream). The fish undergo complex behavioral, physiological, and morphological transformations in transition from one habitat to another. In addition, they must undergo four major migrations: (1) stream to lake, (2) lake to ocean, (3) a feeding migration in the ocean, and (4) a return to the natal stream. Moreover, the timing of each of these events must be precise. For example, migration up and down the freshwater system to the ocean must be timed to correspond with appropriate water flows and availability of food. In some cases, migration into freshwater habitat may precede actual spawning by many months. Several experimental studies have demonstrated that nearly all of these aspects of life history are influenced by genetic differences among individuals and populations (reviewed by Ricker, 1972 and Taylor, 1991).

The best studied aspect of local adaptation in salmonids is the migratory behavior of newly emerged fry. Studies with sockeye salmon, rainbow trout, cutthroat trout, and Arctic grayling have demonstrated innate differences in migratory behavior that correspond to specializations in movement from spawning and incubation habitat in streams to favorable feeding and growth habitat in lakes (Raleigh, 1971; Brannon, 1972; Kelso et al., 1981; Kaya, 1991). Fry emerging from lake-outlet streams typically migrate upstream upon emergence, and fry from inlet streams typically migrate downstream. Quinn (1985) has shown that differences in compass-orientation behavior of

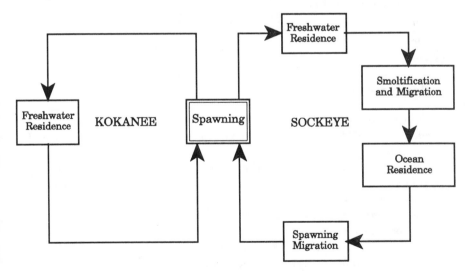

Figure 8.3. Diagram of major life-history events for the non-anadromous (kokanee) and anadromous (sockeye salmon) forms of *Oncorhynchus nerka*.

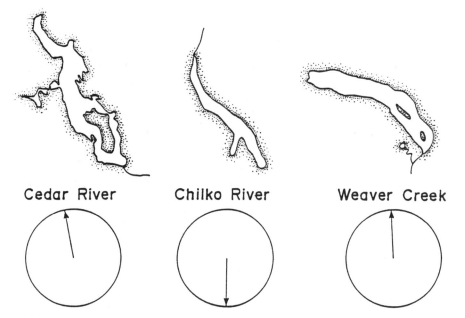

Cedar River   Chilko River   Weaver Creek

Figure 8.4. Mean compass orientations of newly emerged sockeye salmon fry from three populations: Cedar River in Washington, and Chilko River and Weaver Creek in the Fraser River system in British Columbia (Quinn, 1985). The compass orientation of fry reared under experimental conditions corresponds to the direction that will lead them from incubation habitats to their lake feeding areas.

newly emerged sockeye salmon correspond to movements to feeding areas (Figure 8.4).

Conservation

Salmon and Humans in Western North America.   As the glaciers receded some 10,000 years ago, western North America became populated by the ancestral Native Americans. Concurrently, anadromous salmonids evolved a mosaic of populations throughout the Columbia River drainage. Salmon and trout became an important part of the sustenance and culture of native peoples throughout the region.

Euro-Americans first encountered these fish when they arrived in the region some 200 years ago. The first salmon cannery on the West Coast began operation in 1864 near Sacramento, California (Williams et al., 1992). The salmon canning industry on the Columbia River began in 1866 and reached a peak in 1883–1884, with 40 million pounds gross weight of salmon canned. The catch of salmon declined quickly following this peak. By the end of the 19th century, there was great concern about the disappearance of salmon, as evidenced by the U.S. Commissioner of Fisheries report to the U.S. Congress in 1894 (McDonald, 1894):

> There is no reason to doubt — indeed, the fact is beyond question — that the number of salmon now reaching the head waters of streams in the Columbia River basin is insignificant in comparison with the number which some years ago annually visited and spawned in these waters.

In 1892, Livingstone Stone urged the United States to set aside a national park for salmon protection (patterned after the role Yellowstone National Park had served in conservation of bison). Stone prophetically described the demise of salmon in colorful terms:

> The helpless salmon's life is gripped between these two forces — the murderous greed of the fishermen and the white man's advancing civilization — and what hope is there for salmon in the end? Protective laws and artificial breeding are able to hold the first in check, but nothing can stop the last.

It is interesting to note that even early conservationists were aware of differences between local salmon populations and their importance in the success of the fish. R.D. Hume (1893) packed with his canned salmon a booklet stressing the importance to conserve the valuable salmon resource:

> I firmly believe that like conditions must be had in order to bring about like results, and that to transplant salmon successfully they must be placed in rivers where the natural conditions are similar to that from which they have been taken.

W.F. Thompson (1965) provided the first treatment of the importance of population structure and local adaptation in the context of modern evolutionary biology:

> Each stream or lake has its own extremely complex characteristics, and if salmon live in one of them we find that these salmon are adapted in an equally complex way to that environment.

Thompson (1959, 1965) also anticipated and insightfully discussed many current issues faced in salmon conservation (metapopulation structure, misuse of hatcheries to address symptoms rather than causes of decline, and the need for conservation of "subspecific groups of salmon, segregated during spawning").

Challenging issues. Salmonids present a number of unusual and challenging problems that combine to make conservation of these fish a difficult proposition. Among these are the following:

1. *Long-range migrations.* These movements make protection of the habitat of anadromous salmon extremely difficult. Consider salmon that spawn in the Salmon River drainage of Idaho. These fish spend the first years of life in freshwater far from the ocean, at elevations up to 2,000 meters. They then migrate as far as 1,500 km downstream through the Salmon, Snake, and Columbia Rivers before entering the Pacific Ocean. There they spend one, two, or more years, ranging as far north as the coast of Alaska, before returning to spawn in their natal freshwater streams. There is no single habitat or "ecosystem" that encompasses these fish, other than the regional biosphere itself. Protection of salmon requires a combination of habitat measures that takes into account the complex life-history of these fishes.

2. *Natural hybridization.* This phenomenon is much more common in freshwater fish than in other vertebrates for a combination of reasons (e.g., external fertilization, weak ethological isolating mechanisms, and competition for limited spawning habitat; Hubbs, 1955; Campton, 1987). In addition, hybrids between distantly related fish species are sometimes viable, suggesting that fish have less severe developmental incompatibilities than other vertebrate species with comparable levels of genetic divergence (Thorgaard and Allendorf, 1988). This appears to be especially true for salmonids, in which hybrids appear to be developmentally more compatible than other fish, and hybrids between distantly related taxa are often fertile (perhaps because of their polyploid ancestry; Ferguson et al., 1985). The widespread introductions of salmonid species to waters outside their native range have made hybridization and introgression major conservation problems for native trout (Allendorf and Leary, 1988; Leary et al., 1993).

3. *Commercial exploitation.* Local populations of salmon returning to major rivers are often harvested in mixed groups prior to their return to natal streams. An adequate number of individuals for each local reproductive population is needed to ensure persistence of the many reproductive units that make up a fished stock of salmon. In practice, it is extremely difficult to regulate losses to fishing on an individual deme basis. Thousands of demes make up the West Coast salmon fishery, and many of these are likely to be intermingled in any particular catch. The result of regulating fishing on a stock basis, and ignoring the reproductive units that together constitute a stock, is the disappearance or extirpation of some of the demes (Clark, 1984).

4. *Hatchery propagation.* This provides a variety of indirect and direct threats to wild populations of trout and salmon. Perhaps the greatest problem with hatcheries has been indirect; artificial propagation has been promoted as a panacea to the demise of wild salmon from overfishing, loss of access to spawning and rearing habitat (because of dam construction), and

damage to freshwater habitat (Goodman, 1990; Hilborn, 1992; Meffe, 1992). Thus, the real causes of salmon decline were ignored while the public was assured that hatcheries could maintain productivity (Bell, 1937). Hatchery operations have had harmful effects on wild fish through a variety of genetic and environmental mechanisms (Marnell, 1986; Ferguson, 1990; Waples, 1991a; Riddell, 1993).

5. *Legal factors.* Legal arrangements often greatly restrict actions that can be taken for management and conservation of salmon and trout. These include agreements with Native American tribes, international fishing interests, and international hydroelectric development agencies, as specified in national power acts, national fishery management acts, and a variety of state laws (Goodman, 1990). Moreover, extensive migrations of salmon result in their being the "responsibility" of multiple fishery management agencies during different parts of their life cycle. For example, chinook salmon originating in the Lochsa River of Idaho pass through 17 separate management jurisdictions during their lifetime (Goodman, 1990).

POLYPLOIDY AND EVOLUTIONARY GENETICS

Svårdson (1945) first proposed that salmonid species formed a polyploid series because of multivalents in meiotic preparations of several species, and chromosome numbers in some species that seemed to fall in multiples of 10 (e.g., Atlantic salmon have about 60 chromosomes and brown trout have about 80; reviewed in Allendorf and Thorgaard, 1984). This proposal was shown to be incorrect by Rees (1964), who demonstrated that species with different chromosome numbers used by Svårdson to support his hypothesis have similar cellular DNA contents and total chromosome lengths.

Ohno and coworkers (1968) later proposed that salmonids as a group were tetraploid in comparison to related salmonoid and clupeoid fishes, such as smelt and herring. There are several major lines of support for an ancestral tetraploid event in salmonid evolution (Ohno, 1970; Allendorf and Thorgaard, 1984). Salmonid fish have approximately twice the DNA per cell and twice the number of chromosome arms as closely related fish. Multivalents are commonly observed in meiotic preparations. Salmonids show a high incidence of duplicated enzyme loci. Recent studies examining DNA sequences have confirmed the presence of many duplicate genes. For example, Agellon et al. (1988) described two expressed genes that encode growth hormone in rainbow trout, in comparison to the single growth-hormone gene found in almost all other vertebrates. Several other hormones have been found to be encoded by duplicated genes in *Oncorhynchus* species (Hiraoka et al., 1993).

Salmonid Genetics

The genetic analysis of polyploids is extremely complicated (relative to the elegant simplicity of diploid Mendelian systems). Unlike disomic ratios, tetrasomic ratios are affected by differential pairing affinities, multivalent formation, crossovers, and the position of a locus in relation to the centromere. The complexities of tetrasomic inheritance have been discussed by many authors (e.g., Mather, 1936; Little, 1945; Marsden et al., 1987; Soltis and Soltis, 1989); the most comprehensive treatment is that of Burnham (1962).

A new autotetraploid or segmental allopolyploid (Stebbins, 1947) is expected to demonstrate multivalent formation and tetrasomic segregation (Sybenga, 1972). The frequency of non-disjunction is greater when chromosomes associate in multivalents. Therefore, selection for decreased infertility caused by non-disjunction should cause a reduction of multivalent pairing and restoration of disomy. Over time, diploidization of the genome occurs and disomic segregation becomes prevalent (Waines, 1976). The primary mechanism for restoration of disomic inheritance is structural divergence of the four homologs into two pairs of homeologous chromosomes. The two chromosomes within a pair are homologous to each other and are homeologous to the two chromosomes in the other pair.

Patterns of Inheritance

Disomic and Tetrasomic Segregation.   The pattern of segregation for gene loci in salmonids is complex. In females, only disomic segregation ratios have been reported. Strictly disomic inheritance is supported by absence of multivalent formation during oogenesis (Allendorf and Thorgaard, 1984). Most loci in males also are inherited disomically, but some loci show variable patterns of segregation, ranging from disomic ratios in males from some populations to tetrasomic ratios in other populations.

Pairing and recombination between homeologues can lead to partial tetrasomic inheritance and non-independent joint segregation ratios between duplicated loci. The differences between the sexes in multivalent formation and the pairing of homeologues may be explained by a two-stage pattern of pairing in males in which homologous chromosomes pair first, followed by secondary homeologous pairing (such "secondary pairing" was first described in 1960 by Riley in hexaploid wheat). Disjunction such that paired chromosomes pass to opposite poles would ensure that each gamete receives one copy of each homologue. Exchanges between homeologues would produce segregation ratios approaching tetrasomic expectations.

This model produces a mixture of disomic and tetrasomic inheritance, depending upon the map distance between a locus and the centromere. Loci near the centromere would show disomic inheritance and distal loci would show ratios near those expected with tetrasomic inheritance. Exchanges between homeologues would keep the distal part of homologous chromosomes from diverging. Homeologues would maintain their integrity because of divergent sequences near centromeres. The predictions of this model have been tested by mapping gene-centromere distances using half-tetrad analysis via induced gynogenesis. All loci showing an absence of structural divergence and partial tetrasomic inheritance have been found to be distal to the centromere; in contrast, those pairs of loci showing substantial divergence have proved to be proximal (Allendorf et al., 1986).

In interspecific male hybrids between brook and lake trout, Morrison and Wright (1966) reported linkage of the two loci resulting from duplication of the ancestral lactate dehydrogenase-B locus. Further results demonstrated that these inheritance ratios could not be explained by classical linkage because non-parental types were in excess (Morrison, 1970). This phenomenon was first referred to as *pseudolinkage* by Davisson et al. (1973). Johnson et al. (1987) described pseudolinkage at a number of loci in interspecific hybrid males, but never in females. The excess of nonparental gametes is apparently the result of preferential secondary pairing in multivalents of homeologues from the same species, followed by disjunction so that paired chromosomes pass to opposite poles. Analogous patterns of preferential pairing have been described in polyploid plants (Sybenga, 1975).

Recombination and Sex-Linkage. Males are the heterogametic sex in salmonids (reviewed in Hartley, 1987). Linkage studies with salmonids have shown that recombination rates are generally much greater in females than males (May and Johnson, 1990). For example, three pairs of loci that display random joint segregation in females (i.e., 50% recombination) show an average of only 10% recombination in males (May and Johnson, 1990). Wright et al. (1983) suggested that repressed recombination in males is caused by structural constraints imposed on crossing over by multivalent pairing, which occurs only in males.

Allendorf et al. (1994) recently described sex linkage of two enzyme loci in rainbow trout. Joint segregation data from fathers indicated an average of 8.1% recombination between *HEX-2* and the sex-determining locus (*SEX*). The average recombination between *HEX-2* and *sSOD-1* in fathers was 26.8%. No evidence of non-random segregation of *HEX-2* and *sSOD-1* was found in mothers. Unlike the extreme *XX/XY* heteromorphy in mammals, functional alleles for *HEX-2* and *sSOD-1* occur on both *X* and *Y* chromosomes.

Linkage with a sex-determining locus will affect genotypes at another locus if there are differences in allele frequency between males and females.

Such differences between sexes may arise by hybridization, genetic drift in small populations, or natural selection. The rate of decay of non-random associations (i.e., gametic disequilibrium) between the sex locus (SEX) and another locus will depend only upon the rate of recombination in males, because females will always be homozygous (XX) at the sex-determining locus (Allendorf et al., 1994). Allele frequency differences between males and females will result in an excess of heterozygous progeny relative to binomial (i.e., Hardy–Weinberg) expectations. An excess of heterozygotes at sex-linked loci that differ in gender-specific allele frequencies is thus anticipated. Significant nonrandom associations occur between genotypes at HEX-2 and SEX in the hatchery population used for the inheritance study. This gametic disequilibrium has resulted in large changes in allele frequency at HEX-2 from one generation to the next and an excess of heterozygotes in comparison to expected binomial proportions.

The discovery of additional genetic markers on the sex chromosome of salmonid species would be valuable for applying molecular data to conservation questions. Differences in allele frequencies between males and females could be suggestive of hybridization. Such sex-linked markers could be used to provide information about possible hybridization events and introgression (e.g., from hatchery stocks to wild populations) in the recent past.

Interpretation of Molecular Variation.   Extensive gene duplication has made genetic interpretation of isozymes in salmonids more problematic than in species without a polyploid ancestry. Isoloci (genes whose products have identical electrophoretic mobility) are especially difficult because genotypes cannot be determined unambiguously, and there is no way to assign observed variation to a particular locus of the pair (Figure 8.5). Waples (1988) has developed statistical methods for estimating allele frequencies at isoloci, but these methods cannot entirely overcome several problems. First, these methods rely on the ability to score the number of doses and not just presence or absence of electromorphs (e.g., phenotypes 0, I, and II in Figure 8.5). Second, these methods assume that the two loci are genetically discrete, but this is not an appropriate assumption when homeologous exchanges between loci are frequent. Imhof et al. (1980) developed a maximum likelihood method to estimate gametic frequencies at isoloci that avoids the problem of treating the two loci as distinct. However, this method does not consider that gametic frequencies will differ in males and females, even if the two genders have identical genotype frequencies, because of homeologous exchanges in males.

Similar problems are encountered in using DNA methods to study molecular variation in salmonids (Forbes et al., 1994). For example, the general approach of devising "universal" polymerase chain reaction (PCR) primers

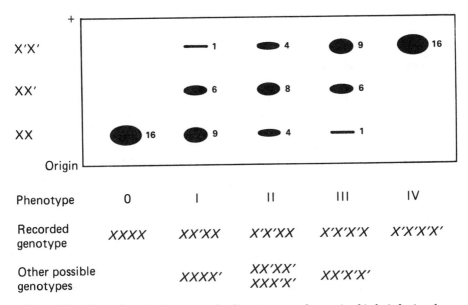

Figure 8.5. Electrophoretic phenotypes of a dimeric enzyme for a pair of isoloci sharing the alleles X and X' (Waples, 1988). The five phenotypes correspond to individuals with 0–4 doses of the variant allele (X'). The expected relative intensities of three electrophoretic bands are indicated by the numbers 1–16.

(Slade et al., 1993) may be less successful in salmonids because of gene duplication. PCR primers designed without detailed knowledge of differences between duplicated loci will likely amplify sequences from both loci, making interpretation difficult. DNA techniques will also have to deal with the complexities of inheritance that have been encountered with isozymes, as discussed previously.

Quantitative Genetics. The genetic system of salmonids will also affect inheritance of polygenic traits as interpreted by methods in quantitative genetics. All published studies of quantitative genetic variation in salmonids (of which we are aware) have used the classic models of quantitative genetics that assume disomic Mendelian inheritance. The fundamental assumptions of these models are not met in salmonids because of residual tetrasomic inheritance and pseudolinkage. Thus, the salmonid genetic system complicates estimation of primary quantitative genetic parameters, such as heritability and genetic components to phenotypic variability (Ehlke and Hill, 1988).

Inbreeding and Outbreeding Depression. Inbreeding depression is caused by two different effects: increase in frequency of deleterious alleles, and loss of allelic variation at loci where heterozygotes are superior. Both effects are the result of genetic drift causing random changes in allele frequencies. Duplicated disomic loci may be somewhat "sheltered" against the harmful effects of an increase in the frequency of a non-functional allele, because of the presence of the second locus. Tetrasomic loci are "buffered" against effects of genetic drift because of the greater number of gene copies. Nonetheless, Kincaid (1976a, 1976b) found in rainbow trout that a 25% loss of heterozygosity due to inbreeding was associated with a pronounced increase in morphological deformities (38%), decreased fry survival (19%), and decreased weight at 364 days (23%).

Outbreeding depression is a reduction in fitness due to mating of genetically divergent individuals. Like inbreeding depression, outbreeding depression can result from two different mechanisms: loss of local adaptation, or breakdown of coadapted genes or chromosomes at different loci (Templeton, 1986; Lynch, 1991). Whereas reductions in fitness due to loss of local adaptation may occur in the F_1 generation, outbreeding depression due to breakdown of coadapted gene complexes is generally not expected until the F_2 generation, because F_1 hybrids retain an entire chromosomal array from each parent.

Pink salmon are an exceptional salmonid species for evolutionary genetic studies because of their rigid two-year semelparous life cycle (Heard, 1991). The temporal isolation has produced genetically distinct odd- and even-year groups in most streams and rivers throughout the natural range of pink salmon. A large number of allozyme studies have revealed relatively large differences between odd- and even-year fish from the same locality, with relatively little divergence between geographic populations within odd- and even-year groups (Beacham et al., 1985; Gagal'chii, 1986). Thus, there appears to be very little genetic exchange between odd- and even-year groups, but there is substantial genetic exchange within these groups throughout their geographic range in western North America and eastern Asia. Gharrett and Smoker (1991) found evidence of severe outbreeding depression in F_2 hybrids between even- and odd-year pink salmon from the same stream in Alaska. The reduction in fitness cannot be attributed to loss of local adaptation since these populations are native to the same stream. Furthermore, the appearance of outbreeding depression in the F_2, but not the F_1 generation, supports the hypothesis that breakdown of gene or chromosomal complexes was involved.

One of the more controversial issues in the conservation genetics of salmon concerns the relative importance of inbreeding and outbreeding depression. These phenomena can be viewed as opposite ends of a continuum of breeding systems (Figure 8.6). Presumably, between these two extremes is a regime of higher fitness associated with an "optimal" level of outbreed-

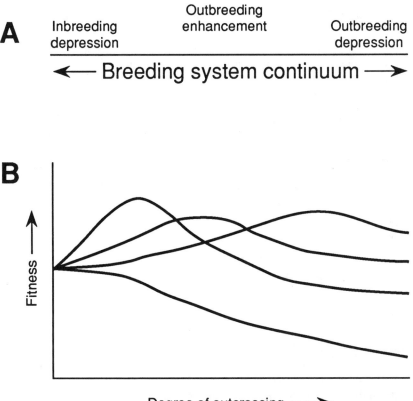

A

| Inbreeding depression | Outbreeding enhancement | Outbreeding depression |

← Breeding system continuum →

B

Fitness ↑

Degree of outcrossing →

Figure 8.6. **A.** Continuum of breeding systems that, at opposite extremes, can lead to inbreeding depression or outbreeding depression. **B.** Several possible forms of the relationship between fitness and the degree of outcrossing. Key questions in evaluating conservation strategies are: (1) What is the form of this relationship for the species in question? and (2) Where on the continuum would the conservation strategy place the population? (Waples, 1995b).

ing. In one view, strong homing fidelity of salmon leads to inbred local populations, which might benefit from infusion of new genes from other populations (e.g., through stock transfers). An alternate view is that natural straying generally provides sufficient levels of outbreeding, and that high levels of genetic exchange risk fitness reductions through loss of adaptive combinations of genotypic variation at multiple loci.

We agree with the latter view, especially with regard to anadromous salmon. The lack of fixed allele frequency differences throughout the range of salmon (see next section) suggests that historically there has been sufficient gene flow to avoid problems of isolation and inbreeding. Moreover, the complex life cycle of salmon makes outbreeding depression due to loss of local adaptation more likely. In addition, the polyploid genetic system of salmonids may minimize effects of inbreeding depression and exacerbate problems of outbreeding depression because of pairing and recombination between homeologous chromosomes in hybrid populations.

Conservation Relevance

The unusual inheritance patterns in salmonids discussed previously must be taken into account when considering conservation of these species. The polyploid ancestry of salmonids provides the genetic architecture upon which the evolutionary forces of mutation, natural selection, genetic drift, and migration have acted during the evolutionary history of these fishes. Moreover, interpretation of genetic experiments and molecular variation in salmonids must reckon with this genetic system.

POPULATION STRUCTURE AND LOCAL ADAPTATION

Population Genetic Structure

Protein electrophoresis was introduced to the study of population genetic structure of salmonids in 1969 and has been widely used for over two decades (reviewed by Utter, 1991). As a result of these efforts, the cumulative amount of allele frequency data for salmonids is huge, perhaps exceeded only by datasets for humans and *Drosophila*. For example, geneticists from a number of state and federal agencies have compiled a dataset for chum salmon that includes allele frequencies for 50–75 gene loci from over 150 populations throughout the Pacific rim (Washington, British Columbia, Alaska, Russia, and Japan). Large datasets are also available for other species of Pacific salmon, several species of trout, and Atlantic salmon. The genetic differences among salmon populations also provide natural markers that have been used to identify the natal origin of fish caught in mixed fisheries (Box 8.3).

DNA techniques have been used increasingly to study population structure of salmonids over the last decade. For example, researchers in at least a dozen laboratories are using DNA assay procedures to study Pacific salmon (Park et al., 1994), and extensive efforts are also underway for other salmonids in North America and Europe. As more informative markers are developed and broadly applied, substantial sets of DNA data will also accumulate.

Although a detailed review of the population structure revealed by allozyme and DNA data is beyond the scope of this chapter, some general

Box 8.3. Genetic Stock Identification

One of the most challenging problems in salmon management is conservation of wild populations harvested incidentally in mixed-stock fisheries that target more abundant (generally hatchery) fish. An approach known as Genetic Stock Identification (GSI), which relies on naturally occurring genetic differences among populations, was developed to address this problem (Grant et al., 1980; Milner et al., 1981).

GSI uses a maximum-likelihood algorithm to provide estimates of stock composition based on multilocus genotypes in a sample from the mixture, and in samples from source populations (the "baseline") that potentially may contribute to the mixture. A number of researchers have examined various statistical aspects of the mixture analysis (e.g., Fournier et al., 1984; Pella and Milner, 1987; Millar, 1987; Wood et al., 1987; Waples, 1990; Waples and Aebersold, 1990; Xu et al., 1994). Smouse et al. (1990) modified the algorithm to allow analysis of mixtures in the absence of complete baseline data.

Most GSI applications have taken advantage of the large baseline datasets that have been generated by protein electrophoresis. In one current application, geneticists from the Washington Department of Fisheries and Wildlife provide weekly, real-time stock composition estimates for a chinook salmon gillnet fishery in the lower Columbia River. The fishery targets relatively abundant, early-returning hatchery fish from the Willamette River, but endangered Snake River chinook salmon begin to move through the fishing area near the end of the run. The fishery is terminated when GSI estimates indicate that fish from the Snake River begin to appear in significant numbers. Other recent applications have included analysis of coastal fisheries involving stocks from the United States and Canada (Utter et al., 1987), analysis of salmon captured illegally in high-seas driftnet fisheries, incorporation of mtDNA data (Marsden et al., 1989), and expansion to regions outside the Pacific Northwest and to species other than salmonids (e.g., Chapter 7).

---

patterns are clear. First, the level of divergence among freshwater-resident populations of salmonids is much higher than for anadromous populations. For example, Gyllensten (1985) summarized indices of gene diversity in marine, anadromous, and freshwater fishes, including a number of salmonid species. For freshwater salmonid species, an average of 20.4% of total genetic diversity was due to differences between populations, whereas the comparable value was only 3.7% for anadromous salmonids. Furthermore, it is not uncommon to find fixed allelic differences between geographically isolated freshwater populations of salmonids (Ryman et al., 1979; Allendorf and Leary, 1988). However, fixed differences are rare in anadromous populations, even if extremes of the range are considered. Thus, anadromous populations within a species are generally characterized by differences in frequency of the same suite of alleles, rather than qualitative differences in the kind of alleles. Fortunately, a large number of variable gene loci can be resolved by protein electrophoresis for most salmonid species, so the overall power to resolve population structure can be quite good in anadromous species, despite the general absence of diagnostic markers.

Second, for most salmonid species studied, there is a strong geographic component to population genetic structure, with geographically proximate populations generally being more similar than those that are more distant (Beacham et al., 1985; Foote et al., 1989; Utter et al., 1989). For example, Waples and Smouse (1990) found that genetic distances between North American chinook salmon populations in different fishery management groups (defined largely by geographic proximity) averaged several times greater than distances between populations from the same management group.

Genetic Structure and Life-History Variation

Two life-history traits are of particular interest with respect to population genetic structure of Pacific salmon: adult run timing and anadromy/nonanadromy. *Run-time designations* refer to the time of year that adults enter freshwater to begin their spawning migration. Thus, in the United States, steelhead are commonly known as summer- or winter-run, and coho salmon populations are often characterized as being early- or late-run. In the Columbia River Basin, three temporal runs of chinook salmon are recognized: spring-, summer-, and fall-run. Allozyme data clearly show that fall-run chinook salmon within the Columbia River Basin are monophyletic, being quite distinct as a group from all spring-run chinook salmon in the basin. In contrast, summer-run chinook salmon are polyphyletic, with summer-run fish from the upper Columbia River showing a high degree of genetic similarity with fall-run chinook salmon, and summer-run fish from the Snake River being genetically much more similar to spring-run chinook salmon (Matthews and Waples, 1991).

Both spring and fall chinook salmon are found in many streams in other regions of the Pacific Northwest, and in many of these a different pattern is observed: populations with different run timing from the same stream are more similar genetically than populations with similar run timing from different areas (Utter et al., 1989). Thus, it is apparent that run-timing differences have evolved independently many times in chinook salmon, and the same appears to be true for other species of anadromous Pacific salmonids (e.g., Utter and Allendorf, 1978; Chilcote et al., 1980; Thorgaard, 1983).

Many salmonid species (e.g., sockeye salmon, coho salmon, rainbow trout, cutthroat trout, brown trout, Atlantic salmon, masu salmon, Arctic char, bull trout, and Dolly Varden) also have both anadromous and resident life histories. For a number of these species, genetic data are available for both anadromous and resident populations, and these data show some consistent patterns. First, almost all studies that have considered anadromous and resident forms from the same area have found evidence of significant reproductive isolation between them (reviewed by Johnson et al., 1994). This can be true even if the two forms spawn at the same time and place (Foote et al., 1989).

Within a single geographic area, multiple subpopulations of each life-history type may exist, with each type forming a genetically coherent group separate from the other (e.g., the multiple subpopulations of sockeye salmon and kokanee in the Babine Lake system in British Columbia; Foote et al., 1989). However, studies have also consistently found that different life-history types from the same area are more similar genetically than either is to the same form from a different geographic area. This pattern is consistent with the hypothesis that the anadromous form has been responsible for colonizing new habitat, and the resident forms have evolved independently several times in each species.

Sympatric and ecologically specialized "morphs" have been described in many salmonid species (e.g., brown trout, Arctic char, and lake whitefish; Behnke, 1972). Normal and "dwarf" sympatric morphs often differ dramatically in body size (Figure 8.7). Perhaps the best studied system is a lake in Iceland, in which four morphs of Arctic char have been described (Box 8.4).

The pattern of genetic relationships among these normal and dwarf morphs generally parallels that for anadromous and resident forms of salmon previously discussed; that is, genetic divergence is greater between populations of the same morph from different lakes than between different morphs from the same lake (e.g., Hindar et al., 1986). Thus, it appears that these morphs have evolved many times by sympatric differentiation, rather than representing sibling species that evolved allopatrically and later invaded the same lakes. However, it is also possible that such morphs evolved allopatrically, and that the present pattern of genetic divergence at allozymes is due to introgression between sympatric populations. If so, differences between the morphs at loci responsible for ecological differentiation must be maintained by natural selection.

Bernatchez and Dodson (1990) have described sympatric normal and dwarf forms of lake whitefish (*Coregonus clupeaformis*) that do not fit the general pattern described previously. Two distinct monophyletic mtDNA genotypes with different geographic distributions were found to exist in northern Maine and eastern Canada. The form with the more western distribution exhibited the normal-size phenotype in all samples, but the eastern

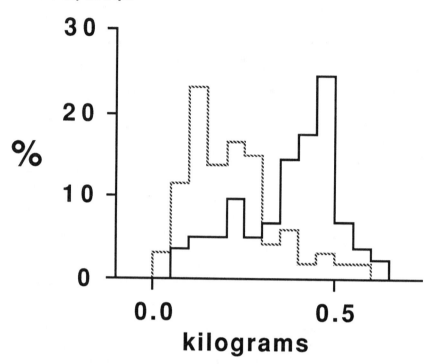

Figure 8.7.   Distribution of weights of five-year-old fish from two sympatric morphs of brown trout identified by allozyme genotypes in a small mountain lake in Sweden (Ryman et al., 1979). Mean weights of the dwarf morph (dashed line) and the normal morph (solid line) are 0.22 and 0.38 kg, respectively.

form exhibited the normal form in allopatry and the dwarf form when found in sympatry with the western form. Thus, the existence of these sympatric forms apparently results from the secondary contact of two monophyletic groups of whitefish that evolved allopatrically during the last glaciation events.

Local Adaptation

Two factors are conducive to the evolution of local adaptations: repro- ductive isolation and variation in selective regimes. Both factors are char- acteristic of salmonid species, many of which occur in a large number of semidiscrete populations over a broad geographic area. Opportunities for strong reproductive isolation are clearly greater for freshwater species than anadromous ones. As noted, this isolation is reflected in substantial genetic differences between freshwater salmonid populations in presumably neutral

Box 8.4. Arctic Char in Thingvallavatn, Iceland

Four sympatric morphs of Arctic char in Thingvallavatn, Iceland, have been described on the basis of morphology, habitat use, diet, parasites, life history, time and place of spawning, and early ontogeny (Sandlund et al., 1992). Thingvallavatn is Iceland's largest lake (83 km^2) and was formed at the end of the last glaciation, approximately 10,000 years ago. The only fish species in Thingvallavatn are Arctic char, brown trout, and threespine stickleback (*Gasterosteus aculeatus*).

Two basic types of char in Thingvallavatn each include two morphs: (1) a benthic type with an overshot mouth and large pectoral fins which has a large (LB-) and small (SB-) morph; (2) a pelagic type with a terminal mouth and small pectoral fins which has a planktivorous (PL-) and piscivorous (PI-) morph. LB-char mature sexually at eight years of age, where as SB-char mature at two years (males) and four years (females). Male and female PL-char mature sexually at four and five years, respectively, and PI-char mature sexually at six years. LB-char spawn in July and August, SB-char from August to November, and PL- and PI-char spawn from September to November.

Controlled laboratory studies of these morphs have shown a partial genetic basis to the morphological differences (Skùlason et al., 1989). However, the four morphs are genetically similar at nuclear and mitochondrial molecular markers. Only the SB-char are distinct on the basis of genetic variation at five allozyme loci. The pattern of mtDNA divergence parallels the allozymes (i.e., SB-char are the most divergent), but the differences are not statistically significant (Danzmann et al., 1991).

characters detected by traditional molecular methods. However, isolated freshwater habitats may share many physical and biotic characteristics, resulting in similar life histories and selective regimes.

Compared to freshwater forms, anadromous salmonids have less potential for isolation, but their complex life histories provide ample opportunity for differences in selective regimes to be manifest. For example, to complete its entire life cycle, an anadromous salmonid must do all of the following: develop as an embryo and emerge at a time appropriate for its natal stream or lake; forage effectively and avoid predators as a juvenile; find cool water for rearing, or have the physiological capability to tolerate elevated water temperatures; complete the complex physiological process of smoltification at a time favorable for outmigration; migrate to sea at an appropriate time to promote survival and growth in the early ocean phase; migrate to oceanic regions with adequate food supply to allow survival for one or more years; reverse the physiological process of smoltification on returning to freshwater as an adult; home accurately to the natal stream and find a mate; and, if female, deposit eggs at an appropriate time, place, and depth to promote survival of offspring. Disruption of even a single link in this chain can lead to a drastic reduction in fitness.

This adaptation to local conditions is demonstrated by the timing of spawning among populations of sockeye salmon in the Fraser River system (Brannon, 1987). Although this timing varies little from year to year within spawning location, there are great and consistent differences among spawning areas because of adaptations to the most favorable conditions for incubation, timing of emergence, and juvenile feeding (Burgner, 1991). The timing of spawning for local populations in the Fraser River is primarily influenced by the temperature regime of the spawning site (Figure 8.8). Spawning is later in the warmer incubation environments. This pattern of differences among populations results in similar emergence times of progeny during the following spring because of the faster rate of development at higher temperatures.

In summary, there is strong evidence that spawning populations of anadromous salmonids exhibit highly specific local adaptations for a number of different traits. These adaptations must be the result of genetic differences between local populations at many loci. On this basis, we would expect it

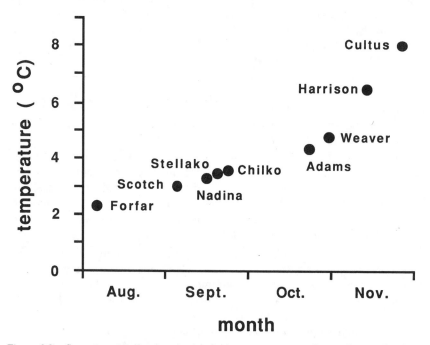

Figure 8.8. Spawning times and mean incubation temperatures of nine demes of sockeye salmon from the Fraser River (Brannon, 1987).

to be difficult to "replace" a local population with transplants from non-local populations. The more complex the life cycle, the more difficult it should be to replace a local population. Complexity would be influenced by the number of migrations and major habitat shifts. A break in any of these links would cause reproductive failure. For a newly introduced population to be successful, at least a few individuals must be able to complete the entire life cycle and reproduce.

The experience with attempted introductions of salmon populations supports these conclusions. Stock transfers within the range of Pacific salmon have been largely unsuccessful, both in North America (Withler, 1982) and Asia (Altukhov and Salmenkova, 1987). For example, efforts to reintroduce sockeye salmon into appropriate habitat in the Fraser River system generally have failed. This has been particularly true of efforts to introduce lower river stocks into upriver areas. Sockeye salmon from the lower Fraser River apparently lack sufficient energy reserves for the long migration into upriver areas that were depleted after a rock slide impeded upstream migration early in this century (Foerster, 1968).

Riddell (1993) has described an example from the Adams River, a tributary of the Fraser River. A logging dam built in 1908 blocked access of sockeye salmon to the Upper Adams River from 1908 to 1921; these runs had been among the largest sockeye runs in the Fraser River system. This area has 1.2 million m^2 of spawning area, which should be sufficient to support six million adult sockeye per year (based on productivity of other sockeye populations in the area). Sixteen attempts between 1949 and 1975 to reintroduce sockeye to these spawning areas have not been successful in reestablishing this run. Today, only a very few fish return to spawn in the Upper Adams River.

In contrast to the lack of success with anadromous sockeye salmon, introductions of kokanee (nonanadromous sockeye salmon) often have been successful (Burgner, 1991). The greater success of introduced kokanee is probably related to their simpler life history in comparison to anadromous sockeye (Figure 8.3). Similarly, introductions of trout to lakes and streams throughout the intermountain west of North America generally have been successful (Allendorf and Leary, 1988).

Conservation Relevance

The sympatric occurrence of ecologically specialized life-history types, or morphs, provides several conservation challenges. The initial problem is the identification of polytypic forms. The amount of genetic divergence at molecular markers is likely a reflection of patterns of genetic exchange between morphs and genetic drift within morphs. Two morphs may be genetically similar at molecular markers that are effectively neutral, or subject to similar selection pressures, but be greatly differentiated at loci under strong natural selection because they are responsible for ecological specializations (Allendorf, 1983).

In evaluating the importance of local adaptation to the conservation of salmonids, we are faced with a paradox. On the one hand, the preponderance of evidence for local adaptation suggests that conservation should focus on small units (locally adapted spawning populations). On the other hand, studies of run timing and anadromy/nonanadromy indicate that these important life-history traits have evolved independently many times within salmon species, and hence are likely to evolve again. The observation that virtually all of the diversity within Pacific salmon species in many parts of the Pacific Northwest has evolved since the last glacial episode ended about 10,000 years ago (McPhail and Lindsay, 1970) is often cited as further evidence for the essentially plastic life history of salmon.

We believe that resolution of this conundrum depends largely on the timescale of concern. In evolutionary time, salmonids have displayed a remarkable capacity for variation in life-history traits. Anadromous forms provide the migratory capability to colonize new habitat, and strong philopatry allows genetically based life-history traits to develop in local populations. However, experience shows that this process cannot be relied upon to occur within a human time frame. The failure of most stock transfers of anadromous salmonids to produce self-sustaining natural populations suggests that if local populations are not conserved, we will be unlikely to reestablish them within years or even decades. Furthermore, recolonization and life-history diversification is a dynamic process that depends heavily on the viability of the species as a whole. A species suffering widespread declines in abundance and loss of local populations is unlikely to provide much impetus in this direction.

The ESA provides a powerful legal mechanism to avoid species extinctions by conserving local populations. In addition to taxonomic species and subspecies, the ESA allows "distinct populations" of vertebrates to be listed as threatened or endangered "species." Unfortunately, the ESA provides no explanation of how to determine which populations are distinct, and this has resulted in confusion and inconsistent application of the provision to list vertebrate populations. Recently, concerns about widespread declines in Pacific salmon (e.g., Nehlsen et al., 1991) raised the question as to which (if any) populations should be considered "distinct" and hence eligible for protection under the ESA. Opinions about the appropriate conservation unit under the ESA vary widely, from local spawning populations at one extreme to the entire biological species at the other.

To provide a consistent, scientifically based framework for interpreting the meaning of "distinct" populations of salmon under the ESA, Waples (1991b) developed an approach based on the "Evolutionarily Significant Unit" (ESU: Ryder, 1986; Dizon et al., 1992), and this approach formed the basis for the policy on this issue adopted by the National Marine Fisheries Service (NMFS). The unifying theme of Waples' ESU concept is conservation of the evolutionary legacy of the biological species; that is, the genetic variability that is a product of past evolutionary events and which represents the reservoir upon which future evolutionary potential depends. The goal is thus to ensure viability of the biological species by conserving enough of its basic components to allow the dynamic process of evolution to proceed largely unaffected by human influence. Waples (1991b) advocated a holistic approach to identifying ESUs and provided specific guidance on how to integrate genetic, phenotypic, life-history, ecological, and geographic information. Waples' ESU concept is inherently hierarchical and applies equally to local populations and groups of populations (metapopulations). The focus is on units that are largely independent over evolutionarily important time frames, and the "bottom line" test for an ESU is whether its demise would represent a significant loss of ecological/genetic diversity to the species as a whole.

The NMFS policy has been used in several ESA listing determinations for Pacific salmon (summarized by Waples, 1995a). The ESU concept has also generated considerable interest and controversy in the broader field of conservation biology (e.g., Dizon et al., 1992; Moritz, 1994; Vogler and DeSalle, 1994), and was the subject of a recent conference on identifying conservation units for aquatic species sponsored by the American Fisheries Society (Anonymous, 1994).

CAPTIVE BREEDING AND HATCHERIES

In comparison to many other vertebrate species of particular interest to conservationists (e.g., mammals, birds, reptiles), fish are highly fecund. Although salmonids are not exceptional among fishes in this respect, the number of eggs spawned in a season by female salmonids ranges from several hundred for freshwater species to several thousand for anadromous species. Conversely, juvenile salmonids typically experience a high natural mortality in early life stages (95% mortality in the first 1–2 years is common). That this combination of high fecundity and high juvenile mortality in nature provides an opportunity to greatly increase production through artificial propagation has long been recognized, and efforts to artificially spawn and propagate salmonids can be traced to at least the 14th century in Europe.

Salmonid Propagation Today

Today, artificial propagation of salmonids is a worldwide enterprise of enormous proportions, with billions of juvenile fish produced each year. Three major types of operation are involved. Trout hatcheries typically raise freshwater species in captivity through an entire life cycle, with the excess production used for food or for stocking lakes and streams. Public salmon hatcheries generally involve a type of sea ranching, in which adults are spawned artificially and their progeny are reared in captivity until release prior to their smolt migration. Adult production in excess of broodstock requirements is generally targeted for harvest. In salmon farming, salmon are raised to maturity in captivity, most commonly in marine net pens, with excess production marketed for food.

Major objectives of salmonid propagation include introductions and stock transfers, and increased production for mitigation or enhancement. These objectives shape the nature and scope of propagation activities, with resulting effects on conservation of wild salmonid resources.

Introductions and Stock Transfers.   As with many plant and animal species worldwide, transplantation efforts for salmonids have a long history. Attempts have been made to introduce salmon into most parts of the globe, with a few notable successes (e.g., New Zealand, Chile, and the North American Great Lakes). Within the native range of salmon species, stock transfers are common even today. Such transfers can contribute to breakdown of population structure and loss of diversity between populations, risk reductions in fitness through outbreeding depression or loss of local adaptation, and deplete wild populations tapped as a source for broodstock (reviewed by Waples, 1991a and 1995b).

Another consequence of repeated introductions and stock transfers of freshwater salmonids is that non-native salmonids have become dominant components of aquatic ecosystems in many areas. For example, brown trout (native to Europe and Asia) are now common in many areas of North America, and rainbow trout have been widely distributed around the world. Brook trout and rainbow trout (native to eastern and western North America, respectively) have both been widely transplanted across the continent. These introductions have led to a wide variety of adverse ecological interactions with native fish species, both salmonid and non-salmonid (e.g., Hebert and Billington, 1991). Adverse effects of these introductions can be genetic as well as ecological. For example, perhaps the greatest danger to conservation of cutthroat trout is introgressive hybridization with rainbow trout (Box 8.5; Allendorf and Leary, 1988). Interbreeding between cutthroat trout and non-native rainbow trout appears to be common and widespread throughout the natural range of cutthroat trout.

Box 8.5. Conservation of Cutthroat Trout

The cutthroat trout is a polytypic species with a wide geographic range that includes the coastal and interior waters of most of the western United States and Canada, from Alaska to the Pecos River of New Mexico (Allendorf and Leary, 1988). As many as 16 subspecies have been recognized in the recent literature; 11 subspecies currently have protected legal status in one or more states, and two subspecies apparently are extinct. The genetic distance between subspecies based on enzyme loci ranges from that usually seen between congeneric species to virtual genetic identity. Subspecies from the western portion of the range of the cutthroat trout are genetically more similar to rainbow trout at enzyme loci than they are to the other subspecies of cutthroat trout.

The greatest danger to the trout of the American West is the introduction of non-native species. Introduced taxa often have serious harmful effects on native taxa through competition or the introduction of pathogenic agents. In cutthroat trout, however, the main impact of introductions has been interbreeding between native and introduced fishes. Hybrid progeny from these matings are often fertile and capable of interbreeding with each other and the parental taxa. This situation destroys the genetic integrity of native populations and results in introgressed populations or hybrid swarms. That is, genes from both parental types are randomly distributed among individuals so that no individual is a representative of either parental taxon. Molecular analysis of isozymes and mtDNA have provided a more reliable and informative means of detecting interbreeding than morphological characters (Allendorf and Leary, 1988). Interbreeding between cutthroat trout and non-native rainbow trout is common throughout the natural range of the cutthroat.

Increased Production. Humans have long recognized their ability to adversely affect natural salmon production through overfishing and habitat degradation. For nearly as long, we have viewed salmon hatcheries as a convenient way to compensate for lost natural production without having to face the difficult measures necessary to correct the problems. This concept of mitigation has been formalized in programs such as the Mitchell Act passed by Congress in 1938, which directed construction and federal funding of over 20 salmon hatcheries to mitigate "in perpetuity" lost production caused by hydroelectric facilities in the Columbia River Basin.

The results of mitigation efforts have generally not lived up to expectations. Enthusiasm for the early efforts at propagating Pacific salmon waned when it was recognized that the survival advantage which hatcheries provide in early life stages does not necessarily translate into increased adult survival (Lichatowich and McIntyre, 1987). Because of essentially nonexistent monitoring programs, however, this realization was slow in coming; in the meantime, many wild stocks were placed at risk from collections of broodstock to support the hatcheries, or from competition with large numbers of juvenile hatchery fish released into natural rearing habitat.

Following improvements in fish health and nutrition, and a shift (beginning in the late 1950s) toward releasing larger juveniles just prior to their smolt migration, many Pacific salmon hatcheries now realize a substantial adult–adult survival advantage over wild populations. Hatchery fish now greatly outnumber wild fish in many areas, and are the mainstay of numerous fisheries (Northwest Power Planning Council, 1987). These successes, however, have not come without cost to wild populations. Although the number of hatchery fish has increased substantially in recent years, total salmon abundance generally has not. Thus, in many cases there has simply been a replacement of wild fish with hatchery fish (Walters and Riddell, 1986; Hilborn, 1992). Furthermore, although salmon spawn in discrete freshwater localities, they are typically harvested at sea, where fish from many geographic areas may commingle. When hatcheries do produce large adult returns, there is pressure to increase harvest rates to take advantage of this resource, with severe but predictable consequences for less abundant wild stocks captured incidentally in mixed-stock fisheries.

Other adverse direct and indirect interactions between hatchery and wild fish include hybridization and breakdown of population structure following the straying of hatchery fish, and random and directional genetic changes associated with fish culture (Waples, 1991a). The recent proliferation of marine net pens for culturing salmon has led to a whole new phase of problems for wild salmon populations. The situation is most acute in Norway, where escaped fish from net pens have represented the majority of the Atlantic salmon spawning in many streams in recent years (e.g., Gausen and Moen, 1991).

Genetic Changes in Hatchery Populations

One objective of many hatchery programs is to maintain the genetic characteristics of the population in order to maximize the probability of success when the population is reintroduced into the wild. Two major types of harmful genetic change can occur during captivity: (1) changes in allele frequencies and loss of genetic variation through genetic drift; (2) adaptation to captivity through natural selection; that is, genotypes best suited for survival and reproduction in captivity will tend to increase in frequency.

Random Changes.   Hatchery practices can reduce effective population size and increase rates of inbreeding and loss of genetic variability, and there are a number of examples in which this has occurred in salmonids (reviewed by Allendorf and Ryman, 1987). However, measures that can minimize random genetic changes in fish culture are fairly straightforward and well recognized, if not always followed. Thus, although this aspect of salmon hatcheries has probably received the greatest amount of attention from conservationists, it is largely a manageable problem (Allendorf and Ryman, 1987).

Selection (Artificial, Natural, or Inadvertent). Darwin (1898) pointed out that selection for tameness and other adaptations to a captive environment is inevitable. Characteristics selected under captivity generally are disadvantageous in the natural environment (Spurway, 1952; Kohane and Parsons, 1988). In the past, it was not uncommon for salmon hatcheries to select broodstock or to cull populations on the basis of phenotypic traits considered desirable in cultured fish (e.g., size, age, sex, appearance, spawn- or run-timing, etc.). There is a growing awareness that phenotypes selected for in the hatchery may not be adaptive in the natural environment, and it is probably accurate to say that there are conscious efforts to avoid artificial selection in most salmon hatcheries today.

However, despite good intentions, selection can also occur inadvertently as a result of fish culture. For example, a common phenomenon in salmon hatcheries is advancement of adult run- and spawn-timing relative to the wild source population. This is likely a result of inadvertent selection due to either or both of the following factors: (1) Egg-take goals may be met by using only the early part of the run for hatchery broodstock; and (2) Progeny of early-spawning hatchery fish may grow faster and survive better in culture. Active measures may be necessary to counteract these trends in cultured populations.

Finally, even the most enlightened hatchery program cannot avoid all selective changes. This can be deduced from first principles: If the hatchery is successful in rearing a high proportion of offspring through early life stages, it will have profoundly altered the mortality profile of the population (Waples, 1991a). The result, inevitably, will be genetic change relative to the wild population. Efforts to alter the hatchery environment to more closely mimic natural conditions may reduce the magnitude of selective changes but cannot escape them altogether.

Conservation Relevance

In the last decade, there has been increasing interest in using salmon hatcheries to aid in conservation of wild populations. Possible roles for artificial propagation in salmonid conservation include avoiding immediate extinction of endangered populations, maintaining depressed populations until habitat problems or other factors limiting survival are addressed, reseeding vacant habitat, and speeding recovery by increasing abundance of naturally spawning fish. If their potential to provide a large survival advantage is realized, salmon hatcheries can play an important role in temporarily reducing risks faced by depleted wild populations. On the other hand, the artificial culture environment also creates risks for the wild population, such as physiological or behavioral changes, domestication selection, erosion of genetic variability, and catastrophic loss due to disease, human error, or technological failure. Therefore, determining the appropriate role of artificial propagation in salmonid conservation involves evaluating relative risks to the wild populations of using versus failing to use artificial propagation (Hard et al., 1992).

In some cases, risks posed by artificial propagation are clearly warranted because of threat of imminent extinction to a wild population. For example, because of high demographic and genetic risks faced by Snake River sockeye salmon (see Box 8.6), it was determined that an aggressive artificial propagation program provided the best chance to avoid immediate extinction or further erosion of genetic variability. A total of 13 adults returned between 1991 and 1993, and these were captured and spawned artificially. The offspring of these matings are being reared to maturity in captivity, and release of their progeny into the natural habitat is scheduled to begin in 1995. Although this captive broodstock program is experimental and involves substantial risk, it is hoped that it will help the population move quickly

---

Box 8.6. Endangered Redfish Lake Sockeye Salmon

Snake River sockeye salmon are considered an ESU, and hence a "species" under the ESA, primarily because of their distinctive habitat (they spawn and rear at an elevation of 2,000 meters) and their remarkable freshwater migration (1,500 km to Redfish Lake in the Sawtooth Mountains of Idaho; Waples et al., 1991). When the population was listed under the ESA as endangered in 1991, a total of less than 30 adults had returned to spawn in the previous five years. One of the key issues in developing a recovery program has been elucidating the relationship between the anadromous sockeye salmon and the resident kokanee populations in Redfish Lake. Genetic data gathered since the listing provide evidence of reproductive isolation and strong genetic divergence between the two forms. However, a third form of *O. nerka*, termed "residual" sockeye salmon, has recently been discovered in the lake. This apparently represents the first time that all three forms (anadromous and residual sockeye salmon, and kokanee) have been found in the same lake system. The residual sockeye salmon appear to have a close genetic and demographic link to the anadromous population and, although not numerous, represent a potentially valuable source to expand the genetic base of the severely depressed anadromous population.

through an extreme bottleneck. Ultimately, of course, the future of this population depends on satisfactory resolution of factors that contributed to its decline in the first place.

In recent years, supplementation (use of hatcheries to increase abundance of naturally spawning populations) has attracted increasing interest as a tool for conservation of Pacific salmon (e.g., Lichatowich and Watson, 1993). Supplementation broodstock are generally drawn from the wild population or from a mix of hatchery and wild fish, and natural spawners also typically include both hatchery and wild fish. In theory, this approach has considerable potential, because a successful salmon hatchery can produce more returning adults than would have resulted from allowing the fish to spawn naturally. However, a number of issues remain to be resolved before it can be determined whether supplementation is a viable conservation strategy for salmon.

First, although considerable efforts have been made to develop strategies to minimize risks of artificial propagation (e.g., Allendorf and Ryman, 1987; Hard et al., 1992; Cuenco et al., 1993), it is not yet clear how effective these efforts will be. Second, broodstock collection to support an unsuccessful hatchery program can result in depletion of the wild population. Third, even if the hatchery produces more adults than would have been produced in the wild, this still may not result in a sustained increase in the abundance of naturally spawning fish. Hatchery fish may have reduced reproductive success in nature (Nickelson et al., 1986; Leider et al., 1990; Fleming and Gross, 1992), and the typical practice of capturing broodstock and releasing progeny at a fixed location often leads to a spatial redistribution of natural spawning, with declines in productive areas upstream of the collection and release sites. We are not aware of a single empirical example in which supplementation has been successfully used as a temporary strategy to increase the long-term abundance of naturally spawning populations of Pacific salmon.

Ryman and Laikre (1991) pointed out that supplementation can lead to reductions in effective population size of the hatchery/wild system as a whole through augmenting only part of the gene pool of a population (i.e., that part represented by broodstock taken into the hatchery). Waples and Do (1994) discussed the applicability of this concept to salmon in greater detail. Although they found that the genetic consequences of supplementation can depend on a number of factors, the single most important consideration was whether increases in abundance attained by supplementation were sustained after supplementation was terminated. This, in turn, will primarily be determined by whether factors responsible for the original declines (or that are preventing recovery) are adequately addressed.

Finally, the legacy of a century of viewing salmon hatcheries as a means of mitigating losses to natural production (rather than fixing the problems that caused the declines) has resulted in a schizophrenic view of many supplementation programs. The stated objectives of supplementation programs typically include production of excess fish for harvest as well as increases in natural populations. Although these two objectives are not necessarily mutually exclusive, it is clear that pursuit of the former may compromise efforts to conserve wild populations. In addition, the intended duration of supplementation programs is often not clearly articulated at the outset. For example, the long-term consequences of a program designed to give a temporary boost in production to a wild population may differ considerably from those of a program designed to provide indefinite support to a wild population that is not self-sustaining.

The NMFS has determined that under the ESA, salmon hatcheries are not a substitute for conservation of natural populations, and that recovery must be evaluated in terms of viability of the natural population rather than numbers of hatchery fish (Hard et al., 1992).

SUMMARY

1. Salmonid fishes provide an unusual and difficult set of conservation challenges because of their complex life histories. In addition, a bewildering and often contradictory assortment of legal agreements associated with the commercial, recreational, and cultural use of salmonid fishes greatly complicates their conservation and management.

2. The polyploid ancestry of salmonids has provided these fishes with an exceptionally complex system of inheritance. The unusual inheritance patterns must be taken into account when considering the genetics of these species.

3. Genetic adaptations to local environmental conditions have evolved in salmon because of their homing behavior and the widely different environmental conditions in which they occur. These adaptations are of great importance, because it has proven to be extremely difficult to reestablish new populations through introductions once local populations have been lost. Salmon should be managed from the premise that local spawning populations are genetically different, and valuable to long-term production of this resource. Managing from this perspective will protect habitat and also protect the salmonid resource over the longer term.

4. Efforts should also be made to identify and protect remaining native wild populations of salmonids and their environments. Populations with unusual genetic adaptations, or those that occupy atypical habitats, are of

special importance. These populations are non-replaceable in human time frames and must be conserved to protect present and future opportunities. This principle seems self-evident, but risks continue to be imposed on such salmonid populations.

5. One of the more controversial issues in conservation genetics of salmon concerns the relative importance of inbreeding and outbreeding depression. In one view, strong homing fidelity of salmon leads to inbred local populations, which could benefit from infusion of new genes from other populations (e.g., through stock transfers). We believe that natural straying generally provides sufficient levels of outbreeding, and that increased genetic exchange through stock transfers is likely to cause loss of adaptive combinations of genotypic variation at multiple loci. In addition, the polyploid genetic system of salmonids may exacerbate problems of outbreeding depression because of observed pairing of homeologous chromosomes in hybrid populations.

6. Ecologically specialized life-history types (e.g., anadromous–non-anadromous, or normal–dwarf), that are at least partially reproductively isolated, occur sympatrically in a wide variety of salmonid populations and species. The conservation of phenotypic and genotypic diversity in salmonid fishes depends upon identification and management of these populations.

7. In the short term, some hatcheries may promote conservation by minimizing risk of extinction for severely threatened populations. However, hatchery propagation of salmon and trout provides a variety of indirect and direct threats to wild populations. In the long term, artificial propagation will serve a legitimate conservation purpose only if it is viewed as a temporary measure (i.e., as a means to an end, but not an end in itself). This concept is explicit in the ESA, which mandates conservation of threatened and endangered species in their native ecosystems.

REFERENCES

AGELLON, L.B., S.L. DAVIES, C.M. LIN, T.T. CHEN, and D.A. POWERS. 1988. Rainbow trout has two genes for growth hormone. *Molec. Reprod. Devel.* 1:11–17.

ALLENDORF, F.W. 1983. Isolation, gene flow, and genetic differentiation among populations. In *Genetics and Conservation*, eds. C. Schonewald-Cox, S. Chambers, B. MacBryde, and L. Thomas, pp. 51–65. Benjamin/Cummings, Menlo Park, CA.

ALLENDORF, F.W., W.A. GELLMAN, and G.H. THORGAARD. 1994. Sex-linkage of two enzyme loci in rainbow trout. *Heredity* 72:498–507.

ALLENDORF, F.W. and R.F. LEARY. 1988. Conservation and distribution of genetic variation in a polytypic species: the cutthroat trout. *Conserv. Biol.* 2:170–184.

ALLENDORF, F.W. and N. RYMAN. 1987. Genetic management of hatchery stocks. In *Population Genetics and Fisheries Management*, eds. N. Ryman and F. Utter, pp. 141–159. University of Washington Press, Seattle.

ALLENDORF, F.W., J.E. SEEB, K.L. KNUDSEN, G.H. THORGAARD, and R.F. LEARY. 1986. Gene-centromere mapping of 25 loci in rainbow trout. *J. Hered.* 77:307–312.

ALLENDORF, F.W. and G.H. THORGAARD. 1984. Polyploidy and the evolution of salmonid fishes. In *The Evolutionary Genetics of Fishes*, ed. B.J. Turner, pp. 1–53. Plenum, New York.

ALTUKHOV, Y.P. and E.A. SALMENKOVA. 1987. Stock transfer relative to natural organization, management, and conservation of fish populations. In *Population Genetics and Fishery Management*, eds. N. Ryman and F. Utter, pp. 333–344. University of Washington Press, Seattle.

ALTUKHOV, Y.P. and E.A. SALMENKOVA. 1991. The genetic structure of salmon populations. *Aquaculture* 98:11–40.

ANONYMOUS. 1994. Symposium explores defining and protecting units of biodiversity. *Fisheries* 19:27–28.

BEACHAM, T.D., R.E. WITHLER, and A.P. GOULD. 1985. Biochemical genetic stock identification of pink salmon (*Oncorhynchus gorbuscha*) in southern British Columbia and Puget Sound. *Can. J. Fish. Aquat. Sci.* 42:1474–1483.

BEHNKE, R.J. 1972. The systematics of salmonid fishes of recently glaciated lakes. *J. Fish. Res. Bd. Can.* 29:639–671.

BEHNKE, R.J. 1992. *Native Trout of Western North America.* American Fisheries Society Monograph 6, Bethesda, MD.

BELL, F.T. 1937. Guarding the Columbia's silver horde. *Nature* 24:43–47.

BERNATCHEZ, L. and J.J. DODSON. 1990. Allopatric origin of sympatric populations of lake whitefish (*Coregonus clupeaformis*) as revealed by mitochondrial-DNA restriction analysis. *Evolution* 44:1263–1271.

BRANNON, E.L. 1972. Mechanisms controlling the migration of sockeye salmon fry. *Int. Pac. Salm. Fish. Comm. Bull.* 21, 86 pp.

BRANNON, E.L. 1987. Mechanisms stabilizing salmonid fry emergence timing. *Can. J. Fish. Aquat. Sci.* Special Issue 96:120–124.

BURGNER, R.L. 1991. Life history of sockeye salmon (*Oncorhynchus nerka*). In *Pacific Salmon Life Histories*, eds. C. Groot and L. Margolis, pp. 1–117. University of British Columbia Press, Vancouver.

BURNHAM, C.R. 1962. *Discussions in Cytogenetics.* Burgess, St. Paul, MN.

CAMPTON, D.E. 1987. Natural hybridization and introgression in fishes: Methods of detection and genetic interpretations. In *Population Genetics and Fishery Management*, eds. N. Ryman and F. Utter, pp. 161–192. University of Washington Press, Seattle.

CAVENDER, T.M. 1970. A comparison of coregonines and other salmonids with the earliest known teleostean fishes. In *Biology of Coregonid Fishes*, eds. C.C. Lindsey and C.S. Woods, pp. 1–32. University of Manitoba, Winnipeg, Canada.

CHILCOTE, M.W., B.A. CRAWFORD, and S.A. LEIDER. 1980. A genetic comparison of sympatric populations of summer and winter steelheads. *Trans. Amer. Fish. Soc.* 109:203–206.

CLARK, C.W. 1984. Strategies for multispecies management: Objectives and constraints. In *Dahlem Conference*, ed. R.M. May, pp. 303–312. Springer, Berlin.

CUENCO, M.L., T.W.H. BACKMAN, and P.R. MUNDY. 1993. The use of supplementation to aid in natural stock restoration. In *Genetic Conservation of Salmonid Fishes*, eds. J.G. Cloud and G.H. Thorgaard, pp. 269–294. Plenum, New York.

DANZMANN, R.G., M.M. FERGUSON, S. SKÚLASON, S.S. SNORRASON, and D.L.G. NOAKES. 1991. Mitochondrial DNA diversity among four sympatric morphs of Arctic charr, *Salvelinus alpinus* (L), from Thingvallavatn, Iceland. *J. Fish Biol.* 39:649–659.

DARWIN, C. 1898. *The Variation of Animals and Plants under Domestication* (vol. 2). Appleton, New York.

DAVISSON, M.T., J.E. WRIGHT, and L.M. ATHERTON. 1973. Cytogenetic analysis of pseudolinkage of LDH loci in the teleost genus *Salvelinus*. *Genetics* 73:645–658.

DIZON, A.E., C. LOCKYER, W.F. PERRIN, D.P. DEMASTER, and J. SISSON. 1992. Rethinking the stock concept— a phylogeographic approach. *Conserv. Biol.* 6:24–36.

EHLKE, N.J. and R.R. HILL, Jr. 1988. Quantitative genetics of allotetraploid and autotetraploid populations. *Genome* 30:63–69.

FERGUSON, M.M. 1990. The genetic impact of introduced fishes on native species. *Can. J. Zool.* 68:1053–1057.

FERGUSON, M.M., R.G. DANZMANN, and F.W. ALLENDORF. 1985. Absence of developmental incompatibility in hybrids between rainbow trout and two subspecies of cutthroat trout. *Biochem. Genet.* 23:557–570.

FLEMING, I.A. and M.R. GROSS. 1992. Reproductive behavior of hatchery and wild coho salmon (*Oncorhynchus kisutch*)— does it differ? *Aquaculture* 103:101–121.

FOERSTER, R.E. 1968. *The Sockeye Salmon. Fish. Res. Bd. Can.*, Bulletin 162.

FOOTE, C.J., C.C. WOOD, and R.E. WITHLER. 1989. Biochemical genetic comparison of sockeye salmon and kokanee, the anadromous and non-anadromous forms of *Oncorhynchus nerka*. *Can. J. Fish. Aquat. Sci.* 46:149–158.

FORBES, S.H., K.L. KNUDSEN, T.W. NORTH, and F.W. ALLENDORF. 1994. One of two growth hormone genes in coho salmon is sex-linked. *Proc. Natl. Acad. Sci. USA* 91:1628–1631.

FOURNIER, D.A., T.D. BEACHAM, B.E. RIDDELL, and C.A. BUSACK. 1984. Estimating stock composition in mixed-stock fisheries using morphometric, meristic, and electrophoretic characteristics. *Can. J. Fish. Aquat. Sci.* 41:400–408.

FRALEY, J. and B. SHEPARD. 1989. Life history and ecology of bull trout (*Salvelinus confluentus*) in a large lake-river system. *Northw. Sci.* 63:133–143.

GAGAL'CHII, N.G. 1986. Biochemical polymorphism of Kamchatka pink salmon *Oncorhynchus gorbuscha* (Walb.). II. Allele frequencies at polymorphic loci in an even-year brood. *Genetika* 22:2839–2846.

GAUSEN, D. and V. MOEN. 1991. Large-scale escapes of farmed Atlantic salmon (*Salmo salar*) into Norwegian rivers threaten natural populations. *Can. J. Fish. Aquat. Sci.* 48:426–428.

GHARRETT, A.J. and W.W. SMOKER. 1991. Two generations of hybrids between even-year and odd-year pink salmon (*Oncorhynchus gorbuscha*): a test for outbreeding depression? *Can. J. Fish. Aquat. Sci.* 48:1744–1749.

GOODMAN, M.L. 1990. Preserving the genetic diversity of salmonid stocks: A call for federal regulation of hatchery programs. *Environ. Law* 20:111–166.

GRANT, W.S., G.B. MILNER, P. KRASNOWSKI, and F.M. UTTER. 1980. Use of biochemical genetic variants for identifications of sockeye salmon (*Oncorhynchus nerka*) stocks in Cook Inlet, Alaska. *Can. J. Fish. Aquat. Sci.* 37:1236–1247.

GYLLENSTEN, U. 1985. The genetic structure of fish: differences in the intraspecific distribution of biochemical genetic variation between marine, anadromous, and freshwater species. *J. Fish Biol.* 26:691–699.

HAAS, G.R. and J.D. MCPHAIL. 1991. Systematics and distributions of Dolly Varden (*Salvelinus malma*) and bull trout (*Salvelinus confluentus*) in North America. *Can. J. Fish. Aquat. Sci.* 48:2191–2211.

HARD, J.J., R.P. JONES, Jr., M.R. DELARM, and R.S. WAPLES. 1992. *Pacific Salmon and Artificial Propagation under the Endangered Species Act.* U.S. Dept. Commer., NOAA Tech. Memo. NMFS-NWFSC-2, 56 pp.

HARTLEY, S.E. 1987. The chromosomes of salmonid fishes. *Biol. Rev. Camb. Philos. Soc.* 62:197–214.

HEALEY, M.C. 1991. Life history of chinook salmon (*Oncorhynchus tshawytscha*). In *Pacific Salmon Life Histories*, eds. C. Groot and L. Margolis, pp. 311–393. University of British Columbia Press, Vancouver.

HEARD, W.R. 1991. Life history of pink salmon (*Oncorhynchus gorbuscha*). In *Pacific Salmon Life Histories*, eds. C. Groot and L. Margolis, pp. 119–230. University of British Columbia Press, Vancouver.

HEBERT, P.D.N. and N. BILLINGTON. 1991. Proceedings of the ecological and genetic implications of fish introductions symposium. *Can. J. Fish. Aquat. Sci.* 48 (Suppl. 1).

HEDRICK, P. and P. MILLER (eds.). 1994. Endangered Pacific salmonids. *Conserv. Biol.* 8:863–894.

HILBORN, R. 1992. Hatcheries and the future of salmon in the Northwest. *Fisheries* 17(1):5–8.

HINDAR, K., N. RYMAN, and G. STÅHL. 1986. Genetic differentiation among local populations and morphotypes of Arctic charr, *Salvelinus alpinus*. *Biol. J. Linn. Soc.* 27:269–285.

HIRAOKA, S., M. SUZUKI, T. YANAGISAWA, M. IWATA, and A. URANO. 1993. Divergence of gene expression in neurohypophysial hormone precursors among salmonids. *Gen. Comp. Endocrin.* 92:292–301.

HUBBS, C.L. 1955. Hybridization between fish species in nature. *Syst. Zool.* 4:1–20.

HUME, R.D. 1893. *Salmon of the Pacific Coast.* Author, Portland, OR.

HUTCHINGS, J.A. and D.W. MORRIS. 1985. The influence of phylogeny, size and behaviour on patterns of covariation in salmonid life histories. *Oikos* 45:118–124.

IMHOF, M., R. LEARY, and H.E. BOOKE. 1980. Population or stock structure of lake whitefish, *Coregonus clupeaformis*, in northern lake Michigan as assessed by isozyme electrophoresis. *Can. J. Fish. Aquat. Sci.* 37:783–793.

JOHNSON, K.R., J.E. WRIGHT, Jr., and B. MAY. 1987. Linkage relationships reflecting ancestral tetraploidy in salmonid fish. *Genetics* 116:579–591.

JOHNSON, O.W., R.S. WAPLES, T.C. WAINWRIGHT, K.G. NEELY, F.W. WAKNITZ, and L.T. PARKER. 1994. *Status Review for Oregon's Umpqua River Sea-run Cutthroat Trout.* U.S. Dept. Commer., NOAA Tech. Memo. NMFS-NWFSC-15, 122 pp.

JORDAN, D.S. and B.W. EVERMAN. 1896. *The Fishes of North and Middle America.* U.S. Government Printing Office, Washington, DC.

KAYA, C.M. 1991. Rheotactic differentiation between fluvial and lacustrine populations of Arctic grayling (*Thymallus arcticus*), and implications for the only remaining indigenous population of fluvial "Montana grayling." *Can. J. Fish. Aquat. Sci.* 48:53–59.

KELSO, B.W., T.G. NORTHCOTE, and C.F. WEHRHAHN. 1981. Genetic and environmental aspects of the response to water current by rainbow trout (*Salmo gairdneri*) originating from inlet to outlet streams of two lakes. *Can. J. Zool.* 59:2177–2185.

KINCAID, H.L. 1976a. Inbreeding in rainbow trout (*Salmo gairdneri*). *J. Fish. Res. Bd. Can.* 33:2420–2426.

KINCAID, H.L. 1976b. Inbreeding depression in rainbow trout. *Trans. Amer. Fish. Soc.* 105:273–280.

KOHANE, M.J. and P.A. PARSONS. 1988. Domestication: Evolutionary change under stress. *Evol. Biol.* 23:31–48.

LAUDER, G.V. and G.F. LIEM. 1983. The evolution and interrelationships of the actinopterygian fishes. *Bull. Mus. Comp. Zool.* 150:95–197.

LEARY, R.F., F.W. ALLENDORF, and S.H. FORBES. 1993. Conservation genetics of bull trout in the Columbia and Klamath River drainages. *Conserv. Biol.* 7:856–865.

LEIDER, S.A., P.L. HULETT, J.J. LOCH, and M.W. CHILCOTTE. 1990. Electrophoretic comparison of the reproductive success of naturally spawning transplanted and wild steelhead trout through the returning adult stage. *Aquaculture* 88:239–252.

LICHATOWICH, J.A. and J.D. MCINTYRE. 1987. Use of hatcheries in the management of anadromous salmonids. *Amer. Fish. Soc. Sympos.* 1:131–136.

LICHATOWICH, J. and B. WATSON. 1993. Use of artificial propagation and supplementation for rebuilding salmon stocks under the Endangered Species Act. *Recovery Issues for Threatened and Endangered Snake River Salmon.* Tech. Rep. to Bonneville Power Administration, Portland, OR.

LITTLE, T.M. 1945. Gene segregation in autotetraploids. *Bot. Rev.* 11:60–85.

LYNCH, M. 1991. The genetic interpretation of inbreeding depression and outbreeding depression. *Evolution* 45:622–629.

MARNELL, L.F. 1986. Impacts of hatchery stocks on wild fish populations. In *Fish Culture in Fisheries Management*, ed. R.H. Stroud, pp. 339–347. American Fisheries Society, Bethesda, MD.

MARSDEN, J.E., C.C. KRUEGER, and B. MAY. 1989. Identification of parental origins of naturally produced lake trout in Lake Ontario: application of mixed-stock analysis to a second generation. *N. Amer. J. Fish. Mgmt.* 9:257–268.

MARSDEN, J.E., S.J. SCHWAGER, and B. MAY. 1987. Single locus inheritance in the tetraploid treefrog *Hyla versicolor* with an analysis of expected progeny ratios in tetraploid organisms. *Genetics* 116:299–311.

MATHER, K. 1936. Segregation and linkage in autotetraploids. *J. Genetics* 32:287–314.

MATTHEWS, G.M. and R.S. WAPLES. 1991. *Status Review for Snake River Spring and Summer Chinook Salmon.* U.S. Dept. Commer., NOAA Tech. Memo. NMFS F/NWC-195, 23 pp.

MAY, B. and K.R. JOHNSON. 1990. Composite linkage map of salmonid fishes (*Salvelinus, Salmo, Oncorhynchus*). In *Genetic Maps— Linkage Maps of Complex Genomes*, ed. S.J. O'Brien, pp. 4.151–4.159. Cold Spring Harbor, New York.

MCDONALD, M. 1894. *The Salmon Fisheries of the Columbia River Basin.* Report of the Commissioner of Fish and Fisheries on investigations in the Columbia River Basin in regard to the salmon fisheries. Misc. Doc. of the Senate of the U.S. 1893–1894, pp. 3–18.

MCPHAIL, J.D. and C.C. LINDSAY. 1970. Freshwater fishes of northwestern Canada and Alaska. *Fish. Res. Bd. Can.*, Bulletin 173.

MEFFE, G.K. 1992. Techno-arrogance and halfway technologies— salmon hatcheries on the Pacific coast of North America. *Conserv. Biol.* 6:350–354.

MILLAR, R.B. 1987. Maximum likelihood estimation of mixed stock fishery composition. *Can. J. Fish. Aquat. Sci.* 44:583–590.

MILNER, G.B., D.J. TEEL, F.M. UTTER, and C.L. BURLEY. 1981. Columbia River stock identification study: validation of method. *Ann. Rep. Res. NOAA*, National Marine Fisheries Service, Northwest and Alaska Fisheries Center, Seattle, WA, 35 pp.

MORITZ, C. 1994. Defining "evolutionarily significant units" for conservation. *Trends Ecol. Evol.* 9:373–375.

MORRISON, W.J. 1970. Nonrandom segregation of two lactate dehydrogenase subunit loci in trout. *Trans. Amer. Fish. Soc.* 99:193–206.

MORRISON, W.J. and J.E. WRIGHT. 1966. Genetic analysis of three lactate dehydrogenase isozyme systems in trout: evidence for linkage of genes coding subunits A and B. *J. Exp. Zool.* 163:259–270.

MULLAN, J. 1987. Status and propagation of chinook salmon in the mid-Columbia River through 1985. *U.S. Fish Wildl. Serv. Biol. Rep.* 87:1–111.

NEHLSEN, W., J.E. WILLIAMS, and J.A. LICHATOWICH. 1991. Pacific salmon at the crossroads: Stocks at risk from California, Oregon, Idaho, and Washington. *Fisheries* 16(2):4–21.

NELSON, J.S. 1994. *Fishes of the World* (3rd ed). Wiley, New York.

NICKELSON, T.E., M.F. SOLAZZI, and S.L. JOHNSON. 1986. Use of hatchery salmon (*Oncorhynchus kisutch*) presmolts to rebuild wild populations in Oregon coastal streams. *Can. J. Fish. Aquat. Sci.* 43:2443–2449.

NORDEN, C.R. 1961. Comparative osteology of representative salmonid fishes, with particular reference to the grayling (*Thymallus arcticus*) and its phylogeny. *J. Fish. Res. Bd. Can.* 18:679–791.

NORTHWEST POWER PLANNING COUNCIL. 1986. Columbia River Basin Fish and Wildlife Program. Appendix D. *Compilation of Information on Salmon and Steelhead Losses in the Columbia River basin.* Northwest Power Planning Council, Portland, OR, 252 pp.

NORTHWEST POWER PLANNING COUNCIL. 1987. *Columbia River Basin Fish and Wildlife Program.* Northwest Power Planning Council, Portland, OR, 246 pp.

OHNO, S. 1970. The enormous diversity in genome sizes of fish as a reflection of nature's extensive experiments with gene duplication. *Trans. Amer. Fish. Soc.* 99:120–130.

OHNO, S., H. WOLF, and N.B. ATKIN. 1968. Evolution from fish to mammals by gene duplication. *Hereditas* 59:169–187.

PARK, L.K., P. MORAN, and R.S. WAPLES (eds.) 1994. Application of DNA technology to the management of Pacific salmon. *U.S. Dept. Commer., NOAA Tech. Memo. NMFS-NWFSC-17.*

PELLA, J.J., and G.B. MILNER. 1987. Use of genetic marks in stock composition analysis. In *Population Genetics and Fisheries Management*, eds. N. Ryman and F. Utter, pp. 247–276. University of Washington Press, Seattle.

PRATT, K.L. 1992. A review of bull trout life history. In *Proceedings of the Gearhart Mountain Bull Trout Workshop*, eds. P.J. Howell and D.V. Buchanan, pp. 5–9. Oregon Chapter of the American Fisheries Society, Corvallis, OR.

QUINN, T.P. 1985. Homing and the evolution of sockeye salmon (*Oncorhynchus nerka*). *Contrib. Mar. Sci.* 27:353–366.

RALEIGH, R.F. 1971. Innate control of migrations of salmon and trout fry from natal gravels to rearing areas. *Ecology* 52:291–297.

REES, H. 1964. The question of polyploidy in Salmonidae. *Chromosoma* 15:275–279.

RICKER, W.E. 1972. Hereditary and environmental factors affecting certain salmonid populations. In *The Stock Concept in Pacific Salmon*, eds. R.C. Simon and P.A. Larkin, pp. 27–160. University of British Columbia Press, Vancouver.

RIDDELL, B.E. 1993. Spatial organization of Pacific salmon: what to conserve? In *Genetic Conservation of Salmonid Fishes*, eds. J.G. Cloud, and G.H. Thorgaard, pp. 23–41. Plenum, New York.

RILEY, R. 1960. The secondary pairing of bivalents with genetically similar chromosomes. *Nature* 185:751–752.

RYDER, O.A. 1986. Species conservation and systematics: the dilemma of subspecies. *Trends Ecol. Evol.* 1:9–10.

RYMAN, N., F.W. ALLENDORF, and G. STÅHL. 1979. Reproductive isolation with little genetic divergence in sympatric populations of brown trout (*Salmo trutta*). *Genetics* 92:247–262.

RYMAN, N. and L. LAIKRE. 1991. Effects of supportive breeding on the genetically effective population size. *Conserv. Biol.* 5:325–329.

SANDLUND, O.T., K. GUNNARSSON, P.M. JONASSON, B. JONSSON, T. LINDEM, K.P. MAGNUSSON, H.J. MALMQUIST, H. SIGURJONSDOTTIR, S. SKÚLASON, and S.S. SNORRASON. 1992. The Arctic charr *Salvelinus alpinus* in Thingvallavatn. *Oikos* 64:305–351.

SKÚLASON, S., D.L.G. NOAKES, and S.S. SNORRASON. 1989. Ontogeny of trophic morphology in four sympatric morphs of arctic charr, *Salvelinus alpinus* in Thingvallavatn, Iceland. *Biol. J. Linn. Soc.* 38:281–301.

SLADE, R.W., C. MORITZ, A. HEIDEMAN, and P.T. HALE. 1993. Rapid assessment of single-copy nuclear DNA variation in diverse species. *Molec. Ecol.* 2:359–373.

SMITH, G.R. 1992. Introgression in fishes— significance for paleontology, cladistics, and evolutionary rates. *Syst. Biol.* 41:41–57.

SMOUSE, P.E., R.S. WAPLES, and J.A. TWOREK. 1990. A genetic mixture analysis for use with incomplete source population data. *Can. J. Fish. Aquat. Sci.* 47:620–634.

SOLTIS, D.E. and P.S. SOLTIS. 1989. Tetrasomic inheritance in *Heuchera micrantha* (Saxifragaceae). *J. Hered.* 80:123–126.

SPURWAY, H. 1952. Can wild animals be bred in captivity? *New Biol.* 13:11–30.

STEARLEY, R.F. 1992. Historical ecology of Salmoninae, with special reference to *Oncorhynchus*. In *Systematics, Historical Ecology, and North American Freshwater Fishes*, ed. R.L. Mayden, pp. 649–658. Stanford University Press, Stanford, CA.

STEARLEY, R.F. and G.R. SMITH. 1993. Phylogeny of the Pacific trouts and salmons (*Oncorhynchus*) and genera of the family Salmonidae. *Trans. Amer. Fish. Soc.* 122:1–33.

STEARNS, S.C. 1976. Life history tactics: a review of the ideas. *Quart. Rev. Biol.* 51:3–47.

STEBBINS, G.L. 1947. Types of polyploids: their classification and significance. *Adv. Genet.* 1:403–429.

STONE, L. 1892. A national salmon park. *Proceedings of the 21st Annual Meeting American Fisheries Society*, pp. 149–162. New York.

SVÄRDSON, G. 1945. Chromosome studies on Salmonidae. *Rep. Swed. St. Inst. Freshw. Res. Drottningh.* 23:1–151.

SYBENGA, J. 1972. *General Cytogenetics*. American Elsevier, New York.

SYBENGA, J. 1975. *Meiotic Configurations*. Springer-Verlag, Berlin.

TAYLOR, E.B. 1991. A review of local adaptation in Salmonidae, with particular reference to Pacific and Atlantic salmon. *Aquaculture* 98:185–207.

TEMPLETON, A.R. 1986. Coadaptation and outbreeding depression. In *Conservation Biology: The Science of Scarcity and Diversity*, ed. M.E. Soulé, pp. 105–116. Sinauer, Sunderland, MA.

THOMPSON, W.F. 1959. An approach to population dynamics of the Pacific red salmon. *Trans. Amer. Fish. Soc.* 88:206–209.

THOMPSON, W.F. 1965. Fishing treaties and salmon of the North Pacific. *Science* 150:1786–1789.

THORGAARD, G.H. 1983. Chromosomal differences among rainbow trout populations. *Copeia* 1983:650–662.

THORGAARD, G.H. and F.W. ALLENDORF. 1988. Developmental genetics of fishes. In *Developmental Genetics of Animals and Plants*, ed. G.M. Malacinski, pp. 369-391. Macmillan, New York.

UTTER, F.M. 1991. Biochemical genetics and fishery management— an historical perspective. *J. Fish Biol.* 39 (Suppl. A):1–20.

UTTER, F.M. and F.W. ALLENDORF. 1978. Determination of the breeding structure of steelhead populations through gene frequency analysis. In *Proceedings of the Genetic Implications of Steelhead Management Symposium*, eds. T.J. Hassler and R.R. Van Kirk, pp. 44–54. Special Report 77-1, Cooperative Fisheries Unit, Arcata, CA.

UTTER, F.M. and F.W. ALLENDORF. 1994. Phylogenetic relationships among species of *Oncorhynchus*: A consensus view. *Conserv. Biol.* 8:864–867.

UTTER, F., G. MILNER, G. STÅHL, and D. TEEL. 1989. Genetic population structure of chinook salmon, *Oncorhynchus tshawytscha*, in the Pacific Northwest. *Fish. Bull.* 85:13–23.

UTTER, F.M., J.E. SEEB, and L.W. SEEB. 1993. Complementary uses of ecological and biochemical genetic data in identifying and conserving salmon populations. *Fish. Res.* 18:59–76.

UTTER, F.M., D.J. TEEL, G.B. MILNER, and D. McISSAC. 1987. Genetic estimates of stock compositions of 1983 chinook salmon, *Oncorhynchus tshawytscha*, harvests off the Washington coast and Columbia River. *Fish. Bull.* 85:13–23.

VOGLER, A.P. and R. DESALLE. 1994. Diagnosing units of conservation management. *Conserv. Biol.* 8:354–363.

WAINES, J.G. 1976. A model for the origin of diploidizing mechanisms in polyploid species. *Amer. Nat.* 110:415–430.

WALTERS, C. and B. RIDDELL. 1986. Multiple objectives in salmon management: The chinook sport fishery in the Strait of Georgia, B.C. *Northw. Environ. J.* 2:1–15.

WAPLES, R.S. 1988. Estimation of allele frequencies at isoloci. *Genetics* 118:371–384.

WAPLES, R.S. 1990. Temporal changes of allele frequency in Pacific salmon populations: implications for mixed-stock fishery analysis. *Can. J. Fish. Aquat. Sci.* 47:968–976.

WAPLES, R.S. 1991a. Genetic interactions between hatchery and wild salmonids— lessons from the Pacific Northwest. *Can. J. Fish. Aquat. Sci.* 48(Suppl. 1):124–133.

WAPLES, R.S. 1991b. Pacific salmon, *Oncorhynchus* spp., and the definition of "species" under the Endangered Species Act. *Mar. Fish. Rev.* 53:11–22.

WAPLES, R.S. 1995a. Evolutionarily significant units and the conservation of biological diversity under the Endangered Species Act. In *Evolution and the Aquatic Ecosystem*, eds. J.L. Nielsen and D.A. Powers. American Fisheries Society, Bethesda, MD, *in press*.

WAPLES, R.S. 1995b. Genetic effects of stock transfers of fish. In *Protection of Aquatic Biodiversity, Proceedings of the World Fisheries Congress, Theme 3*, ed. D. Philipp, pp. 51–69. Oxford and IBH, New Delhi.

WAPLES, R.S. and P.B. AEBERSOLD. 1990. Treatment of data for duplicated gene loci in mixed-stock fishery analysis. *Can. J. Fish. Aquat. Sci.* 47:2092–2098.

WAPLES, R.S. and C. DO. 1994. Genetic risk associated with supplementation of Pacific salmonids. I. Captive broodstock programs. *Can. J. Fish. Aquat. Sci.* 51 (Suppl. 1):310–329.

WAPLES, R.S., O.W. JOHNSON, and R.P. JONES, Jr. 1991. Status review for Snake River sockeye salmon. *U.S. Dept. Commer., NOAA Tech. Memo. NMFS F/NWC-195*, 23 pp.

WAPLES, R.S. and P.E. SMOUSE. 1990. Gametic disequilibrium analysis as a means of identifying mixtures of salmon populations. *Amer. Fish. Soc. Sympos.* 7:439–458.

WILLIAMS, J.E., J.A. LICHATOWICH, and W. NEHLSEN. 1992. Declining salmon and steelhead populations: New endangered species concerns for the west. *Endang. Spec. Upd.* 9:1–8.

WILSON, M.V.H. 1977. Middle Eocene freshwater fishes from British Columbia. *Roy. Ontar. Mus. Life Sci. Contrib.* 113:1–61.

WITHLER, F. 1982. Transplanting Pacific salmon. *Can. Tech. Rep. Fish. Aquat. Sci.* 1079, 27 pp.

WOOD, C.C. and C.J. FOOTE. 1990. Genetic differences in the early development and growth of sympatric sockeye salmon and kokanee (*Oncorhynchus nerka*), and their hybrids. *Can. J. Fish. Aquat. Sci.* 47:2250–2260.

WOOD, C.C., S. MCKINNELL, T.J. MULLIGAN, and D.A. FOURNIER. 1987. Stock identification with the maximum-likelihood mixture model: Sensitivity analysis and application to complex problems. *Can. J. Fish. Aquat. Sci.* 44:866–881.

WORLD CONSERVATION UNION. 1993. *1994 IUCN Red List of Threatened Animals*. World Conservation Union, Gland, Switzerland.

WRIGHT, J.E., K. JOHNSON, A. HOLLISTER, and B. MAY. 1983. Meiotic models to explain classical linkage, pseudolinkage and chromosome pairing in tetraploid derivative salmonids. In *Isozymes: Current Topics in Biological and Medical Research*, eds. M.C. Rattazzi, J.G. Scandalios, and G.S. Whitt, pp. 239–260. Liss, New York.

XU, S., C.J. KOBAK, and P.E. SMOUSE. 1994. Constrained least squares estimation of mixed population stock composition from mtDNA haplotype frequency data. *Can. J. Fish. Aquat. Sci.* 51:417–425.

9

CONSERVATION
GENETICS
OF ENDEMIC
PLANT SPECIES

J.L. Hamrick
and Mary Jo W. Godt

INTRODUCTION

The role of population genetics in plant conservation biology has been the subject of considerable discussion during the last decade. Some researchers (e.g., Hamrick, 1983; Falk and Holsinger, 1991; Ellstrand and Elam, 1993; Loeschcke et al., 1994) have argued that knowledge of a species' genetic composition is essential for any comprehensive conservation plan, whereas others (e.g., Lande, 1988; Schemske et al., 1994) doubt that genetic diversity plays a decisive role in the survival of populations or species. These latter authors argue that populations go extinct for ecological reasons (e.g., habitat destruction or environmental changes) rather than because they lack genetic variation. Most biologists agree that ecological factors and demographic characteristics of populations almost certainly play a dominant role in the short-term survival of populations and species. However, a lack of concern for the preservation of natural levels of genetic diversity by conservation agencies is shortsighted, at best. There are examples in the plant literature that describe the potential loss of populations (e.g., Lakeside daisy; DeMauro, 1993) or of species (e.g., American chestnut; see Box 9.1) because they lacked genetic diversity to reproduce or to survive environmental challenges.

Box 9.1. American Chestnut

At one time, American chestnut (*Castanea dentata* Mill, Fagaceae) was the dominant tree in the upper Piedmont and lower Appalachian slopes of the eastern United States (Stephenson et al., 1993). Over most of its range (New England to Alabama), American chestnut accounted for 40% of the overstory canopy (Keever, 1953). In addition to its dominant ecological role, the American chestnut was the most economically valuable hardwood species in the United States. It yielded a great variety of products, including high-grade timber, paper, fiber board, and tannins (Boyce, 1961). It was highly prized as a shade tree, and its nuts were a valuable food resource for a wide variety of wildlife.

In 1904, chestnut blight (*Endothia parasitica* Murr.), a fungal disease, was introduced into the New York Zoological Park on Asiatic chestnut nursery stock (Merkel, 1906). The disease attacks the stem of living chestnuts and kills the above ground portions of the tree, but does not affect the root system. American chestnut sprouts vigorously from rootstocks after the stem dies, but sprouts are killed by the fungus once they reach a few centimeters in diameter.

Because the spores of chestnut blight are wind-dispersed, the disease spread rapidly, and by the 1950s chestnut blight was present in every natural American chestnut stand. One estimate places chestnut losses at more than 3.6 million hectares (Anonymous, 1954). No resistance to the disease has been found in American chestnut, although Asian chestnuts are naturally resistant, presumably because they evolved with the disease. Natural hybrids between American and Asian chestnuts have intermediate levels of resistance (Anonymous, 1954).

Interestingly, the Holocene paleontological record indicates that eastern hemlock (*Tsuga canadensis* [L.] Carr; Pinaceae) may have experienced an environmental challenge similar to that of the American chestnut. Pollen profiles taken throughout the range of eastern hemlock indicate that it disappeared rather suddenly about 4,800 years BP (Davis, 1981). The rapidity of its disappearance while other plant species within the community remained relatively constant suggests that its dramatic decrease was not due to climatic factors, but rather to an epidemic of an unknown disease (Davis, 1981). For about 1,000 years, eastern hemlock pollen was absent or in low abundance in pollen profiles. Eastern hemlock pollen then gradually increased, and the species reached its former abundance 2,000 years after the initial decline (Davis, 1981, 1983). Evidently, eastern hemlock differed from American chestnut in that its populations maintained or acquired genetic variation that allowed it to evolve resistance to its pathogen. Currently, eastern hemlock maintains a low level of allozyme diversity (Zabinski, 1992), perhaps as a consequence of the genetic bottleneck passed through 4,800 years ago.

Ecological and genetic studies of populations complement each other, and the contributions of both disciplines enable conservation biologists to develop a better understanding of the biology of endangered species. Furthermore, there are several concerns of conservation biologists that can only be addressed by detailed population genetics studies. For example, the loss of genetic diversity in *ex situ* or *in situ* conservation programs can only be measured if genetic diversity in natural populations is known. Also, the choice of populations as sources of propagules for the restoration of extinct or demographically threatened populations should depend, in part, on genetic diversity within donor populations and the genetic similarity of these populations to those in need of restoration.

Little is known of the genetic structure of the vast majority of plant species. Ideally, for every species of conservation concern, transplant or common garden experiments would be performed to determine the genetic similarity of populations for morphological, physiological, and reproductive traits. For most plant species such information is unavailable, and it is unlikely to become available for the majority of species in danger of extinction. Therefore, many conservation biologists have turned to biochemical and molecular markers to describe the genetic composition of plant species. The assumption that forms the basis for such analyses is that genetic structure as measured by "neutral" genes reflects evolutionary processes (e.g., inbreeding, genetic drift, and gene flow) that affect the entire genome. However, it is unreasonable to expect that even the most straightforward genetic analyses (i.e., allozyme studies) can be carried out on a significant proportion of the endangered plant species in the United States, much less those in underdeveloped nations.

As a result, there have been attempts (e.g., Crawford, 1983; Hamrick, 1983; Ellstrand and Roose, 1987; Hamrick et al., 1991) to utilize the existing plant population genetics literature (largely based on allozymes) to make predictions concerning the level and distribution of genetic variation within plant species. Typically, mean measures of genetic diversity within and among populations are obtained for species that have been classified for various characteristics (e.g., island vs. continental location; selfing vs. outcrossing mode of reproduction). If significant differences in genetic features are shown among species' classifications, it is then argued that these generalizations permit characterization of species for which direct genetic information is unavailable. It has become apparent, however, that a relatively low proportion of the variation among species for genetic diversity can be explained by such categorical descriptions. In this chapter, we summarize generalizations developed from reviews of the plant allozyme literature and point out the inadequacies of this approach. We then use three case studies of endangered or threatened plant species from the southeastern United

States to illustrate how empirical knowledge of the species' genetic composition can provide additional insights into their ecological and evolutionary histories.

GENERALIZATIONS FROM THE PLANT ALLOZYME LITERATURE

The allozyme literature provides a rich database with which to examine hypotheses regarding levels and distribution of genetic diversity in plant species. The most recent and comprehensive review is that of Hamrick and Godt (1989), who classified species for eight ecological and life-history traits and calculated standard population genetic parameters (see Box 9.2). Genetic diversity was calculated within species and populations, and the proportion of the total genetic variation found among populations was estimated. The values of these parameters were then summarized by category for each of the eight traits. This analysis differed from previous reviews (Brown, 1978; Hamrick et al., 1979; Gottlieb, 1981; Loveless and Hamrick, 1984) in that the same dataset was used to estimate genetic diversity at each of the three levels.

We summarize the results of this analysis for the categories "geographic range" and "breeding system," two traits of interest and importance to plant conservation biologists. With respect to geographic range, endemic species tend to have fewer polymorphic loci and less genetic diversity (H_{es}) than more widespread species (Table 9.1). Interestingly, genetic differentiation among populations (G_{ST}) of endemic species was similar to that found among populations of more widespread species. This result illustrates one difficulty in the interpretation of such data. Because population differentiation is assumed to be largely a function of gene flow among populations (Slatkin and Barton, 1989), interpopulation heterogeneity should reflect isolation and spatial scale. Thus, if widespread species were sampled on the same limited spatial scale as localized endemics, they should have less among-population differentiation than endemic species. This prediction is partially supported by the analysis of Loveless and Hamrick (1984).

Breeding systems were correlated with the level and distribution of genetic variation, with predominantly selfing species having less genetic diversity overall and dramatically greater differentiation among populations (Table 9.1). Outcrossing taxa, especially wind-pollinated species, have more genetic diversity within species and much lower heterogeneity among populations. Although it is apparent why species with limited pollen movement should display less within-population genetic diversity, it is less obvious why selfing species have less genetic diversity overall. Hamrick and Nason (1996)

Box 9.2. Measures of Genetic Diversity

Several standard measures of genetic diversity for single-gene markers (such as allozymes) are employed in this chapter. These parameters estimate diversity within species, within populations, and among populations. Within-species measures of genetic diversity estimate total genetic diversity available to the species, and are not confounded by how diversity is partitioned among populations. Species with the same overall genetic variation can have quite different levels of within-population genetic diversity, depending on how the variation is distributed.

Measures of Diversity within Populations or Species

P = percent of loci polymorphic.

AP = mean number of alleles per polymorphic locus.

A = mean number of alleles per locus (monomorphic as well as polymorphic loci included).

A_e = effective number of alleles per locus ($1/\Sigma p_i^2$, where p_i is the frequency of the ith allele). This measure is influenced by the proportion of polymorphic loci, the number of alleles per polymorphic locus, and allele frequencies.

H_o = observed proportion of loci heterozygous per individual.

H_e = expected proportion of loci heterozygous per individual in a random mating population ($1 - \Sigma p_i^2$, where p_i is the frequency of the ith allele). This measure is a function of the proportion of polymorphic loci, the number of alleles per polymorphic locus, and allele frequencies. This parameter is also called genic diversity, and is the most comprehensive measure of genetic diversity for allozyme data.

D_G = genotypic diversity ($1 - \Sigma\{[n_i(n_i - 1)]/[N(N - 1)]\}$, where n_i is the number of individuals of genotype i in a population sample of size N. (In this chapter, values of parameters at the species level are subscripted by an s, and those at the population level by a p.)

Measures of Genetic Diversity among Populations

F_{ST} = standardized interpopulational variance in allele frequencies ($\sigma_p^2/\bar{p}_i(1 - \bar{p}_i)$, where σ_p^2 is the variance in allele frequencies among populations and $\bar{p}_i$ is the mean frequency of the ith allele). F_{ST} values are calculated for individual loci and alleles.

G_{ST} = the proportion of the total genetic diversity that occurs among populations (D_{ST}/H_T, where D_{ST} is the genetic diversity among populations; $D_{ST} = H_T - \bar{H}_S$, where $\bar{H}_S$ is the mean heterozygosity within populations and $H_T = 1 - \Sigma\bar{p}_i^2$). G_{ST} is also calculated for each locus and is equilvalent to a multiallelic F_{ST}.

Genetic Identity and Genetic Distance

I = the genetic identity or genetic similarity value ($\Sigma x_i y_i/(\Sigma x_i^2 \Sigma y_i^2)^{0.5}$ [Nei, 1972], where x_i and y_i are the frequencies of the ith alleles in populations x and y). For populations that share no alleles, $I = 0.0$, whereas for populations with identical allele frequencies, $I = 1.0$.

Box 9.2. *(cont.)*

D = genetic distance $(-\log_e I)$.

I and D values are usually presented as a matrix of pairwise identity and distance values, with each matrix element representing a comparison between a pair of populations or species. Values presented usually represent mean I and D values over all loci.

Indirect Estimates of Gene Flow

Nm = the estimated number of migrants per generation $[(1 - F_{ST})/4 F_{ST}$ (Wright, 1931), where N is the recipient population size and m is the rate of gene flow].

$\log_{10}(\bar{p}(1)) = a \log_{10} Nm + b$ (Barton and Slatkin, 1986), where a and b are variables related to population size and $\bar{p}(1)$ is the mean frequency of "private alleles" (i.e., alleles that occur in a single population).

---

reasoned that selfing species are often annuals or short-lived perennials, species whose populations commonly undergo large fluctuations in population numbers and, as a result, are prone to extinction. Furthermore, in selfing species, an allele novel to one population is relatively unlikely to be introduced into other populations. Thus, when the population of origin goes extinct, novel alleles are lost to the species. Furthermore, in populations of predominantly selfing species, relatively new alleles are more likely to be in a homozygous state and may be exposed to more intense selection. By contrast, novel alleles in outcrossing species are more likely to be introduced into several populations via gene flow where they reside in heterozygotes. The more widespread distribution of these novel alleles effectively buffers them against loss. The overall inference is that outcrossing species should have a higher percentage of polymorphic loci, more alleles per polymorphic locus and, thus, more genetic diversity.

Significant differences were also seen among categories of several other traits (Table 9.2). For example, wind-pollinated conifer trees from temperate regions tend to have higher levels of genetic diversity and less genetic heterogeneity among populations than species with other combinations of traits. Generalizations from such analyses may prove useful to conservation biologists faced with developing management plans for species lacking empirical genetic data. However, an important point is that much of the variation in genetic parameters among species was *not* explained by the eight life-history traits considered. This was particularly true for genetic diversity within populations (H_{ep}) and total genetic diversity within species (H_{es}) (R^2 values = 0.21 and 0.17, respectively; Table 9.2), and somewhat less so for among-population differentiation (G_{ST}; R^2 = 0.45, Table 9.2). Fortunately, it

Table 9.1. Mean Values of Genetic Diversity and Population Structure Derived from the Allozyme Literature for Plant Species Categorized by Geographic Range and Breeding System [1]

| | *Within species* | | *Within populations* | | *Among populations* |
	P_s	H_{es}	P_p	H_{ep}	G_{ST}
A. *Geographic range*					
Endemic	40.0c	0.096c	26.3c	0.063c	0.248a
Narrow	45.1bc	0.137b	30.6bc	0.105b	0.242a
Regional	52.9ab	0.150b	36.4ab	0.118b	0.216a
Widespread	58.9a	0.202a	43.0a	0.159a	0.210a
B. *Breeding system* [2]					
Selfing	41.8b	0.124b	20.0c	0.074d	0.510a
Mixed-animal	40.0b	0.120b	29.2bc	0.090cd	0.216a
Mixed-wind	73.5a	0.194a	54.4a	0.198a	0.100c
Outcrossing-animal	50.1b	0.167ab	35.9b	0.124bc	0.197b
Outcrossing-wind	66.1b	0.162ab	49.7a	0.148b	0.099c

[1] After Hamrick and Godt, 1989.

[2] Mixed breeding systems involve both selfing and outcrossing; also indicated is whether pollination occurs by wind or by animal vectors.

Note: P is the percent of polymorphic loci, H_e is the Hardy–Weinberg expected level of individual heterozygosity, and G_{ST} is the proportion of the total genetic diversity found among populations (see Box 9.2). Values at the species level are subscripted by an *s*; those at the population level by a *p*. Means followed by different letters in a column are significantly different at the five percent probability level.

is often more important for conservation biologists to understand how genetic variation is distributed among populations than it is to estimate genetic diversity within species.

Schoen and Brown (1991; Brown and Schoen, 1992) have also argued that the uncritical application of such generalizations could lead to serious conservation errors, particularly with selfing species. They demonstrate that genetic diversity within populations varies more among populations of selfing species than among populations of predominantly outcrossing species. Some populations of selfers are polymorphic at several loci and have high levels of genetic diversity, whereas other populations are largely monomorphic. Thus, as differentiation among populations increases, the need for specific empirical data becomes more important to the conservation biologist whose goal is to preserve genetic diversity.

Why is it so difficult to predict the level of genetic diversity, even when much is known about the present-day biology and ecology of the species? One possibility is that generalizations from the allozyme literature are affected by differences among studies in terms of loci examined, sample sizes, and the number and spatial distribution of the populations sampled. Another, more biologically interesting likelihood is that the genetic diversity maintained by a species is a function of its ecological and evolutionary history. Thus, factors such as fluctuations in the number and size of popu-

Table 9.2. Summary of an Analysis of Allozyme Variation in Plants (Hamrick and Godt, 1989)

A. *Genetic diversity within species* (H_{es})

	H_{es}	
	Low	High
Characteristic		
Taxonomic status (ts)	Dicots	Gymnosperms
Geographic range (gr)	Endemic	Widespread
Regional distribution (rd)	Temperate	Boreal-temperate
Life form (lf)	Short-lived woody	Long-lived woody
Mode of reproduction (mr)	Not significant	
Breeding system (bs)	Selfing	Outcrossed-wind
Seed dispersal (sd)	Explosive	Ingested
Stage of succession (ss)	Not significant	

Order of importance for H_{es}: lf > bs > gr > sd > ts
$$R^2 = 0.171$$

B. *Genetic diversity within populations* (H_{ep})

	H_{ep}	
	Low	High
Characteristic		
Taxonomic status (ts)	Dicots	Gymnosperms
Geographic range (gr)	Endemic	Widespread
Regional distribution (rd)	Not significant	
Life form (lf)	Long-lived herbaceous	Long-lived woody
Mode of reproduction (mr)	Not significant	
Breeding system (bs)	Selfing	Outcrossed-wind
Seed dispersal (sd)	Explosive	Wind
Stage of succession (ss)	Not significant	

Order of importance for H_{ep}: lf > bs > gr > ts > rd > sd
$$R^2 = 0.215$$

C. *Proportion of variation among populations* (G_{ST})

	G_{ST}	
	Low	High
Characteristic		
Taxonomic status (ts)	Gymnosperms	Dicots
Geographic range (gr)	Not significant	
Regional distribution (rd)	Boreal-temperate	Temperate
Life form (lf)	Long-lived woody	Annual
Mode of reproduction (mr)	Not significant	
Breeding system (bs)	Outcrossed-wind	Selfing
Seed dispersal (sd)	Wind	Gravity
Stage of succession (ss)	Late	Early

Order of importance for G_{ST}: bs > lf > ts > sd > rd
$$R^2 = 0.446$$

Note: R^2 values indicate the total fraction of variation among species that is explained by the eight characteristics studied.

lations, biogeography, and the speciation process itself may have played important roles in determining the present-day genetic composition of species.

THREE CASE STUDIES
FROM THE SOUTHEASTERN UNITED STATES

The southeastern United States is one of the most floristically diverse areas outside of the tropics. Nearly 5,000 plant species occur in the region, with approximately 300 species endemic (P. White, personal communication). At least 83 of these species have been classified as threatened or endangered by the U.S. Fish and Wildlife Service. In the following sections, we review case histories involving three endangered or threatened species that illustrate how empirical knowledge of contemporary population genetic structure can provide insights into evolutionary history. These insights for particular species permit more informed management and conservation decisions.

Spreading Avens, *Geum radiatum*

Biological and Genetic Background. Because historical data is usually scarce or nonexistent, predictions concerning genetic diversity are often based on the present-day size and vigor of populations, despite the fact that current status may not reflect population demographic history. A typical assumption of evolutionary biologists is that small populations, especially those in demographic decline, should contain less genetic diversity than large, vigorous populations.

An example in which the association between population size and genetic diversity seems to hold involves *Geum radiatum* Gray (Rosaceae), commonly called spreading avens. *Geum radiatum* is a rare perennial herb endemic to a few mountaintops in western North Carolina and eastern Tennessee. It prefers full sun and occurs predominantly in shallow, acidic soils on high-elevation cliffs, outcrops, steep slopes, and gravelly talus. Sixteen populations of *G. radiatum* were documented (U.S. Fish and Wildlife Service, 1991), of which five have gone extinct. Observations of four currently declining populations suggest that the extinct populations may have been destroyed by heavy recreational use of the areas by hikers, climbers, and sightseers. Because of the vulnerability of this rare species to human impacts, *G. radiatum* has been federally listed as an endangered species.

One population (CGG) of *G. radiatum* that is presently in danger of extirpation is in the Blue Ridge Parkway in North Carolina, on a heavily trampled scenic overlook. The National Park Service has identified this site as requiring special conservation attention and is attempting to divert tourist traffic from the most fragile areas. There is considerable interest in restoring *G. radiatum* and several other co-occurring plant species to their predisturbance population numbers at this site. The management question associated with this goal is: Which *G. radiatum* population is most appropriate as a source of propagules to restore population CGG?

Five populations of *G. radiatum* were analyzed for 25 allozyme loci (Table 9.3). Seven loci (28%) were polymorphic within this sample, and on average approximately six loci (23%) were polymorphic per population. There were 2.57 alleles per polymorphic locus within the species, and overall genetic diversity was 0.098. Since other *Geum* species have not been analyzed, it is not known whether *G. radiatum* has lower levels of genetic diversity than more widespread congeners, but in any event the level of genetic diversity measured for *G. radiatum* is somewhat lower than the means for endemic species and for short-lived herbaceous plants (unpublished data, updated from Hamrick and Godt, 1989).

The majority (84%) of the allozyme diversity within *G. radiatum* resides within its populations, which have a mean genetic identity (see Box 9.2) of $I = 0.958$. There are, however, rather large differences in the level of genetic diversity displayed by individual populations. Population CGG, the site with the fewest estimated plants (< 100, B. Johnson, personal communication), had only four polymorphic loci, whereas the larger populations (RMT and PMT) were polymorphic for seven assayed loci. Of the five sites sampled,

Table 9.3. Allozyme Genetic Diversity in *Geum radiatum*[1]

Population	Sample size	P	AP	H_o	H_e	I
PMT	60	28.0	2.14	0.056	0.091	0.969
RMT	72	28.0	2.43	0.050	0.086	0.958
GMT	48	24.0	2.33	0.049	0.066	0.947
CTP	48	20.0	2.20	0.054	0.064	0.966
CGG	48	16.0	2.00	0.050	0.061	0.947
Population means	55.2	23.2	2.22	0.052	0.074	0.958
Total within *G. radiatum*	–	28.0	2.57	–	0.098	–
Other endemic species[2] (n = 154)	–	44.0	2.99	–	0.110	–
Short-lived herbaceous species[2] (n = 185)	–	43.4	2.73	–	0.125	– .

[1] From data in Godt, Johnson, and Hamrick, unpublished.

[2] Unpublished data; updated from Hamrick and Godt (1989).

Note: P is the proportion of loci polymorphic; *AP* is the mean number of alleles per polymorphic locus; H_o and H_e are the observed and expected heterozygosities; and *I* is the mean genetic identity of the population to all others surveyed.

population CGG also had the lowest genetic diversity (H_e = 0.061) and the fewest alleles per polymorphic locus (Table 9.3). Population PMT had the highest genetic diversity (H_e = 0.091) but had fewer alleles per polymorphic locus than three other populations. Overall, there was a close positive association between current population size and genetic diversity.

Conservation Implications. When a decision is made to restore a threatened or declining population, several factors should be considered. First, will the collection of propagules adversely affect the donor population? Second, does the potential donor population contain a representative sample of the species' genetic diversity? And, finally, which of the more genetically diverse donor populations has the highest genetic similarity to the recipient population? This last question is difficult to answer with single-gene traits because there is usually no direct association between such loci and adaptive characters (Hamrick, 1989). Ideally, common garden or transplant studies established in habitats similar to the threatened population could provide this information, but the rarity of endangered species and their habitats may prohibit such transplant studies for most endangered species. For the restoration of population CGG, populations PMT and RMT would be the most appropriate, because both have large population numbers and high levels of genetic diversity.

Conservation organizations must frequently choose which populations to acquire and protect. Although many factors usually go into these decisions, the preservation of genetically diverse populations should be a priority. For *G. radiatum*, a positive association appears to exist between genetic diversity and population size. Three other plant species (*Calamagrostis cainii, Carex misera* and *Trichophorum cespitosum*) that co-occur with *G. radiatum* were also analyzed electrophoretically (Godt, Johnson, and Hamrick, unpublished data). In general, the same association was seen, with the largest populations maintaining the highest levels of genetic diversity. However, as will be shown, historical factors that influence the genetic diversity of populations may in some cases lead to a decoupling of the usual associations between contemporary population size and extant genetic variation.

Swamp Pink, *Helonias bullata*

Biological and Genetic Background. Many species have experienced pronounced changes in geographic ranges. Certainly, the distributions of North American plant species have shifted several hundred kilometers since the peak of the Wisconsin glacial epoch, about 20,000 years ago. During the last glacial epoch, species that characterize the current vegetation zones of North America were forced into refugia considerably south of their present ranges (Davis, 1983; Delcourt et al., 1983). Following the retreat of the glaciers (beginning about 12,000 years BP), many dominant species rapidly expanded northward, and by approximately 4,000 years ago, present day vegetation zones were largely in place (Davis, 1983). The migration of species across the continent has almost certainly had genetic consequences. For example, the low levels of genetic diversity found in red pine (*Pinus resinosa*) may have resulted from a series of genetic bottlenecks during several glacial epochs (Box 9.3). Even within species that

---

Box 9.3. The Genus *Pinus*

Pinaceae is the most extensively investigated plant family in allozyme surveys, with at least 274 studies published. In the genus *Pinus* alone, 169 such studies have appeared. Other genera of Pinaceae that have been widely assayed (and the number of studies) include *Abies* (15), *Cedrus* (4), *Larix* (13), *Picea* (60), and *Pseudotsuga* (13). The genus *Pinus* comprises about 100 species worldwide, of which 56 are represented in the allozyme literature (Hamrick and Godt, unpublished data). The numerous studies of pines may be due in part to the fact that the group is primarily north temperate, and includes many species of economic value.

Pinus is a particularly appropriate genus in which to examine the effects of biogeographic history on the levels and distribution of genetic diversity, because members of the genus share most life-history traits. All are wind-pollinated, long-lived, woody, and most distribute their seeds by wind. The chief differences among pines involve geographic range, degree of spatial isolation of populations, and successional stage of their habitats.

Pines, like many other woody species, maintain relatively high levels of genetic variability and display little genetic differentiation among populations (mean $H_{es} = 0.157$; $H_{ep} = 0.136$; $G_{ST} = 0.065$

[Hamrick et al., 1992]). Nonetheless, both genetic diversity and population structure vary considerably within the genus. For example, genetic diversity within species ranges from 0.000 for *P. resinosa* (Fowler and Morris, 1977; Simon et al., 1986) and 0.017 for *P. torreyana* (Ledig and Conkle, 1983) to 0.261 for *P. jeffreyi* (Conkle, 1981) and 0.275 for *P. lambertiana* (Conkle, 1981). *Pinus torreyana* is a local endemic comprised of two disjunct populations in California, whereas *P. resinosa* ranges from Minnesota to Nova Scotia. *Pinus jefferyi* and *P. lambertiana* are montane species that range from Oregon to southern California.

The level of genetic heterogeneity among pine populations also varies considerably among species. Pines with geographically disjunct populations tend to display higher G_{ST} values (e.g., 0.161 for *P. brutia* [Conkle et al., 1988], 0.164 for *P. longaeva* [Hamrick et al., 1994], 0.199 for *P. muricata* [Millar, 1983], and 0.300 for *P. halepensis* [Schiller et al., 1985]), whereas more continuously distributed species tend to have much lower G_{ST} values (e.g., 0.039 for *P. rigida* [Guries and Ledig, 1982], 0.036 for *P. contorta* [Wheeler and Guries, 1982], and 0.039 for *P. monophylla* [Hamrick et al., 1994]).

have maintained genetic diversity during periods of range contraction, the present-day distribution of genetic variation may not represent a stable condition. For example, recently founded populations may be in the process of acquiring genetic variation from populations in more centrally located or ancestral regions.

The present distribution of genetic diversity within species is influenced by both natural history and evolutionary history. For example, populations of species with long-distance seed dispersal may have been founded by several individuals and, thus, are likely to contain a representative sample of genetic diversity. Furthermore, species with continuous distributions or the potential for long-distance pollen and seed dispersal may subsequently acquire genetic diversity from neighboring populations. Such species usually have relatively little population-to-population variation in genetic diversity (Box 9.3). In contrast, species with limited dispersal capabilities, or those in discontinuously distributed or isolated habitats, are less likely to display comparable levels of within-population genetic variation. For example, North American populations far from glacial refugia may tend to display low levels of genetic variation (e.g., Schwaegerle and Schaal, 1979).

Helonias bullata L. (Liliaceae) is a perennial herb that occurs in wetland habitats characterized by consistently saturated soils with a low frequency of inundation (Rawinski and Cassin, 1986). Historically, the range of *H. bullata* (swamp pink) extended from New York to northern Georgia (U.S. Fish and Wildlife Service, 1991). There are currently 112 known populations, ranging from New Jersey to northern Georgia. The four southern Appalachian populations are isolated from the Virginia populations by approximately 400 km. Populations vary in size from several individuals to more than 10,000 rosettes. New Jersey populations are generally larger and less isolated than their southern counterparts. *Helonias bullata* is currently listed as a threatened species (U.S. Fish and Wildlife Service, 1988). The loss and alteration of wetland habitats are purportedly the most significant threats to the species.

Low levels of genetic variation were indicated by an allozyme analysis of 15 populations representing the geographic range of *H. bullata* (Table 9.4). Eleven of 33 assayed loci (33.3%) were polymorphic within the species but, on average, only four loci (12.8%) were polymorphic within a population. A relatively high proportion (30.6%) of the total genetic variation resided among populations. For the six loci with the highest levels of genetic diversity, the proportion of genetic diversity among populations (50.4%) was higher yet. A positive correlation (0.30) existed between genetic distance and geographic distance. Interestingly, when populations were pooled within regions (New Jersey, Virginia, and southern Appalachians), a trend of decreasing genetic variation with increasing latitude was observed (Table 9.4).

Table 9.4. Allozyme Genetic Diversity in *Helonias bullata*[1]

Geographic location	P	AP	H_e
S. Appalachian region (4)	24.2	2.38	0.061
Virginia region (6)	33.3	2.18	0.045
New Jersey region (5)	18.2	2.17	0.033
Population means (15)	12.8	2.09	0.029
Species total	33.3	2.36	0.053

[1] From Godt et al., 1995.
Note: P is the proportion of loci polymorphic; AP is the mean number of alleles per polymorphic locus; and H_e is the genetic diversity. Regional values were obtained by pooling across the populations (numbers in parentheses) in each region.

Notably, the northernmost New Jersey population was polymorphic for only two loci (one with two alleles and the other with three). This New Jersey population had an expected heterozygosity of 0.002, and only three multilocus genotypes were observed among the 48 individuals sampled.

The fixation of alternate alleles in different populations, and the substantial differentiation in allele frequencies among *H. bullata* populations, suggest that genetic drift has played a major role in the evolutionary history of the species. This conclusion is supported by an indirect estimate of gene flow that indicates that fewer than one individual per generation migrates between populations ($Nm = 0.56$). Seeds of *H. bullata* are not particularly well adapted for long-distance dispersal, which suggests that new sites are colonized by few individuals. In addition, pollen movement among populations is unlikely, because swamp pink habitats are spatially restricted and isolated. Once established, *H. bullata* populations may experience numerous population bottlenecks caused by adverse environmental conditions such as extended droughts. Such fluctuations in population numbers should lead to the further loss of alleles. The relatively large differences in genetic diversity seen among populations within the same geographic region also support the idea that genetic drift has played a major role in shaping the genetic structure of *Helonias bullata*.

Prior to the empirical genetic surveys, the New Jersey region (which contains the majority of known populations of *H. bullata*) might have been expected to harbor the most genetic diversity, whereas the disjunct southern Appalachian populations might be genetically depauperate. The observed pattern is contrary to these expectations. These results suggest that the southern Appalachians contain relict populations from the last glacial epoch, and that populations within this region may have been the source of founders for northern populations. The New Jersey populations did not exist during the last glacial period and may have been founded relatively recently.

It appears that loss of a significant proportion of this species' genetic diversity accompanied these introductions.

Conservation Implications. The study of *Helonias bullata* provides two exceptions to commonly held generalizations. First, disjunct marginal populations can in some cases be more diverse genetically than more centrally located populations. Second, large and demographically vigorous populations can in some cases be less diverse genetically than small populations. Thus, for *H. bullata*, current population sizes and their distribution are not indicative of the genetic patterns observed.

These results illustrate the value of empirical measures of genetic diversity, and also carry relevance for conservation programs. Based on the *a priori* assumption (now recognized to be incorrect) that southern Appalachian populations of *H. bullata* are ecologically marginal and genetically depauperate, they would be unlikely candidates for protection. But based on the genetic findings, as interpreted in a historical biogeographic framework, these southern populations should be given a high conservation priority.

False Poison Sumac, *Rhus michauxii*

Biological and Genetic Background. Low levels of genetic diversity in endangered species may be attributed to small population sizes, population extinctions, and habitat loss during recent times. It is also possible that endemic species may have captured only a small fraction of the genetic diversity present in its progenitor(s). These alternatives can most easily be resolved by comparing the endangered species with closely related congeners. If the genus is uniformly characterized by low levels of genetic diversity, it is likely that the endangered species was always genetically depauperate. However, if the endemic species displays only a fraction of the alleles of a sister taxon, one explanation is that it was recently derived from a small subset of the ancestral species.

An example of a proposed progenitor-derived species complex is the rare and endangered false poison sumac (*Rhus michauxii* Sargent; Anacardiaceae), and its closely related congener the smooth sumac (*R. glabra* L.). *Rhus glabra*, the most widespread woody species in North America, ranges from southern Canada to northern Mexico and is native to all 48 contiguous states. *Rhus michauxii* is a rare endemic confined to the Piedmont of central North Carolina and one recently discovered site in Virginia (Hardin and Phillips, 1985; J. Jacobs, personal communication). (*R. michauxii* previously occurred in Georgia and Florida also, but today only two Georgia individuals remain, and the Florida population has not been relocated [Georgia Nature Conservancy, personal communication].) Based on several morphological traits, as well as results of experimental and natural hybridizations, Hardin and Phillips (1985) proposed a close evolutionary relationship between *R. michauxii* and *R. glabra*.

Sherman-Broyles et al. (1992) used allozymes to estimate genetic diversity in six populations of *R. glabra* (Table 9.5) and nine populations of *R. michauxii*. Of 17 loci analyzed, 14 (82.4%) were polymorphic for *R. glabra*, whereas only six (35.3%) were polymorphic for *R. michauxii*. *Rhus glabra* also had more alleles per polymorphic locus than *R. michauxii* (2.67 vs. 2.38), and levels of individual heterozygosity (H_e) were more than two times higher in *R. glabra* (0.23 vs. 0.10). A higher proportion (33.5%) of the genetic diversity of *R. michauxii* was found among its populations as compared to *R. glabra* (19.8%). Finally, compared to *R. glabra*, populations of *R. michauxii* usually displayed fewer genotypes, often with one genotype predominating. Also, two unisexual populations consisted of a single multilocus genotype. It appears that the successful establishment of seedlings is a relatively rare event in *R. michauxii* populations, and that most reproduction is clonal.

Genetic identities among the nine *R. michauxii* populations ($\overline{I} = 0.94$) were within the range expected of conspecific populations (Crawford, 1983), as were identity values for *R. glabra* ($\overline{I} = 0.93$). Genetic identities between the two taxa were lower ($\overline{I} = 0.87$) but within the range normally observed for closely related congeners. Usually, *R. michauxii* displayed the common alleles characteristic of *R. glabra*. A more distantly related congener (*R. copallina* L.) showed $\overline{I} < 0.60$ with both *R. glabra* and *R. michauxii*.

In summary, the genetic evidence suggests that *R. michauxii* is a derivative of *R. glabra*. The high genetic similarity of the two species, the reduced genetic variation in *R. michauxii*, and the presence of nearly all of the common alleles of *R. glabra* at high frequencies in *R. michauxii* populations support the view that the progenitors of *R. michauxii* went through a

Table 9.5. Allozyme Genetic Diversity in *Rhus michauxii* and *R. glabra*[1]

Species	P	AP	H_e	D_G	G_{ST}
R. michauxii					
Within species	47.0	2.38	0.10	0.41	0.335
Within populations	18.0	2.08	0.05	0.00–0.76	–
R. glabra					
Within species	88.0	2.67	0.23	0.95	0.198
Within populations	48.0	2.29	0.15	0.85–0.99	–

[1] From Sherman-Broyles et al., 1992.

Note: P is the proportion of loci polymorphic; AP is the mean number of alleles per polymorphic locus; H_e is the genetic diversity; D_G is the genotypic diversity; and G_{ST} is the proportion of genetic variation among populations.

genetic bottleneck during the speciation process. Similar examples in the literature indicate that endemic species often have considerably less variation than the more widespread congeners from which they may have been derived (Karron, 1987; Loveless and Hamrick, 1988; Pleasants and Wendel, 1989).

Conservation Implications. The low genetic diversity within *R. michauxii* is likely the result of a speciation bottleneck rather than habitat destruction or modification. However, genetic drift in the small, isolated *R. michauxii* populations may have led to the further loss of genetic variation within the species, and increased differentiation among populations. Also, the predominantly clonal reproduction exhibited by *R. michauxii* populations no doubt serves to further restrict genetic diversity within (and magnify differentiation among) populations.

Conservation efforts should focus on *R. michauxii* populations that maintain the most genotypic diversity as sources of genetic variation for reestablishment of extirpated sites, or for the genetic enrichment of monotypic unisexual populations. When such restoration efforts are undertaken, propagules should consist of as many genotypes as is feasible. Propagules that are the product of sexual reproduction should be more genotypically diverse than cuttings taken from the limited number of established genotypes. The introduction of genotypically diverse progeny arrays should improve the chances of seedling survival in habitats whose microenvironmental conditions differ slightly from those of the donor populations (Vrijenhoek, 1994).

CONCLUDING REMARKS

Two goals of any conservation program should be the insurance of the long-term survival of species and the maintenance of ecological and evolutionary processes. Both goals require the preservation of genetic diversity. Biologists who argue for the priority of ecologic and demographic considerations over genetic considerations (e.g., Lande, 1988; Schemske et al., 1994) appear to have missed this point. Few would dispute that the short-term survival of a species is largely dependent on the ecological health of its populations, and that few populations or species have gone extinct for purely genetic reasons (but see Box 9.1). Nevertheless, the maintenance of evolutionary processes is also an important conservation goal and one that is dependent on the maintenance of genetic diversity.

The short-term preservation of plant species can be accomplished by the *in situ* conservation of a sample of populations, by maintaining representatives of the species in gardens, or by the storage of seed or other propagules. Each strategy requires that decisions be made concerning which populations to preserve or to sample (Brown and Briggs, 1991; Hamrick et al., 1991), and it is here that empirical genetic information can play a useful role. An understanding of the total level and distribution of genetic diversity is required to design a suitable preservation scheme. This information can also serve as a baseline to determine whether genetic diversity is lost during sampling and propagation for *ex situ* conservation. In the absence of empirical information, one can only assume (based on generalities developed from literature reviews) that the distribution of genetic diversity within a threatened species of interest resembles that of a congener, or of species with similar life-history traits. The case studies described here illustrate the risks of predicting the distribution of genetic diversity in the absence of empirical data on the species in question.

As illustrated by examples in this book, many decisions concerning the development of conservation strategies are based on biochemical or molecular markers. Although such markers provide insights into evolutionary patterns and processes that were unthinkable even a decade ago (Avise, 1994), the question has been raised whether such traits provide appropriate information for the conservation of genetic diversity. One issue is whether the distribution of variation in single-gene traits is representative of quantitative genetic variation in adaptive characters sensitive to natural selection (Chapter 15). At best, the plant genetics literature is contradictory in this regard (Hamrick, 1989), although there is some indication that selfing species tend to have associations between single-gene and quantitative traits (Price et al., 1984). Nevertheless, for most plant species there is little evidence that the distribution of variation for single-gene traits is indicative of patterns of variation for other attributes (Hamrick, 1989). Well-designed common garden and/or transplant studies could provide information to help conservation biologists decide which populations to preserve and whether individuals could be successfully translocated among sites. Unfortunately, such studies are difficult logistically, and when implemented at an advanced demographic stage (e.g., seedlings) may neglect the life stages most critical to successful colonization and establishment.

It is our opinion that data from single-gene traits, if employed properly, can be valuable to conservation decisions. For example, such data can be used to identify populations with high levels of genetic variation that are genetically similar to threatened populations, and they can provide insights into the evolutionary and ecological processes that have shaped a species' genetic structure. Nevertheless, conservation biologists should be aware of the limitations of studies based on presumably neutral traits. Furthermore, they should be willing to support studies that permit the placement of the genetic analyses of rare and endangered species into the proper biogeographical, phylogenetic, and evolutionary context.

A question currently under discussion in conservation circles is whether propagules from potential donor populations should be used to restore extirpated or declining populations. Some biologists argue that the restoration of populations by "foreign" propagules will disrupt the adaptation of recipient populations and ultimately increase the probabilities of extinction. Most natural populations are not static, however, but rather are ecologically and evolutionarily dynamic entities that change in size and genetic composition in response to selection, drift, inbreeding, and gene flow. It is not unusual for populations of some species to go through cycles of extinction and recolonization. Many species (e.g., those associated with disturbance) track habitats spatially and temporally. Although most plant populations appear to be reasonably well adapted to their local environments, the genetic basis of this adaptation is probably constantly changing in response to variation in the physical and biotic environment (an exception is if there is insufficient genetic variation available to the population). Hence, populations of most species are generally in a state of flux, both genetically and with respect to size.

One factor that has been given relatively little consideration is that gene flow among populations of many endangered species has undoubtedly decreased in recent decades due to habitat fragmentation. Indeed, decreasing population sizes (leading to genetic drift and inbreeding) and diminished levels of gene flow may be contributing factors in the decline of many endangered species. Although the introduction of propagules from a "foreign" to a threatened population is likely to produce changes in the genetic composition of the recipient population, these changes are not necessarily dire. For populations severely affected by drift and inbreeding, such introductions may be "life saving." Several generations may be required to sort out favorable genotypes, but selection will accomplish this, and presumably has done so in the past when the populations experienced gene flow or were reestablished via colonization.

The introduction of materials from genetically diverse, healthy populations into threatened populations of endangered species may be warranted when empirical data indicate that the populations are in decline. If species endangerment is due to human alteration or destruction of the species' habitat, the case for management may be even stronger. However, population restoration efforts should not be undertaken lightly. Serious consideration should be given to the ecology and genetics of the population, the mating system of the species, the source of propagules, and the goals of the restoration effort. For example, the reestablishment of a plant population using small numbers of individuals (e.g., a few maternal plants) may seriously undermine the restoration effort by producing a population that suffers from immediate founder effects, and/or has limited evolutionary potential because of a restricted genetic base. For species with self-incompatibility systems, special care should be taken to include multiple genetic types so as to enhance the prospects for population self-maintenance. Finally, managers should be aware that the clustering of related individuals within the population can result in near-neighbor pollination and subsequent inbreeding or lack of seed-set. Clearly, restoration efforts require consideration of a complex set of factors, the neglect of which may doom the projects to failure.

SUMMARY

1. Generalizations from the plant allozyme literature can be used to predict the levels and distribution of genetic diversity in unstudied species, but the accuracy of such predictions is low. High variation in estimates of genetic diversity among species with similar life-history traits is undoubtedly due in part to the phylogenetic, biogeographical, and evolutionary histories of particular species.

2. One generalization from the allozyme literature is that endemic plant species have less genetic variation than more widespread species. Comparisons of genetic variation in localized endemics and their widespread congeners may provide insights into this difference. For example, a genetic analysis of the endangered shrub *Rhus michauxii* suggested that the low level of genetic variation in this species is a legacy of its phylogenetic history. *Rhus michauxii* contains a subset of the alleles of its most closely related congener (*R. glabra*), and may have acquired only a small fraction of the genetic variation in its presumed ancestor during the speciation process. Furthermore, the genetic analysis indicated that many populations of *R. michauxii* consist of few genetic clones, suggesting that sexual recruitment is rare in this species relative to its progenitor.

3. Small endemic populations often lose genetic variation by genetic drift. Thus, in some cases, current population sizes may be closely associated with levels of genetic variation. Such appears to be the situation within *Geum radiatum*, an endangered perennial herb endemic to high-elevation rocky outcrops and cliff faces in the southern Appalachians. As gauged by allozyme analyses, larger populations of *G. radiatum* contained higher levels of genetic diversity than smaller populations. Five of the sixteen known populations of *G. radiatum* have been extirpated, and several extant populations are in decline. Thus, knowledge of the recent history and ecology of this species suggests that genetic drift has played a dominant role in the current genetic structure of the species.

4. Present-day genetic structures of species may also reflect past bio-geographical events, a finding illustrated by patterns of genetic variation within *Helonias bullata*. In this threatened wetland species, a latitudinal trend in genetic diversity was observed, with decreasing levels of genetic variation associated with increasing latitude. Although northern populations (in the New Jersey area) have been considered to represent the geographic and genetic stronghold for the species, these populations displayed much lower genetic variation than did smaller "marginal" populations in the southern Appalachians. The observed pattern can be explained by post-glacial recolonization of the northern areas from southern refugia.

5. Such case studies provide insights into the ecology, biogeography, and phylogenetic history of particular species, and have conservation impli-cations. For example, ecological and genetic evidence for population declines in *Geum radiatum* suggest that some populations may benefit demo-graphically from increased genetic variation. In *Helonias bullata,* populations that had been considered geographically and demographically marginal ap-pear upon examination to be especially important for the conservation of the genetic diversity within this species. Finally, low levels of genetic vari-ation within *Rhus michauxii* may be attributable in large part to the phylogenetic history of the species. These studies demonstrate why empirical data are necessary for the establishment of effective genetic conservation programs. Without such information, the distribution of genetic diversity in a threatened species can only be assumed to resemble that of a congener or other species with similar life-history traits.

6. Information garnered from genetic markers, if carefully applied, can serve as a basis for the development of genetic conservation programs. However, conservation biologists should also be aware of the limitations of such studies, and where possible should support research efforts that aim to place genetic results into a better biogeographical, phylogenetic, and evol-utionary context.

7. Under natural conditions, plant populations tend to be genetically and demographically dynamic, changing in size and genetic composition in response to a variety of ecological and evolutionary factors. Human activ-ities often exacerbate the genetic and demographic challenges experienced by endangered species by altering the ecological and evolutionary processes that determine population and metapopulation structure. In this context, the restoration of threatened or extirpated populations by the introduction of propagules from outside sources can be viewed as a management strategy that attempts to preserve the evolutionary processes of gene flow and col-onization.

REFERENCES

ANONYMOUS. 1954. Chestnut blight and resistant chestnuts. *U.S. Dept. Agr. Farm. Bull.* 2068:1–21.

AVISE, J.C. 1994. *Molecular Markers, Natural History and Evolution.* Chapman & Hall, New York.

BARTON, N.H. and M. SLATKIN. 1986. A quasi-equilibrium theory of the distribution of rare alleles in a subdivided population. *Heredity* 56:409–415.

BOYCE, J.S. 1961. *Forest Pathology.* McGraw-Hill, New York.

BROWN, A.H.D. 1978. Isozymes, plant population genetic structure, and genetic conservation. *Theor. Appl. Genet.* 52:145–157.

BROWN, A.H.D. and J.D. BRIGGS. 1991. Sampling strategies for genetic variation in *ex situ* collections of endangered plant species. In *Genetics and Conservation of Rare Plants,* eds. D.A. Falk and K.E. Holsinger, pp. 99–119. Oxford University Press, New York.

BROWN, A.H.D. and D.J. SCHOEN. 1992. Plant population genetic structure and biological conservation. In *Conservation of Biodiversity for Sustainable Development,* eds. O.T. Sandlund, K. Hindar, and A.H.D. Brown, pp. 88–104. Scandinavian University Press, Oslo.

CONKLE, M.T. 1981. Isozyme variation and linkage in six conifer species. In *Proceedings of the Symposium on Isozymes of North American Forest Trees and Forest Insects,* ed. M.T. Conkle, pp. 11–17. United States Department of Agriculture Forest Service Pacific Southwest Forest and Range Experiment Station General Technical Report PSW 48, Berkeley, CA.

CONKLE, M.T., G. SCHILLER, and C. GRUNWALD. 1988. Electrophoretic analysis of diversity and phylogeny of *Pinus brutia* and closely related taxa. *Syst. Bot.* 13:411–424.

CRAWFORD, D.J. 1983. Phylogenetic and systematic inferences from electrophoretic studies. In *Isozymes in Plant Breeding, Part A,* eds. S.D. Tanksely and T.J. Orton, pp. 257–287. Elsevier, Amsterdam.

DAVIS, M.B. 1981. Outbreaks of forest pathogens in Quaternary history. *Proc. 4th Int. Palynolog. Conf.* 3:216–227.

DAVIS, M.B. 1983. Holocene vegetational history of the eastern United States. In *Late-Quaternary Environments of the United States: Vol. 2. The Holocene,* ed. H.E. Wright, Jr., pp. 166–181. University of Minnesota Press, Minneapolis.

DELCOURT, P.A., H.R. DELCOURT, and J.L. DAVIDSON. 1983. Mapping and calibration of modern pollen-vegetation relationships in the southeastern United States. *Rev. Paleobot. Palynol.* 39:1–45.

DEMAURO, M.M. 1993. Relationship of breeding system to rarity in the Lakeside Daisy (*Hymenoxys acaulis* var. *glabra*). *Conserv. Biol.* 7:542–550.

ELLSTRAND, N.C. and D.R. ELAM. 1993. Population genetic consequences of small population size. Implications for plant conservation. *Annu. Rev. Ecol. Syst.* 24:217–242.

ELLSTRAND, N.C. and M.L. ROOSE. 1987. Patterns of genotypic diversity in clonal plant species. *Amer. J. Bot.* 74:132–135.

FALK, D.A. and K. HOLSINGER. 1991. *Genetics and Conservation of Rare Plants.* Oxford University Press, New York.

FOWLER, D.P. and R.W. MORRIS. 1977. Genetic diversity in red pine: evidence for low genic heterozygosity. *Can. J. For. Res.* 7:343–347.

GODT, M.J.W., J.L. HAMRICK, and S. BRATTON. 1995. Genetic diversity in a threatened wetland species, *Helonias bullata* (Liliaceae). *Conserv. Biol.* 9:596–604.

GOTTLIEB, L.D. 1981. Electrophoretic evidence and plant populations. *Prog. Phytochem.* 7:1–46.

GURIES, R.P. and F.T. LEDIG. 1982. Genetic diversity and population structure in pitch pine (*Pinus rigida* Mill.). *Evolution* 36:387–402.

HAMRICK, J.L. 1983. The distribution of genetic variation within and among natural plant populations. In *Genetics and Conservation*, eds. C.M. Schonewald-Cox, S.M. Chambers, B. MacBryde, and W.L. Thomas, pp. 335–348. Benjamin/Cummings, Menlo Park, CA.

HAMRICK, J.L. 1989. Isozymes and analyses of genetic structure of plant populations. In *Isozymes in Plant Biology*, eds. D. Soltis and P. Soltis, pp. 87–105. Dioscorides Press, Portland, OR.

HAMRICK, J.L. and M.J.W. GODT. 1989. Allozyme diversity in plant species. In *Plant Population Genetics, Breeding and Genetic Resources*, eds. A.H.D. Brown, M.T. Clegg, A.L. Kahler, and B.S. Weir, pp. 43–63. Sinauer, Sunderland, MA.

HAMRICK, J.L., M.J.W. GODT, D.A. MURAWSKI, and M.D. LOVELESS. 1991. Correlations between species traits and allozyme diversity: Implications for conservation biology. In *Genetics and Conservation of Rare Plants*, eds. D. Falk and K. Holsinger, pp. 75–86. Oxford University Press, New York.

HAMRICK, J.L., M.J.W. GODT, and S.L. SHERMAN-BROYLES. 1992. Factors influencing levels of genetic diversity in woody plant species. *New Fores.* 6:95–124.

HAMRICK, J.L., Y.B. LINHART, and J.B. MITTON. 1979. Relationships between life history characteristics and electrophoretically detectable genetic variation in plants. *Annu. Rev. Ecol. Syst.* 10:173–200.

HAMRICK, J.L. and J.D. NASON. 1996. Consequences of dispersal in plants. In *Population Dynamics in Space and Time*, eds. O.E. Rhodes, R.H. Chesser, and M.H. Smith. University of Chicago Press, Chicago, *in press*.

HAMRICK, J.L., A. SCHNABEL, and P.V. WELLS. 1994. Distributions of genetic diversity within and among populations of Great Basin Conifers. In *Natural History of the Colorado Plateau and Great Basin*, eds. K.T. Harper, L.L. St.Clair, K.H. Thorne, and W.M. Hess, pp. 147–161. University Press of Colorado, Niwot, CO.

HARDIN, J.W. and L.L. PHILLIPS. 1985. Hybridization in eastern North American *Rhus* (Anacardiaceae). *Assoc. Southeast. Biol. Bull.* 32:99–106.

KARRON J.D. 1987. A comparison of levels of genetic polymorphism and self-compatibility in geographically restricted and widespread plant congeners. *Evol. Ecol.* 1:47–58.

KEEVER, C. 1953. Present composition of some stands of the former oak-chestnut forest in the southern Blue Ridge Mountains. *Ecology* 34:44–54.

LANDE, R. 1988. Genetics and demography in biological conservation. *Science* 241:1455–1460.

LEDIG, F.T. and M.T. CONKLE. 1983. Gene diversity and genetic structure in a narrow endemic, Torrey pine, (*Pinus torreyana* Parry ex Larr). *Evolution* 37:79–85.

LOESCHCKE, V., J. TOMIUK, and S.K. JAIN. 1994. Introductory remarks: Genetics and conservation biology. In *Conservation Genetics*, eds. V. Loeschcke, J. Tomiuk, and S.K. Jain, pp. 3–8. Birkhäuser Verlag, Basel, Switzerland.

LOVELESS, M.D. and J.L. HAMRICK. 1984. Ecological determinants of genetic structure in plant populations. *Annu. Rev. Ecol. Syst.* 15:65–95.

LOVELESS, M.D. and J.L HAMRICK. 1988. Genetic organization and evolutionary history in two North American species of *Cirsium. Evolution* 42:254–265.

MERKEL, H.W. 1906. A deadly fungus on the American chestnut. *N.Y. Zool. Annu. Rep.* 10:97–103.

MILLAR, C.T. 1983. A steep cline in *Pinus muricata. Evolution* 37:311–319.

NEI, M. 1972. Genetic distance between populations. *Amer. Nat.* 106:283–292.

PLEASANTS, J.M. and J.F. WENDEL. 1989. Genetic diversity in a clonal narrow endemic *Erythronium propullans*, and in its widespread progenitor, *Erythronium albidum. Amer. J. Bot.* 76:1136–1151.

PRICE, S.C., K.N. SCHUMAKER, A.L. KAHLER, R.W. ALLARD, and J.E. HILL. 1984. Estimates of population differentiation obtained from enzyme polymorphisms and quantitative characters. *J. Hered.* 75:141–142.

RAWINSKI, T. and J.T. CASSIN. 1986. Range-wide status summary of *Helonias bullata*. Unpublished report to the U.S. Fish and Wildlife Service.

SCHEMSKE, D.W., B.C. HUSBAND, M.H. RUCKELHAUS, C. GOODWILLIE, I.M. PARKER, and J.G. BISHOP. 1994. Evaluating approaches to the conservation of rare and endangered plants. *Ecology* 75:584–606.

SCHILLER, G., M.T. CONKLE, and L. GRUNWALD. 1985. Local differentiation among Mediterranean populations of Aleppo pine in their isozymes. *Silv. Genet.* 35:11–19.

SCHOEN, D.J. and A.H.D. BROWN. 1991. Intraspecific variation in population gene diversity and effective population size correlates with the mating system in plants. *Proc. Natl. Acad. Sci. USA* 88:4494–4497.

SCHWAEGERLE, K.E. and B.A. SCHAAL. 1979. Genetic variability and founder effect in the pitcher plant *Sarracenia purpurea* L. *Evolution* 33:1210–1218.

SHERMAN-BROYLES, S.L., J.P. GIBSON, J.L. HAMRICK, M.A. BUCHER, and M.J. GIBSON. 1992. Comparisons of allozyme diversity among rare and widespread *Rhus* species. *Syst. Bot.* 17:551–559.

SIMON, J.P., Y. GERGERON, and D. GAGNON. 1986. Isozyme uniformity in populations of red pine (*Pinus resinosa*) in the Abitibi Region, Quebec. *Can. J. For. Res.* 16:1133–1135.

SLATKIN, M. and N.H. BARTON. 1989. A comparison of three indirect methods for estimating average levels of gene flow. *Evolution* 43:1349–1368.

STEPHENSON, S.L., A.W. ASH, and D.F. STOUFFER. 1993. Appalachian oak-forests. In *Biodiversity of the Southeastern United States: Upland Terrestrial Communities*, eds. W.H. Martin, S.G. Boyce, and A.C. Echternacht, pp. 255–304. Wiley, New York.

U.S. FISH and WILDLIFE SERVUCE. 1988. Endangered and threatened wildlife and plants; Proposal to determine *Helonias bullata* (swamp pink) to be a threatened species. *Fed. Reg.* 53(175):5740–5743.

U.S. FISH and WILDLIFE SERVICE. 1991. *Swamp Pink (Helonias bullata) Recovery Plan.* United States Fish and Wildlife Service, Newton Corner, MA.

VRIJENHOEK, R.C. 1994. Genetic diversity and fitness in small populations. In *Conservation Genetics*, eds. V. Loeschcke, J. Tomuik, and S. K. Jain, pp. 37–53. Birkhäuser Verlag, Basel, Switzerland.

WHEELER, N.C. and R.P. GURIES. 1982. Population structure, genetic diversity and morphological variation in *Pinus contorta*. *Can. J. For. Res.* 12:595–606.

WRIGHT, S. 1931. Evolution in Mendelian populations. *Genetics* 16:97–159.

ZABINSKI, C. 1992. Isozyme variation in eastern hemlock. *Can. J. For. Res.* 22:1838–1842.

10

CONSERVATION GENETICS OF ENDANGERED ISLAND PLANTS

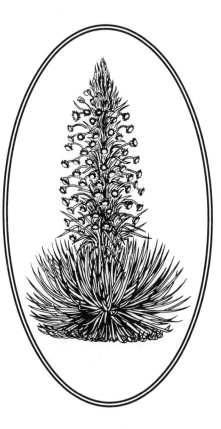

Loren H. Rieseberg and Susan M. Swensen

INTRODUCTION

Perhaps the most striking observation arising from geographical studies of rarity is the high proportion of endangered species on islands. For example, of 680 native U.S. plants identified as being at risk of extinction in the next decade, 232 (34%) are from the Hawaiian Islands or Puerto Rico (Center for Plant Conservation, 1988), yet these islands comprise less than 0.3% of the total geographic area of the United States. Thus, the number of endangered island plants is greater than 100 times that expected based on geographic area alone.

Two primary differences between continental land masses and islands may contribute to the high frequency of endangered island species: restricted land area and isolation. The importance of land area to species extinction rates was theoretically predicted by MacArthur and Wilson (1967), who concluded that smaller land areas suffer higher extinction rates than larger ones. However, as pointed out by Carlquist (1974), human activities are currently the primary cause of extinction, rather than "natural" processes. Nonetheless, restricted land area clearly contributes to endangerment in a variety of ways. First, island species typically comprise fewer populations, and island populations are often numerically small. Second, island organisms often must adapt to and survive within a more limited range of habitats, although the physical and ecological diversity of some islands is remarkable. Third, the effects of introduced predators, herbivores, competitors, parasites, and pathogens are "geometrically greater on small land areas than larger ones" (Carlquist, 1974). Finally, disturbance is more likely to lead to hybridization and genetic assimilation among island species because of small niche size, extensive sympatry, and the often close genetic relationships among species (Rieseberg et al., 1989).

Isolation also contributes to the susceptibility of island species to extinction. Island species are highly susceptible to introduced parasites, predators, and herbivores, because they often have evolved in the absence of such species. For example, the introduction of feline and rodent predators has led to the extinction of large numbers of flightless or ground-nesting birds on oceanic islands (Diamond, 1989). Likewise, introduced mammalian herbivores, such as goats and pigs, have devastated native island vegetation. Of course, goats and pigs devastate mainland vegetation as well, and mainland organisms are subject to many of the coevolutionary pressures faced by island species. However, Carlquist (1974) suggested that the adaptation of immigrants to island conditions is often irreversible, perhaps making it more difficult for them to adapt to biotic and abiotic changes in their environment.

The origin of many island taxa via long-distance dispersal may also contribute to their vulnerability to biotic and abiotic change. Founding populations may consist of only one or several individuals, often resulting in lower levels of genetic diversity in island versus continental species. Furthermore, the tendency for many plant or animal groups to undergo adaptive radiation in open habitats (Box 10.1) often leads to species that are ecologically isolated, but not significantly diverged at the gene level—potentially leading to extensive hybridization if ecological barriers to mating are broken down.

Although small land area and isolation are clearly critical to species endangerment on oceanic islands, the role of these factors on islands derived from continental land masses (continental islands) is less clear. In the latter situation, the geological history of the island, the distance from the adjacent continent, and the dispersal capability of the organisms in question appear critical to interpreting the role of land area and isolation. Nonetheless, the biota of continental islands appear more vulnerable to extinction than continental organisms. For example, 23% of the species presumed extinct in California and nearly 10% of species listed in the California Native Plant Society Inventory of Rare and Endangered Species (Smith and Berg, 1988) are from the California Islands (Box 10.2), which comprise only 0.2% of the

Box 10.1. Adaptive Radiation

The diversification of species derived from a single common ancestor into many ecological niches is termed *adaptive radiation* (Futuyma, 1986). Although there undoubtedly have been many such radiations on continents, they appear to be more frequent and obvious on islands. The primary factor responsible for these bursts of rapid evolutionary change appears to be the release of species from competition. This allows evolutionary changes in resource use that are usually constrained by the presence of competing species. The lack of many mainland plant and animal groups on islands creates numerous vacant niches or ecological opportunities for island immigrants, leading to numerous spectacular radiations on islands (Carlquist, 1974). Impressive examples include the Galápagos Island finches, the Hawaiian honeycreepers, and the Hawaiian silversword alliance. The peculiarity and richness of these groups have become defining features of these island biotas.

land area of the state. Moreover, as will be apparent later in this chapter, these plants are often endangered by factors similar to those acting on oceanic islands.

The rapid loss of native island floras and faunas is disturbing not only because of the richness and peculiarity of insular organisms (Box 10.1), but also because of the major role island biology has played and continues to play in understanding evolutionary processes, and in the development of evolutionary theory. Islands have long been considered natural experiments in evolutionary biology because they represent discrete microcosms, yet vary in shape, size, and degree of isolation (Darwin, 1859; Wallace, 1880; MacArthur and Wilson, 1967). Thus, it is perhaps not surprising that studies of variation among populations of birds from the Galápagos Islands led Darwin to doubt the fixity of species and ultimately to develop the theory of natural selection. Even today, the efforts of evolutionary biologists disproportionately focus on the biology of islands because of the clarity of evolutionary patterns and the utility of island systems for the study of complex biological phenomena.

This chapter employs genetic data to interpret the natural history of several rare island plants representing both continental (California Channel) and oceanic (Hawaiian) islands. The conservation implications of these data will also be considered, given the current vulnerability of these species to extinction and the severity of endangerment of island biotas in general.

Box 10.2. The California Channel Islands

These comprise a group of eight islands occurring off the coast of Southern California (Power, 1980; Figure 10.1). Because they occur along the edge of the continent, these islands are typically classified as continental rather than oceanic. The islands vary considerably in land area, ranging from 2.6 km² (Santa Barbara) to 249 km² (Santa Cruz). They also differ in degree of isolation, with Anacarpa only 20 km from the mainland, compared to 98 km between San Nicolas and the California coast. The Mediterranean climate of the islands is moderated relative to mainland Southern California, due to ocean influence.

As with most island systems, the biota of the Channel Islands has been devastated by human activities, beginning with extirpation of the dwarf mammoth shortly after the arrival of Native Americans approximately 8,000 years ago (Johnson, 1980). Following the appearance of European explorers in 1542, the islands served as refuges for English pirates, Yankee smugglers, and Russian fur traders (Thorne, 1967). During the 19th century, Native Americans still remaining on the islands were moved to the mainland by government order and were replaced on several islands by white squatters, who arrived in response to a brief mining boom (gold in California was first discovered on Santa Catalina), or to raise domestic livestock.

Goats were disastrous to the Channel Island flora (Coblentz, 1980; Hobbs, 1980) because they overgrazed vegetation, and destroyed the soil by removing plant cover. Vegetation loss was exacerbated by large herds of sheep, by feral pigs' rooting activities, and by humans who cut timber for firewood. Large numbers of species were thus eliminated from the islands, including many endemic plants (Thorne, 1967). Since 1950, the National Parks Service, the Catalina Island Conservancy, and the Nature Conservancy (Santa Cruz) have greatly reduced population sizes of the introduced herbivores, probably saving numerous native species from extinction. Nonetheless, the problem has not been eliminated entirely, and non-native herbivores remain the greatest single threat to the Channel Island flora.

CATALINA MAHOGANY

The Catalina mahogany (*Cercocarpus traskiae* Eastwood; Rosaceae) is a member of a small genus of shrubs and small trees widely distributed throughout western North America. The genus has been monographed several times (Schneider, 1905; Rydberg, 1913; Martin, 1950; R. Lis in Hickman, 1993), and in each treatment, leaves have provided the majority of characters used to delimit taxa (Lis, 1992). Nonetheless, the number of species recognized varies dramatically among treatments, ranging from as few as six in Martin's treatment to as many as 21 in Rydberg's classification scheme. The most recent classification for *Cercocarpus* (R. Lis in Hickman, 1993) recognizes ten species.

The Catalina mahogany is one of the most distinctive species in the genus due to its thick, coriaceous leaves and woolly pubescence on the leaf undersurface. The species is extremely rare, restricted to Wild Boar Gully on the southwest side of Santa Catalina Island in the Channel Islands off the coast of California (Figures 10.1 and 10.2). First collected at this locality in 1897 by B. Trask (Thorne, 1967), the population has declined from more than 40 individuals to six adult plants today. Catalina mahogany is listed as "endangered" by the state of California and thus is afforded protection by state law. Inexplicably, the species is not listed as endangered or threatened by the U.S. Fish and Wildlife Service, although it is under consideration for listing.

The decline of the Catalina mahogany is directly related to Santa Catalina Island's recent history (Box 10.2). Santa Catalina Island (194 km^2) is the largest of the southern group of California Channel islands (Figure 10.1). The introduction of goats to the islands during the 19th century was disastrous for the Catalina mahogany and other taxa within the island flora. During the early 1950s, the Santa Catalina Island Company (a private company that later founded the Catalina Island Conservancy) made a concerted effort to reduce the number and impact of introduced herbivores, probably saving the Catalina mahogany from immediate extinction and reducing the threat to other species. Nevertheless, grazing and rooting by goats, pigs, and the more recently introduced bison and mule deer remain a serious threat to the Catalina mahogany and to the entire Santa Catalina Island flora.

Active management of the Catalina mahogany began in the late 1970s (Box 10.3), and the detailed inventories resulting from these efforts led to the discovery of several individuals in Wild Boar Gully that resembled mountain mahogany (C. *betuloides* var. *blanchae*) in key morphological features such as leaf pubescence and leaf thickness, or were intermediate for these characters. Mountain mahogany is the more abundant species on the island and was known to occur in adjacent canyons, so hybridization between the two species seemed possible. However, the morphological features differentiating *Cercocarpus* species are sometimes controlled by ecological rather than genetic factors (Searcy, 1969; Mortenson, 1973), and it is generally understood that morphological intermediacy can result from evolutionary processes other than hybridization (Rieseberg and Ellstrand, 1993).

Classification of Adult Trees

In an attempt to clarify the identity of morphologically ambiguous individuals, Rieseberg et al. (1989) undertook a study of allozymic variation and

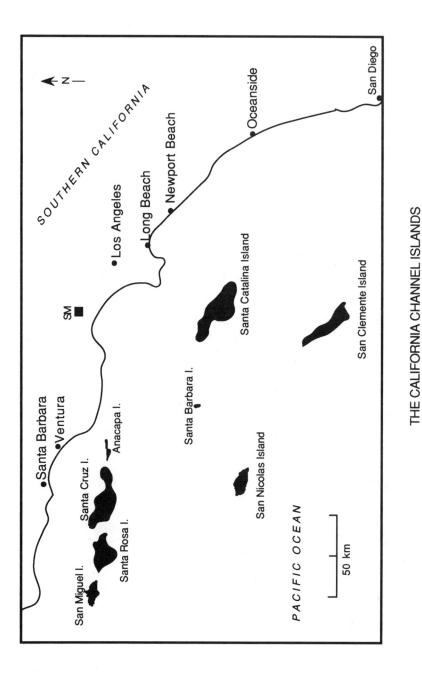

Figure 10.1. Map of the California Channel Islands (modified from Power, 1980).

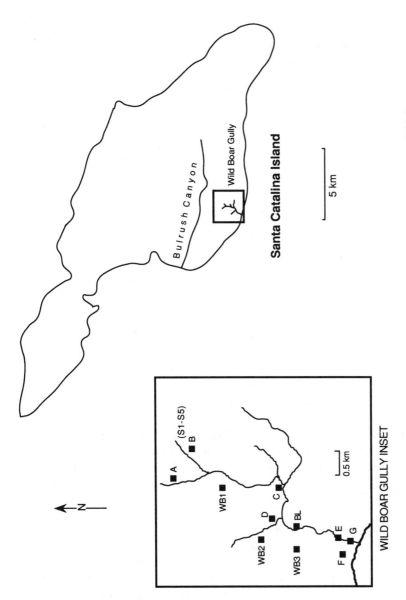

Figure 10.2. Map of Santa Catalina Island showing distribution of *Cercocarpus* in Wild Boar Gully (updated and corrected from Rieseberg et al., 1989). Trees A–D, WB1, WB2 = Catalina mahogany; trees E–G, BL, WB3 and seedlings S1–S5 are hybrids between Catalina and mountain mahogany.

Box 10.3. Management of the Catalina Mahogany

Active management of this species began with the fencing of two trees (individuals A and B; Figure 10.2) in the late 1970s by the Santa Catalina Island Conservancy (Martin, 1984). In 1985, the Conservancy constructed a larger fence encompassing about an acre of suitable habitat around these trees. The effort was extremely successful, and by 1988 approximately 70 seedlings were observed within the fenced area. Isozyme analyses (Rieseberg et al., 1989) revealed that most seedlings were "pure" Catalina mahogany. Although seedling mortality and turnover have been high, the total number of seedlings has only declined slightly over the past six years, with approximately 50 seedlings currently alive.

Several cuttings from the "pure" adult trees have also been propagated at the Rancho Santa Ana Botanic Garden,

and 16 of these specimens have been returned to test plots on the island (J. Takara, personal communication). Five more plants remain at the Island nursery. Survival rates of outplanted trees have been high (close to 90%). Furthermore, several trees have produced viable fruit, and seedlings have been observed at one of the sites. Potential problems (which are recognized by the Conservancy) include the low number of individuals (1–5) planted at each site, the lack of herbivore protection at most locales, the dissimilarity between the habitat of the test-plot sites and the original population, and the loss of identification tags from many trees. Nonetheless, the test plots are well isolated from the original population, and have clearly reduced the immediate likelihood of the species' extinction.

leaf anatomy of all the adult mahogany trees that had been discovered in Wild Boar Gully. Of the eight adult trees in the Gully, three were anatomically intermediate between Catalina and mountain mahogany, suggestive of hybridization. Analysis of 17 isozyme loci in the Catalina mahogany from Wild Boar Gully, as well as in 30 plants of mountain mahogany from a nearby population, revealed one diagnostic isozyme polymorphism differentiating the two species. Based on this single isozyme polymorphism, five of the eight trees were diagnosed as Catalina mahogany, one as mountain mahogany, and two as hybrids.

The results from this study were not entirely satisfactory, however, because the classification of trees as parental versus hybrid was based on a single locus. The use of a single diagnostic locus can ensure the correct identification of first-generation hybrids (barring intraspecific polymorphism), but F_2 or later generation hybrid or backcross progeny could easily be misdiagnosed as belonging to one of the parental classes. Thus, it was not surprising that two of the anatomically intermediate plants lacked a hybrid enzyme genotype. Rieseberg et al. (1989) were unable to distinguish between phenotypic plasticity and hybridization as explanations for the discordance between the anatomical and isozyme datasets.

In the five years following the publication of Rieseberg et al. (1989), three additional *Cercocarpus* trees were discovered in Wild Boar Gully (M. Gay, personal communication). In addition, a single tree from the Santa Monica Mountains was discovered that resembled the Catalina mahogany morphologically (D. Carroll, personal communication). The discovery of these additional trees, combined with the prior uncertainty regarding the correct classification of several trees from the previous study, prompted a more detailed genetic analysis of the mahogany situation (Rieseberg and Gerber, 1995) using random amplified polymorphic DNAs (RAPDs; Williams et al., 1990). In addition, the original diagnostic isozyme locus was assayed in the newly discovered trees.

The results of the RAPD and isozyme analysis (Table 10.1) generated the following classification of hybrid and parental classes of *Cercocarpus* trees, which agrees remarkably well with morphological observations. First, six of the 11 adult mahogany trees in Wild Boar Gully appear to be "true" Catalina Mahogany (trees A–D, WB1, WB2; Figure 10.2). Second, the remaining five adult trees in Wild Boar Gully appear to be of hybrid origin. Of these, trees E, BL, and WB3 appear to represent later generation hybrids. Trees F and G may be later generation hybrids as well, but the possibility that they represent F_1 hybrids cannot be ruled out. Third, the individual from the Santa Monica Mountains is most similar genetically to Catalina mahogany. However, it lacks one RAPD marker diagnostic for Catalina mahogany and carries allozymes of both Catalina and mountain mahogany. Possibly, this individual represents the last remaining descendent of an ancestral or sister population of Catalina mahogany. Alternatively, it represents a mainland relict of Catalina mahogany that has since hybridized with the mainland form of mountain mahogany, *C. betuloides* var. *betuloides*. Finally, the possibility that it represents a recent human introduction from Wild Boar Gully cannot be ruled out. Phylogenetic analysis of the California *Cercocarpus* species will likely be required to distinguish among these hypotheses.

Classification of Juvenile Trees

An almost immediate result of constructing a fence around trees A and B in 1985 (Box 10.3; Figure 10.2) was the emergence of over 70 *Cercocarpus* seedlings in the enclosed area. Electrophoretic analysis of the 28 largest seedlings in 1988 revealed that all were parented by "true" Catalina mahogany— probably trees A and B (Rieseberg et al., 1989). Unfortunately, juvenile mortality and turnover has been high, and reasons for the survival

Table 10.1. RAPD Data for the Catalina and Mountain Mohagany

Taxon/Genotype	Primer:		470	290		188	458	437	173	155	6pgd1
	Fragment (kb): 450										
	0.55	0.70	0.80*	0.80	0.95	1.4	1.0	0.7*	0.9*	0.7*	
Mountain mahogany (20 trees):	0.71	0.82	0.35	0	0	0.73	0	0.10	0.15	0.55	aa
Catalina mahogany:											
A	−	−	+	+	+	−	+	+	+	+	bb
B	−	−	+	+	+	−	+	−	+	+	bb
C	−	−	−	−	−	−	+	−	+	+	bb
D	−	−	−	−	−	−	+	+	+	+	bb
WB1	−	−	+	+	+	−	+	+	+	+	bb
WB2	−	−	+	+	+	−	+	+	+	+	bb
Hybrid trees:											
BL	−	−	−	+	+	−	−	+	+	+	aa
E	−	+	−	−	−	−	−	−	+	−	bb
F	+	+	−	−	−	+	−	+	+	−	ab
G	+	+	+	+	+	−	−	+	+	−	ab
WB3	−	−	+	−	−	+	+	−	+	+	aa
Juvenile trees:											
S1	+	−	−	+	+	+	+	−	−	+	bb
S2	−	−	+	+	+	−	−	−	+	+	ab
S3	−	−	+	+	+	−	+	−	+	+	ab
S4	−	−	−	+	+	−	+	−	+	+	ab
S5	−	+	+	+	+	+	+	−	−	+	bb
Mainland unknown:											
SM	−	−	−	−	+	−	−	+	+	+	ab

Note: The presence or absence of fragments is indicated by a "+" and "−", respectively. Asterisks indicate fragments that are not specific to either parental taxon, but potentially informative regarding the parentage of juvenile trees.
Source: Expanded from Rieseberg and Gerber, 1995.

314

of some seedlings and loss of others is not clear. Nonetheless, the total number of seedlings has declined only slightly, with approximately 50 individuals currently extant. In the summer of 1993, five seedlings were observed with some hybrid morphological characteristics. RAPD and isozyme analysis (Rieseberg and Gerber, 1995) revealed that none of the five seedlings could have resulted from mating between trees A and B, and that at least one of the parents must have been a hybrid tree. If we assume that tree B is the maternal parent of the seedlings (a reasonable assumption given that the seedlings occur beside tree B), then the probable father(s) of each seedling can be determined (Table 10.1): S1 and S5 must have been sired by tree F; and S2–S4 could have been sired by trees F, G, BL, or WB3. It is not clear whether these seedlings existed but were too small to be sampled in the 1989 study, or whether they emerged only recently.

Conservation Relevance

Biological Issues. Hybridization between rare and common species has two potential consequences that are important to conservation biology. If F_1 or later generation hybrids are partly sterile or have reduced vigor, then the rare species may be endangered by outbreeding depression (Price and Waser, 1979; Templeton, 1986); that is, rare populations may have reduced fitness due to gamete wastage in the formation of unfit hybrid individuals. Because only a small fraction of the pollen produced by plants is actually required to fertilize ovules, the primary cost of outbreeding depression in plants appears to be reduced seed set by the maternal parent (Ellstrand and Elam, 1993). For example, seed abortion rates of over 50% and 90% have been reported for interspecific matings in *Gilia* (Grant, 1964) and *Helianthus* (Heiser et al., 1969), and outbreeding depression may account for occasional reports (California Fish and Game, 1992) of unusually low seed set in endangered populations that are sympatric with a common relative (Ellstrand and Elam, 1993). However, outbreeding depression is unlikely to be a problem in long-lived perennial plants, such as *Cercocarpus*, in which the number of seeds produced greatly exceeds the number of seedlings that can become established.

On the other hand, if hybrids are fertile and vigorous, hybridization may lead to the genetic assimilation of the rare taxon by a numerically larger one. Island plants are particularly susceptible to genetic assimilation through hybridization because of small population size, the general lack of strong genetic barriers to hybridization, the invasion and colonization of islands by closely related exotics, and the increasing loss and disturbance of island habitat due to human activities (Rieseberg, 1991). Habitat disturbance tends to increase the frequency of hybrid matings and seedling establishment (Anderson, 1949), particularly in island systems in which ecological barriers to hybridization are significant. For example, a recent survey of the hybrid flora of Hawaii (Ellstrand and Rieseberg, 1995) revealed the occurrence of hybridization in nearly 40 genera and 23 plant families. The majority of the hybrid combinations involved endemic, often rare, species of *Cyrtandra* (67), *Dubautia* (24), *Bidens* (10), and *Clermontia* (8). Hybridization also occurs frequently on the California Channel Islands, often involving endangered taxa such as the federally listed San Clemente Island endemic, *Lotus scoparius* ssp. *traskiae* (Liston et al., 1990) and the example discussed previously, *Cercocarpus traskiae*. Nonetheless, it must be recognized that only in that fraction of situations in which the frequency of interspecific mating is high and the population at risk is numerically smaller than its congener, does hybridization pose a significant threat.

When these conditions do occur, such as in the Catalina mahogany, there are two possible management solutions: elimination of the less-desired species from the area of hybridization, and/or transplantation of the rare population to a remote location where the other hybridizing taxon does not occur (Rieseberg, 1991). In some instances it may also be necessary to eliminate all hybrid and introgressive individuals as well as the less-desired species (e.g., Allendorf and Leary, 1988). However, none of the *Cercocarpus* in Wild Boar Gully are now classified as mountain mahogany (Table 10.1). Furthermore, nearly half of the global genetic diversity in the Catalina mahogany would be lost by the removal of hybrid or introgressed adult and juvenile individuals. Thus, there appears to be little to gain and much to lose by eliminating hybrid or introgressed individuals from Wild Boar Gully.

The other potential solution (recommended by Rieseberg et al., 1989) involves the propagation and transplanting of cuttings of "pure" Catalina mahogany individuals to other locations where hybridization is unlikely. Although this approach is currently being implemented by the Catalina Conservancy (J. Takara, personal communication; Box 10.3), potential problems with transplantations should be recognized. For example, it may be difficult to duplicate the habitat or interspecific interactions (e.g., pollinators) of the original population, which could lead to low survival rates (Rieseberg, 1991). In addition, the possible transfer of diseases or parasites from the nursery to wild populations and the potential loss of the "rich historical genetic records of populations" must be considered in the development of management strategies that involve transplantations (Avise, 1994). Nonetheless, the Catalina mahogany transplant program represents a situation in which transplantations seem necessary, and to date, appears to have been successful (Box 10.3).

Legal Issues. Hybridization involving rare plant taxa also has legal ramifications (O'Brien et al., 1990; O'Brien and Mayr, 1991; Whitham et al., 1991; Leary et al., 1993; Avise, 1994). Because isolating barriers are critical to the formation and delimitation of species, the legal status of rare species that happen to hybridize is unclear. The irony here is that most plant species hybridize at least occasionally in nature, and rare taxa are more likely to hybridize than abundant taxa because of the scarcity of potential mates in small populations (Grant and Grant, 1992). A second question involves the legal status of the hybrid products, which may include F_1 hybrids, hybrid swarms, introgressive races, and hybrid species.

The "hybrid policy" interpretation of the Endangered Species Act (see Avise, 1994 for further details) indicates that hybrids between endangered populations, subspecies, or species should not be legally protected. The effect of this policy could be to exclude F_1 hybrids, hybrid swarms, and perhaps even introgressive races from legal protection. Indeed, the hybrid policy formed the basis for unsuccessful attempts to remove hybridizing rare species from the endangered species list (e.g., Fergus, 1991; see also Chapter 3) because of such populations' lack of "genetic integrity." (However, the law does not appear to exclude hybridizing rare species, such as the California black walnut [McGranahan et al., 1988] and Catalina mahogany.) Conversely, others have argued for the importance of extending protection to hybrid swarms for the sake of preserving important pools of genetic diversity (e.g., Whitham et al., 1991).

Our view is that stretching the endangered species act to cover first-generation hybrids, hybrid swarms, or hybrid zones would damage the credibility of the endangered species program and could waste the limited resources of various management agencies. One possible exception to this general rule might be cases such as the red wolf (Wayne and Jenks, 1991; Wayne, 1992; see Chapter 4), in which the introgressed races represent the only remaining populations of these distinctive taxa (see also O'Brien and Mayr, 1991). The protection of hybrid taxa such as *Helianthus paradoxus* (Rieseberg et al., 1990) and *Iris nelsonii* (Arnold et al., 1991) seems appropriate also, and perhaps the criteria of Jolly (1989) should be employed to evaluate these situations. He suggested protection for the hybrid taxon if: (1) its evolution has proceeded beyond the point at which crossing of the parental stock would recreate the plant considered; (2) it is taxonomically distinct; and (3) it is sufficiently rare or imperiled.

SANTA CRUZ ISLAND BUSH MALLOW

The genus *Malacothamnus* Nutt. (Malvaceae) is an example of a taxonomically difficult plant group. This genus was established by Greene (1906) as a receptacle for the shrubby California elements in *Malvastrum*. Since then,

Box 10.4. Taxonomy of *Malacothamnus*

During the last century, the Southern California bush mallows have been classified as *Malvastrum* (Estes, 1925; Eastwood, 1939), *Sphaeralcea* (Jepson, 1936) and *Malacothamnus* (Kearney, 1951; Hickman, 1993). Two major groups of closely allied species occur within *Malacothamnus*— one typified by *M. fremontii* and the other by *M. fasciculatus*. The number of species recognized within each of these groups varies dramatically, however, depending on the treatment. Nonetheless, the Santa Cruz bush mallow has consistently been treated as one of several *M. fasciculatus* varieties. The other more common varieties of *M. fasciculatus* (sensu Munz and Keck, 1974) occur on the mainland or Santa Catalina Island, and include varieties *catalinensis*, *fasciculatus*, *laxiflorus*, and *nuttallii*. These varieties can usually be distinguished by a number of vegetative and floral features, but morphological intergradation among them, possibly due to intraspecific hybridization, makes varietal delimitation otherwise difficult. Furthermore, several varieties of *M. fasciculatus* are thought to hybridize with other species of both the *fasciculatus* and *fremontii* groups, and may be the source of locally variable and indistinct forms (Kearney, 1951; D. Bates in Hickman, 1993).

it has had a rather remarkable taxonomic history, including three different generic names and species numbers ranging from eight to 27 (Box 10.4). The constant taxonomic flux is unfortunate because the group includes two taxa presumed to be extinct and five additional forms that are very rare and local (Smith and Berg, 1988). Two of these rare forms occur on the California Channel Islands: the San Clemente Island bush mallow (*M. clementinus* [Munz & Johns.] Kearn), and the Santa Cruz Island bush mallow (M. *fasciculatus* [Nutt.] Greene var. *nesioticus* [Rob.] Kearn).

The Santa Cruz Island bush mallow is a weakly self-compatible taxon (D. Wilken, personal communication), known from only two sites on Santa Cruz Island (Figure 10.3). The populations at both sites consist of 25–30 bushes growing on dry chaparral canyon slopes. However, because the species appears to be clonal, the actual number of genetic individuals is likely to be much lower (to be discussed). Although seed set is low (< 10%; D. Wilken, personal communication), some seed is produced. The history of Santa Cruz Island largely parallels that of other Channel Islands (Box 10.2). It comes as no surprise, therefore, that overgrazing by feral animals, particularly goats and sheep, coupled with continued habitat deterioration due to feral pigs, erosion, and drought, have reduced both populations and threaten their continued existence (Swensen et al., 1995). Thus, the Santa Cruz Island bush mallow is listed as "endangered" by the state of California and is under consideration for federal listing.

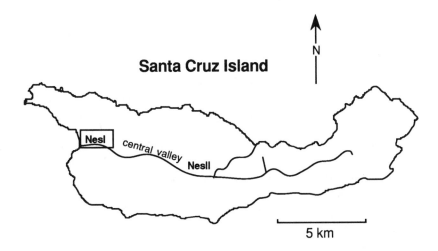

Santa Cruz Island

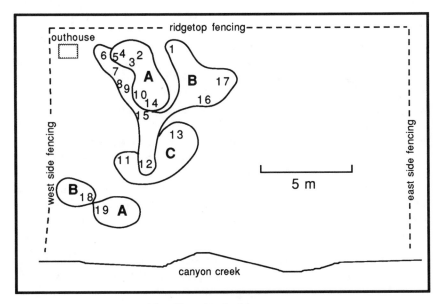

Nes(I) population

Figure 10.3. Map of Santa Cruz Island showing distribution of the Santa Cruz Island bush mallow. The inset shows the clonal population structure of the Nes(I) population. Bushes sampled are indicated by numbers; bold letters indicate clonal individuals as determined by rDNA analysis.

An important consideration that has hampered federal listing and delayed implementation of recovery plans for the Santa Cruz Island bush mallow has been its ambiguous taxonomic status. For example, Kearney (1951) writes: "This extremely local variety differs from all other forms of *M. fasciculatus* in the fastigiate character of the loosely many-flowered panicle, the numerous rather rigid branches mostly erect or strictly ascending, and the flowers cymosely disposed on the branchlets." In contrast, David Bates in Hickman (1993) argued that "Santa Cruz Island var. *nesioticus* is essentially indistinguishable from mainland var. *nuttallii*." In addition to taxonomic ambiguity, the potential clonal structure of the two populations has led to questions regarding the relationship and numbers of genetic individuals in the two populations of Santa Cruz Island bush mallow.

Genetic Distinctness

Empirical patterns.   In an attempt to clarify the taxonomic ambiguity surrounding the Santa Cruz Island bush mallow, Swensen et al. (1995) genetically compared the two known populations of var. *nesioticus* to the other varieties of *M. fasciculatus*. Several types of genetic markers were employed because molecules with different patterns of evolutionary divergence, inheritance, and linkage may reveal different aspects of the evolutionary history of the populations being compared (Avise, 1989a; Doyle, 1992).

The results from this study were unambiguous. The Santa Cruz Island bush mallow had unique alleles at two enzyme loci for all individuals examined, whereas only three other unique alleles were detected among the other four varieties. A similar pattern was observed for RAPD loci, where a total of seven unique alleles was observed in the two populations of var. *nesioticus*. Moreover, the two populations of var. *nesioticus* differed from the other four varieties in terms of restriction endonuclease profiles of the internal transcribed spacer (ITS) of ribosomal genes. Thus, Swensen et al. concluded: "The genetic distinctness of the two *nesioticus* populations is indicated by the genetic analysis, and as such, their protection from herbivory and erosion is not only warranted, but critical to the continued existence of this taxon."

Conservation relevance.   Biological classification provides the fundamental basis for management decisions (Avise and Nelson, 1989; Daugherty et al., 1990). Examples in which poor taxonomy has delayed management efforts, often with disastrous results, are common (e.g., Daugherty et al., 1990; Rieseberg, 1991). Unfortunately, classical taxonomic methods are often inadequate to differentiate groups along phylogenetic lines (e.g., Avise and Nelson, 1989; Beckstrom-Sternberg et al., 1991), or to provide precise identification or delimitation of closely related species or intraspecific taxa.

The latter problem is particularly common in taxonomically difficult plant groups, in which the number of species, subspecies, or varieties can vary by an order of magnitude depending on the philosophical stance of the author. For example, the increasingly widespread view among plant taxonomists (e.g., Hickman, 1993), that species are never real entities, can have serious implications for plant classification (Edwards and Clinnick, 1993). If species are not real, then utility and convenience become the guiding principles of taxonomy, and there is little incentive to account for and describe the often complex and bewildering patterns of variation observed in the field. As pointed out by Edwards and Clinnick, this has the potential to lead to severe lumping and to the dismissal of intraspecific taxa merely for expediency. This appears to be the general trend for many groups (Hickman, 1993), and has immediate implications for conservation biology due to the critical role taxonomy plays in the implementation of the U.S. Endangered Species Act. Our view is that the burden of proof in situations in which the lumping or dismissal of species or intraspecific taxa is considered should lie with the advocates of the proposed taxonomic change, not with its detractors. That is, a taxon such as the Santa Cruz Island bush mallow should be considered genetically distinct unless proven otherwise, and judgments such as these often will require the use of modern genetic methods.

Clonal Structure

Empirical Patterns. To determine the number of genetic individuals in each of the two Santa Cruz Island bush mallow populations, Swensen et al. (1995) tested leaf samples from each discrete mallow bush for RAPD and ITS variation, as well as for length variation in the non-transcribed spacer (NTS) of ribosomal genes. No variation was found within either population for RAPD or ITS variation, but three different NTS length variants were observed for one of the two populations (designated Nes(I) in Figure 10.3). The almost complete identity among individuals within each population is probably due to clonal population structure, with NTS length variation perhaps arising by somatic mutation or a few instances of sexual reproduction. However, the possibility that the populations are genetically homogeneous due to small effective population size and genetic drift, rather than clonal growth, cannot be ruled out.

The three genetic individuals from the Nes(I) population differ from the single genetic individual in the Nes(II) population. Thus, the global diversity of the Santa Cruz Island bush mallow appears to consist of four genetic individuals.

Conservation Relevance. The existence of only four remaining genetic individuals of the Santa Cruz Island bush mallow might be used to justify the establishment of new populations of the variety consisting of all four genotypes. However, there are numerous problems associated with the mixing of genetic stocks within species, including outbreeding depression (e.g., Leberg, 1993), the loss of local adaptations, and the spread of pathogens, particularly if the introduced populations come into contact with existing populations (Avise, 1994). In addition, theoretical genetic models indicate that although small isolated populations rapidly lose within-population variation, overall genetic diversity may be retained better than in a single panmictic population (Lacy, 1987). Thus, management efforts that focus on the protection of existing populations (e.g., from herbivory), and the discovery and protection of new populations of the variety, would probably represent a more intelligent strategy than the creation of artificial hybrid populations of unknown ecological behavior. Nonetheless, the initiation of a captive breeding program for the Santa Cruz Island bush mallow away from the originating populations is critical in order to maintain a germ-plasm resource in the event that one or both of the originating populations is extirpated. In addition, these plants could be used to assess the ecological performance of selfed progeny relative to progeny from within- and among-population matings.

THE HAWAIIAN SILVERSWORD ALLIANCE

This plant alliance consists of 28 species and three genera (*Dubautia, Argyroxiphium,* and *Wilksia*) endemic to Hawaii (Box 10.5). Although all of these species appear to have arisen from a single mainland ancestor (Carlquist, 1959; Baldwin et al., 1991), they exhibit spectacular morphological and ecological diversity (Robichaux et al., 1990). The group includes rosette plants, cushion plants, subshrubs, shrubs, trees, and vines. Moreover, the species occur in nearly all of the island habitats, ranging from sea level to alpine, and from wet tropical to desert-like. So impressive is the broad array of morphological, anatomical and ecophysiological traits required for success in these habitats, that Carlquist (1974) referred to the silversword alliance as "undoubtedly the most outstanding example of adaptive radiation among Hawaiian angiosperms." Unfortunately, several members of this unique group are in danger of extinction (Box 10.6).

Population Genetic Structure and Genetic Differentiation

Analysis of isozyme variation among members of the Hawaiian silversword alliance (Witter and Carr, 1988; Witter, 1990) revealed two patterns

Box 10.5. The Hawaiian Islands

The Hawaiian archipelago comprises 17 volcanic islands (Figure 10.4) that vary dramatically in land area (1–8,000 km^2), elevation (near sea level to 3,700 m), and rainfall (0.3–11.4 m). The islands occur about 3,500 km from the nearest continent, and are arranged in a line of descending age from the northwest to southeast. The chain terminates in the southeast with the youngest and largest island, Hawaii, which is less than 500 thousand years old (MacDonald et al., 1983).

The physical diversity and isolation of these islands has led to the evolution of a diverse and spectacular biota through multiple invasions and explosive adaptive radiations (Box 10.1). Thus,

the majority of species that occur on the islands are found nowhere else. For example, more than 90% of the 1,100 plant species inhabiting the islands are endemic (Mehrhoff, 1993).

Colonization of the Hawaiian archipelago by humans, and subsequent over-hunting, habitat loss and deterioration, and the introduction of pests, herbivores, and predators, has led to the extirpation of numerous native species, including many endemic landbirds (Chapter 6) and 97 plants (Carlquist, 1990; Mehrhoff, 1993). In addition, another 267 plant species (24%) are thought to be in immediate danger of extinction (Mehrhoff, 1993).

---

of genetic differentiation. Cytologically derived species of *Dubautia* with $n = 13$ pairs of chromosomes (Carr, 1985) had extremely high interspecific genetic identities ($I > 0.90$). These values are similar to those observed for conspecific populations of mainland species and suggest that the $n = 13$ species (which occur on the younger islands of Maui and Hawaii) are of relatively recent origin. In contrast, the $n = 14$ species of *Dubautia*, most of which are endemic to the oldest major island of Kauai, had genetic identities similar to interspecific comparisons from other plant groups ($I \approx 0.67$; Crawford, 1983), indicating that given sufficient time (up to six million years) island plants will accumulate significant genetic differences.

In addition to the extremely high genetic identities observed among some members of the silversword alliance, the species were genetically depauperate, displaying less than half the genetic variability ($P = 0.24$; $H = 0.14$) characterizing an average plant species (Hamrick et al., 1991). These two patterns (high genetic identities and low genetic diversity) appear to be typical of island plants in general (e.g., Rick and Fobes, 1975; Crawford et al., 1985, 1993; Helenurm and Ganders, 1985; Lowrey and Crawford, 1985; Wendel and Percy, 1990). Witter and Carr (1988) suggested that low levels of genetic variation in island populations are due to founder effects and genetic drift. Furthermore, because the introductions are usually recent, there has been insufficient time for variation to accumulate. Thus, these

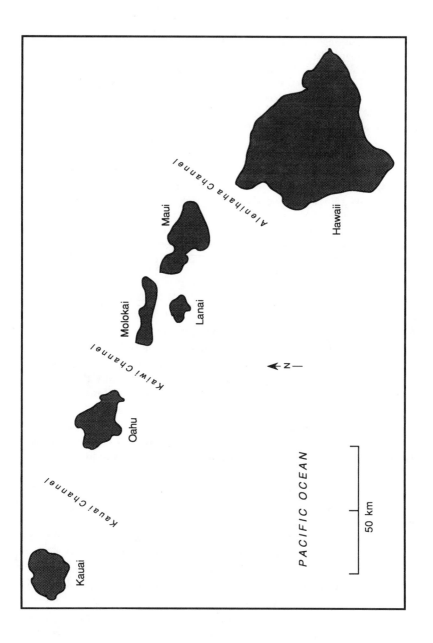

Figure 10.4. Map of the Hawaiian Islands.

Box 10.6. Conservation Status of Species in the Hawaiian Silversword Alliance

As is the case for most speciose groups of endemic Hawaiian plants, a significant proportion (25%) of the species of the silversword alliance are listed as threatened or endangered by the U.S. Fish and Wildlife Service, including:

Argyroxiphium kauense (Rock & M. Neal) Degener & I. Degener
Argyroxiphium sandwicense DC. subsp. *macrocephalum* (A. Gray) Meyrat
Argyroxiphium DC. *sandwicense* subsp. *sandwicense*
Dubautia herbstobatae G. Carr
Dubautia latifolia (A. Gray) Keck

Dubautia pauciflorula H. St. John & G. Carr
Wilkesia gymnoxiphium A. Gray

Primary threats to members of the silversword alliance are destruction by non-native mammalian herbivores such as goats, sheep and feral pigs, and competition from exotic weeds (Carlquist, 1990; Robichaux et al., 1992). The impact of herbivory has recently been reduced, however, by the construction of fences around many of the *Argyroxiphium* populations.

---

authors suggest that allozyme divergence is likely to occur only as new mutants arise, rather than by divergence in the frequencies of multiple ancestral alleles at a single locus. Evidence for this type of divergence is particularly strong in the $n = 13$ *Dubautia* species that share a common allele at most loci. In this situation, genetic divergence can only occur by the production and fixation of new mutations.

The most detailed analysis of within-population genetic variation has been conducted in the last remaining natural population of the Mauna Kea silversword from Waipahoehoe, Hawaii, *Argyroxiphium sandwicense* subsp. *sandwicense* (Box 10.6). The population consists of fewer than 50 widely spaced adult individuals. The population is variable genetically, with polymorphism observed at 28.6% of 63 RAPD loci assayed (Friar et al., 1994). However, seed set is very low (< 1%), apparently due to pollen limitations (Powell, 1992). Silverswords are self-incompatible, and most plants flower in isolation with little chance of cross-pollination.

In an attempt to reduce the risk of extinction for the Mauna Kea silversword, several hundred individuals of the species have been outplanted (Powell, 1992). The three surviving outplant populations comprise groups of three, 10–25, and 300–400 individuals, respectively (Robichaux, personal communication). Unfortunately, the vast majority of the outplants appears to be descended from the same female parent, with one or two possible male parents. Because of sporophytic self-incompatibility (Carr et al., 1986), sexual reproduction can only occur between individuals carrying different "s" alleles. As a result, seed set in outplanted individuals is typically less than 20% (Powell, 1992), although it increases to over 60% when the outplants are pollinated by individuals from the natural population (Robichaux and Friar, personal communication). Other consequences of using outplants from such a limited parental source include differences in morphology and genetic diversity (Robichaux and Friar, personal communication). Over 70% of the individuals from the natural population are monocarpic (Powell, 1992). In contrast, the outplants were derived from the less frequent multiple-stemmed, polycarpic individuals in the natural population and, thus, over 80% of the outplants are branched. Likewise, the outplants display less than one third the RAPD variation of the natural population ($P = 0.079$).

Molecular Phylogeny

The Hawaiian silversword alliance has been the subject of extensive phylogenetic studies over the past five years (Baldwin et al., 1990, 1991; Baldwin, 1992; Baldwin and Robichaux, 1995). The phylogenetic hypotheses derived from these studies have been compared to relationships proposed on the basis of cytogenetic, morphological, and allozyme studies of these species and will be briefly summarized here.

Analysis of chloroplast (cp) DNA and ITS variation in the Hawaiian silversword alliance and putative mainland relatives resulted in two major findings (Baldwin et al., 1990, 1991; Baldwin, 1992), namely that the Hawaiian silversword alliance is: (1) monophyletic, indicating a single introduction to Hawaii and subsequent adaptive radiation; and (2) derived from the California tarweeds (as first hypothesized by Carlquist, 1959), indicating extreme long-distance dispersal.

The phylogenetic data also have several important implications regarding evolution within the Hawaiian silversword alliance. The generic partitions based on morphological and cytogenetic data are largely supported by the molecular phylogenetic data. One major exception is that several species of *Dubautia* are more closely related to species of *Wilkesia* than to other congeners. There are several minor discrepancies between the two molecular-based phylogenetic hypotheses, as well as between the molecular phylogenies and those based on morphology, cytogenetics, and ITS variation. At least some of the phylogenetic discordance appears to result from hybridization and introgression (Baldwin et al., 1990; Baldwin, 1992). With regard to the phylogenetic distribution of rarity, both the species-depauparate *Wilkesia* (5–7 species) and *Argyroxiphium* (four species), as well as the speciose *Dubautia* (16 species), contain endangered species. Thus, there is no apparent phylogenetic component to the distribution of rarity.

The phylogenetic data also suggest several general trends, including the colonization of taxa from the oldest to the youngest islands, and repeated shifts within lineages between wet and dry habitats. The basal position of *Argyroxiphium*, and the fact that these species are known only from the younger islands, confounds the contention that the Hawaiian silversword alliance originated on the oldest island, Kauai (or an older submerged island; Baldwin and Robichaux, 1995). Assumption of an extinct Kauaian *Argyroxiphium* ancestor may solve this apparent discrepancy. Assuming that there was a wet-adapted ancestor of the Hawaiian silversword alliance, at least five ecological shifts from wet to dry habitats are suggested by the resolution in both cpDNA and ITS phylogenies, whereas under the assumption of a dry-habitat ancestor, at least six shifts of habitat are required (Baldwin and Robichaux, 1995).

As was observed for the isozyme dataset, cpDNA and ITS sequence divergence values among the Hawaiian silverswords were much lower on average than those observed for the mainland perennial tarweeds (Baldwin et al., 1991; Baldwin, 1992; Baldwin and Robichaux, 1995). The cpDNA divergence values were also lower than those observed for most other mainland plant genera that have been analyzed to date (Soltis et al., 1992). Unfortunately, there is insufficient data for other mainland plants to allow similar comparisons for ITS.

Conservation Relevance

Until recently, most management strategies have focused on the maintenance of maximum levels of genetic diversity or heterozygosity within populations (Falk and Holsinger, 1991). The reasoning underlying this approach is that populations with high levels of diversity are thought to be capable of adapting more successfully to biotic and abiotic changes in their environment (Huenneke, 1991). Furthermore, a positive correlation between heterozygosity and fitness has been observed in some organisms (e.g., Wildt et al., 1987; see also Chapters 3 and 12). However, as pointed out by Avise (1989b), the relationship between fitness and heterozygosity is not a clear one. For example, in many plant groups, spatial structuring of populations and mating system (e.g., autogamy) appear to promote low levels of polymorphism within populations. Moreover, outbreeding depression and the disruption of locally adapted gene combinations are sometimes a consequence of mating between individuals from different source populations (e.g., Leberg, 1993). Thus, attempts to maximize diversity in historically depauparate populations (such as many island populations) by transplant or reintroduction may lead to the loss of critical adaptive features and perhaps to outbreeding depression. This seems particularly likely in a group such as the Hawaiian silversword alliance, in which populations exhibit such a remarkable array of morphological, anatomical, and ecophysiological adaptations.

Nonetheless, if outplantings are undertaken, genetic factors should be considered (Huenneke, 1991). In particular, care should be taken to maintain the genetic characteristics involved in an organism's habitat specialization. Although there are a number of reasons to avoid mixing material from different source populations (Avise, 1994; although see Barrett and Kohn, 1991), the transplant population should include the full range of ecologically relevant variation from a single site in order to increase the chances of preserving that population's full ecological amplitude. Additionally, in self-incompatible plants such as the Mauna Kea silversword, it is critical to include progeny from as many different individuals in the source population as possible so as to maximize compatibility among matings and, thus, ensure high levels of seed set. The low levels of seed set and different growth form of current outplants are the unfortunate results of outplanting descendants from only one or two maternal plants.

One final comment concerns the conservation relevance of the phylogenetic data generated for the Hawaiian silversword alliance. Because of the existence of an excellent taxonomic treatment for this group (G. Carr in Wagner, 1990), the phylogenetic data did not result in major new taxonomic insights that would impact management decisions, nor was there any obvious phylogenetic component to the evolution of rarity. It has been suggested recently that the allocation of limited resources for conservation biology should be guided by phylogenetic criteria (e.g., Vane-Wright et al., 1991). The basic reasoning is that organisms representing phylogenetically unique or distinctive lineages should receive priority, whereas those representing more speciose clades or with common sibling taxa would be of lower conservation priority. In the Hawaiian silverswords, this reasoning would place a priority on preserving rare species of *Argyroxiphium* and *Wilkesia*, whereas rare taxa in the more speciose *Dubautia* lineage would be of lower conservation priority. We do not believe that phylogenetic data are necessarily a good predictor of the biological uniqueness and significance of taxa and, thus, we are not supportive of attempts to allocate resources using phylogenetic criteria. In fact, we find little that is commendable in general about placing priorities on the preservation of unique and irreplaceable entities such as endangered species.

SUMMARY

1. Due to the restricted land area and isolation of islands, island biotas typically have a much higher proportion of endangered species than do their mainland counterparts. This is unfortunate not only because of the richness

and peculiarity of insular organisms, but also because of the important role islands have played in the development of evolutionary theory. Threats to the long-term preservation of island plants and animals include overhunting, habitat destruction and deterioration, the introduction of non-native pests, parasites, herbivores, and competitors, and hybridization with numerically larger congeners. In some instances, genetic data can be relevant to the preservation of rare island species, such as the Santa Catalina Island mahogany, Santa Cruz Island bush mallow, and members of the Hawaiian silversword alliance.

2. Isozyme and RAPD analyses of the last remaining population of the Catalina mahogany revealed that it was threatened due to hybridization with a more common congener, the mountain mahogany. In fact, five of the 11 remaining adult trees are first- or later-generation hybrids, and at least five juvenile trees are of hybrid origin. The correct identification of these trees is essential for a transplant program currently underway on Santa Catalina Island.

3. The Santa Cruz Island bush mallow is known from only two populations of 25–30 bushes on Santa Cruz Island. A recent taxonomic treatment has suggested that these populations are not different from a mainland bush mallow variety. However, analysis of island and mainland bush mallow populations revealed several allozymes, RAPD fragments, and ITS sequence variants unique to the Santa Cruz bush mallow. Thus, this population is genetically distinct and appears to provide a good example of the general rule that taxa should be considered genetically distinct unless proven otherwise. Furthermore, the species appears to be clonal, consisting of a total of only four genetic individuals. Although we do not advocate a transplantation program at this time, knowledge of clonal distribution in these populations will be critical if it becomes necessary to implement a transplantation option at a later date.

4. The Hawaiian silversword alliance is considered the most outstanding example of adaptive radiation in flowering plants. Observations of low levels of isozyme and cpDNA divergence among species in the silversword alliance, and low levels of polymorphism within populations, are typical of island plant groups in general and may be partly related to their susceptibility to biotic and abiotic change. Analysis of the phylogenetic distribution of endangered Hawaiian silversword species revealed no obvious phylogenetic component to rarity, although some taxa represented "phylogenetically more unique" lineages. However, we see little justification for prioritizing endangered species based on phylogenetic criteria.

5. Because of the uniqueness and rarity of members of the silversword alliance, transplantation programs have been initiated for several species, including the endangered Mauna Kea silversword. This species demonstrates the hazards of implementing a transplantation program without first considering genetic factors. All of the outplanted individuals are descendants of one to two maternal plants, with the consequences that the outplants: (1) represent a form that is rare in the natural population; (2) retain only one fourth the genetic diversity characteristic of the remaining natural population; and (3) have a seed set of less than 20%, due to sporophytic self-incompatibility.

REFERENCES

ALLENDORF, F.W. and R.F. LEARY. 1988. Conservation and distribution of genetic variation in a polytypic species, the cutthroat trout. *Conserv. Biol.* 2:170–184.

ANDERSON, E. 1949. *Introgressive Hybridization.* Wiley, New York.

ARNOLD, M.L., C.M. BUCKNER, and J.J. ROBINSON. 1991. Pollen mediated introgression and hybrid speciation in Louisiana irises. *Proc. Natl. Acad. Sci. USA* 88:1398–1402.

AVISE, J.C. 1989a. Gene trees and organismal histories: a phylogenetic approach to population biology. *Evolution* 43:1192–1208.

AVISE, J.C. 1989b. A role for molecular genetics in the recognition and conservation of endangered species. *Trends Ecol. Evol.* 4:279–281.

AVISE, J.C. 1994. *Molecular Markers, Natural History and Evolution.* Chapman & Hall, New York.

AVISE, J.C. and W.S. NELSON. 1989. Molecular relationships of the extinct dusky seaside sparrow. *Science* 243:646–648.

BALDWIN, B.G. 1992. Phylogenetic utility of the internal transcribed spacers of nuclear ribosomal DNA in plants: an example from the Compositae. *Molec. Phylogenet. Evol.* 1:3–16.

BALDWIN, B.G., D.W. KYHOS, and J. DVORAK. 1990. Chloroplast DNA evolution and adaptive radiation in the Hawaiian silversword alliance (Asteraceae–Madiinae). *Ann. Missouri Bot. Gard.* 77:96–109.

BALDWIN, B.G., D.W. KYHOS, J. DVORAK, and G.D. CARR. 1991. Chloroplast DNA evidence for a North American origin of the Hawaiian silversword alliance (Asteraceae). *Proc. Natl. Acad. Sci. USA* 88:1840–1843.

BALDWIN, B.G. and R.H. ROBICHAUX. 1995. Historical biogeography and ecology of the Hawaiian silversword alliance (Compositae): New molecular phylogenetic perspectives. In *Hawaiian Biogeography: Evolution on a Hot Spot Archipelago,* eds. W.L. Wagner and V.A. Funks. Smithsonian Institution Press, Washington, DC, *in press.*

BARRETT, S.C.H. and J.R. KOHN. 1991. Genetic and evolutionary consequences of small population size in plants: implications for conservation. In *Genetics and Conservation of Rare Plants,* eds. D.A. Falk and K.E. Holsinger, pp. 3–30. Oxford University Press, New York.

BECKSTROM-STERNBERG, S., L.H. RIESEBERG, and K. DOAN. 1991. Gene lineage analysis of populations of *Helianthus niveus* and *H. petiolaris. Plant Syst. Evol.* 175:125–138.

CALIFORNIA FISH AND GAME. 1992. *RAREFIND.* Natural Heritage Division/Natural Diversity Data Base, Sacramento, CA.

CARLQUIST, S. 1959. Studies on Madiinae: anatomy, cytology, and evolutionary relationships. *Aliso* 4:171–236.

CARLQUIST, S. 1974. *Island Biology.* Columbia University Press, New York.

CARLQUIST, S. 1990. Worst case scenarios for island conservation: the endemic biota of Hawaii. In *Ecological Restoration of New Zealand Islands*, eds. D.R. Towns, C.H. Daugherty and I.A.E. Atkinson, pp. 207–212. Conservation Sciences Publication No. 2., Department of Conservation, Wellington, New Zealand.

CARR, G.D. 1985. Monograph of the Hawaiian Madiinae (Asteraceae): *Argyroxiphium, Dubautia*, and *Wilkesia*. *Allertonia* 4:1–123.

CARR, G.D., E.A. POWELL, and D.W. KYHOS. 1986. Self-incompatibility in the Hawaiian Madiinae (Compositae): an exception to Baker's Rule. *Evolution* 40:430–434.

CENTER FOR PLANT CONSERVATION. 1988. CPC survey reveals 680 native U.S. plants may become extinct within 10 years. *Cent. Plant Conserv.* 4:1–2.

COBLENTZ, B.E. 1980. Effects of feral goats on the Santa Catalina Island ecosystem. In *The California Islands: Proceedings of a Multidisciplinary Symposium*, ed. D.M. Power, pp. 167–172. Haagden Printing, Santa Barbara, CA.

CRAWFORD, D.J. 1983. Phylogenetic and systematic inferences from electrophoretic studies. In *Isozymes in Plant Genetics and Breeding, Part A*, eds. S.D. Tanksley and T.J. Orton, pp. 257–287. Elsevier, New York.

CRAWFORD, D.J., T.F. STUESSEY, R. RODRIGUEZ, and M. RONDINELLI. 1993. Genetic diversity in *Raphithamnus venustus* (Verbenaceae), a species endemic to the Juan Fernandez Islands. *Bull. Torr. Bot. Club* 120:23–28.

CRAWFORD, D.J., T.F. STUESSEY, and O. SILVA. 1985. Allozyme divergence and evolution of *Dendroseris* (Compositae), a progenitor-derivative species pair. *Evolution* 36:379–386.

DARWIN, C. 1859. *On the Origin of Species by Means of Natural Selection, or the Preservation of Favored Races in the Struggle for Life*. Murray, London.

DAUGHERTY, C.H., A. CREE, J.M. HAY, and M.B. THOMPSON. 1990. Neglected taxonomy and continuing extinctions of tuatara (*Sphenodon*). *Nature* 347:177–179.

DIAMOND, J. 1989. The past, present and future of human-caused extinctions. *Phil. Trans. Roy. Soc. Lond.* B325:469–477.

DOYLE, J.J. 1992. Gene trees and species trees: molecular systematics as one-character taxonomy. *Syst. Bot.* 17:144–163.

EASTWOOD, A. 1939. New Californian plants. *Leafl. West. Bot.* 1:213–220.

EDWARDS, S.W. and R.W. CLINNICK. 1993. Species denial. *Four Seas.* 9:4–20.

ELLSTRAND, N.C. and D.R. ELAM. 1993. Population genetic consequences of small population size: Implications for plant conservation. *Annu. Rev. Ecol. Syst.* 24:217–242.

ELLSTRAND, N.C. and L.H. RIESEBERG. 1995. Distribution of spontaneous plant hybrids. *Proc. Natl. Acad. Sci. USA* (in review).

ESTES, F.W. 1925. The shrubby malvastrums of Southern California. *Bull. S. Calif. Acad. Sci.* 24:81–87.

FALK, D.A. and K.E. HOLSINGER (eds.). 1991. *Genetics and Conservation of Rare Plants*. Oxford University Press, New York.

FERGUS, C. 1991. The Florida panther verges on extinction. *Science* 251:1178–1180.

FRIAR, E., R.H. ROBICHAUX, and D.W. MOUNT. 1994. Loss of genetic variability in the endangered Mauna Kea silversword (*Argyroxiphium sandwicense* ssp. *sandwicense*) and its implications for population recovery. Abstract of paper presented at July 1994 Hawaii Conservation Conference, Honolulu.

FUTUYMA, D.J. 1986. *Evolutionary Biology*. Sinauer, Sunderland, MA.

GRANT, P.R. and B.R. GRANT. 1992. Hybridization of bird species. *Science* 256:193–197.

GRANT, V. 1964. Genetic and taxonomic studies of *Gilia*. XII. Fertility relationships of the polyploid cobwebby gilias. *Aliso* 4:435–481.

GREENE, E.L. 1906. Certain malvaceous types. *Leafl. Bot. Obs.* 1:207–208.

HAMRICK, J.L., M.J.W. GODT, D.A. MURAWSKI, and M.D. LOVELESS. 1991. Correlations between species traits and allozyme diversity: Implications for conservation biology. In *Conservation of Rare Plants: Biology and Genetics*, eds. D.A. Falk and K.E. Holsinger, pp. 75–86. Oxford University Press, Oxford.

HEISER, C.B., D.M. SMITH, S. CLEVENGER, and W.C. MARTIN. 1969. The North American sunflowers (*Helianthus*). *Mem. Torr. Bot. Club* 22:1–218.

HELENURM, K. and F.R. GANDERS. 1985. Adaptive radiation and genetic differentiation in Hawaiian *Bidens*. *Evolution* 39:753–765.

HICKMAN, J. 1993. *The Jepson Manual, Higher Plants of California*. University of California Press, Los Angeles.

HOBBS, E. 1980. Effects of grazing on the northern populations of *Pinus muricata* on Santa Cruz Island, California. In *The California Islands: Proceedings of a Multidisciplinary Symposium*, ed. D.M. Power, pp. 159–166. Haagden Printing, Santa Barbara, CA.

HUENNEKE, L.F. 1991. Ecological implications of genetic variation in plant populations. In *Genetics and Conservation of Rare Plants*, eds. D.A. Falk and K.E. Holsinger, pp. 31–44. Oxford University Press, New York.

JEPSON, W.L. 1936. *A Flora of California*, vol. 2. University California Press, Berkeley.

JOHNSON, D.L. 1980. Episodic vegetation stripping, soil erosion, and landscape modification in prehistoric and recent historical time, San Miguel Island, California. In *The California Islands: Proceedings of a Multidisciplinary Symposium*, ed. D.M. Power, pp. 103–122. Haagden Printing, Santa Barbara, CA.

JOLLY, D. 1989. Letter on USFS Southwestern Region Policy on protection of rare plants. *Memo 2670, U.S.F.S. Region 3*, Albuquerque, NM.

KEARNEY, T.H. 1951. The genus *Malacothamnus*, Greene (Malvaceae). *Leafl. West. Bot.* 6:113–140.

LACY, R.C. 1987. Loss of genetic diversity from managed populations: Interacting effects of drift, mutation, immigration, selection and population subdivision. *Conserv. Biol.* 1: 143–158.

LEARY, R.F., F.W. ALLENDORF, and S.H. FORBES. 1993. Conservation genetics of bull trout in the Columbia and Klamath River drainages. *Conserv. Biol.* 7:856–865.

LEBERG, P.L. 1993. Strategies for population reintroduction: effects of genetic variability on population growth and size. *Conserv. Biol.* 7:194–199.

LIS, R. 1992. Leaf architectural survey of *Cercocarpus* (Rosaceae) and its systematic significance. *Int. J. Plant Sci.* 153:258–272.

LISTON, A.L., L.H. RIESEBERG, and O. MISTRETTA. 1990. Ribosomal DNA evidence for hybridization between island endemic subspecies of *Lotus. Biochem. Syst. Ecol.* 18:239–244.

LOWREY, T.K. and D.J. CRAWFORD. 1985. Allozyme divergence and evolution in *Tetramolopium* (Compositae: Astereae) on the Hawaiian Islands. *Syst. Bot.* 10:64–72.

MACARTHUR, R.H. and E.O. WILSON. 1967. *Island Biogeography*. Princeton University Press, Princeton, NJ.

MACDONALD, G.A., A.T. ABBOTT, and F.L. PETERSON. 1983. *Volcanoes in the Sea*, 2nd ed. University of Hawaii Press, Honolulu.

MARTIN, F.L. 1950. A revision of *Cercocarpus. Brittonia* 7:91–111.

MARTIN, T. 1984. Status report for *Cercocarpus traskiae*. *California Natural Diversity Data Base*. Element Code SPLC.EB3.

MCGRANAHAN, G.H., J. HANSEN, and D.V. SHAW. 1988. Inter- and intraspecific variation in the California black walnuts. *J. Amer. Soc. Hort. Sci.* 113:760–765.

MEHRHOFF, L. 1993. Rare plants in Hawaii: A status report. *Plant Conserv.* 7:1–2.

MORTENSON, T.H. 1973. Ecological variation in the leaf anatomy of selected species of *Cercocarpus. Aliso* 8:19–48.

MUNZ, P.A. and D.D. KECK. 1974. *A California Flora.* University of California Press, Los Angeles.

O'BRIEN, S.J. and E. MAYR. 1991. Bureaucratic mischief: recognizing endangered species and subspecies. *Science* 251:1187–1188.

O'BRIEN, S.J., M.E. ROELKE, N. YUHKI, K.W. RICHARDS, W.E. JOHNSON, W.L. FRANKLIN, A.E. ANDERSON, O.L. BASS, JR., R.C. BELDEN, and J.S. MARTENSON. 1990. Genetic introgression in the Florida panther *Felis concolor coryi. Natl. Geog. Res.* 6:485–494.

POWELL, E.A. 1992. Life history, reproductive biology, and conservation of the Mauna Kea silversword, *Argyroxiphium sandwichense* DC (Asteraceae), an endangered plant of Hawaii. PhD dissertation, University of Hawaii, Honolulu.

POWER, D.M. 1980. Introduction. In *The California Islands: Proceedings of a Multidisciplinary Symposium,* ed. D.M. Power, pp. 1–6. Haagden Printing, Santa Barbara, CA.

PRICE, M.V. and N.M. WASER. 1979. Pollen dispersal and optimal outcrossing in *Delphinium nelsoni. Nature* 277:294–297.

RICK, C.M. and J.F. FOBES. 1975. Allozymes of Galapagos tomatoes: polymorphism, geographic distribution, and affinities. *Evolution* 29:443–457.

RIESEBERG, L.H. 1991. Hybridization in rare plants: insights from case studies in *Helianthus* and *Cercocarpus.* In *Conservation of Rare Plants: Biology and Genetics,* eds. D.A. Falk and K.E. Holsinger, pp. 171–181. Oxford University Press, Oxford.

RIESEBERG, L.H., R. CARTER, and S. ZONA. 1990. Molecular tests of the hypothesized hybrid origin of two diploid *Helianthus* species (Asteraceae). *Evolution* 44:1498–1511.

RIESEBERG, L.H. and N.C. ELLSTRAND. 1993. What can morphological and molecular markers tell us about plant hybridization? *Crit. Rev. Plant Sci.* 12:213–241.

RIESEBERG, L.H. and D. GERBER. 1995. Hybridization in the Catalina Island mountain mahogany (*Cercocarpus traskiae*): RAPD evidence. *Conserv. Biol.* 9:199–203.

RIESEBERG, L.H., S. ZONA, L. ABERBOM, and T. MARTIN. 1989. Hybridization in the island endemic, Catalina mahogany. *Conserv. Biol.* 3:52–58.

ROBICHAUX, R.H., G.D. CARR, and B.G. BALDWIN. 1992. Diversity and conservation of the Hawaiian silversword alliance (Compositae). *Bull. Ecol. Soc. Amer.* 73:325.

ROBICHAUX, R.H., G.D. CARR, M. LIEBMAN, and R.W. PEARCY. 1990. Adaptive radiation of the Hawaiian silversword alliance (Compositae-Madiinae): ecological, morphological, and physiological diversity. *Ann. Missouri Bot. Gard.* 77:64–72.

RYDBERG, P.A. 1913. *Cercocarpus. N. Amer. Flora* 22:418–424.

SCHNEIDER, C.K. 1905. Beitrag zur Kenntnis der Arten und Formen der Gattung *Cercocarpus* Kunth. *Mitt. Deutsch. Dendrol. Ges.* 14:125–129.

SEARCY, K.B. 1969. Variation in *Cercocarpus* in Southern California. *New Phytol.* 68:829–839.

SMITH, J.P. and K. BERG. 1988. *Inventory of Rare and Endangered Vascular Plants of California.* California Native Plant Society, Sacramento, CA.

SOLTIS, P.S., D.E. SOLTIS, and J.J. DOYLE (eds.). 1992. *Molecular Systematics of Plants.* Chapman & Hall, New York.

SWENSEN, S.M., G.J. ALLAN, M. HOWE, W.J. ELISENS, S.A. JUNAK, and L.H. RIESEBERG. 1995. Genetic analysis of the endangered island endemic *Malacothamnus fasciculatus* (Nutt.) Greene var. *nesioticus* (Rob.) Kern. (Malvaceae). *Conserv. Biol.* 9:404–415.

TEMPLETON, A.R. 1986. Coadaptation and outbreeding depression. In *Conservation Biology: The Science of Scarcity and Diversity,* ed. M.E. Soulé, pp. 105–115. Sinauer, Sunderland, MA.

THORNE, R. 1967. A flora of Santa Catalina Island, California. *Aliso* 6:1–77.

VANE-WRIGHT, R.I., C.J. HUMPHRIES, and P.H. WILLIAMS. 1991. What to protect— systematics and the agony of choice. *Biol. Conserv.* 55:235–254.

WAGNER, W.L. 1990. *Manual of the Flowering Plants of Hawaii.* University of Hawaii Press, Honolulu.

WALLACE, A.R. 1880. *Island Life*. Macmillan, London.

WAYNE, R.K. 1992. On the use of morphologic and molecular genetic characters to investigate species status. *Conserv. Biol.* 6:590–592.

WAYNE, R.K. and S.M. JENKS. 1991. Mitochondrial DNA analysis implying extensive hybridization of the endangered red wolf *Canus rufus. Nature* 351:564–568.

WENDEL, J.F. and R.G. PERCY. 1990. Allozyme diversity and introgression in the Galapagos islands endemic *Gossypium darwinii* and its relationship to continental *G. barbadense. Biochem. Syst. Ecol.* 18:517–528.

WHITHAM, T.G., P.A. MORROW, and B.M. POTTS. 1991. Conservation of hybrid plants. *Science* 254:779–780.

WIDLT, D.E., M. BUSH, K.L. GOODROWE, C. PACKER, A.E. PUSEY, J.L. BROWN, P. JOSLIN, and S.J. O'BRIEN. 1987. Reproductive and genetic consequences of founding isolated lion populations. *Nature* 329:328–331.

WILLIAMS, J.G.K., A.R. KUBELIK, K.J. LIVAK, J.A. RAFALSKI, and S.V. TINGEY. 1990. DNA polymorphisms amplified by arbitrary primers are useful as genetic markers. *Nucl. Acids Res.* 18:6531–6535.

WITTER, M.S. 1990. Evolution in the Madiinae: Evidence from enzyme electrophoresis. *Ann. Missouri Bot. Gard.* 77:110–116.

WITTER, M.S. and G.D. CARR. 1988. Genetic differentiation in the Hawaiian silversword alliance (Compositae: Madiinae). *Evolution* 42:1278–1287.

11

CONSERVATION GENETICS OF FISHES
IN THE PELAGIC MARINE REALM

John E. Graves

INTRODUCTION

The upper waters of the world's oceans comprise the largest continuous environment on the surface of the planet. A wide array of fishes occur within the pelagic realm, encompassing a variety of unique adaptations and life histories. Several of these species, notably the tunas and billfishes, are the targets of extensive international fisheries and have been exploited for generations. As a result of intense fishing pressure, major decreases in abundance have been noted for many of these fishes (Box 11.1).

The vast area of the epipelagic realm, coupled with the large population sizes of most pelagic fishes, would tend to preclude the possibility of biological extinction by overfishing. Furthermore, the economics of most commercial fisheries are such that the fishery would collapse well before the target species became threatened or endangered. However, with certain pelagic fisheries, the remarkably high value of individual fish could promote commercial harvest even when numbers become extremely low. For example, a giant bluefin tuna in good condition can be worth more than $30,000. If the price for bluefin continues to rise as the supply diminishes (as has been the case), the economic collapse of the fishery could be coincident with biological extinction.

Box 11.1. Decreasing Trends in Abundance

Because it is difficult to directly estimate the number of fish available to a fishery, stock abundances are most commonly determined from analyses of catch-per-effort data. Over the last 30 years, almost all pelagic fisheries have shown dramatic decreases in catch per effort, reflecting concomitant declines in biomass. The trends in catch-per-unit effort (which are similar for most species of tuna and billfish) are as follows. During the late 1950s and early 1960s, indices of catch-per-unit effort remained fairly constant, while total catches of tunas and billfishes dramatically increased. Following peak harvests in the 1960s, catch-per-unit effort values continually de-creased to the present time (current values of some fisheries are less than 4% of those 30 years ago). As a consequence, most stocks of tunas and billfishes are considered fully exploited or overexploited, especially in the Atlantic Ocean (ICCAT, 1993). Long-term trends in abundance are not available for pelagic sharks but may be inferred from trends in catch per effort of large, coastal sharks that have also experienced high fishing pressure. Although few such time series are available, the data depict steep declines in abundance over the last 20 years (Musick et al., 1993). Nonetheless, at the present time, no pelagic fishes are listed as threatened or endangered.

The stocks of most pelagic fishes are considered to be fully exploited or overexploited, with annual catches meeting or exceeding the estimated maximum sustainable yields. As such, most of the annual production of the population is removed as catch, and the stability of the system is highly dependent on consistent recruitment of juveniles into the fishery. However, if recruitment should fail due to environmental variations, the current levels of intense fishing pressure would deplete the spawning stock, and successive recruitment failures could lead to the collapse of the fishery. A major objective of fishery management is to maintain a spawning biomass that is large enough to prevent a fishery collapse due to recruitment failure.

Fishery management is a difficult proposition on any level. Managers must address a wide range of concerns, including biological, economic, social and political issues. The inability to satisfy competing interests and maintain a viable fishery is evidenced by the collapse or severe reduction of most major national fisheries. The problems of fishery management are exacerbated for pelagic fishes, due to the conflicting objectives of various nations competing for the same, limited resource.

The major pelagic fisheries are managed by a variety of international treaty organizations, each of which oversees a specific geographic area. For example, in the Atlantic Ocean the harvest of tunas and billfishes is monitored and regulated by member countries of the International Commission for the Conservation of Atlantic Tunas (ICCAT), and similar organizations exist in the eastern Pacific, western Pacific, South Pacific, and Indian Oceans. Member countries of these treaty organizations typically meet annually to compile catch and effort data, determine the status of stocks, allocate future catches, and impose restrictions to the fisheries as needed. However, the effectiveness of many treaty organizations is compromised because not all major fishing nations are members of the various treaty organizations. Consequently, there are concerns over the accuracy of catch-and-effort reporting by non-member countries, and there are no effective means to ensure compliance to the quotas and size limits set by treaty organizations.

In addition to problems with reporting and enforcement, the management of most pelagic fisheries is hindered by a lack of biological information. Little is known of the genetic basis of stock structure for any pelagic fish. As a consequence, management models incorporate stocks that are defined primarily by limited tag-and-recapture results and spatial/temporal trends in catch-per-effort data. Only recently have pelagic fishes been examined with molecular genetic techniques, and although some species demonstrate a high degree of genetic homogeneity, others reveal an unanticipated level of population structure. The distribution and nature of genetic variation within these species reflects both historical and recent population structuring, and dictates that smaller stocks (management units) be incorporated into management plans to conserve genetic variation in the face of intense fishing pressure.

This chapter is a review of the genetic population structures of pelagic fishes. It emphasizes what is known of the spatial and temporal distribution of genetic variation of selected pelagic fishes, and how this information relates to conservation and management practices. Although a variety of molecular genetic techniques has been used to survey pelagic fishes, this review focuses primarily on studies employing restriction fragment length polymorphisms (RFLPs) in mitochondrial (mt) DNA. Results from these studies are compared with those from allozyme analyses, and with recent investigations employing direct sequence analyses of mitochondrial genes.

POPULATION GENETIC STRUCTURE OF TUNAS

The family Scombridae (Perciformes, suborder Scombroidei) comprises the mackerels, bonitos, Spanish mackerels, and tunas (Figure 11.1.). Of these, the tunas (tribe Thunnini) are pelagic and support large, international fisheries (Box 11.2). Despite the economic importance of tunas, until recently little genetic information has been available to evaluate the genetic structure of any species.

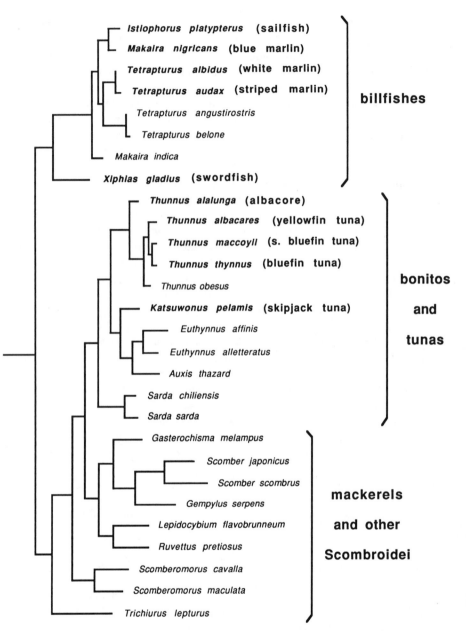

Figure 11.1. Phylogenetic relationships among representative species of the suborder Scombroidei, as estimated from sequence comparisons of a portion of the cytochrome *b* gene in mtDNA (after Block et al., 1993). Species which have been assayed for geographic variation in molecular genetic characters are indicated in boldface and by common names (see text).

Box 11.2. Tunas

The tunas and tuna-like fishes belong to the tribe Thunnini of the family Scombridae. They comprise four genera, and a total of 11 species (Collette and Nauen, 1983). More than 3.25 million metric tons of these fishes are harvested each year. The species and total catches for 1990 are listed below (Food and Agriculture Organization of the United Nations, 1992).

	Catch (10^3 metric tons)
Genus *Thunnus*	
Bluefin tuna (*T. thynnus*)	29
Southern bluefin tuna (*T. maccoyii*)	16
Albacore (*T. alalunga*)	233
Bigeye tuna (*T. obesus*)	258
Yellowfin tuna (*T. albacares*)	987
Blackfin tuna (*T. atlanticus*)	3
Longtail tuna (*T. tonggol*)	163
Genus *Katsuwonus*	
Skipjack tuna (*K. pelamis*)	1,239
Genus *Euthynnus*	
Black skipjack tuna (*E. lineatus*)	<1
Spotted tunny (*E. alletteratus*)	21
Kawakawa (*E. affinis*)	124
Genus *Auxis*	
Bullet tuna (*A. rochei*)	*
Frigate tuna (*A. thazard*)	*
* combined catch	195

Skipjack Tuna

This species occurs throughout the world's tropical and subtropical oceans. Relative to other tunas, skipjack are small and short-lived, with a maximum lifespan of about six years (Forsbergh, 1980). They grow rapidly and reach sexual maturity in their second year (Cayre and Farrugio, 1986). Spawning is restricted to warm waters and occurs over a broad geographic range (Nishikawa et al., 1985).

Skipjack tuna are capable of extensive movements (Figure 11.2). A large-scale tagging study by the South Pacific Commission revealed that although most fish were recaptured near the release area (Argue, 1981), many individuals moved long distances. The results of this and other studies

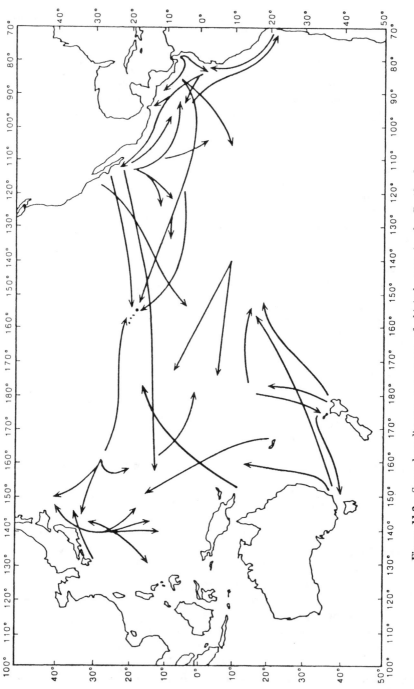

Figure 11.2. Some long-distance movements of skipjack tuna in the Pacific Ocean as documented from tag-and-recapture studies (from IATTC, 1993).

also indicated that skipjack schools were not cohesive units. In some cases, fish tagged from the same school were recovered several months later in different schools hundreds of kilometers apart (Argue, 1981; Bayliff, 1988; Hilborn, 1991). The mixing of skipjack tuna from different schools and the high dispersal capabilities of individuals are factors that should facilitate effective gene flow and make strong population genetic structure seem unlikely.

Skipjack tuna population genetic structure has been investigated with a variety of molecular genetic techniques. Allozyme analyses revealed a slight cline for two loci (an esterase and guanine deaminase) across the Pacific Ocean, and considerable heterogeneity of frequencies from samples collected in the same region (Argue, 1981; Fujino et al., 1981). Argue (1981) reviewed the data from several allozyme analyses of skipjack tuna within the Pacific Ocean and concluded that Pacific fish do not comprise a single panmictic population. It was suggested that the results were better explained by either an isolation-by-distance model or the presence of distinct breeding groups in the central and western Pacific. Electrophoretic analysis of samples of Atlantic and Pacific skipjack tuna demonstrated slight frequency differences between ocean populations (Fujino, 1969), implying some limit to interocean gene flow. However, the level of genetic divergence was more than an order of magnitude below that found between shorefishes separated by the uplift of the Panama land bridge three million years ago (Vawter et al., 1980), suggesting either a much more recent isolation of skipjack populations, or a low level of gene flow between oceans.

The lack of substantial allozymic divergence between Atlantic and Pacific skipjack tuna was supported by RFLP analysis of mtDNA (Graves et al., 1984). Analysis of nine Pacific and seven Atlantic skipjack revealed variation within the pooled samples. All mtDNA haplotypes were closely related, differing by only one assayed restriction-site gain or loss. No consistent restriction-site differences were found between ocean populations, nor were there significant differences in haplotype frequencies between the two small samples (Table 11.1). Thus, the null hypothesis that the samples came from a common gene pool could not be rejected. The lack of major genetic divergence between Atlantic and Pacific skipjack is consistent with an alternate hypothesis that there is sufficient gene flow between ocean populations to prevent the accumulation of genetic divergence. Such gene flow would be facilitated by either larval transport or the dispersal abilities of vagile adults. Movement between ocean basins is possible around Africa, because suitable environmental conditions exist below the Cape of Good Hope during the austral summer, and skipjack have been collected in those waters (Talbot and Penrith, 1962).

Table 11.1. Distributions of mtDNA Haplotypes (as Revealed in RFLP Analyses) between Atlantic (Atl) and Pacific (Pac) Samples of Four Tuna Species

Haplotype	Skipjack tuna Atl	Pac
1	5	7
Unique	2	2
Totals	7	9

Haplotype	Albacore Atl	Pac
1	10	8
Unique	1	4
Totals	11	12

Yellowfin tuna

Haplotype	Atl	Pac Ecu	Mex	Haw	Aus	Png
1	7	7	6	7	8	8
2	3	2	5	3	4	3
3	1	3	0	1	0	1
4	2	1	1	0	2	0
5	2	0	0	0	1	1
6	1	1	0	1	0	0
7	0	2	1	0	0	0
8	0	0	2	0	1	0
Unique	4	4	5	8	4	7
Totals	20	20	20	20	20	20

Haplotype	Bluefin tuna Atl	Pac
1	0	31
2	16	1
Totals	16	32

Note: Samples of yellowfin tuna from the Pacific Ocean were collected from Ecuador (Ecu), Mexico (Mex), Hawaii (Haw), Australia (Aus), and Papua New Guinea (Png). Unique haplotypes denote single occurrences, and no haplotypes were shared among species.

Source: Data are from Graves et al. (1984), Graves and Dizon (1989), Scoles and Graves (1993), and Chow and Inoue (1993).

Albacore

The albacore has a distribution and life history quite different from that of the skipjack tuna (Foreman, 1980). Spawning occurs seasonally in the tropics, and is temporally displaced in the Northern and Southern Hemispheres. Young albacore move through gyre current systems to temperate waters, where they typically spend several years, undertaking distinct seasonal movements that may cross ocean basins. As albacore reach sexual maturity, they return to tropical waters to spawn.

There have been only limited genetic analyses of albacore population structure. A preliminary RFLP analysis of mtDNA from samples of 12 individuals from San Diego, CA and 11 individuals from Cape Town, South Africa revealed little genetic divergence (Graves and Dizon, 1989). The 13 restriction enzymes employed revealed modest amounts of variation (a pooled nucleon diversity of 0.60), and variant haplotypes were all related to the common haplotype by the gain or loss of a single restriction site. A common haplotype predominated in both collections (Table 11.1), and because of limited sample sizes, the occurrence of unique genotypes was not informative. However, analysis of larger sample sizes revealed interoceanic genetic divergence. Chow and Ushiama (1995) surveyed the mitochondrial ATPase gene of more than 600 albacore from the Atlantic and Pacific Oceans with two polymorphic restriction enzymes. Although there was no evidence of within-ocean population structuring, a slight but statistically significant difference in haplotype distribution was found between the pooled Atlantic and Pacific samples. These results are consistent with gene flow around the Cape of Good Hope, but suggest that it is limited.

Yellowfin Tuna

This species is distributed throughout the tropical and subtropical oceans, and supports the largest of the tuna fisheries. Yellowfin tuna grow rapidly, with most individuals reaching sexual maturity at a length of 100 cm at two years of age (Inter-American Tropical Tuna Commission, 1991), and have a longevity of at least eight years. Like skipjack tuna, individual yellowfin tuna are known to move long distances, although most tagged fish are recovered within several hundred kilometers of the site of release (Hunter et al., 1986). Within the Atlantic, several transoceanic migrations have been reported (International Commission for the Conservation of Atlantic Tunas, 1993). Spawning occurs in broad areas throughout the tropical Atlantic, Pacific, and Indian Oceans, typically in waters above 24°C (Nishikawa et al., 1985).

Allozyme analyses of yellowfin tuna stock structure in the Pacific Ocean have revealed little heterogeneity. Barrett and Tsuyuki (1967) and Fujino and Kang (1968) both analyzed a transferrin polymorphism and reported no population structuring, whereas Sharp (1978) reported differences for the glucose phosphate isomerase (GPI) locus between samples of yellowfin tuna from the eastern and western Pacific. The difference in GPI allele frequencies within the Pacific Ocean remained evident in a recent allozyme analysis of the species (Ward et al., 1994).

Population structuring of yellowfin tuna was not evident in RFLP analyses of mtDNA from samples taken across the Pacific Ocean. Scoles and Graves (1993) employed 12 enzymes to survey 100 yellowfin tuna from five Pacific locations, and one sample of 20 individuals from the Atlantic, and found no significant differences in the distribution of haplotypes among samples within or between oceans (Table 11.1). Similarly, Ward et al. (1994) examined 118 yellowfin tuna with eight restriction enzymes, and an additional 316 fish with two variable enzymes, finding no difference in the distribution of haplotypes among samples. As with the skipjack tuna and albacore, these data are consistent with the hypothesis that there is sufficient gene flow within and between oceans to prevent the accumulation of large genetic differences.

Bluefin Tuna and Southern Bluefin Tuna

These species are distinguished from the other tunas by their longevity, large size, and discrete spawning areas— characteristics which make them vulnerable to overexploitation. It is therefore not surprising that their stocks are the most depleted of any tuna species.

The bluefin tuna is the largest of the tunas, with some individuals exceeding 700 kg. Individuals do not reach sexual maturity until about five years of age and may live beyond 20 years (Bayliff, 1991). Spawning appears to be spatially and temporally restricted. Within the Atlantic there are two major spawning areas, one in the Mediterranean Sea and the other in the Gulf of Mexico (Richards, 1976). It is known from tag-and-recapture studies that individuals make trans-Atlantic movements, and such movements account for 3% of tag recoveries (Suzuki, 1991). Such findings led a recent panel of the National Research Council (1994) to conclude that considerable mixing occurs between eastern and western Atlantic stocks. However, it is not known whether the two stocks interbreed.

Within the Pacific, spawning appears to be restricted to the area between the Philippines and Japan, and the Sea of Japan (Bayliff, 1991). Some juveniles make trans-Pacific migrations, and the bluefin tuna taken in the eastern Pacific are all believed to be derived from spawning events in the western Pacific.

Although bluefin tuna represent an important fishery resource, little is known about the population genetic structure of the species. The discrete nature of spawning areas could promote the accumulation of intraspecific genetic divergence both within and between oceans. Chow and Inoue (1993) surveyed small samples of Atlantic and Pacific bluefin tuna with RFLP analysis of several mitochondrial genes and found major differences in the occurrence of haplotypes between ocean populations (Table 11.1). These data suggest that unlike other species of tuna, Atlantic and Pacific bluefin tuna do not share a common gene pool. Whether the discrete spawning areas of bluefin tuna within the Atlantic represent independent reproductive units remains unknown. Efforts are currently underway to collect larvae from each of the spawning sites for genetic analyses.

The southern bluefin tuna is morphologically similar to the bluefin tuna, but is genetically distinct (Bartlett and Davidson, 1991; Smith et al. 1994; see also Figure 11.1). The southern bluefin tuna reaches sexual maturity at 3–5 years, is long-lived, and attains a large size (Caton, 1991). All spawning in southern bluefin tuna appears to take place south of Java in the Indian Ocean, with juveniles making a distinct migration south of Australia, and subsequently dispersing into the southern ocean.

Southern bluefin tuna are assumed to comprise one stock because spawning is apparently restricted to a single geographical location, little morphological variation exists, and there is considerable mixing (movement) of individuals (Caton, 1991). Consequently, there has been little impetus to survey molecular characters for intraspecific population structure. Rather, attention has been directed on rebuilding the population of spawning adults, which has decreased dramatically over the past 30 years.

Conservation Relevance

The tunas encompass a diversity of life histories, from the short-lived skipjack that spawns over wide oceanic expanses, to the long-lived bluefin and southern bluefin tunas with discrete spawning areas. In general, both allozyme analyses and RFLP analysis of mtDNA reveal little intra- or interocean genetic divergence within most species. These data are consistent with the hypothesis that there has been sufficient historical connectedness and gene flow within and between oceans to prevent the accumulation of significant genetic differences. In theory, this exchange can be on the order of individuals per generation (Hartl and Clark, 1989).

Essential to effective fishery management is a quantitative understanding of the magnitude of exchange between populations (i.e., the degree of demographic interdependence of fisheries in different geographic locations). Although inferences are limited by small sample sizes, RFLP analyses of mtDNA in yellowfin, albacore, and skipjack tuna indicate that for these species, there has been at least sufficient exchange within and between oceans to prevent the accumulation of pronounced genetic divergence. Whether this exchange represents a few or several thousand genomes per generation cannot be evaluated from these data, but the results do indicate that the collapse of a fishery in one area would not likely result in the appreciable loss of unique genetic variation.

POPULATION GENETIC STRUCTURE OF BILLFISHES

The billfishes, including the swordfish, marlins, sailfish and spearfishes, comprise two families (Box 11.3). As a group, these fishes represent an important commercial and recreational fisheries resource, with average annual landings about 10% that of tunas of the genus *Thunnus*. Billfishes are targeted by longline fisheries in some regions, and represent a major bycatch of tuna fishing operations throughout most tropical and temperate waters. Like the tunas, the abundances of most billfish species have dropped precipitously over the past three decades (Box 11.1).

Little is known of intraspecific population structure for any of the billfishes. It was generally assumed that due to their vagility and continuous distribution, billfishes, like tunas, would exhibit little or no population subdivision within or between oceans. These assumptions have been supported by results of tag-and-recapture analyses, which indicate that several billfish species are capable of undertaking transoceanic movements (Scott et al., 1990). However, because tag-recovery rates are typically very low, ranging from 0.5–2.0% for the various species, and the distribution of tagging and recovery efforts are not equally distributed (billfish are usually tagged by the inshore recreational fishery, and are often recovered by the offshore commercial fishery), delineations of stocks based on tagging data were considered tentative. Only recently have molecular genetic studies been used to examine billfish population structure, and surprisingly, these analyses have revealed considerable intraspecific genetic differentiation.

Striped Marlin

This species occurs throughout the Pacific and Indian Oceans and is closely related to the white marlin of the Atlantic. Striped marlin exhibit a more temperate distribution than blue marlin or sailfish, with ranges extending to 45°N in the North Pacific and 45°S in the western South Pacific (Nakamura, 1985). Striped marlin are capable of extended movements, and a few individuals have been recovered more than 1,000 km from the site of

Box 11.3. Billfishes

The billfishes comprise two families, Istiophoridae and Xiphiidae. There is considerable confusion regarding the numbers of species within all three genera of the Istiophoridae. For example, although Nakamura (1985) recognized two species of blue marlin and two of sailfish, genetic data suggest that Atlantic and Indo–Pacific populations of each comprise a single species. The specific status of various morphotypes of *Tetrapturus* is also problematic: species listed by Nakamura (1985) are presented here, as are their world catches in 1990 (Food and Agriculture Organization of the United Nations, 1992).

	Catch (10^3 metric tons)
Family Istiophoridae	
Genus *Makaira*	
Blue marlin (*M. nigricans*)	26.8
Black marlin (*M. indica*)	4.3
Genus *Istiophorus*	
Sailfish (*I. platypterus*)	8.5
Genus *Tetrapturus*	
White marlin (*T. albidus*)	1.0
Striped marlin (*T. audax*)	13.6
Longbill spearfish (*T. pfluegeri*)	*
Shortbill spearfish (*T. angustirostris*)	*
Mediterranean spearfish (*T. belone*)	*
Roundscale spearfish (*T. georgei*)	*
Family Xiphiidae	
Genus *Xiphias*	
Swordfish (*X. gladius*)	77. 5

* The 1990 catch of spearfishes and unidentified sailfish and marlins was 15.2 thousand metric tons.

first capture (Squire, 1987). On the basis of seasonal analyses of catch-per-effort data, morphometric studies, tag recoveries, and reproductive studies, Squire and Suzuki (1990) concluded that striped marlin in the Pacific Ocean probably represent a single stock.

However, genetic analysis of striped marlin within the Pacific Ocean suggests the presence of more than one genetic population. Graves and McDowell (1994) employed RFLP analysis of mtDNA to survey samples of approximately 40 striped marlin from each of four Pacific locations. Levels of variation were high, with sample nucleon diversities ranging from 0.69 to 0.84. In contrast to the tunas, there was evident partitioning of this genetic variation (Table 11.2). The distribution of haplotypes was significantly

Table 11.2. Distributions of mtDNA Haplotypes (as Revealed in RFLP Analyses) within Striped Marlin, White Marlin, and Blue Marlin

Haplotype	Striped marlin			
	Mex	Ecu	Haw	Aus
1	12	20	14	12
2	6	6	7	3
3	8	9	4	1
4	0	2	10	8
5	0	1	2	7
6	0	1	0	11
7	6	0	0	0
Unique	4	1	6	5
Totals	36	40	43	47

Haplotype	White marlin		
	NJ	Dom	Bra
1	3	7	5
2	3	2	4
3	0	2	0
4	4	0	2
Unique	4	3	4
Totals	14	14	15

Haplotype	Blue marlin	
	Atl	Pac
1	29	53
2	11	16
3	5	8
4	12	0
5	8	2
6	2	3
7	2	3
8	3	1
9	4	0
10	1	2
11	1	2
12	3	0
13	3	0
14	0	3
15	2	0
16	2	0
17	2	0
Unique	48	13
Totals	138	106

Note: Unique haplotypes denote single occurrences, and no haplotypes were shared among species. Locale abbreviations as in Table 11.1, and as follows: NJ, New Jersey, U.S.; Dom, Dominican Republic; Bra, Brazil.
Source: Data are from Graves and McDowell (1994, 1995, and unpublished data).

different between all samples in pairwise comparisons, and temporal analysis of the Australian collection demonstrated that the haplotypes characteristic of that sample were present in consecutive years. Thus, in contrast to many of the tunas, the null hypothesis that the striped marlin samples represent random draws from a single gene pool was rejected.

Typically, mtDNA haplotypes in the striped marlin differed by one or two restriction-site changes, and nucleotide sequence divergences were small, ranging from 0.16 to 1.11%. Phylogeographic analysis of the relationships and distribution of haplotypes revealed only a shallow structuring (Figure 11.3). Such a distribution is consistent with evolutionarily recent restrictions to gene flow (Avise et al., 1987).

The RFLP analyses of striped marlin mtDNA can be compared to similar analyses of yellowfin tuna described earlier (Scoles and Graves, 1993). The two studies surveyed about the same proportion of the mtDNA genome, resulting in comparable pooled nucleon diversities (0.82 and 0.84, respectively) and mean sequence diversities (0.30 and 0.28, respectively). Within both species, haplotypes were closely related, typically differing by one or two restriction-site changes. However, within the yellowfin tuna, haplotypes were homogeneously distributed across all sampling locations, whereas in striped marlin certain haplotypes demonstrated significant spatial partitioning (compare Tables 11.1 and 11.2).

Both striped marlin and yellowfin tuna maintain large populations, and individuals are highly vagile and migratory. What then accounts for the apparent difference in stock structures? The limited gene flow within striped marlin might be attributed to unsuccessful spawning after migration (movement but no gene flow), to an unsuspected social or behavioral tendency for related fish to remain together, or to some degree of spawning-site fidelity. Because little is known of the temporal and geographic distribution of striped marlin spawning (Nakamura, 1985), these hypotheses cannot be adequately evaluated at present. However, the moderate degree of stock structure within the striped marlin suggests that current management of the species based on basin-wide stocks is inappropriate, and could result in the loss of unique genetic variation.

White Marlin

This species is similar in many respects to the striped marlin, but is confined to the Atlantic Ocean. The ranges of the two species marginally overlap off South Africa, and individuals are distinguished by morphological differences of the median and paired fins (Nakamura, 1985).

Preliminary analyses suggest that white marlin may not display as much population structure as striped marlin. Edmunds (1972) analyzed variation

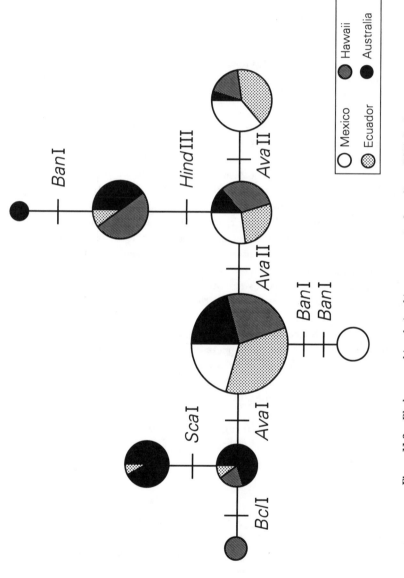

Figure 11.3. Phylogeographic relationships among striped marlin mtDNA haplotypes occurring in two or more individuals. Included are haplotypes of 166 striped marlin generated with 11 restriction endonucleases (Graves and McDowell, 1994). Circle areas are proportional to the overall abundance of a haplotype, with the smallest circle representing two individuals.

of seven blood and tissue proteins within more than 100 white marlin from the U.S. mid-Atlantic Bight, Gulf of Mexico, and Caribbean Sea, but found no significant differences in the distribution of alleles among collection sites. Similarly, RFLP analysis of white marlin mtDNA revealed no significant heterogeneity among three samples of approximately 15 fish each from three sites in the western Atlantic (although differences in the distributions of haplotypes were noted; Table 11.2; Graves and McDowell, 1995; Finnerty and Block, 1995).

White marlin and striped marlin have extremely similar mitochondrial genomes as indicated both by cytochrome *b* sequences (Figure 11.1), and by RFLP analysis (Graves and McDowell, 1995). In the latter study, two haplotypes were common to both species, accounting for 87% of the white marlin and 25% of the striped marlin samples. The remaining white marlin and striped marlin haplotypes also were similar, typically differing by only one or two assayed restriction-site changes (Figure 11.4). These data suggest that if white and striped marlin are valid species, they either separated recently, or introgression has occurred since the time of speciation.

Blue Marlin

Whereas the specific status of white and striped marlin has not been considered problematic, the relationship of Atlantic and Indo–Pacific populations of blue marlin has. Some researchers consider blue marlin to be a single, circumtropical species (e.g., Briggs, 1960; Robins and de Sylva, 1960; Rivas, 1975), but Nakamura (1985) recognized Atlantic and Indo–Pacific populations as distinct species based on morphological differences of the lateral line. Until recently, molecular genetic data were not available to evaluate the taxonomic status of the putative species, or to survey population structure.

Blue marlin exhibit considerable genetic variation. Preliminary evidence for the utility of allozyme analysis to elucidate the population structure of blue marlin was presented by Shaklee et al. (1983), whose electrophoretic survey of 35 loci from 95 Hawaiian blue marlin revealed sufficient variation for intra- and interocean analyses of stock structure. RFLP analysis of blue marlin mtDNA also revealed considerable variation, and analyses of Atlantic and Indo–Pacific samples demonstrated both recent and historical phylogeographic patterning (Graves and McDowell, 1995). Analysis of 244 blue marlin (138 Atlantic, 106 Pacific) with 11 restriction enzymes revealed 78 genotypes, an overall nucleon diversity of 0.84, and an overall mean nucleotide sequence diversity of 0.54%. Variation was significantly higher in the Atlantic sample than in the Indo–Pacific sample.

The distribution of blue marlin haplotypes was significantly different

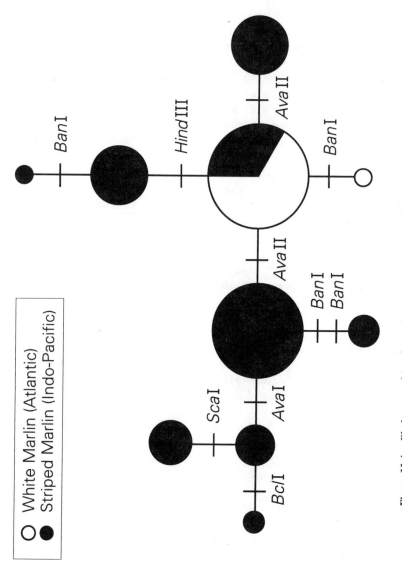

O White Marlin (Atlantic)
● Striped Marlin (Indo-Pacific)

Figure 11.4. Phylogeographic relationships among striped marlin and white marlin mtDNA haplotypes occurring in two or more individuals. Included are haplotypes of 166 striped marlin and 52 white marlin generated with 10 restriction endonucleases (Graves and McDowell, 1995). See caption to Figure 11.3.

between samples from the Atlantic and Pacific Oceans (Table 11.2), indicating a severe restriction to gene flow between ocean basins (a result that contrasts sharply with the homogeneity demonstrated for several species of tunas). Furthermore, blue marlin, like striped marlin, appear to exhibit significant within-ocean population structure. For example, one blue marlin mtDNA haplotype apparently confined to the Atlantic Ocean was found in 11 of 74 Jamaican blue marlin (five of 25 in 1991, six of 49 in 1992), but only in one of 74 blue marlin collected outside of Jamaica (Graves and McDowell, 1995).

The considerable level of population structuring in the blue marlin is enigmatic, because tagging data demonstrate that individuals are capable of extended movements. One tenth of the recoveries of blue marlin tagged in the western Atlantic have been from the eastern Atlantic, and recently a fish tagged off the U.S. mid-Atlantic coast was recovered in the Indian Ocean (Scott et al., 1990; E. Prince, personal communication). As was true for the striped marlin, the spatial partitioning of genetic variation within the highly mobile blue marlin raises the possibility of fidelity to spawning sites.

Phylogeographic analysis of the blue marlin mtDNA revealed two groups of haplotypes that differed by several consistent restriction-site differences and a mean nucleotide sequence divergence of 1.35% (Figure 11.5). One group of haplotypes was ubiquitous, occurring in approximately 55% of Atlantic blue marlin and all Indo–Pacific individuals. Ten of the haplotypes within this group were present in both Atlantic and Indo–Pacific fish (Table 11.2). The other group of blue marlin haplotypes was found only in Atlantic blue marlin, accounting for 45% of the individuals.

A similar phylogeographic patterning of mtDNA haplotypes among Atlantic and Pacific blue marlin was reported by Finnerty and Block (1992). Their direct sequence analysis of a 612-bp region of the cytochrome *b* gene of 26 blue marlin, 12 from the Atlantic Ocean and 14 from the Pacific Ocean, revealed seven haplotypes that clustered into two groups that differed by a minimum of nine base substitutions. One group was found within both Atlantic and Pacific samples, whereas the other was restricted primarily to the Atlantic Ocean, accounting for six of the 12 Atlantic blue marlin samples. Direct sequence analysis of the mitochondrial control region of 100 blue marlin also demonstrated two major groups of mtDNA haplotypes in Atlantic blue marlin, only one of which was present in the Pacific sample (A. Meyer, personal communication). A substantial amount of variation was revealed in the analysis of the control region, with each of the 100 blue marlin possessing a unique sequence.

The presence of two genetically divergent groups of mtDNA haplotypes within blue marlin suggests that historical isolation has occurred within the

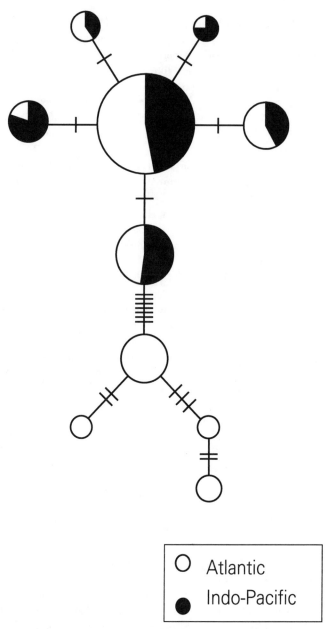

Figure 11.5.   Phylogeographic relationships among blue marlin mtDNA haplotypes present in more than two individuals. Included are haplotypes from 138 Atlantic blue marlin and 68 Indo–Pacific blue marlin (Graves and McDowell, 1995, and unpublished data). See caption to Figure 11.3.

species. A similar phylogeographic distribution of mtDNA haplotypes is also evident within the sailfish, as described next.

Sailfish

The taxonomic status of Atlantic and Indo–Pacific sailfish, like blue marlin, has been problematic. Nakamura (1985) recognized two species on the basis of morphological differences, although several researchers recognize a single, circumtropical species (e.g., Morrow and Harbo, 1969).

RFLP analyses of mtDNA from 56 Atlantic and 33 Indo–Pacific sailfish revealed a phylogeographic pattern similar to that found in blue marlin (Graves and McDowell, 1995, and unpublished data). A higher diversity of mtDNA haplotypes was found within Atlantic sailfish, which was accounted for by the presence of two distinct groups of mtDNA haplotypes, of which only one was present in the Indo–Pacific sailfish (Figure 11.6). The two groups of mtDNA haplotypes were separated by three consistent restriction-site differences, and a mean nucleotide sequence divergence of 0.65%. In contrast to the blue marlin, the ubiquitous sailfish haplotypes were not well represented in Atlantic samples. Only two haplotypes were shared between oceans, together accounting for 76% of the Indo–Pacific samples, and 9% of Atlantic individuals (Table 11.3).

The phylogeographic distributions of haplotypes between Atlantic and Indo–Pacific populations of blue marlin and sailfish are similar: Atlantic samples comprise two distinct groups of haplotypes (which differ by nucleotide sequence divergences of 1.35% and 0.65%, respectively), of which only one is present in the Indo–Pacific samples. The observed distribution of haplotypes could have arisen from the extinction of intermediate haplotypes, but such events would be unlikely to occur in both species. Rather, the phylogeographic distribution of haplotypes is consistent with a long period of isolation between conspecific populations, coupled with recent unidirectional exchange between oceans. This isolation may have occurred between Atlantic Ocean and Indo–Pacific populations during a period of tropical compression, when surface temperatures south of the Cape of Good Hope were too cold for interoceanic movement. The current distribution of haplotypes suggests that interocean gene flow of blue marlin and sailfish has been predominately from the Indo–Pacific to the Atlantic.

Swordfish

This species has the largest range of any billfish, occurring in tropical, temperate, and occasionally even cold (5°C) waters (Nakamura, 1985). Swordfish are the target of a large fishery, and the annual swordfish catch is greater than that of all istiophorid billfish landings combined (Box 11.3).

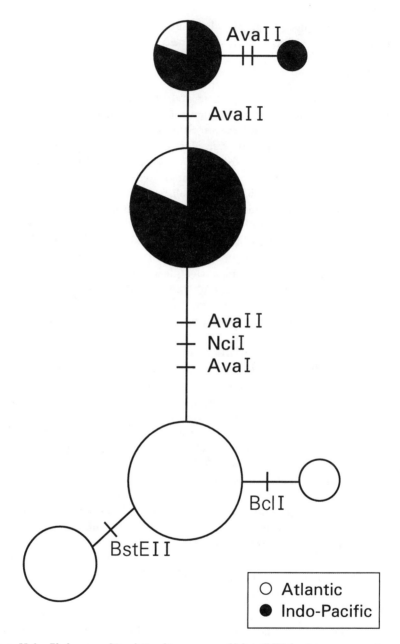

Figure 11.6.   Phylogeographic relationships among sailfish mtDNA haplotypes present in two or more individuals. Included are haplotypes from 56 Atlantic sailfish and 33 Indo–Pacific sailfish (Graves and McDowell, 1995, and unpublished data). See caption to Figure 11.3.

Table 11.3. Distributions of mtDNA Haplotypes (as Revealed in RFLP Analyses) within Sailfish, Swordfish and Mako Shark

Sailfish				
Haplotype	N. Atl	S. Atl	N. Pac	Ind
1	1	3	19	2
2	10	18	0	0
3	7	3	0	0
4	0	1	0	4
5	2	1	0	0
6	0	0	0	2
7	1	1	0	0
Unique	2	6	1	5
Totals	23	33	20	13

Swordfish				
Haplotype	Gre	Ita	Spa	Tar	Gog
1	13	23	23	7	8
2	4	7	7	5	30
3	12	16	22	10	35
4	20	20	22	12	6
Totals	49	66	64	34	79

Shortfin mako				
Haplotype	N. Atl	S. Atl	N. Pac	S. Pac
1	0	0	4	3
2	8	5	10	4
3	28	2	3	5
4	4	3	3	1
5	0	0	3	0
6	3	0	0	1
7	0	1	1	2
8	0	0	0	2
9	0	1	2	0
Unique	2	5	4	4
Totals	45	17	30	22

Note: Locale abbreviations as in Table 11.1, and as follows: Ind, Indian Ocean; Gre, Greece; Ita, Italy; Spa, Spain; Tar, Tarifa; Gog, Gulf of Guinea. Unique haplotypes denote single occurrences, and no haplotypes were shared among species.

Source: Data are from Magoulas et al. (1993), Graves and McDowell (1995), and Heist et al. (1995a).

Like the blue and black marlins, swordfish demonstrate a sexual dimorphism in size, with males rarely exceeding 140 kg, whereas females may grow to 540 kg (Nakamura, 1985).

The population structure of swordfish has recently received considerable attention. Finnerty and Block (1992) used direct sequencing of a 612-bp fragment of the mitochondrial cytochrome *b* to survey variation among six swordfish from the Mediterranean Sea, western North Atlantic, and central North Pacific. Only two of the 612 base pairs varied, revealing three genotypes. One genotype was common to swordfish from the Pacific and western North Atlantic, whereas the two Mediterranean swordfish possessed variant haplotypes.

Divergence between North Atlantic and Mediterranean swordfish has been supported by recent analyses of large samples. Magoulas et al. (1993) surveyed 377 swordfish from three sites in the Mediterranean and two sites in the eastern North Atlantic using two informative restriction endonucleases (Table 11.3). The four observed haplotypes were closely related, differing by the gain or loss of a restriction site. Although all haplotypes were present in the Mediterranean and North Atlantic samples, there was significant heterogeneity in the distribution of haplotypes among samples. Mediterranean samples were not significantly different from the geographically proximate sample in the Atlantic, indicating that there is some exchange through the Strait of Gibraltar. However, the distributions of haplotypes within these samples were significantly different from those of swordfish collected in the Gulf of Guinea.

Similarly, Alvarado-Bremer (1994) directly sequenced 280 bp of the swordfish mtDNA control region from 50 individuals. The 33 genotypes were grouped into two clades that differed by an average of 3.8% sequence divergence. The distribution of the clades was significantly different between the North Atlantic and Mediterranean.

Grijalva-Chon et al. (1994) surveyed mtDNA variation among 150 swordfish collected from the eastern, central and western North Pacific Ocean. RFLP analyses, which employed 14 informative restriction endonucleases, revealed 27 haplotypes, of which 14 were unique occurrences. One haplotype was found in more than 50% of the individuals from each sample, three others were common to the three samples, and ten haplotypes were shared by two of the sample locations. The distribution of haplotypes was not statistically different among locations.

Conservation Relevance

In contrast to the tunas, most billfishes demonstrate considerable spatial partitioning of genetic variation. Thus, a regional fishery collapse could result in the loss of unique genetic variation. Because many of the billfish stocks recognized by international fishery management agencies cover large geographic areas (e.g., North Pacific striped marlin, western Atlantic sailfish), they comprise heterogeneous genetic assemblages. To properly conserve the genetic variation within the billfishes, the possibility of smaller, regional stocks should be included in future management discussions.

Before closing this section, it should also be emphasized that the magnitudes of phylogeographic population structure in several of the pelagic billfishes remain small relative to those characterizing many widely distributed coastal, freshwater and terrestrial populations in other vertebrate groups (e.g., Chapter 14). The quasi-continuous nature of the pelagic realm, and the remarkable vagilities of many of its inhabitants, clearly have conspired to restrain the phylogeographic structure of conspecific populations to relatively low levels.

POPULATION GENETIC STRUCTURE
OF OTHER PELAGIC FISHES

Although tunas and billfishes comprise the major fraction of pelagic fishery landings, other fishes with divergent life histories are also harvested. Important among these are the dolphin fish (*Coryphaena hippurus*) and the shortfin mako shark (*Isurus oxyrinchus*), which exhibit quite different life histories.

Dolphin Fish

This species is a keen visual predator, occupying surface waters of the pelagic realm. The dolphin fish is characterized by rapid growth, with analyses of daily increments of otoliths revealing growth rates of four mm/day over the first year of life (Uchiyama et al., 1986). Dolphin fish reach sexual maturity during their first year and are short-lived, with a lifespan of only two or three years (Bohlke and Chaplin, 1993).

Like many tunas, the dolphin fish exhibits little genetic divergence within or between oceans. Allozyme analysis of 36 presumptive gene loci of 15 Atlantic and 10 Pacific dolphin revealed no significant genetic differentiation (Rosenblatt and Waples, 1986). Similarly, direct sequence analysis of 258 bp of the mitochondrial cytochrome *b* gene of 55 dolphin from 11 geographically distant sampling locations provided no evidence of significant intra- or interocean population structure (C. Reeb, personal communication). Thirteen unique haplotypes which differed by no more than two mutational steps from each other were encountered. One haplotype was common to most samples, occurring in 55% of the individuals. Although conclusions are limited by small sample sizes, spatial partitioning of the genetic variation was not evident within or between ocean samples. Direct sequence analysis of the mitochondrial control region is now being employed to determine if any fine-scale population structure exists in this pelagic species.

Shortfin Mako Shark

Many species of elasmobranchs have life histories that are considerably different from most bony fishes, but not unlike some cetaceans. Sharks are typically long-lived, mature at a late age, and have long gestation periods and low fecundities. As a result, they are highly susceptible to overexploitation. Dramatic decreases in the abundance of large sharks have been noted in several regions (Box 11.1), but in general, long-term catch records have not been maintained to infer such decreases (Compagno, 1984).

The shortfin mako maintains an elevated body temperature (Carey and Teal, 1969) and is found throughout tropical and warm temperate pelagic waters. This species grows rapidly relative to other sharks, and has a life span of approximately 11.5 years (Pratt and Casey, 1983). Sexual maturity is reached at 2+ years for males, and 6+ years for females (Pratt and Casey, 1983; Stevens, 1983). Females have a gestation period of 8–12 months (Pratt and Casey, 1983) and bear 4–16 oviphagus pups that are approximately 70 cm at birth (Bass et al., 1975; Stevens, 1983). The shortfin mako is highly vagile, and undertakes extensive movements. Analysis of 231 tag recoveries demonstrated that 31 individuals had traveled distances greater than 1,600 km (Casey and Kolher, 1992).

Despite the vagility of adults, the shortfin mako exhibits genetic population structure. RFLP analysis (nine enzymes) of mtDNA samples from approximately 20 mako sharks from each of five locations in the Atlantic and Pacific Oceans (Heist et al., 1995a; Graves, unpublished data) revealed the following: (1) substantial variation, with sample nucleon diversities ranging from 0.49 to 0.91 (these genetic diversities contrast to the low values previously reported for shark mtDNA [Martin et al., 1992; Heist et al., 1995b]); and (2) highly significant spatial heterogeneity in the distribution of haplotypes (Table 11.3). In particular, the two North Atlantic samples were similar to each other, but significantly different from the South Atlantic, South Pacific, and North Pacific samples, which in turn, were similar to each other. These results suggest severe restrictions to gene flow across the equatorial Atlantic, a finding with important management implications. The shortfin mako is a common bycatch of swordfishing operations, and consequently experiences high harvest pressure in the North Atlantic. It would appear from the genetic data that animals from the North Atlantic comprise a self-sustaining population that does not experience extensive genetic immigration from the south.

Conservation Relevance

In the dolphin fish, a regional fishery collapse would not likely result in the loss of unique genetic variation, given the homogeneous distribution of mtDNA haplotypes among samples from different oceans. Furthermore, the long-distance movements, high fecundity, and rapid growth rates of these fish would presumably ensure relatively rapid repopulation of depleted areas. A different scenario exists for the shortfin mako. The dissimilarity of mtDNA haplotype distributions between populations in the North versus South Atlantic suggests severe limitations to gene flow across the equator. Quite likely, therefore, local depletion would not be rapidly relieved by long-distance migration. Furthermore, repopulation by local individuals would be slow due to the shortfin mako's late age of sexual maturity and low fecundity.

SUMMARY

1. Several pelagic fishes, including skipjack tuna, albacore, yellowfin tuna, and dolphin fish, exhibit little spatial partitioning of genetic variation within or between ocean basins. This observation is consistent with the hypothesis that these species have sufficient historical connectedness and gene flow to have prevented the accumulation of significant genetic divergence. Such exchange, no doubt, is promoted by the high vagility of the fishes, the occurrence of a continuous, circumtropical pelagic environment, and a wide range of suitable spawning conditions.

2. Little is known of the genetic population structure of bluefin tuna and southern bluefin tuna. Of all the tunas, these species are the most susceptible to overexploitation due to slower growth rates, later ages of maturation, and the staggering market value of individual fish. Both bluefin tuna and southern bluefin tuna have temporally and spatially discrete spawning grounds. Unlike other tunas, distributions of mtDNA haplotypes are significantly different between Atlantic and Pacific bluefin tuna. The genetic relationships of eastern and western Atlantic bluefin tuna have not been determined, but are critical for effective management of the fishery.

3. Although billfishes and tunas appear to have similar life histories, they exhibit rather different levels of population structure. Evolutionarily distant groups of mtDNA haplotypes are found within several species of billfish, and these lineages are heterogeneously distributed between sword-

fish from the Mediterranean Sea and North Atlantic Ocean, as well as between Atlantic and Indo–Pacific populations of blue marlin and sailfish. The magnitude of nucleotide sequence divergence between mtDNA haplotypes is consistent with relatively long periods of isolation between ocean populations (perhaps resulting from compression of the tropical pelagic environment during Pliocene or Pleistocene cooling), coupled with more recent unidirectional movement of genes from the Indo–Pacific into the Atlantic.

4. A more recent (shallower) population structure is also evident within several billfishes. Striped marlin in the Pacific Ocean demonstrate significant spatial partitioning of closely related mtDNA haplotypes. Similar within-ocean population structuring is also evident in sailfish and blue marlin. Tag-and-recapture analyses demonstrate that all of these species undertake extended movements. How these vagile organisms maintain significant population structure is not known, but spawning-site fidelity is suggested. To prevent the loss of unique genetic variation due to regional overfishing, these fisheries must be managed with units smaller than those currently applied.

5. Of all pelagic fishes, the sharks may be the most vulnerable to overfishing, due to their late age of maturity and low fecundities. Little is known of the biology of pelagic sharks, and historical records are not available to evaluate the status of the fisheries. Tag-and-recapture studies indicate that some pelagic sharks are quite vagile, yet preliminary mtDNA analysis of the shortfin mako shark demonstrated significant spatial partitioning of genetic variation.

6. Notwithstanding the presence of statistically significant genetic differentiation among conspecific populations of at least some pelagic marine fishes, the magnitudes and depths of the phylogeographic separations normally remain extremely shallow in comparison to those typifying most widely distributed species inhabitating coastal, freshwater, and terrestrial environments. The physical nature of the pelagic realm and the biology of its inhabitants have clearly constrained the magnitudes of genetic separation.

REFERENCES

ALVARADO-BREMER, J.R. 1994. Assessment of morphological and genetic variation of the swordfish (*Xiphias gladius* Linnaeus): evolutionary implications of allometric growth and of the patterns of nucleotide substitution in the mitochondrial genome. PhD dissertation, University of Toronto, Canada.

ARGUE, A.W. (ed.). 1981. *Report of the Second Skipjack Survey and Assessment Programme Workshop to Review Results from Genetic Analysis of Skipjack Blood Samples.* South Pacific Commission, Skipjack Survey and Assessment Programme, Technical Report No. 6, 39 pp.

AVISE, J.C., J. ARNOLD, R.M. BALL, E. BERMINGHAM, T. LAMB, J.E. NEIGEL, C.A. REEB, and N.C. SAUNDERS. 1987. Intraspecific phylogeography: the mitochondrial DNA bridge between population genetics and systematics. *Annu. Rev. Ecol. Syst.* 18:489–522.

BARRETT, I. and H. TSUYUKI. 1967. Serum transferrin polymorphism in some scombroid fishes. *Copeia* 1967:551–557.

BARTLETT, S.E. and W.S. DAVIDSON. 1991. Identification of *Thunnus* species by the polymerase chain reaction and direct sequence analysis of their mitochondrial cytochrome *b* genes. *Can. J. Fish. Aquat. Sci.* 48:309–317.

BASS, A.J., J.D. D'AUBREY, and N. KISTNASAMY. 1975. *Sharks of the East Coast of Southern Africa. IV. The families Odontaspididae, Scapanorhynchidae, Isuridae, Cetorhinidae, Alopiidae, Orectolobidae and Rhiniodontidae.* Oceanographic Research Institute (Durban), Investigative Report No. 39.

BAYLIFF, W.H. 1988. Integrity of schools of skipjack tuna, *Katsuwonus pelamis*, in the eastern Pacific Ocean, as determined from tagging data. *Fish. Bull.* 86:631–643.

BAYLIFF, W.H. 1991. Status of northern bluefin tuna in the Pacific Ocean. In *World Meeting on Stock Assessment of Bluefin Tunas: Strengths and Weaknesses*, eds. R.B. Deriso and W.H. Bayliff, pp. 29–88. Inter-American Tropical Tuna Commission, Spec. Rep. No. 7. IATTC, La Jolla, CA.

BLOCK, B.A., J.R. FINNERTY, A.F.R. STEWART, and J. KIDD. 1993. Evolution of endothermy in fish: mapping physiological traits on a molecular phylogeny. *Science* 260:210–214.

BOHLKE, J.E. and C.C.G. CHAPLIN. 1993. *Fishes of the Bahamas and Adjacent Tropical Waters*, 2nd ed. University of Texas Press, Austin.

BRIGGS, J.C. 1960. Fishes of worldwide (circumtropical) distribution. *Copeia* 1960:171–180.

CAREY, F.G. and J.M. TEAL. 1969. Mako and porbeagle: warm-bodied sharks. *Comp. Biochem. Physiol.* 28:199–204.

CASEY, J.G. and N.E. KOHLER. 1992. Tagging studies on the shortfin mako shark (*Isurus oxyrinchus*) in the western North Atlantic. *Aust. J. Mar. Freshw. Res.* 43:45–60.

CATON, A. (ed.). 1991. Review of aspects of southern bluefin tuna: biology, population and fisheries. In *World Meeting on Stock Assessment of Bluefin Tunas: Strengths and Weaknesses*, eds. R.B. Deriso and W.H. Bayliff, pp. 181–350. Inter-American Tropical Tuna Commission, Spec. Rep. No. 7. IATTC, La Jolla, CA.

CAYRE, P. and H. FARRUGIO. 1986. Biologie de la reproducion du listao (*Katsuwonus pelamis*) de l'Ocean Atlantique. In *Proceedings of the ICCAT Conference on the International Skipjack Year Program*, eds. P.E.K. Symons, P.M. Miyake, and G.T. Sakagawa, pp. 242–251. International Commission for the Conservation of Atlantic Tunas, Madrid.

CHOW, S. and S. INOUE. 1993. Intra- and interspecific restriction fragment length polymorphism in mitochondrial genes of *Thunnus* tuna species. *Bull. Nat. Res. Inst. Far Seas Fish.* 30:207–225.

CHOW, S. and S. USHIAMA. 1995. Global population structure of albacore (*Thunnus alalunga*) infered by RFLP analysis of the mitochondrial ATPase gene. *Mar. Biol.* 123:39–45.

COLLETTE, B.B. and C.E. NAUEN. 1983. FAO species catalogue. Vol 2. Scombrids of the world: an annotated and illustrated catalogue of tunas, mackerels, bonitos and related species known to date. *FAO Fish. Synop.* 125, 137 pp.

COMPAGNO, L.J.V. 1984. Species catalogue. Vol. 4. Sharks of the world: an annotated and illustrated catalogue of shark species known to date. Part 1. Hexanchiformes to Lamniformes. *FAO Fish. Synop.* 125, pp. 237–249.

EDMUNDS, P.H. 1972. *Genic Polymorphism of Blood Proteins from White Marlin*. National Marine Fisheries Service, NOAA, Res. Rep. No. 77, 15 pp.

FINNERTY, J.R. and B.A. BLOCK. 1992. Direct sequencing of mitochondrial DNA detects highly divergent haplotypes in blue marlin (*Makaira nigricans*). *Molec. Mar. Biol. Biotech.* 1:206–214.

FINNERTY, J.R. and B.A. BLOCK. 1995. Evolution of cytochrome *b* in the Scombroidei (Teleostei): molecular insights into billfish (Istiophoridae and Xiphiidae) relationships. *Fish. Bull.*, 93:78–96.

FOOD AND AGRICULTURE ORGANIZATION OF THE UNITED NATIONS. 1992. Yearbook of fisheries statistics. Catches and landings, 1990. *Yearb. Fish. Stats.*, Vol. 70.

FOREMAN, T.J. 1980. Synopsis of biological data on the albacore tuna, *Thunnus alalunga* (Bonnaterre, 1788) in the Pacific Ocean. In *Synopsis of Biological Data on Eight Species of Scombrids*, ed. W.H. Bayliff, pp. 17–70. Inter-American Tropical Tuna Commission, Spec. Rep. No. 2. IATTC, La Jolla, CA.

FORSBERGH, E.D. 1980. Synopsis of biological data on the skipjack tuna, *Katsuwonus pelamis* (Linnaeus, 1758) in the Pacific Ocean. In *Synopsis of Biological Data on Eight Species of Scombrids*, ed. W.H. Bayliff, pp. 295–360. Inter-American Tropical Tuna Commission, Spec. Rep. No. 2. IATTC, La Jolla, CA.

FUJINO, K. 1969. Atlantic skipjack tuna genetically distinct from Pacific specimens. *Copeia* 1969:626–629.

FUJINO, K. and T. KANG. 1968. Serum esterase groups of Pacific and Atlantic tunas. *Copeia* 1968:56–63.

FUJINO, K., K. SASAKI, and S. OKUMURA. 1981. Genetic diversity of skipjack tuna in the Atlantic, Indian, and Pacific Oceans. *Bull. Jap. Soc. Sci. Fish.* 47:215–222.

GRAVES, J.E. and A.E. DIZON. 1989. Mitochondrial DNA sequence similarity of Atlantic and Pacific albacore tuna. *Can. J. Fish. Aquat. Sci.* 46:870–873.

GRAVES, J.E., S.D. FERRIS and A.E. DIZON. 1984. High genetic similarity of Atlantic and Pacific skipjack tuna demonstrated with restriction endonuclease analysis of mitochondrial DNA. *Mar. Biol.* 79:315–319.

GRAVES, J.E. and J.R. MCDOWELL. 1994. Genetic analysis of striped marlin *Tetrapturus audax* population structure in the Pacific Ocean. *Can. J. Fish. Aquat. Sci.* 51:1762–1768.

GRAVES, J.E. and J.R. MCDOWELL. 1995. Inter-ocean genetic differentiation of istiophorid billfishes. *Mar. Biol.* 122:193–203.

GRIJALVA-CHON, J.M., K. NUMACHI, O. SOSA-NISHIZAKI, and J. DE LA ROSA-VELEZ. 1994. Mitochondrial DNA analysis of north Pacific swordfish (*Xiphias gladius*) population structure. *Mar. Ecol. Prog. Ser.* 115:15–19.

HARTL, D.L. and A.G. CLARK. 1989. *Principles of Population Genetics*, 2nd ed. Sinauer, Sunderland, MA.

HEIST, E.J., J.E. GRAVES, and J.A. MUSICK. 1995a. Population genetics of the cosmopolitan shortfin mako (*Isurus oxyrinchus*). *Can. J. Fish. Aquat. Sci.*, in press.

HEIST, E.J., J.E. GRAVES, and J.A. MUSICK. 1995b. Population genetics of the sandbar shark (*Carcharhinus plumbeus*) in the Gulf of Mexico and mid-Atlantic bight. *Copeia*, in press.

HILBORN, R. 1991. Modeling the stability of fish schools: exchange of individual fish between schools of skipjack tuna (*Katsuwonus pelamis*). *Fish. Bull.* 48:1081–1091.

HUNTER, J.R., A.W. ARGUE, W.H. BAYLIFF, A.E. DIZON, A. FONTENEAU, D. GOODMAN, and G.R. SECKEL. 1986. *The Dynamics of Tuna Movements: an Evaluation of Past and Future Research*. Food and Agriculture Organization of the United Nations, Fish., Tech. Pap. No. 277, Rome.

INTER-AMERICAN TROPICAL TUNA COMMISSION. 1991. *1990 Annual Report of the Inter-American Tropical Tuna Commission*. IATTC La Jolla, CA.

INTER-AMERICAN TROPICAL TUNA COMMISSION. 1993. *1992 Annual Report of the Inter-American Tropical Tuna Commission.* IATTC, La Jolla, CA.

INTERNATIONAL COMMISSION FOR THE CONSERVATION OF ATLANTIC TUNAS. 1993. Report of the standing committee on research and statistics, yellowfin tuna. In *International Commission for the Conservation of Atlantic Tunas, Report for Biennial Period 1992-3,* pp. 135–139. ICCAT, Madrid.

MAGOULAS, A., G. KOTOULAS, J.M. DE LA SERNA, G. DE METRIO, N. TSIMENIDES, and E. ZOUROS. 1993. Genetic structure of swordfish (*Xiphias gladius*) populations of the Mediterranean and the eastern side of the Atlantic: analysis by mitochondrial DNA markers. ICCAT, *Coll. Vol. Sci. Pap.* 40:126–136.

MARTIN, A.P., G.J.P. NAYLOR, and S.R. PALUMBI. 1992. Rates of mitochondrial DNA evolution in sharks are slow compared with mammals. *Nature* 257:153–155.

MORROW, J.E. and S.J. HARBO. 1969. A revision of the sailfish genus *Istiophorus. Copeia* 1969: 34–44.

MUSICK, J.A., S. BRANSTETTER, and J.A. COLVOCORESSES. 1993. Trends in shark abundance from 1974 to 1991 for the Chesapeake Bight region of the U.S. mid-Atlantic coast. In *Conservation Biology of Elasmobranchs,* ed. S. Branstetter, pp. 1–18. NOAA Tech. Rep. No. 115.

NAKAMURA, I. 1985. FAO species catalogue. Vol. 5. Billfishes of the world: an annotated and illustrated catalogue of marlins, sailfishes, spearfishes and swordfishes known to date. *FAO Fish. Synop.* 125, Rome.

NISHIKAWA, Y., M. HONMA, S. UEYANAGI, and S. KIKAWA. 1985. Average distribution of larvae of oceanic species of scombroid fishes, 1956–1981. *Far Seas Fish. Res. Lab. S Ser.* No. 12, 99 pp.

NATIONAL RESEARCH COUNCIL. 1994. *An Assessment of Atlantic Bluefin Tuna.* Commission on Geosciences, Environment and Resources. National Academy Press, Washington, DC.

PRATT, H.L., Jr. and J.G. CASEY. 1983. Age and growth of the shortfin mako, *Isurus oxyrinchus,* using four methods. *Can. J. Fish. Aquat. Sci.* 40:1944–1957.

RICHARDS, W.J. 1976. Spawning of bluefin tuna (*Thunnus thynnus*) in the Atlantic Ocean and adjacent seas. ICCAT, *Coll. Vol. Sci. Pap.* 5:267–278.

RIVAS, L.R. 1975. Synopsis of biological data on blue marlin, *Makaira nigricans* Lacepede, 1802. In *Proceedings of the International Billfish Symposium,* eds. R.S. Shomura and F. Williams, pp. 1–16. Kailua-Kona, Hawaii. Part 3. U.S. Department of Commerce, NOAA Tech. Rep. NMFS SSRF-675.

ROBINS, C.R. and D.P. DE SYLVA. 1960. Description and relationships of the longbill spearfish, *Tetrapturus belone,* based on western North Atlantic specimens. *Bull. Mar. Sci.* 10:385–413.

ROSENBLATT, R.H. and R.S. WAPLES. 1986. A genetic comparison of allopatric populations of shore fish species from the eastern and central Pacific Ocean: dispersal or vicariance? *Copeia* 1986:275–284.

SCOLES, D.R. and J.E. GRAVES. 1993. Genetic analysis of the population structure of yellowfin tuna *Thunnus albacares* in the Pacific Ocean. *Fish. Bull.* 91:690–698.

SCOTT, E.L., E.D. PRINCE, and C.D. GOODYEAR. 1990. History of the cooperative game fish tagging program in the Atlantic Ocean, Gulf of Mexico, and Caribbean Sea, 1954–1987. *Amer. Fish. Soc. Symp.* 7:841–853.

SHAKLEE, J.B., R.W. BRILL, and R. ACERRA. 1983. Biochemical genetics of Pacific blue marlin, *Makaira nigricans,* from Hawaiian waters. *Fish. Bull.* 81:85–90.

SHARP, G.D. 1978. Behavioral and physiological properties of tunas and their effects on vulnerability to fishing gear. In *The Physiological Ecology of Tunas,* eds. G.D. Sharp and A.E. Dizon, pp. 397–449. Academic Press, New York.

SMITH, P.J., A.M. CONROY, and P.R. TAYLOR. 1994. Biochemical genetic identification of northern bluefin tuna *Thunnus thynnus* in the New Zealand fishery. *New Zeal. J. Mar. Freshw. Res.* 28:113–118.

SQUIRE, J.L. 1987. Striped marlin, *Tetrapturus audax*, migration patterns and rates in the northeast Pacific Ocean as determined by a cooperative tagging program: its relation to resource management. *Mar. Fish. Rev.* 49:26–43.

SQUIRE, J.L. and Z. SUZUKI. 1990. Migration trends of striped marlin (*Tetrapturus audax*) in the Pacific Ocean. In *Planning the Future of Billfishes: Research and Management in the 90s and Beyond*, ed. R.H. Stroud, pp. 67–80. National Coalition for Marine Conservation, Savannah, GA.

STEVENS, J.D. 1983. Observations on reproduction in the shortfin mako *Isurus oxyrinchus*. *Copeia* 1983:126–130.

SUZUKI, Z. 1991. Migration—western Atlantic. In *World Meeting on Stock Assessment of Bluefin Tunas: Strengths and Weaknesses*, eds. R.B. Deriso and W.H. Bayliff, pp. 129–130. Inter-American Tropical Tuna Commission, Spec. Rep. No. 7. IATTC, La Jolla, CA.

TALBOT, F.H. and M.J. PENRITH. 1962. Tunnies and marlins of South Africa. *Nature* 193:558–559.

UCHIYAMA, J.H., R.K. BURCH and S.A. KRAUL, Jr. 1986. Growth of dolphins, *Coryphaena hippurus* and *C. equiselis*, in Hawaiian waters as determined by daily increments on otoliths. *Fish. Bull.* 84:186–191.

VAWTER, A.T., R.H. ROSENBLATT, and G.C. GORMAN. 1980. Genetic divergence among fishes of the eastern Pacific and Caribbean: support for the molecular clock. *Evolution* 34:705–711.

WARD, R.D., N.G. ELLIOT, P.M. GREWE, and A.J. SMOLENSKI. 1994. Allozyme and mitochondrial DNA variation in yellowfin tuna (*Thunnus albacares*) from the Pacific Ocean. *Mar. Biol.* 118:531–539.

12

CONSERVATION GENETICS
OF NORTH AMERICA DESERT FISHES

Robert C. Vrijenhoek

INTRODUCTION

Fishes in the desert— that can seem almost a contradiction in terms since by definition fishes inhabit water and deserts have little. ...But conflicting expectations to the contrary, there is water in the desert and fishes do live there— no contradiction but quite a marvel. (Rolston, 1991, p. 93).

Despite the paradox of "desert fishes," the Great Basin, Mohave, Sonoran, and Chihuahuan deserts of North America are homes to a remarkable variety of stream-dwelling fish. These fish populations are remnants of earlier and wetter (pluvial) times (Minckley et al., 1986). Natural drying during the last 8,000 to 12,000 years has led to many isolated populations in tiny springs and disconnected stream segments (Miller, 1958). Recently, the plight of these remnant populations has been exacerbated by habitat loss due to the voracious human appetite for water. Of the 224 fish species currently listed as threatened or endangered in the United States, approximately 50% live west of the Continental Divide, and many of these occur in desert habitats (Minckley and Douglas, 1991). More than 20 species from the North American West have become extinct during the past 100 years.

ENVIRONMENTAL BACKGROUND

Natural Processes

Attempts to preserve and restore these fish must recognize the tight interplay between physico-chemical and biological processes in desert eco-systems. Strong seasonality characterizes the aquatic habitats. In the Sonoran Desert region, a long dry season is followed by intense summer rains (Figure 12.1). In some years, El Niño conditions bring midwinter storms. Farther south in Sinaloa, Mexico, summer rainfall increases and streams become more permanent. In contrast, total rainfall decreases dramatically north of Sonora, in the Mohave and Great Basin deserts. Some regions, like Death Valley in California, may receive only four cm of rain annually (Soltz and Naiman, 1978). Rainy seasons in the desert are characterized by flash floods, such that stream segments and springs that were isolated during the dry season may be temporarily reconnected. At such times, fish can disperse to reclaim temporary habitats and mix with populations from other refugia. However, temporary habitats disappear again during the dry season, and

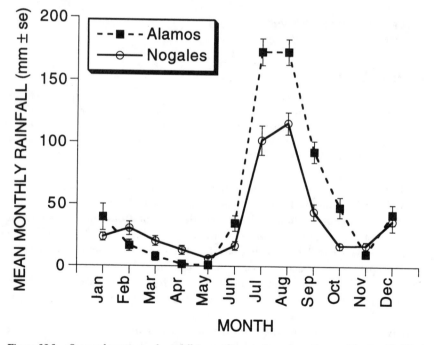

Figure 12.1.   Seasonal patterns of rainfall in northern and southern Sonora, Mexico. Modified from Vrijenhoek (1994a).

the fish either die locally or retreat to permanent refugia. The North American deserts are geologically young, and perhaps as a consequence their fishes have not evolved special adaptations like aerial respiration, aestivation, and desiccation-resistant embryos that characterize some fish in other arid regions of the world (Cole, 1981).

Residual pools that serve as refugia during the dry season can be harsh environments. Often, fish are extremely crowded, a situation that also concentrates parasite vectors and leads to increased disease risk. Water temperatures can elevate to lethal levels during the day, and dissolved oxygen can plummet, especially at night near the detrital interface (Deacon and Minckley, 1974). Chemical properties of these pools may vary immensely across short intervals. Salinity and hardness of the water may range from high levels in desiccating pools to very low levels after a rainstorm.

Food supplies also vary widely across the seasons. Insects must oviposit in residual pools during the dry season, concentrating their larvae in habitats with accumulated detritus and aquatic plants. In southern Sonora, Mexico, desert topminnows of the genus *Poeciliopsis* prey on these larvae and eat the detritus. These fish are relatively well nourished when crowded in springs and residual pools, and are able to produce eggs (Thibault, 1974; Schenck and Vrijenhoek, 1989). During the rainy season, water and space are abundant, but the streams are scoured of plants and detritus, and insect larvae become super-dispersed. With food less available, these fish rapidly lose the ability to reproduce. A sudden abundance of water is not necessarily advantageous for these desert fish.

In contrast, desert springs may provide very stable habitats. For an individual spring, the discharge rate, water level, temperature, salinity, and oxygen level can be nearly constant over long periods (Naiman, 1981). Many springs, like those in the Death Valley region, are fed by ancient aquifers with water that is 10,000–20,000 years old. Individual springs may issue waters ranging from fresh to highly mineralized, and ranging in temperature from cold to warm geothermal. Hot and cold springs are interspersed in some areas, creating opportunities for disruptive selection of warm- and cold-water adaptations (Vrijenhoek et al., 1992).

During the Pleistocene, the Great Basin and Mohave regions contained many large lakes. For example, the salt encrusted desert of Death Valley (Figure 12.2) is all that remains of Lake Manly, which at its height was 160 km long, 10–17 km wide, and 185 m deep. Springs that fed this intermontane basin along the California–Nevada border still house remnant fish populations, primarily cyprinodonts and a few minnows. The Great Salt Lake of Utah represents another stage in the drying process. Its hypersaline waters presently do not support freshwater fish, whereas the deep waters of Pyramid Lake in northern Nevada still do. In recent historical times, Pyra-

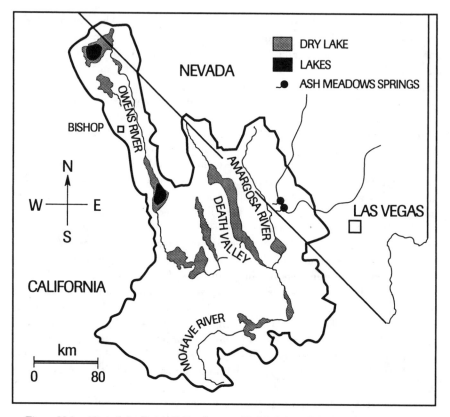

Figure 12.2.   Map of the Death Valley System. Modified from Soltz and Naiman (1978).

mid Lake had a population of cutthroat trout (*Oncorhynchus clarki*) that sustained a commercial fishery, but overfishing led to the demise of both (LaRivers, 1962). The postpluvial period has exposed desert fish populations to intense selection for living in these difficult and extreme environments.

Anthropogenic Disturbance

Loss of aquatic habitats and fragmentation of populations due to post-pluvial drying has been exacerbated by excessive ground water pumping that has lowered aquifers and dried natural springs. Diversion of surface waters has eliminated natural streams and marshes. Entire habitats have disappeared in the Death Valley system, and endemic pupfish (genus *Cyprinodon*) and killifish (*Empetrichthys*) have been lost (Box 12.1).

Construction of dams on major rivers of the North American West has altered downstream flow and consequently changed temperature and

Box 12.1. Water Mining, Habitat Loss, and the Desert Fishes Council

Agricultural development has posed a serious threat to fish of the Death Valley system, particularly the Amargosa River and associated springs (Figure 12.2). Pumping actually "mines" water because underground aquifers are not recharged as fast as they are pumped (Dudley and Larson, 1976). Pumping has eliminated all major springs of the Pahrump Valley and along with them the Pahrump killifish (*Empetrichthys latos pahrump*; Soltz and Naiman, 1978). Similarly, the Ash Meadows pupfish (*Cyprinodon nevadensis mionectes*) disappeared when Jackrabbit Spring was drained by an irrigation pump (Deacon and Bunnell, 1970). In the period from 1967–1972, pumping in the Ash Meadows area intensified to such a degree that it lowered the aquifer and threatened isolated habitats like Devil's Hole. The threat of extinction of the endemic Devil's Hole pupfish (*Cyprinodon diabolis*) resulted in a flurry of conservation activities and litigation. Representatives from government agencies, conservation organizations, academic scientists, and private citizens met in 1969 at Death Valley National Monument headquarters and formed the Pupfish Taskforce that became the nucleus for the Desert Fishes Council (Pister, 1991).

---

sedimentation patterns. For example, prior to impoundment, the Colorado River exhibited large seasonal changes in water volume and temperature. Water management practices have modified aquatic habitats and resulted in the serious decline and near extinction of several large species endemic to the drainage, including the Colorado squawfish (*Ptychocheilus lucius*), humpback chub (*Gila cypha*), bonytail (*G. elegans*), and the razorback sucker (*Xyrauchen texanus*).

Introduction of exotic species has been the *coup de grace* for many native fish. Presently, only 40% of Colorado River fish are natives; nearly 30 species are introduced (Richardson, 1976). Construction of dams aided the establishment of several exotic species, such as the African tilapiines which flourish in impoundments and outcompete native species for food and nesting sites. The extraordinary success of tilapiines also results from a remarkable plasticity in life-history traits. When tilapiines escape from impoundments, they are capable of reproducing at small sizes in surrounding streams, and can quickly replace local cyprinids and poeciliids in riverine habitats. Mosquitofish (*Gambusia* spp.) too have benefited from construction of dams. Unlike stream-dwelling native fish that are adapted to huge seasonal fluctuations in water flow, *Gambusia* tend to be washed away during the episodic flash floods that dams control (Meffe, 1984). Introduced throughout the American West for insect control, mosquitofish have devastated native species because they also prey on larval and juvenile fish (Meffe, 1985).

Introduced species pose an additional threat if they harbor exotic parasites. Mosquitofish and guppies (*Poecilia reticulata*) may expose native
poeciliids to exotic gyrodactylids (Trematoda: Monogenea; Rogers and
Wellborn, 1965; Harris, 1986), and some Sonoran *Poeciliopsis* (Poeciliidae)
are known to be susceptible to guppy trematodes (Leberg and Vrijenhoek,
1994). For example, *Gyrodactylus turnbulli* infections are lethal for a clonal
strain of the unisexual fish *P. monacha-lucida* (Box 12.2). However, a
coexisting clone is resistant to this parasite. If introduced to Sonora, *G.
turnbulli* could alter the dynamic balance between native *P. monacha-lucida*
clones and their sexual relatives. The potential impact of exotic helminths,
bacteria, and viruses should be assessed before government agencies become involved in the transport and release of exotic fish.

Box 12.2. All-Female Fish and the Mother of Clonal Diversity, *Poeciliopsis monacha*

Among the oddities of desert fish are
all-female forms that arose as hybrids
between sexually reproducing species of
the genus *Poeciliopsis* in Sonora and
Sinaloa, Mexico (Schultz, 1969, 1977).
Hybridization between *P. monacha* females and *P. lucida* males produced a
diploid biotype *P. monacha-lucida* that
reproduces by hybridogenesis. These females transmit only the *monacha* chromosomes to each egg, but true fertilization by males of *P. lucida* restores the
monacha-lucida genotype in each generation. *P. monacha-latidens* and *P. mon-
acha-occidentalis* also transmit only their
monacha genomes to ova, but rely on
males of *P. latidens* and *P. occidentalis*,
respectively, for insemination. Three
triploid biotypes (*P. 2 monacha-lucida*,
P. monacha-2 lucida, *P. monacha-
viriosa-lucida*) reproduce by gynogenesis,
a form of reproduction that transmits
the entire triploid genome, without recombination, to ova. Sperm from sexual
males is needed to activate development
of triploid ova, but paternal genes are
not expressed in the offspring. Inheritance is strictly clonal.

Although the all-female forms are generally abundant in their respective rivers
and habitats, the maternal ancestor of all
these biotypes, *P. monacha*, is seriously endangered (Quattro et al., 1991). It was
known from a few isolated headwater
tributaries in the Ríos Mayo, Fuerte, and
Sinaloa (Miller, 1960), and during the past
25 years, apparently has disappeared from
the Río Sinaloa and declined in headwaters
of the Río Mayo (where both *Gambusia*
and *Tilapia* were introduced). *P. monacha*
remains in isolated headwaters of the Río
Fuerte, but the construction of small impoundments and introduction of exotics
also threaten the species there. If *P. monacha* occurred in the United States, it
would undoubtedly be listed as endangered. What a shame it would be to lose
this extraordinary fish— the mother of a
remarkably diverse assemblage of fish
clones that have served so well as a model
system for evolutionary and ecological studies (Vrijenhoek, 1994b).

Fish hatcheries and sport fishing also have had devastating effects on native fish (Meffe, 1992). In many streams on the western slope of the Rockies, hatchery-reared rainbow trout (*Oncorhynchus mykiss*) have replaced native cutthroat trout (*O. clarki*). In other cases, cutthroat populations are thoroughly introgressed with rainbow trout genes (Allendorf and Leary, 1988). Introduced rainbow trout have hybridized with and replaced native Apache (*O. apache*) and Gila trout (*O. gilae*; Loudenslager et al., 1986). Released bait fish also have had deleterious effects on native fish. For example, sheepshead minnows (*Cyprinodon variegatus*) used as bait have hybridized with native Pecos River pupfish (*Cyprinodon pecosensis*; Echelle and Conner, 1989).

GENETIC DIVERSITY
AND POPULATION STRUCTURE

Efforts to preserve desert fish are meaningless unless accompanied by attempts to prevent further deterioration of their physical and biological habitats. With sound land- and water-management practices, habitat restoration and reclamation can create opportunities to restore endangered species throughout their recent historical ranges. In the meantime, husbandry must be provided for native fish stocks. Success at husbandry will depend on the ability to maintain healthy fish populations in nature, in preserves, in hatcheries, and sometimes in aquaria. One such facility is the Dexter National Fish Hatchery and Technology Center (DNFHTC) in New Mexico, which is devoted to preserving and breeding native desert fishes. It presently maintains 14 endangered species and subspecies of fish intended for restoration throughout the desert Southwest (Table 12.1). The DNFHTC recently added a geneticist to its staff, an important step that should help avoid further losses of genetic diversity and inbreeding in its precious stocks (see razorback sucker, to follow).

Loss of Genetic Diversity
and Inbreeding Depression

Considerable evidence exists for an association between increased homozygosity of individuals and inbreeding depression (Falconer, 1960). Decreased fertility, survival, and disease resistance are associated with homozygosity and inbreeding in captively propagated zoo animals (Ralls et al., 1986a). Inbreeding depression has been inferred from low heterozygosity and decreased vigor in natural populations of birds and mammals (Ralls et

Table 12.1. Fishes Held at Dexter National Fish Hatchery, New Mexico (June, 1994)

Scientific name	Common name	Natural distribution[1]
Cyprinidae		
Gila elegans	Sonoran chub	AZ, CO, NV, UT, WY
G. nigrescens	Chihuahua chub	NM, CHI
G. robusta jordani	Paranagat roundtail chub	NV
G. r. seminuda	Virgin roundtail chub	AZ, NV, UT
Cyprinella f. formosa	Beautiful shiner	NM, CHI
Plagopterus argentissimus	Woundfin	AZ, CA, NV, UT, SON(?), BCN(?)
Ptychocheilus lucius	Colorado squawfish	AZ, CA, CO, NM, NV, UT, WY, SON, BCN
Catastomidae		
Xyrauchen texanus	Razorback sucker	AZ, CA, CO, NM, NV, UT, WY, SON, BCN
Catostomus warnerensis	Warner's sucker	CA, OR
Ictaluridae		
Ictalurus pricei	Yaqui catfish	AZ, SON
Cyprinodontidae		
Cyprinodon bovinus	Leon Springs pupfish	TX
C. m. macularis	Desert pupfish	AZ, CA, SON, BCN
Poeciliidae		
Gambusia gaigei	Big Bend gambusia	TX
Poeciliopsis o. occidentalis	Gila topminnow	AZ, CA(?), NM, SON

[1] Abbreviations for Mexican states: BCN = Baja California; CHI = Chihuahua; SON = Sonora; (?) = questionable occurrence.
Source: Updated from Johnson and Jensen (1991), courtesy of B.L. Jensen.

al., 1986b), and has been verified in hatchery stocks of salmonid fishes (Gall, 1987). Conversely, multilocus heterozygosity as revealed by allozyme surveys is often associated with increased survival, fertility, and general vigor in a wide variety of plants and animals (Mitton and Grant, 1984). Similar associations have been documented in natural and experimental populations of fish (Allendorf and Leary, 1986).

The mechanisms underlying correlations between multilocus heterozygosity and fitness are not well understood, and thus a debate exists over the effectors of such associations (Turelli and Ginzburg, 1983; Zouros and Foltz, 1987). Is the correlation due to a superiority of heterozygous enzyme loci *per se*? Is it due to transitory linkage disequilibrium with "phantom loci" that are true effectors of fitness (Ohta, 1981)? Is it due to a relationship between allozyme homozygosity and an increased probability of being inbred (Ledig et al., 1983)? Despite the persistence of this debate among population geneticists, conservationists should not be blinded to the potential usefulness of heterozygosity–fitness correlations in the management of remnant populations (Vrijenhoek, 1994a).

Genetic diversity may decline rapidly in captive and natural fish populations as a consequence of severe bottlenecks in the number of breeding adults or founder events associated with local extinction and recolonization events in desert streams. Mating between relatives also increases the inbreeding coefficient of individuals and risks exposing deleterious recessive traits. Permanently small populations lose genetic diversity at a rate that is inversely related to the population's genetically effective size, N_e (Box 12.3). Under most circumstances, N_e is smaller than the actual number of breeding adults. It is reduced by uneven sex ratios, fluctuations in population size over time, excess (i.e., non-Poisson) reproduction of a small number of families, and non-random mating. Managers of aquatic preserves and hatchery stocks should be aware of factors that reduce N_e, increase homozygosity, and thereby increase the risk of inbreeding depression.

In hatchery populations, artificial spawning through hormone induction and stripping of eggs and sperm can substantially reduce N_e. For example, equalizing the numbers of males and females that contribute to the spawning would maximize N_e, but this is based on the assumption that eggs and sperm from different individuals are equally viable and fertile at the time of

---

Box 12.3. Effective Population Size

Population geneticists recognize three different kinds of genetically effective population size (Gliddon and Gowdet, 1994):

1. The *inbreeding effective size* (N_{ef}) predicts the rate of loss of heterozygosity in individuals due to homozygosity for genes that are identical by descent (Wright, 1931).

2. The *variance effective size* (N_{ev}) predicts the rate of loss of genetic diversity in a population due to genetic drift. It is "the size of an ideal population that has the same rate of genetic drift as the actual population" (Crow and Denniston, 1988).

3. The *extinction (or eigenvalue) effective size* ($N_{e\lambda}$) predicts the rate at which alleles are lost from a population (Ewens, 1982; Haldane, 1939).

Under most circumstances, one cannot distinguish among the kinds of N_e in a randomly mating natural population, because they are essentially the same. However, they should be distinguished if pedigree information is available, if non-random mating is suspected, if a population is highly subdivided, or if populations are rapidly growing or declining (Templeton and Read, 1994). With inbreeding, N_{ef} may be substantially smaller than N_{ev} and $N_{e\lambda}$. In a subdivided population, N_{ev} plays a significant role in determining the total genetic diversity. In declining populations, N_{ev} and $N_{e\lambda}$ may be much smaller than N_{ef}, and *vice versa* in growing populations.

spawning. Often, embryos and juveniles from different families are not equally viable, and in reality only a few males and females are responsible for producing the next generation. Bartley et al. (1992) used genetic techniques to infer N_e in hatchery stocks of Chinook salmon (*Oncorhynchus tshawytscha*) and rainbow trout (*O. mykiss*). Despite the use of large numbers of adults, genetically effective sizes were limited. For example, mass strip-spawning of 2,000 rainbow trout was genetically equivalent to reproduction by only 88.5 individuals, and 10,000 adult Chinook salmon were equivalent to that of only 132.5 individuals. An example involving razorback suckers (*Xyrauchen texanus*) is given later.

Clearly, continuous monitoring is necessary to assure the maintenance of genetic diversity in captive and natural stocks of endangered fish. It is best to keep stock sizes in natural refugia and preserves as large as possible to avoid genetic drift and inbreeding depression. In most cases, population sizes large enough to avoid demographic stochasticity and extinction also are large enough to avoid substantive losses of genetic diversity (Lande, 1988). However, assessing the risk of demographic stochasticity and extinction requires detailed life-history information that is rarely available for most species.

Diversity among Populations

Management of desert fish also requires knowledge of the distribution of genetic diversity between remnant populations. Total genetic diversity in a fragmented species can be partitioned into variation that exists within and among subpopulations: $H_T = H_S + D_{ST}$, where H_S and D_{ST} are the within- and among–subpopulation components of variance, respectively (Nei, 1977; Wright, 1978). Differences among subpopulations may result from random genetic drift or differential adaptation to diverse environments. Partitioning of genetic diversity can be extended hierarchically to more complex situations. For example, for a riverine fish, $H_T = H_S + D_{SR} + D_{RT}$, where D_{SR} is the variance among subpopulations within a river system, and D_{RT} is the variance among populations in different river systems, and so on. The proportion of the total genetic diversity found among subpopulations, $G_{ST} = D_{ST}/H_T$ (Nei, 1977), is informative regarding the degree of isolation and gene flow among subpopulations. For neutral polymorphisms, low G_{ST}'s are a product of high gene flow, and high G_{ST}'s are a product of relative isolation. Values of G_{ST} can be estimated in hierarchical populations as well (Nei, 1977; Slatkin and Voelm, 1991).

Protein electrophoresis and mitochondrial (mt) DNA analysis provide relatively easy methods for obtaining data on allele frequency variances within and among natural populations. Sensitive techniques based on the polymerase chain reaction (PCR) permit nondestructive sampling from rare and endangered species. As little as a drop of blood, a fin clip, or a scale can provide sufficient material for the amplification of DNA. Nondestructive sampling and PCR-based techniques such as those recently employed by Karl et al. (1992) allow multi-locus genetic analyses that heretofore have been intractable with allozyme and mtDNA techniques.

Echelle (1991) recently summarized allozyme studies of genetic diversity in North American desert fish. He considered 27 taxa that were recognized as threatened or endangered in the United States and Mexico. The total genetic diversity found in these taxa (mean $H_T = 0.04$; range $= 0.01-0.07$) was lower than that in marine fishes, but typical of other freshwater fish. On average, heterozygosity within local subpopulations (H_S) of desert fish also was lower than in marine fish. Overall, desert fish did not differ greatly from other non-migratory freshwater fish with respect to genetic diversity; however, comparisons within some taxa were revealing. For example, eight *Gambusia* species from the Chihuahuan Desert had significantly lower heterozygosity (mean $H_S = 0.016$) than did 14 species from less arid regions (mean $H_S = 0.047$; Echelle et al., 1989). The only cases of $H_S = 0.000$ occurred in geographically restricted desert populations. Echelle attributed lower heterozygosity in desert *Gambusia* to an increased likelihood of population bottlenecks and an absence of gene flow due to geographical isolation. Despite the opportunity for losses of heterozygosity in presently isolated populations, most of the 27 taxa reviewed by Echelle had 50% or more of the total genetic diversity represented within subpopulations— an indicator, perhaps, of substantial gene flow in their recent past.

Only a few studies have examined hierarchical partitioning of genetic diversity between subpopulations within drainage systems and between populations inhabiting different drainages (Echelle, 1991). In rainbow trout (*Oncorhynchus mykiss*), a migratory species, diversity within populations is high ($H_S = 85.0\%$ of the total) with the remainder evenly split between samples within drainages (7.7%) and between drainages (7.3%). In the nonmigratory Yellowstone cutthroat trout (*O. clarki lewisi*), diversity within populations is lower ($H_S = 67.6\%$ of the total), and again the remainder is evenly split between samples within drainages (15.7%) and between drainages (16.7%). In the small nonmigratory poeciliid (*Poeciliopsis occidentalis*), diversity within populations is much lower ($H_S = 21.3\%$ of the total), and less diversity exists between samples within drainages (25.5%) than between distinct drainages (53.2%; Vrijenhoek et al., 1985).

The fragmented population structures of desert fish are probably best described by the metapopulation concept first introduced by Levins (1970). Subpopulations are connected through a history of colonization, subsequent dispersal, and patch turnover due to local extinction and recolonization events. The demographic processes that define metapopulation dynamics also have an impact on genetic structure (McCauley, 1991). Depending on patterns of colonization and dispersal, genetic divergence among subpopulations will either be increased or decreased (Slatkin, 1977). To manage fragmented metapopulations, much more information must be gathered on local turnover rates of subpopulations, patterns of dispersal, and sources of colonists. Only then will it be possible to properly assess alternative management strategies (Hanski and Gilpin, 1991; Gliddon and Goudet, 1994).

MODELS OF DISPERSAL AND GENE FLOW

Higher dispersal rates result in more diversity within, and less diversity between, subpopulations of desert fish species. The proportion of genetic diversity between populations, G_{ST}, is inversely proportional to the effective rate of dispersal (Figure 12.3). The pseudoparameter Nm can be interpreted as the effective number of migrants exchanged between populations per

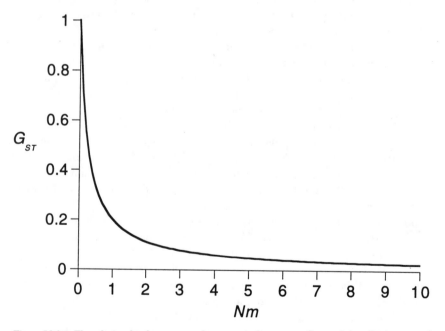

Figure 12.3.   The relationship between random genetic divergence, G_{ST}, and the effective rate of gene flow, Nm, among subpopulations under an island model of population structure. In the present case, G_{ST} is substituted for F_{ST} in Wright's (1951) relationship, $F_{ST} \approx 1/(1 + 4Nm)$.

Box 12.4. Estimation of Gene Flow

Several methods exist for inferring rates of gene flow from geographical surveys of allozymes and molecular sequence variation (Slatkin and Barton, 1989; Slatkin and Maddison, 1990; Slatkin, 1993). They are generally based on estimation of the migration parameter Nm, which can be interpreted as the effective number of migrants randomly exchanged between populations per generation, under an *island model* of population structure (Wright, 1951). The island model represents a type of null model of discrete population structure; dispersal is random with respect to the distance between subpopulations. A linear *stepping-stone model* may be a better approximation for riverine populations.

Stepping-stone models assume that dispersal occurs between neighboring subpopulations; long-distance dispersal is assumed to be rare. Stepping-stone dispersal results in a correlation of gene frequencies between subpopulations that decreases as a function of the number of steps between subpopulations (Kimura and Weiss, 1964). A similar decrease in genetic correlation with increasing geographical distance occurs as a consequence of "isolation by distance" in continuously distributed populations (Wright, 1943). The slope of these decreasing correlations can provide information about historical patterns and rates of dispersal (Sokal and Wartenberg, 1983; Slatkin, 1993).

generation (Box 12.4). Under an "island" model of population structure, genetic drift (a function of N) is balanced by gene flow (a function of m). Low migration rates ($Nm < 1$) permit substantial divergence among populations (Wright, 1951), whereas values of $Nm \gg 1$ are sufficient to counteract substantial divergence among subpopulations. Note however, that Nm is a compound parameter. An $Nm = 4$ might be due to large subpopulations with little migration ($N = 10,000$ and $m = 0.0004$), or to small subpopulations with much migration ($N = 20$ and $m = 0.20$). With $N = 10,000$, genetic drift will be negligible, so that little gene flow is necessary to balance it; but when $N = 20$, drift will be substantial, and high gene flow is necessary to prevent divergence in neutral characters. Nevertheless, even high rates of gene flow ($Nm > 1$) may result in statistically significant allele frequency variances among subpopulations, although the subpopulations will tend to be similar in allelic composition.

Meffe and Vrijenhoek (1988) proposed different management criteria that relate to ends of the spectrum regarding natural opportunities for gene flow in desert fish. These scenarios, the "Death Valley" and "Stream Hierarchy" models, will be described next.

The Death Valley Model

Pupfish (*Cyprinodon*) of the Death Valley System exemplify one end of the gene flow spectrum (Box 12.5). During pluvial times, these pupfish were distributed throughout an interconnected system of rivers and lakes situated in intermontane basins near the present borders of southern California and Nevada (Figure 12.2). Through natural drying of watercourses, the remnant populations of these fish are now completely isolated. Genetic drift and, presumably, adaptation during the past 20,000 years to the unique physico-chemical properties of individual springs, streams, and marshes, has resulted

---

Box 12.5. Death Valley Pupfish, the *Cyprinodon nevadensis* Species Complex

The *C. nevadensis* complex comprises four species and seven extant subspecies that are endemic to closed basins in the Death Valley system, the Owens River Valley, and Ash Meadows/Death Valley (Figure 12.2). One member of this complex, *C. diabolis*, may be the most geographically restricted vertebrate species in the world (Soltz and Naiman, 1978). It exists naturally in an isolated depression called Devil's Hole near Ash Meadows, CA. The entire habitat of Devil's Hole pupfish consists of a few square meters of a shallow limestone shelf suspended above an extensive subterranean cavern system. It took a Supreme Court decision (426 U.S. 128 [1976]) to stop further groundwater pumping and protect this species (U.S. Fish and Wildlife Service, 1983). To date, *C. diabolis* has not been reared in captivity, but a stock was established in a cement pond near Hoover Dam. The Devil's Hole pupfish was placed on the Federal Endangered Species List in 1966.

Although substantial differences exist among members of the *C. nevadensis* species complex with respect to morphology, life history, physiology, and behav-ior, they are strikingly similar with respect to allozymes (Turner, 1974; Echelle and Echelle, 1993) and mtDNA (Echelle and Dowling, 1992). Apparently, their brief isolation since pluvial times has allowed divergence in quantitative traits, but has not been sufficiently long to allow an accumulation of mutations that would be revealed in mtDNA or proteins. However, the Owens Valley pupfish (*C. radiosus*) has a mtDNA variant that links it to members of a related species complex (*C. macularis*) that occurs in isolated basins formerly connected to the lower Colorado River system. Thus, Echelle and Dowling (1992) suggested that the *C. nevadensis* complex may be an artificial taxonomic grouping (based on geography rather than phylogeny), or alternatively, that secondary contact during an earlier pluvial period resulted in introgressive hybridization between the *C. nevadensis* and *C. macularis* complexes. Although this problem remains unresolved, it exemplifies the risks associated with using present-day geographical distribution as the sole criterion for taxonomic distinction and hence conservation management.

in remarkable interlocality divergence with respect to morphology, physiology, behavior, and life-history traits— a level of divergence that rivals the classical example of Darwin's finches (Soltz and Hirshfield, 1981). Genetic diversity in this and other such species complexes can be represented simply as that which occurs within versus among remnant populations.

Although historical patterns of connectivity might be recovered from genetic analysis of such fish, these patterns have little bearing on the management of current populations. Translocation of pupfish between the isolated springs and marshes of the Death Valley system might seem to be a good way to supplement depleted populations and simultaneously combat inbreeding depression. However, translocation would be unnatural, and probably inappropriate for this and similar systems, because the current isolation of distinct populations is a consequence of natural climatic and geological processes rather than anthropogenic disturbance. Artificial introgression between isolated pupfish populations will no doubt alter the traits that define these native strains. Furthermore, indiscriminate mixing might disrupt local adaptations and lead to *outbreeding depression* (Templeton, 1986). An alternative strategy, currently adopted by the conservation agencies involved with these fish, is to maintain ecologically and genetically healthy populations within their natural locales. Translocation of fish may be desirable, however, in cases where recent anthropogenic factors have severed natural connections between extant subpopulations, as discussed under the subsequent model.

The Stream Hierarchy Model

Desert fish that inhabit dendritic rivers exemplify a more complex population structure based on hierarchical connectivity within and between river systems. Subpopulations may be disconnected during the dry season and reconnected when the rains resume. The structuring of genetic diversity will be influenced by local subpopulation sizes, natural barriers to gene flow, and the life-history characteristics of individual species. However, rates of gene flow between localities are not constant. Instead they vary across generations depending on seasonal rains and the extent of periodic droughts. Over the long term, the genetic variability in subpopulations will be affected by a mean rate of gene flow defined in part by periodic connectivity with other sites, and in part by the generation time of fish relative to these periods. At equilibrium, total genetic diversity in a dendritic population system could be partitioned as $H_T = H_S + D_{SR} + D_{RT}$ (and so on), as mentioned earlier.

Not surprisingly, in empirical studies of desert fish based on molecular markers, less genetic divergence typically is observed between subpopulations within river systems than between populations in different rivers, but exceptions exist (Vrijenhoek et al., 1985). Nonetheless, inferred rates of gene flow (Box 12.4) can be difficult to interpret in riverine systems. Interdrainage dispersal might occur through coastal flooding, anastomosis of meandering streams in coastal deltas, connections between estuarine regions, or through headwater or lateral stream captures. Hierarchical analysis of Nm modeled on present hydrologic structure might not accurately reflect the history of fish movements. Different species with unique migratory behaviors might share the same physical history but have different histories of dispersal events. Also, fish with different generation lengths might respond differently to periodic droughts and seasonal floods. It would be interesting in this regard to examine whether desert fish species with parallel distributions and similar life histories also maintain parallel genetical population structures.

The complex hierarchical branching patterns of riverine systems and the potential for isolation between rivers calls for models that can take this complexity into account. Slatkin (1993) developed a robust approach for estimating Nm between all possible pairs of populations, regardless of the underlying model of population structure. Pairwise Nm values can be examined for correlations with various geographical distance metrics. For example, one could test for genetic correlations with a matrix of geographical distances between populations measured along river beds. Alternatively, if headwater stream capture is suspected, one could test for genetic correlations with a matrix of shortest (straight-line) distances between populations. Finally, it might be informative to test for correlations with a connectivity matrix that simply counts the number of hierarchical transitions in stream order among sites. The Mantel test can be used to test the significance of correlations between pairwise gene flow estimates and these alternative models for dispersal (Smouse et al., 1986).

For most rivers of the North American West, groundwater pumping, impoundments, and stream diversion have greatly reduced the connectivity between remnant fish populations. One goal of species preservation plans should be to restore pre-fragmented levels of dispersal and gene flow in anthropogenically fragmented populations. Fish ladders that circumvent dams and manual translocations are examples of methods that can accomplish this goal, but rates of transport ideally should be based on knowledge of historical patterns of gene flow. Otherwise, local divergence might be swamped by novel genes.

Management decisions regarding translocation and mixing of desert fish populations should be based on sound models of historical connectivity inferred from a knowledge of genetic structure, geography, and geology. For example, when extensive genetic analysis reveals no substantive divergence between populations, and geographic evidence suggests recent population fragmentation, interlocality translocations might be warranted as a means of supplementing local population sizes and avoiding inbreeding. Conversely, when populations are strongly divergent genetically, and geographically long separated, it is best to keep native fish populations within their respective basins. In all cases, the genetic inferences ideally should be based on multiple genes, because it is known from empirical experience that variances in estimates of H_S and G_{ST} can be high across loci (Nei, 1978). Differential selection and sex-biased patterns of gene flow are examples of evolutionary processes that can produce divergent results in independent genetic systems of a single species (Avise, 1989, 1994). Thus, multiple independent loci are needed to test for a consensus regarding historical processes that may have impinged on natural populations (Karl and Avise, 1992), and to produce unbiased and comprehensive estimates of population genetic parameters.

GENETIC DIVERSITY AND FITNESS: CASE STUDIES

The Razorback Sucker (*Xyrauchen texanus*)

Prior to the construction of dams, razorback suckers were widespread and abundant throughout the Colorado River basin. Today, the largest remnant populations exist in the Upper Green River (CO–UT) and Lake Mohave (AZ–NV), and small populations remain in the Gila River (AZ). Following construction of the major dams on the Colorado River system, reproductive recruitment in the razorback sucker has been limited or nonexistent (Marsh and Minckley, 1989). Natural populations are composed almost entirely of adults that probably were born prior to 1954. The razorback sucker was federally listed as endangered in 1991 (U.S. Fish and Wildlife Service, 1991).

Analysis of mtDNA variation in the Lake Mohave population of razorback suckers identified numerous haplotypes distinguishable by restriction fragment length polymorphisms (Dowling and Minckley, 1993). Haplotype diversity ($\hat{h}$ of Nei and Tajima, 1981) is high when mtDNA variants are evenly represented, and low when only a few haplotypes predominate. In the present case, a sample of 48 razorback suckers revealed 32 different haplotypes and an extraordinarily high haplotype diversity ($\hat{h} = 0.98$). Such high diversity in the Lake Mohave razorback suckers suggested that these fish are "direct descendants of an exceedingly large, diverse, panmictic population inhabiting the lower Colorado basin in pre-development times" (Dowling and Minckley, 1993).

Razorback suckers from Lake Mohave were used as broodstock for a restoration program run by the Dexter National Fish Hatchery. The number of founders used to establish the DNFHTC population was unknown, but surveys of mtDNA diversity in the current stock provided an excellent example of the decline of diversity in a hatchery setting (Dowling and Minckley, 1993). Surveys of the year-classes for 1987 and 1989 revealed 6–7 haplotypes, far fewer than the 32 haplotypes found in Lake Mohave (nevertheless, because mtDNA haplotypic frequencies were relatively even, haplotype diversity remained high [$\hat{h} \approx 0.90$]). However, the 1990 year-class (which was produced from only 17 females) retained only four mtDNA haplotypes, and $\hat{h}$ had declined to 0.71. Dowling and Minckley (1993) suggested that the 1990 offspring were inbred, and that low viability of some families reduced N_e, concomitantly reducing gene diversity. An allozyme analysis of this stock is warranted and apparently under contract (H. Williamson, personal communication).

Mass stripping and spawning should be discontinued in hatchery programs aimed at maintaining genetic diversity in endangered species. Paired matings and separate cultures of resulting families should be used to reduce variance in family size and thereby to maximize N_e. Furthermore, wild fish from the Lake Mohave population should probably be introduced into the DNFHTC stock to reverse the inbreeding trend. It is ironic that wild fish of an endangered species are needed to rescue a hatchery population intended for restocking.

The risks associated with captive breeding programs are evident from this example. In most cases, *in situ* breeding would be a preferable alternative, but unfortunately, this approach is not successful with razorback suckers. Although natural spawning does occur and hatchlings are abundant, they quickly disappear, apparently as a consequence of predation by non-native fish. Thus, hatchlings must be reared in predator-free isolation facilities and protected sites. Dowling and Minckley (1993) favor predator-free rearing at protected sites, and strongly recommend against reintroduction of the genetically depauperate DNFHTC stock to Lake Mohave. Whether or not this advice is followed, sound genetic designs and continuous monitoring are needed for programs like this. An important message is that captive propagation and stock supplementation of endangered species cannot always be accomplished with the traditional methods of mass strip-spawning used by sport and commercial fish hatcheries.

The Sonoran Topminnow (*Poeciliopsis occidentalis*)

This livebearing poeciliid was once distributed throughout the lower Colorado River drainage of Arizona and New Mexico. As recently as the 1940s, it was one of the most abundant fish species in the Gila River basin of southern Arizona (Minckley and Deacon, 1968). Its decline over the past 50 years is due primarily to loss of habitat, and introduction of *Gambusia*, which are voracious predators on newborn *P. occidentalis* (Meffe, 1985; Meffe et al., 1983). The subspecies *P.o. occidentalis* presently persists in nine natural localities— all natural springs, seeps, or streams in the Gila River drainage of southern Arizona. *P.o. occidentalis* was placed on the Federal Endangered Species List in 1973 (Deacon and Bunnell, 1970); however, the species is still abundant in Sonora, Mexico.

With the goal of restoring *P.o. occidentalis* throughout its former range in Arizona, the U.S. Fish and Wildlife Service established a stock at the Dexter National Fish Hatchery. The DNFHTC stock derived from Monkey Spring (MS), a thermally stable, spring-fed pool in southern Arizona. The 1984 Topminnow Recovery Plan called for introductions of this stock to 191 sites within the historical range of this species (Simons et al., 1989). Downlisting of the species was planned when 20 stocked populations had survived for at least three years, and delisting was planned when at least 50% of the natural and reclaimed habitats were free of mosquitofish. The downlisting criteria were met by 1985, but only 23 of these populations survived at least five years to 1987. In 1990, a severe drought eliminated all but the nine natural populations.

Even the best of genes will not save these fish when habitats dry completely. Nevertheless, based on genetic criteria, Vrijenhoek et al. (1985) suggested that the MS stock was not the best choice for this restoration effort. The source population derived from a thermally stable spring, and it lacked variation at 25 protein-determining gene loci. A genetically variable population from a thermally fluctuating stream, Sharp Spring (SS), was recommended instead. The SS population is more likely to adapt to novel restored habitats than the homozygous MS stock.

Experimental studies subsequently revealed that fitness of the wild MS population indeed was compromised (Quattro and Vrijenhoek, 1989). Under controlled laboratory conditions, fitness-related traits of the homozygous MS fish ($H = 0.000$) were compared with those of genetically variable SS fish ($H = 0.037$) and TS (Tule Spring) fish ($H = 0.015$). Laboratory-born progeny from 10 wild females of each stock were tested for survival and growth rate during a 12-week grow-out period, after which the fish were sacrificed and measured. Fecundity was determined and adjusted with regression techniques to account for differences in fish size. Finally, the fish were examined for fluctuating asymmetry in seven bilateral characters. Increased fluctuating asymmetry often is interpreted as a sign of diminished developmental stability (Van Valen, 1962).

The homozygous MS fish performed poorly for each fitness-related trait. Of the three stocks, the MS fish exhibited the highest mortality, slowest growth rate, poorest fecundity, and weakest developmental stability. Their poor performance under benign laboratory conditions suggested inbreeding depression. In contrast, the genetically most variable SS stock exhibited the highest growth rate, fecundity, and developmental stability. It did not differ from the TS stock with respect to survival. The TS stock had intermediate heterozygosity and it also was intermediate with respect to growth rate, fecundity, and developmental stability.

The fitness differences among the MS, SS, and TS stocks may have been coincidentally correlated with multilocus heterozygosity levels. Allozyme heterozygosity may reflect little more than the history of population bottlenecks at each locality, and the differences in fitness observed in the laboratory may independently reflect the history of dealing with natural environmental fluctuations versus stabilities at these localities. Constantz (1979) argued that the MS fish have life-history characteristics (i.e., low fecundity) expected for fish living in a thermally stable habitat with few predators. If true, the data of Quattro et al. (1991) do not necessarily reflect fitness differences in nature. Nevertheless, the inability of MS fish to perform well in a novel laboratory environment designed to favor growth and reproduction in *Poeciliopsis* bodes poorly for the survival of these fish in novel reclaimed habitats. Apparently, the SS stock also performs better than the MS stock under culture conditions at the DNFHTC (B.L. Jensen, personal communication). Recently, the U.S. Fish and Wildlife Service has adopted aspects of our proposal to use the more diverse, robust, and fecund fish from Sharp Spring for their *Gila* Topminnow Recovery Plan (Abarca et al., 1994).

Relative Fitness in *Poeciliopsis monacha*

Recently, a geneticist argued in a conservation journal that "loss of diversity should not be a cause for concern because the vast majority of genetic polymorphisms are selectively neutral" (Hughes, 1991). Although this point is highly debatable (Vrijenhoek and Leberg, 1991), molecular polymorphisms may provide information about correlated variability at other loci that do affect fitness in small endangered populations (Vrijenhoek, 1994a). A long-term field study of the Mexican poeciliid, *Poeciliopsis monacha*, provides considerable insight regarding heterozygosity-fitness correlations in small populations. Fitness of sexually reproducing *P. monacha* can be judged relative to that of the coexisting clonal biotype *P. 2 monacha-lucida* (abbreviated *MML*). For comparative purposes, the clones are treated as genotypically uniform controls. The population dynamics and genetics of this fish complex have been monitored in a small headwater tributary, the Arroyo de los Platanos and neighboring streams, since 1975 (Vrijenhoek, 1989).

An Extinction–Recolonization Event and Fitness Lost. The Platanos descends a steep montane gradient with many little waterfalls and barriers to upstream dispersal (Figure 12.4). A natural seep in the midportion of this stream (Log Pool) supplies water even in years of extreme drought. Other populations occur in the mainstream of the Arroyo de Jag-

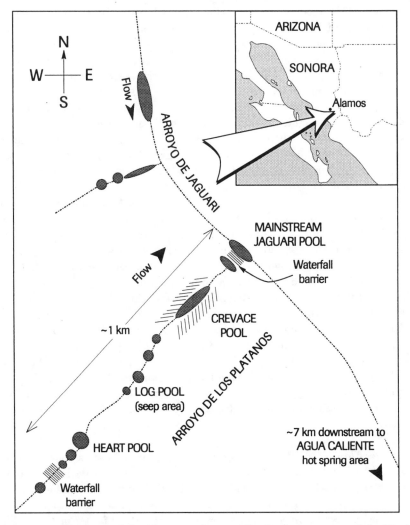

Figure 12.4. The Arroyo de los Platanos and surrounding headwater streams of the Río del Fuerte. Fish have never been found upstream from the steep waterfall above Heart Pool. The Log Pool area is a permanent seep. Other pools illustrated here have been temporary during the past 23 years. Modified from Vrijenhoek et al. (1992).

uari, near the mouth of the Platanos. In 1975, *P. monacha* constituted 76% of the fish in upper Platanos (Heart Pool); the rest were *MML* (Figure 12.5). The Heart Pool population of *P. monacha* was polymorphic at four enzyme-determining gene loci, and mean gene diversity across these four loci was high. In 1976, a severe drought eliminated all fish above the Log Pool seep, including the Heart Pool population. By the Spring of 1978, *P. monacha* and *MML* recolonized Heart Pool, probably from the Log Pool region downstream. In 1978, the *P. monacha* founder population in Heart Pool had essentially no genetic diversity, and it was significantly outnumbered by *MML*, which now constituted 95% of the fish. By 1983, subsequent gene flow had not replenished genetic diversity in *P. monacha*, and the sexual form still was rare relative to the clones.

Apparently, the loss of heterozygosity in *P. monacha* during founder events that accompanied upstream colonization also resulted in diminished competitive ability with respect to *MML*. No similar shifts in the relative abundance of *P. monacha* and *MML* occurred in the mainstream of the Arroyo de Jaguari, where gene diversity in *P. monacha* remained relatively high and stable (Figure 12.5). The following independent lines of evidence also exist for inbreeding depression in the *P. monacha* founder population: (1) Developmental instability, as assessed by fluctuating asymmetry, was significantly greater in the founder population than in *MML* (Vrijenhoek and Lerman, 1982); (2) Survival under hypoxic stress was severely compromised in the founder population of *P. monacha*, relative to *MML* (Box 12.6); and (3) The homozygous founder population of *P. monacha* had significantly greater parasite loads than coexisting clones (Lively et al., 1990). These results contrasted markedly with genetically variable *P. monacha* populations, which typically had: (1) a level of developmental instability that did not differ markedly from that of *MML*; (2) a similar ability to survive hypoxic stress as *MML*; and (3) lower parasite loads than the predominant *MML* clone at a given locality. Thus, the homozygous founder population of *P. monacha* in Heart Pool exhibited numerous signs of inbreeding depression and poor health that debilitated it relative to coexisting clones.

Gene Flow and Fitness Regained.   In March 1983, a local rancher planned to build a dam across the mouth of the Arroyo de los Platanos. Later that summer, the rancher abandoned the dam project, and fortunately for the fish, the dam remains unbuilt. A few years earlier, an impoundment was constructed on an adjacent tributary of the Río Fuerte, in the neighboring village of Guiricoba (SON), and stocked with African *Tilapia* provided by the state of Sonora. Native poeciliids below the dam subsequently were essentially eliminated.

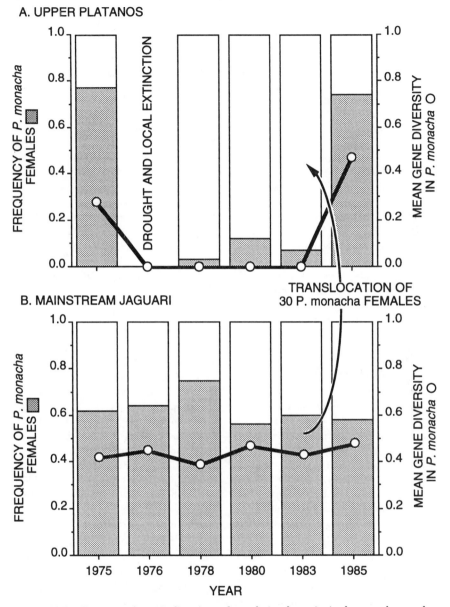

Figure 12.5. Ten years of genetic diversity and population dynamics in the *monacha* complex of the Arroyo de los Platanos and the Arroyo de Jaguari. Histogram bars represent the frequency of *P. monacha* (gray) and *MML* (white) in each year. The black line represents the mean heterozygosity averaged across four polymorphic loci (*Ldh-C*, *Idh-2*, *Pgd*, and *Ck-A*) in *Poeciliopsis monacha*. Modified from Vrijenhoek (1989).

Box 12.6. Heterozygosity and Survival During Stress

Desert fish that live in residual pools and stream segments may experience extreme spatial and temporal fluctuations in water temperature and levels of dissolved oxygen. For example, *Poeciliopsis monacha* lives in streams that are interspersed with geothermal springs and cold-water seeps. For this species, survival of extreme environmental conditions is associated with genetic diversity marked by allozymes (Vrijenhoek et al., 1992). In contrast to a homozygous founder population in an upstream portion of the Arroyo de los Platanos, genetically variable *P. monacha* from the Arroyo de Jaguari survived hypoxic stress about as well as coexisting *MML* clones, which are highly heterozygous as a consequence of their hybrid nature (Vrijenhoek, 1994a). The Jaguari *P. monacha*'s were polymorphic at four enzyme-determining gene loci, and at each locus, homozygotes for the common allele ($+/+$) exhibited higher survival during hypoxic stress than the alternative homozygotes for a local variant allele (v/v). Application of heat stress to a new sample of *P. monacha* produced results for all four loci that were nearly identical to those obtained earlier under hypoxic stress. The parallel results were not surprising, however, as hot water and low dissolved-oxygen levels are coupled in nature, and the fish must respond to both stresses simultaneously.

Independent cold stress trials with Jaguari *P. monacha* produced strikingly opposite results. Now, homozygotes for the local variant allele (v/v) at each of the four loci survived best, and the $+/+$ homozygotes exhibited poor survival. Heterozygotes generally exhibited intermediate survival under stress (e.g., the *Ldh-C* locus in Figure 12.6), but most important, heterozygotes never had the worst survival when exposed to these opposing forms of stress. In this spatially and temporally heterogeneous environment, such heterozygotes would have the highest geometrical mean fitness, and thereby produce a balanced polymorphism.

Significant correlations in allelic frequencies across these polymorphic loci suggested that the allozymes marked larger blocks of genes that may have contained the true effectors of survival (Vrijenhoek, 1994a). Alternatively, one or more of the allozymes may be directly involved in survival, but that remains to be tested. Although the specific mechanisms underlying these genotype–environment interactions remain unknown, the significance of allozyme polymorphisms (and the hidden variation with which it may be correlated) should not be discounted in small, highly structured populations such as these.

In any event, a similar fate was anticipated for the Platanos and Jaguari fish. Thus, a small experiment involving these fish was initiated in March 1983 (Vrijenhoek, 1989). Thirty genetically variable *P. monacha* females were collected from the mainstream Jaguari near the mouth of the Platanos and moved one km upstream to Heart Pool. The 30 translocated individuals probably constituted no more than 10% of the *P. monacha* females in Heart

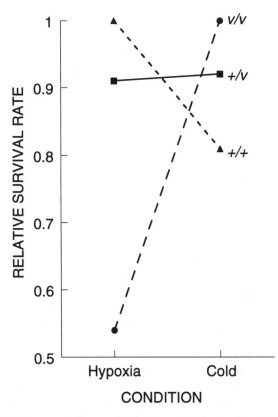

Figure 12.6. Relative survival rates of *Ldh-C* genotypes in *Poeciliopsis monacha* during hypoxic and cold stress. Relative survival is determined as the absolute survival rate for that genotype divided by the maximal survival rate under a particular type of stress (from results reported in Vrijenhoek et al., 1992).

Pool. The subsequent recovery of *P. monacha* in the upper Platanos was dramatic. By 1985, gene diversity rose to a level comparable to that in the mainstream Jaguari population (Figure 12.5). *P. monacha* also recovered its competitive ability and constituted nearly 80% of the fish in Heart Pool. Its relative abundance and genetic diversity have remained stable in four subsequent yearly collections (1987–1991). Parasite loads also decreased in the reconstituted Heart Pool population of *P. monacha*, to levels below that of coexisting clones (Lively et al., 1990).

Thus, a single transplant of 30 fish into a population estimated to contain about 300 adult *P. monacha* females resulted in the restoration of genetic variability and relative fitness as manifested by competitive ability and disease susceptibility. An *Nm* of 30 would be large if repeated in each generation, and would quickly swamp the local gene pool. However, spread across the intervening 10 years (roughly 30 generations for these fish), this level of introduction amounted to only $Nm \approx 1$ per generation. In principle, such rates of gene flow will produce qualitative similarity in allelic composition between source and recipient populations, but will not be sufficient to overwhelm local selective pressures (Allendorf, 1983).

SUMMARY

1.   Management of desert fish populations should take into account the causes of present-day population fragmentation. When natural postpluvial drying is the cause, as in the case of Death Valley pupfishes, the populations should remain isolated in their respective refugia. Mixing of populations to elevate heterozygosity and avoid possible inbreeding depression is not warranted because the isolated state of these fish, inbred or not, is *natural*. Rather, conservation efforts for these fish should be directed at protecting the water supplies that feed these habitats, and in removing exotic predators and competitors. Geneticists can help in devising breeding plans and monitoring the genetic diversity and integrity of captive populations that are being raised to supplement native stocks of these fish.

2.   In anthropogenically fragmented water systems such as those occupied by many riverine species, analyses of genetic population structure can provide information about patterns of gene flow and historical connectivity between populations. This information can be used to restore natural patterns of dispersal, to design programs aimed at maintaining heterozygosity within formerly connected subpopulations, and simultaneously to preserve the differences that exist among evolutionarily divergent populations.

3.   Evidence is reviewed for fitness benefits associated with genetic diversity in desert fish. For two well-studied fish of the genus *Poeciliopsis*, rapid losses of heterozygosity in small, isolated populations are associated with a decline in fitness that is manifested as poor competitive ability, growth rate, developmental stability, and resistance to parasites. In these cases, fitness associations with heterozygosity are likely to be a consequence of small population size itself, because genetic drift and inbreeding probably lead to temporary associations between allozyme markers and other genes that can have manifold effects on fitness. Such associations are likely to persist or recur in populations that are subject to frequent population bottlenecks and founder events.

4.   Most North American desert fish exist as small and fragmented subpopulations. Remnant genetic variation that exists within and among these

populations should be considered precious. It is an integral part of the overall biological diversity that conservationists must strive to preserve.

REFERENCES

ABARCA, F.J., B.E. BAGLEY, D.A. HENDRICKSON and J.R. SIMMS. 1994. *January Draft:* Gila Topminnow *(*Poeciliopsis occidentalis occidentalis*) Revised Recovery Plan.* U.S. Fish and Wildlife Service, Region 2, Albuquerque, NM.

ALLENDORF, F.W. 1983. Gene flow, and genetic differentiation among populations. In *Genetics and Conservation: A Reference for Managing Wild Animal and Plant Populations,* eds. C.M. Schonewald-Cox, S.M. Chambers, F. McBryde and L. Thomas, pp. 241–261. Benjamin/ Cummings, Menlo Park, CA.

ALLENDORF, F.W. and R.F. LEARY. 1986. Heterozygosity and fitness in natural populations of animals. In *Conservation Biology: The Science of Scarcity and Diversity,* ed. M.E. Soulé, pp. 57–76. Sinauer, Sunderland, MA.

ALLENDORF, F.W. and R.F. LEARY. 1988. Conservation and distribution of genetic variation in a polytypic species, the cutthroat trout. *Conserv. Biol.* 2:170–184.

AVISE, J.C. 1989. Gene trees and organismal histories: A phylogenetic approach to population biology. *Evolution* 43:1192–1208.

AVISE, J.C. 1994. *Molecular Markers, Natural History and Evolution.* Chapman & Hall, New York.

BARTLEY, D., M. BAGLEY, G. GALL and B. BENTLY. 1992. Use of linkage disequilibrium data to estimate effective size of hatchery and natural fish populations. *Conserv. Biol.* 6:365–375.

COLE, G.A. 1981. Habitats of North American Desert Fishes. In *Fishes of the North American Deserts,* eds. R.J. Naiman and D.L. Soltz, pp. 477–493. Wiley, New York.

CONSTANTZ, G.D. 1979. Life history patterns of a livebearing fish in contrasting environments. *Oecologia* 40:189–201.

CROW, J.F. and C. DENNISTON. 1988. Inbreeding and variance effective population numbers. *Evolution* 42:482–495.

DEACON, J.E. and S. BUNNELL. 1970. Man and pupfish, a process of destruction. *Cry Calif.* 5:14–21.

DEACON, J.E. and W.L. MINCKLEY. 1974. Desert fishes. In *Desert Biology,* ed. G.W. Brown, Jr., pp. 385–488. Academic Press, New York.

DOWLING, T.E. and W.L. MINCKLEY. 1993. *Genetic Diversity of Razorback Sucker as Determined by Restriction Endonuclease Analysis of Mitochondrial DNA.* Final Report to the U.S. Bureau of Reclamation, Award #0-FC-40-09530-004.

DUDLEY, W.W., Jr. and J.D. LARSON. 1976. Effect of irrigation pumping on desert pupfish habitats in Ash Meadows, Nye County, Nevada. *U.S. Geol. Surv. Prof. Pap.* 927:1–52.

ECHELLE, A.A. 1991. Conservation genetics and genic diversity in freshwater fishes of North America. In *Battle Against Extinction: Native Fish Management in the American West,* eds. W.L. Minckley and J.E. Deacon, pp. 141–153. University of Arizona Press, Tucson.

ECHELLE, A.A. and P.J. CONNER. 1989. Rapid, geographically extensive genetic introgression after secondary contact between two pupfish species (*Cyprinodon*, Cyprinodontidae). *Evolution* 43:717–727.

ECHELLE, A.A. and T.E. DOWLING. 1992. Mitochondrial DNA variation of the Death Valley pupfishes (*Cyprinodon*, Cyprinodontidae). *Evolution* 46:193–206.

ECHELLE, A.A. and A.F. ECHELLE. 1993. Allozyme perspective on mitochondrial DNA variation and evolution of the Death Valley pupfishes (Cyprinodontidae: *Cyprinodon*). *Copeia* 1993:275–287.

ECHELLE, A.A., D.M. WILDROCK and A.F. ECHELLE. 1989. Allozyme studies of genetic variation in poeciliid fishes. In *The Ecology and Evolution of Poeciliid Fishes*, eds. G.A. Meffe and F.F. Snelson, Jr., pp. 217–233. Prentice-Hall, Englewood Cliffs, NJ.

EWENS, W.J. 1982. On the concept of effective population size. *Theoret. Pop. Biol.* 21:373–378.

FALCONER, D.S. 1960. *Introduction to Quantitative Genetics*. Ronald Press, New York.

GALL, G.A.E. 1987. Inbreeding. In *Population Genetics and Fishery Management*, eds. N. Ryman and F. Utter, pp. 47-87. University of Washington Press, Seattle.

GLIDDON, C. and J. GOUDET. 1994. The genetic structure of metapopulations and conservation biology. In *Conservation Genetics*, eds. V. Loeschcke, J. Tomiuk and S. K. Jain, pp. 107–114. Birkhäuser Verlag, Basel, Switzerland.

HALDANE, J.B.S. 1939. The equilibrium between mutation and random extinction. *Ann. Eugen.* 9:400–405.

HANSKI, I. and M. GILPIN. 1991. Metapopulation dynamics: brief history and conceptual domain. *Biol. J. Linn. Soc.* 42:3–16.

HARRIS, P.D. 1986. Species of *Gyrodactylus* von Nordmann 1932 (Monogenea Gyrodactylidae) from poeciliid fishes, with a description of *G. turnbulli* sp. nov. from the guppy, *Poecilia reticulata* Peters. *J. Nat. Hist.* 20:183–191.

HUGHES, A.L. 1991. MHC polymorphism and the design of captive breeding programs. *Conserv. Biol.* 5:249–251.

JOHNSON, J.E. and B.L. JENSEN. 1991. Hatcheries for Endangered Freshwater Fishes. In *Battle Against Extinction: Native Fish Management in the American West*, eds. W.L. Minckley and J.E. Deacon, pp. 199-217. University of Arizona Press, Tucson.

KARL, S.A. and J.C. AVISE. 1992. Balancing selection at allozyme loci in oysters: implications from nuclear RFLPs. *Science* 256:100–102.

KARL, S.A., B.W. BOWEN and J.C. AVISE. 1992. Global population genetic structure and male-mediated gene flow in the green turtle (*Chelonia mydas*): RFLP analysis of anonymous nuclear loci. *Genetics* 131:163–173.

KIMURA, M. and W.H. WEISS. 1964. The stepping stone model of genetic structure and the decrease of genetic correlation with distance. *Genetics* 49:561–576.

LANDE, R. 1988. Genetics and demography in biological conservation. *Science* 241:1455–1460.

LA RIVERS, I. 1962. *Fishes and Fisheries of Nevada*. Nevada State Printing Office, Carson City.

LEBERG, P. and R.C. VRIJENHOEK. 1994. Variation among desert topminnows in their susceptibility to attack by exotic parasites. *Conserv. Biol.* 8:419–424.

LEDIG, F.T., R.P. GURIES and B.A. BONEFIELD. 1983. The relation of growth to heterozygosity in pitch pine. *Evolution* 37:1227–1238.

LEVINS, R. 1970. Extinction. In *Some Mathematical Questions in Biology*, ed. M. Gerstenhaber, pp. 75–108. American Mathematical Society, Providence, RI.

LIVELY, C.M., C. CRADDOCK and R.C. VRIJENHOEK. 1990. The Red Queen hypothesis supported by parasitism in sexual and clonal fish. *Nature* 344:864–866.

LOUDENSLAGER, E.J., J.N. RINNE, G.A.E. GALL and R.E. DAVID. 1986. Biochemical genetic studies of native Arizona and New Mexico trout. *Southw. Nat.* 31:221–234.

MARSH, P.C. and W.L. MINCKLEY. 1989. Observations on recruitment and ecology of razorback suckers. *Env. Biol. Fish.* 21:59–67.

MCCAULEY, D.E. 1991. Genetic consequences of local population extinction and recolonization. *Trends Ecol. Evol.* 6:5–8.

MEFFE, G.K. 1984. Effects of abiotic disturbance on coexistence of predator–prey fish species. *Ecology* 65:1525–1534.

MEFFE, G.K. 1985. Predation and species replacement in American southwestern fishes: a case study. *Southw. Nat.* 30:173–187.

MEFFE, G.K. 1992. Techno-arrogance and halfway technologies: Salmon hatcheries on the Pacific Coast of North America. *Conserv. Biol.* 3:350–354.

MEFFE, G.K., D.A. HENDRICKSON, W.L. MINCKLEY and J.N. RINNE. 1983. Factors resulting in decline of the endangered Sonoran topminnow *Poeciliopsis occidentalis* (Atheriniformes: Poeciliidae) in the United States. *Biol. Conserv.* 25:135–159.

MEFFE, G.K. and R.C. VRIJENHOEK. 1988. Conservation genetics in the management of desert fishes. *Conserv. Biol.* 2:157–169.

MILLER, R.R. 1958. Origin and affinities of the freshwater fish fauna of western North America. *Zoogeography* 51:187–222.

MILLER, R.R. 1960. Four new species of viviparous fishes, genus *Poeciliopsis* from northwestern Mexico. *Occ. Pap. Mus. Zool. Univ. Mich.* 433:1–9.

MINCKLEY, W.L. and J.E. DEACON. 1968. Southwestern fishes and the enigma of 'endangered species'. *Science* 159:1424–1432.

MINCKLEY, W.L. and M.E. DOUGLAS. 1991. Discovery and extinction of western fishes: a blink in the eye of geological time. In *Battle Against Extinction: Native Fish Management in the American West*, eds. W.L. Minckley and J.E. Deacon, pp. 7–18. University of Arizona Press, Tucson.

MINCKLEY, W.R., D.A. HENDRICKSON and C.E. BOND. 1986. Geography of western North American freshwater fishes: Description and relationships to intercontinental tectonism. In *Zoogeography of North American Freshwater Fishes*, eds. C.H. Honeycutt and E.O. Wiley, pp. 519–613. Wiley, New York.

MITTON, J.B. and M.C. GRANT. 1984. Associations among protein heterozygosity, growth rate, and developmental homeostasis. *Annu. Rev. Ecol. Syst.* 15:479–499.

NAIMAN, R.J. 1981. An ecosystem overview: Desert fishes and their habitats. In *Fishes of the North American Deserts*, eds. R.J. Naiman and D.L. Soltz, pp. 493–531. Wiley, New York.

NEI, M. 1977. F-statistics and analysis of gene diversity in subdivided populations. *Ann. Hum. Genet.* 41:225–233.

NEI, M. 1978. Estimation of average heterozygosity and genetic distance from a small number of individuals. *Genetics* 89:583–590.

NEI, M. and F. TAJIMA. 1981. DNA polymorphism detectable by restriction endonucleases. *Genetics* 97:145–163.

OHTA, T. 1981. Associative overdominance caused by linked detrimental mutations. *Genet. Res.* 18:277–286.

PISTER, E.P. 1991. The Desert Fishes Council: catalyst for change. In *Battle Against Extinction: Native Fish Management in the American West*, eds. W.L. Minckley and J.E. Deacon, pp. 55–68. University of Arizona Press, Tucson.

QUATTRO, J.M., J.C. AVISE and R.C. VRIJENHOEK. 1991. Molecular evidence for multiple origins of hybridogenetic fish clones (Poeciliidae: *Poeciliopsis*). *Genetics* 127:391–398.

QUATTRO, J.M. and R.C. VRIJENHOEK. 1989. Fitness differences among remnant populations of the Sonoran topminnow, *Poeciliopsis occidentalis*. *Science* 245:976–978.

RALLS, K., J. BALLOU and A. TEMPLETON. 1986a. Estimates of lethal equivalents and the cost of inbreeding in mammals. *Conserv. Biol.* 2:185–193.

RALLS, K., P.H. HARVEY and A.M. LYLES. 1986b. Inbreeding in natural populations of birds and mammals. In *Conservation Biology: The Science of Scarcity and Diversity*, ed. M.E. Soulé, pp. 35–56. Sinauer, Sunderland, MA.

RICHARDSON, W.M. 1976. Fishes known to be in the Colorado River drainage. In *Technical Committee Minutes*, pp. 12–18. Colorado River Wildlife Council, Las Vegas, NV.

ROGERS, W.A. and T.L. WELLBORN, Jr. 1965. Studies of *Gyrodactylus* (Trematoda: Monogenea) with descriptions of five new species from the southeastern U.S. *J. Parasit.* 51:977–982.

ROLSTON, H. 1991. Fishes in the desert: paradox and responsibility. In *Battle Against Extinction: Native Fish Management in the American West*, eds. W.L. Minckley and J.E. Deacon, pp. 93–108. University of Arizona Press, Tucson.

SCHENCK, R.A. and R.C. VRIJENHOEK. 1989. Coexistence among sexual and asexual forms of *Poeciliopsis*: foraging behavior and microhabitat selection. In *Evolution and Ecology of Unisexual Vertebrates*, eds. R. Dawley and J. Bogart, pp. 39–48. New York State Museum, Albany.

SCHULTZ, R.J. 1969. Hybridization, unisexuality and polyploidy in the teleost *Poeciliopsis* (Poeciliidae) and other vertebrates. *Amer. Nat.* 103:605–619.

SCHULTZ, R.J. 1977. Evolution and ecology of unisexual fishes. *Evol. Biol.* 10:277–331.

SIMONS, L.H., D.A. HENDRICKSON and D. PAPOULIAS. 1989. Recovery of the Gila topminnow: A success story? *Conserv. Biol.* 3:11–15.

SLATKIN, M. 1977. Gene flow and genetic drift in a species subject to frequent local extinctions. *Theor. Pop. Biol.* 12:253–262.

SLATKIN, M. 1993. Isolation by distance in equilibrium and non-equilibrium populations. *Evolution* 47:264–279.

SLATKIN, M. and N.H. BARTON. 1989. A comparison of three indirect methods for estimating average levels of gene flow. *Evolution* 43:1349–1368.

SLATKIN, M. and W.P. MADDISON. 1990. Detecting isolation by distance using phylogenies of genes. *Genetics* 126:249–260.

SLATKIN, M. and L. VOELM. 1991. F_{ST} in a hierarchical island model. *Genetics* 127:627–629.

SMOUSE, P.E., J.C. LONG and R.R. SOKAL. 1986. Multiple regression and correlation extensions of the Mantel Test of matrix correspondence. *Syst. Zool.* 35:627–632.

SOKAL, R.R. and D.E. WARTENBERG. 1983. A test of spatial autocorrelation analysis using an isolation-by-distance model. *Genetics* 105:219–237.

SOLTZ, D.L. and M.F. HIRSHFIELD. 1981. Genetic differentiation of pupfishes (genus *Cyprinodon*) in the American Southwest. In *Fishes of the North American Deserts*, eds. R.J. Naiman and D.L. Soltz, pp. 291–333. Wiley, New York.

SOLTZ, D.L. and R.J. NAIMAN. 1978. The Natural History of Native Fishes in the Death Valley System. *Nat. Hist. Mus. L.A. Co. Sci. Ser.* 30:1–76.

TEMPLETON, A. 1986. Coadaptation and outbreeding depression. In *Conservation Biology: The Science of Scarcity and Diversity*, ed. M.E. Soulé, pp. 105–116. Sinauer, Sunderland, MA.

TEMPLETON, A.R. and B. READ. 1994. Inbreeding: one word, several meanings, much confusion. In *Conservation Genetics*, eds. V. Loeschcke, J. Tomiuk and S.K. Jain, pp. 91–105. Birkhäuser Verlag, Basel, Switzerland.

THIBAULT, R.E. 1974. The ecology of unisexual and bisexual fishes of the genus *Poeciliopsis*: a study in niche relationships. PhD dissertation, University of Connecticut, Storrs.

TURELLI, M. and L.R. GINZBURG. 1983. Should individual fitness increase with heterozygosity? *Genetics* 104:191–209.

TURNER, B.J. 1974. Genetic divergence of Death Valley pupfish species: biochemical versus morphological evidence. *Evolution* 28:281–294.

U.S. FISH AND WILDLIFE SERVICE. 1983. *Gila and Yaqui Topminnow Recovery Plan.* U.S. Fish and Wildlife Service, Albuquerque, NM.

U.S. FISH AND WILDLIFE SERVICE. 1991. Endangered and threatened wildlife and plants: the razorback sucker (*Xyrauchen texanus*) determined to be an endangered species. *Fed. Reg.* 56:54957–54967.

VAN VALEN, L. 1962. A study of fluctuating asymmetry. *Evolution* 16:125–142.

VRIJENHOEK, R.C. 1989. Genotypic diversity and coexistence among sexual and clonal forms of *Poeciliopsis*. In *Speciation and Its Consequences*, eds. D. Otte and J. Endler, pp. 386–400. Sinauer, Sunderland, MA.

VRIJENHOEK, R.C. 1994a. Genetic diversity and fitness in small populations. In *Conservation Genetics*, eds. V. Loeschcke, J. Tomiuk and S.K. Jain, pp. 37–53. Birkhäuser Verlag, Basel, Switzerland.

VRIJENHOEK, R.C. 1994b. Unisexual fish: Models for studying ecology and evolution. *Annu. Rev. Ecol. Syst.* 25:71–96.

VRIJENHOEK, R.C., M.E. DOUGLAS and G.K. MEFFE. 1985. Conservation genetics of endangered fish populations in Arizona. *Science* 229:400–402.

VRIJENHOEK, R.C. and P.L. LEBERG. 1991. Let's not throw the baby out with the bathwater: a comment on management for MHC diversity in captive populations. *Conserv. Biol.* 5:252–253.

VRIJENHOEK, R.C. and S. LERMAN. 1982. Heterozygosity and developmental stability under sexual and asexual breeding systems. *Evolution* 36:768–776.

VRIJENHOEK, R.C., E. PFEILER and J. WETHERINGTON. 1992. Balancing selection in a desert stream-dwelling fish, *Poeciliopsis monacha*. *Evolution* 46:1642–1657.

WRIGHT, S. 1931. Evolution in Mendelian populations. *Genetics* 16:97–159.

WRIGHT, S. 1943. Isolation by distance. *Genetics* 28:114–138.

WRIGHT, S. 1951. The genetical structure of populations. *Ann. Eugen.* 15:323–354.

WRIGHT, S. 1978. *Evolution and the Genetics of Populations. Vol. 4: Variability Within and Among Natural Populations.* The University of Chicago Press, Chicago.

ZOUROS, E. and D.W. FOLTZ. 1987. The use of allelic isozyme variation for the study of heterosis. In *Isozymes: Current Topics in Biological and Medical Research*, eds. M.C. Rattazi, J.G. Scandalios and G.S. Whitt, pp. 1–59. Liss, New York.

13

A LANDSCAPE APPROACH TO CONSERVATION GENETICS: CONSERVING EVOLUTIONARY PROCESSES IN THE AFRICAN BOVIDAE

Alan R. Templeton and Nicholas J. Georgiadis

INTRODUCTION

Conservation agencies and efforts are increasingly adopting a landscape (ecosystem or bioreserve) approach. There are many reasons for this change in emphasis from the more traditional endangered species perspective (Grumbine, 1990). The endangered species approach focuses on situations that, by definition, are already critical, have limited chances of success, are constrained by a reduced number of management options, and often depend upon laborious and expensive interventions. Moreover, the implementations of this approach have been quite discriminatory, focusing primarily on higher plants and animals and ignoring the vast majority of the biodiversity on this planet. Finally, any endangered species program is destined for ultimate failure unless it can discover and correct the underlying causes of endangerment, which in turn often have to be addressed at the landscape or ecosystem level. The landscape or bioreserve strategy directly focuses attention on natural processes (and their disruptions) that result in the endangerment of not only rare or charismatic fauna and flora, but also of all species in the landscape and the communities into which they are assembled. For many species, the present situation is not yet at the critical stage, so the initial starting conditions in terms of genetic diversity, population sizes, and so on, are often much better than those associated with endangered species programs.

Given this change in emphasis from endangered species to ecosystems and landscapes, what role will genetics play in landscape management? Conservation genetics, with the exception of the use of genetics as a systematic tool, has for the most part had an intraspecific focus (Templeton, 1991). The landscape strategy focuses more on the processes that have historically operated on the ecosystem, which is reflected in the recent evolutionary histories of the species living within the landscape. Hence, by reconstructing or estimating recent evolutionary histories of several species that inhabit the landscape, it should, in principle, be possible to make inferences about some of the landscape-level processes that have shaped the current assemblage of species and communities. This approach does not require each species to have the same recent evolutionary history, but is predicated upon the assumption that generalities will emerge as more and more case histories are developed. The variation from species to species can then provide significant insights into how species differ when faced by the same or similar environmental conditions and may suggest hypotheses for these differences.

Genetics can play an important role in such studies because historical information is encoded in genes and gene pools, and modern molecular survey techniques are particularly effective at extracting and estimating this evolutionary, historical information (Avise, 1994; Templeton, 1993, 1994a, 1994b). Paradoxically, it is those species that have been least affected by human activities that are likely to be the more valuable ones to include in such genetic surveys. The pattern of genetic variation in these species should still reflect the processes that have been operating over evolutionary time. In contrast, most endangered species are endangered precisely because they have been disrupted by human activities, including those designed to aid in their recovery (e.g., translocations). Hence, the landscape perspective increases the importance of genetic surveys on non-endangered species as a tool for process assessment and management.

In this chapter, this approach will be illustrated by focusing upon a landscape that is rich in biodiversity: the savanna ecosystems in Tanzania, Botswana, and Zimbabwe. These savannas are increasingly surrounded by human habitation and cultivation, and the areas protected in National Parks are becoming increasingly isolated from each other by human development (Borner, 1987; Newmark et al., 1994). To understand the significance of these human-induced alterations, it is critical to understand what processes operated on the savanna species in the past. The impact of these alterations in the future may then be anticipated under various management strategies. More important, it may become possible to manage savanna communities in such a manner that natural processes are better maintained.

The savanna ecosystems of Tanzania, Botswana, and Zimbabwe are characterized by similar, large herbivore communities, which constitute one of the major biotic components of this landscape (Lamprey, 1963; Hansen et al., 1985). Accordingly, the use of genetic surveys to reveal historical information contained within mitochondrial (mt) DNA will be illustrated from analyses conducted on three widespread bovid species that are characteristic of African savannas: buffalo (*Syncerus caffer*), impala (*Aepyceros melampus*) and wildebeest (*Connochaetes taurinus*). The basic approach to extracting this historical information is to overlay geography upon the estimated haplotype evolutionary tree. Such an approach is an example of intraspecific *phylogeography* (Avise et al., 1987). The phylogeographic approach has yielded great insights into the recent evolutionary history of many species (e.g., see Avise, 1994), but most such analyses have not provided a rigorous statistical assessment of the strength of association between the geographic data and the evolutionary history. By using recently developed permutational testing statistics (Templeton, 1993, 1994b; Templeton and Sing, 1993), this historical information can be extracted with a rigorous statistical design and objective criteria for making inferences about historical processes and events. This approach also has the advantage of indicating when samples are inadequate to make an inference, and thereby can guide the researcher's future efforts.

NATURAL HISTORY BACKGROUND

Within a relatively small region in northern Tanzania, sharp disjunctions in geology, relief, and forest habitat separate the Serengeti, Ngorongoro, and Tarangire savannas (Figure 13.1). These protected areas include exposed parts of the ancient African shield in the northwest and the more recent (<5 mya) volcanic Ngorongoro highlands. The Gregory Rift Valley (formed >5 mya) bisects the region. Annual rainfall varies from a minimum of 250 mm at Olduvai (elevation 1,000 m) to 1,500 mm in the eastern Crater Highlands (3,000 m), only 20 km to the east. An equally dramatic variety of edaphic and vegetation types is associated with this geological diversity, from expansive short, medium, and long grass plains, through mixed savannas, to dense riverine and highland forests (McNaughton, 1983). In particular, there is a sharp disjunction between the forests of the Crater Highlands

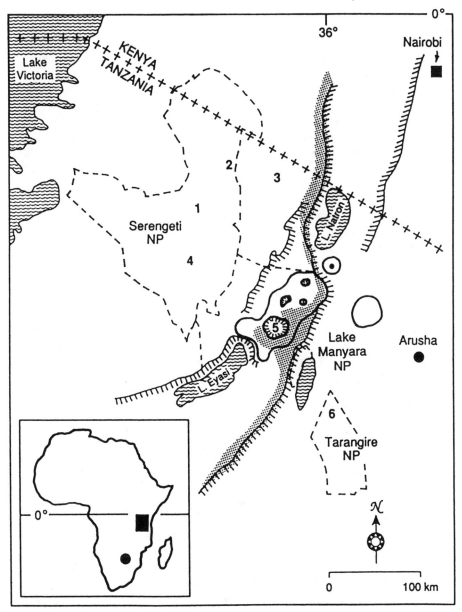

Figure 13.1. Map of the sampling locations in northern Tanzania: (1) Seronera, (2) Lobo, (3) Loliondo, (4) Naabi Hill, (5) Ngorongoro Crater, and (6) Tarangire. Sampling locations in Botswana and Zimbabwe are marked by the filled circle in the inset. Volcanic Highlands are marked by a heavy line, step faults of the Rift Valley by feathered lines, and closed canopy forest and woodland by shading.

and the savanna habitats on either side. The savannas in northeastern Botswana (Chobe) and Zimbabwe (Hwange) are contiguous (Figure 13.1) and encompass much lower geographical, edaphic, and ecological diversity than is true for the Serengeti–Ngorongoro–Tarangire area.

All three species of bovids chosen for this study are sympatric over large areas in eastern and southern Africa and are characteristic of savanna ecosystems. All are polygynous, live in herds, have male-biased dispersal, and depend on frequent access to water (Estes, 1991). Of the three species, the buffalo is the most widespread, occurring primarily in savannas with high grass biomass, but also being found in dense bush and montane forests. Buffalo form large aggregations that stay within persistent home ranges. Females tend to remain in their natal herd, and males form a dominance hierarchy. Impala are found in wooded savannas with relatively open patches where they browse, or graze, according to seasonal forage availability. Females aggregate into groups year-round, but group membership rarely remains constant. Collections of local female groups form "clans" with stable, overlapping home ranges but with 2% per annum dispersal between clans (Murray, 1982). Males disperse from their natal range to breed, fighting for exclusive but temporary ownership of territories, where most mating occurs. Wildebeest are strict grazers, preferring short green grass in open habitats. They actively avoid dense cover, where they are vulnerable to predators. Social organization is based on unstable female aggregations. Where water and nutritional requirements are met year round, local "resident" populations may persist over many generations. Alternatively, massive aggregations migrate vast distances, following seasonal rainfall patterns.

All three species are physically capable of moving within a few days a distance equivalent to the largest linear transect between the sampling locations in Tanzania (200 km). For all species, lifetime cumulative travel distances of individuals are far greater than the distance separating Serengeti from the Chobe/Hwange savanna (> 2,000 km). Thus, the potential for dispersal over ecological time is large across the entire species range. However, it is possible that the sharp disjunctions in relief or habitat that characterize the savannas in northern Tanzania will have an impact on genetic exchange. In this context, the important contrasts among these species are as follows: Buffalo are strongly philopatric, yet they can occupy and disperse through virtually all terrestrial habitat types except deserts; wildebeest are less philopatric and can be exceptionally mobile, yet are restricted to open habitats; and impala are restricted to mixed savannas and display much local dispersal among groups and more limited dispersal among clans, which as an aggregate tend to be relatively sedentary.

SAMPLING DESIGN
AND STATISTICAL ANALYSES

Samples and Genetic Assays

The particular sampling locations (Figure 13.1) were chosen because they allow us to examine many different factors that could affect genetic exchange. To reveal patterns of gene flow within a continuous savanna habitat, each species was sampled at two locations within the Serengeti (buffalo and impala from Seronera and Lobo, migratory wildebeest from Naabi Hill, and resident wildebeest from Loliondo; Figure 13.1). To test whether geographical relief affects genetic exchange within a savanna habitat, wildebeest and buffalo from the Ngorongoro Crater were compared to those from the Serengeti (impala are not found within the crater, presumably due to lack of appropriate habitat). The Serengeti and Ngorongoro Crater are separated by marked geographical relief, including steep (but not unscaleable) crater walls, but otherwise show no marked habitat discontinuities. To test for the effects of habitat discontinuities on genetic exchange, populations of all three species were sampled from the Tarangire savannas, which are separated from the Serengeti–Ngorongoro systems by about 50 km of forest habitat associated with the eastern Ngorongoro highlands and the dense woodland of the Rift Valley. Finally, to assess genetic exchange across the continent, samples were obtained in southern Africa (buffalo and impala were sampled within and near Chobe National Park in northeastern Botswana, and wildebeest from Hwange National Park in western Zimbabwe).

The skin-biopsy dart method (Karesh et al., 1989) was used to collect 17 samples from each species in all locations except two: The wildebeest samples from Loliondo were cut from muscle tissue of culled animals, and the buffalo samples from Botswana were cut from dried skins. After total genomic DNA extraction (Maniatis et al., 1989), DNA samples were transferred to the United States (under permit from the U.S. Dept. of Agriculture). There, two mtDNA segments (2.0 kb of the displacement loop and 2.4 kb between the Glu- and Leu- tRNAs in the ORF5/6 regions) were separately amplified in each sample by the polymerase chain reaction (PCR), using flanking primers designed from mitochondrial sequences conserved across mammals and amphibians. Variation was detected by cutting the amplified DNA with a battery of restriction enzymes (details are given in Georgiadis et al., 1994).

Cladogram Estimation and Confidence Sets

For each species, an unrooted cladogram, or haplotype network, was estimated from the mtDNA haplotype data using the algorithm of Templeton et al. (1992). This algorithm is specific for intraspecific gene trees and involves an evaluation of the limits of parsimony. In evaluating these limits, a DNA evolution model was assumed with an extreme bias in favor of transitions over transversions ($b = 1$ in the parameterization of Templeton et al., 1992), as is standard for mtDNA. Using the resulting estimates of when parsimony can be assumed and when (and by how many mutational steps) it is likely to be violated, a 95% plausible set of cladograms was constructed; that is, all linkages among haplotypes, starting with the most probable, are included in this set of plausible cladograms until the cumulative probability of all shown linkages exceeds 0.95. In some cases, the set of probable linkages forms a loop, thereby indicating ambiguity in the correct evolutionary topology. Probabilities were assigned to the various ways of breaking such loops using the techniques given in Crandall et al. (1994). Once again, an extreme transition bias is assumed in calculating these probabilities. If only one of the alternatives for breaking a loop had a probability greater than or equal to 0.95, the loop was regarded as resolved; otherwise, all possibilities were retained in subsequent analyses.

Statistical Analysis of Geographical Data

The estimated 95% plausible sets of haplotype networks were converted into a nested statistical design using the algorithm of Templeton et al. (1987) and, where necessary, using the nesting rules given in Templeton and Sing (1993) that deal with cladogram ambiguity. These procedures convert the haplotype tree for each species into a series of nested categories, with haplotypes nested into 1-step clades, 1-step clades nested into 2-step clades, and so on (Templeton et al., 1987).

Geography is overlaid upon this nested set of evolutionary hierarchies by using two types of distance measures: the clade distance and the nested-clade distance (Figure 13.2) . The clade distance is calculated by first determining the geographical center of all observations of a clade (haplotypes, or a higher step clade) and then calculating the average distance (great circle distances were used in all the analyses given here) of the observations from their geographical center. The clade distance therefore measures the extent to which individuals bearing haplotypes that are members of that clade are geographically dispersed. The second distance measure is the nested-clade distance. In this case, the geographical center is determined of all observations of the clade of interest, as well as all other clades with which it is nested into a higher step nesting clade (that is, the geographical center of the nesting clade). Next, the average distance of bearers of haplotypes from the clade of interest is calculated from the geographical center of its nesting clade. The nested-clade distance measures how far bearers of haplotypes

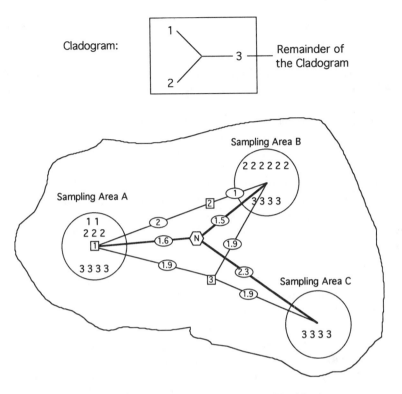

$D_c(1) = 0$ $D_c(2) = (1/3)(2) + (2/3)(1) = 1.33$ $D_c(3) = 1.9$

$D_n(1) = 1.6$ $D_n(2) = (1/3)(1.6) + (2/3)(1.5) = 1.53$ $D_n(3) = (1.6 + 1.5 + 2.3)/3 = 1.8$

$$D_c(\text{Interior}) - D_c(\text{Tip}) = 1.9 - (0 + 1.33)/2 = 1.23$$

$$D_n(\text{Interior}) - D_n(\text{Tip}) = 1.8 - (1.6 + 1.53)/2 = 0.23$$

Figure 13.2.   A hypothetical example of clade and nested-clade distances. Three sampling areas on an island are indicated by the letters A, B, and C. Within each sampling area, the number of times haplotypes of type 1, 2, and 3 are sampled is indicated by the number of times their respective haplotype numbers appear within the circle centered at the sampling location. These three haplotypes are all within a common nested clade, as indicated at the top of the figure. Note that haplotypes 1 and 2 are tips and that haplotype 3 is an interior haplotype. The geographical center of a particular haplotype is indicated by a box containing the number of the haplotype. The geographical center of the entire nested clade is indicated by the hexagon enclosing the letter N. The great circle distances between these geographical centers and the sampling locations are indicated by the numbers enclosed on an oval on the line connecting the geographical locations under consideration. The bottom part of the figure illustrates how these great circle distances are combined to calculate the clade (D_c) and nested-clade (D_n) distances, and the average difference between tip- and interior-haplotype clade and nested-clade distances.

from a clade are from individuals that bear evolutionarily close haplotypes (including haplotypes from the clade of interest).

The null hypothesis of no significant geographical associations is tested by randomly permuting 1,000 times in a nested fashion the categorical variables of clade type versus sampling location, as described in Templeton and Sing (1993). The distances are recalculated after each random permutation, and the observed distances are compared to this set of randomly generated distances under the null hypothesis of no association to determine which clade and nested-clade distances are significantly small or large at the 5% level of significance.

Another statistic calculated is the average difference of these distance measures between interior versus tip clades within the same nesting clade. Although the haplotype networks are not rooted (accurate rooting of gene trees is extremely difficult; see Castelloe and Templeton, 1994), tip clades (clades connected to the rest of the cladogram by only one mutational pathway) strongly tend to be younger than the clades that lie immediately interior to them (interior clades are connected to the rest of the cladogram by two or more mutational pathways; that is, they lie on an internal node of the network) (Castelloe and Templeton, 1994). Hence, these statistics tend to contrast the older versus younger members of a clade. The statistical significances of the average interior minus tip clade and nested-clade distances are also evaluated through the random permutation procedure. In those cases in which the tip–interior status of a clade is uncertain, the averages are determined by weighting the tip versus interior possibilities by the probabilities calculated using the techniques given in Crandall et al. (1994).

Temporal polarity is also given by the nested design itself. The age of a nesting clade must be greater than or equal to that of all clades found within it, so as clade level increases in a nested series, the average age of the clades must be nondecreasing.

The advantage of using these statistics is not only that they allow one to test rigorously the null hypothesis of no association between the haplotype tree and geography (and hence prevent making conclusions when the sample size is too small), but when the null hypothesis is rejected, the pattern of rejection can often be informative about the likely underlying biological causes. The expected patterns under restricted but non-zero gene flow, range expansion, and population fragmentation are discussed elsewhere (Templeton, 1993, 1994b; Templeton et al., 1995), and the Appendix gives a key indicating how to interpret the results when the null hypothesis is rejected. For the most part, these interpretations simply represent a translation of how intraspecific phylogeographic inferences have been made in the past in terms of the distance statistics mentioned earlier, but the key ensures that such inferences are made in an objective, reproducible fashion that also warns the researcher when unambiguous inference cannot be made because the cladogram is not resolved or detailed enough, or when the sampling scheme is too sparse geographically.

Conservation Relevance

Although one does not normally think of analytical techniques as having conservation relevance, conservation biology is a science, and the accuracy and strength of conclusions in any scientific endeavor depend as much upon analytical methodology as upon the data. This is particularly true for conservation biology, because the questions in this area often lie at the population level and above, and usually in a non-controlled situation that is not amenable to traditional experimental science. Hence, use must be made of "natural experiments" to justify scientific inference. Rigorous data and statistical analyses are essential in such studies in defining the *natural experimental design* and in using the inferred design to test hypotheses.

The natural experimental design that will be used is the nested, hierarchical design provided by the evolutionary history of the DNA region under investigation. This design in turn depends upon estimating an intraspecific gene tree, including an assessment of statistical confidence in that estimate. Despite the fact that gene trees are widely employed in intraspecific studies, it is surprising how little effort has been made to quantify the accuracy of intraspecific gene tree estimates. Typically, gene trees are estimated using some criterion, such as maximum parsimony or neighbor-joining, with no evaluation of the validity of these criteria for the specific dataset being analyzed. Occasionally, some statistical-confidence technique is used that was originally developed for interspecific trees, such as bootstrapping, despite the fact that these techniques sometimes lack statistical power at this level (Templeton et al., 1992; Crandall and Templeton, 1993; Crandall, 1994; Crandall et al., 1994). By using the estimation and confidence-assessment algorithm of Templeton et al. (1992) and by using the nesting algorithm of Templeton and Sing (1993) that allows for uncertainty in the natural design defined by the estimated gene tree, it becomes possible to avoid inappropriate conclusions based upon an estimated gene tree that is treated in an unjustifiably accurate manner.

For the same reason, it is essential that the analyses of data using the gene trees as a natural statistical design also be done in a statistically rigorous fashion with explicit, objective criteria for inference. This goal is achieved in this chapter by using the nested-permutation testing procedure given in Templeton and Sing (1993), coupled with the explicit inference structure given in the Appendix. By using this procedure, it is possible not only to statistically justify conclusions, but also to identify those situations in which the sampling scheme does not justify a conclusion. Recognizing what is not known with available samples is critical for evaluating the merits or strengths of management recommendations and for guiding future research endeavors.

GENETIC OBSERVATIONS

Table 13.1 gives the number of distinct mtDNA haplotypes and their frequencies at each geographical location as revealed by genetic surveys for each of the three bovid species. Following are the analyses of these data species by species.

Buffalo

Haplotypes separated by up to three restriction-site changes lie within the 95% confidence bounds of parsimony. All haplotypes but two (numbers 22 and 25) are separated by three or fewer restriction-site changes from one or more other haplotypes, allowing a parsimonious network to be constructed (Figure 13.3). No loops are found in this parsimonious network, so the topology of this portion of the cladogram is unambiguous. Haplotype 25 differs from haplotype 11 by a minimum of five mutational steps, and the probability of this branch length deviating from parsimony by no more than one additional step is 0.995. There are three possible ways of connecting haplotype 25 to the parsimonious network that involve branch lengths of five or six mutational events, as is shown in Figure 13.3. Haplotype 22 is minimally 10 mutational steps away from one of the intermediate haplotypes inferred to exist on the branch between haplotype 25 and the parsimonious network. In this case, 95% confidence is not achieved until branch lengths up to 12 steps are included in the 95% plausible set. This means that haplotype 22 could be connected to haplotype 25 or any of four intermediate haplotype states on the branch between haplotype 25 and the parsimonious network. Hence, there are a total of 15 cladograms in the 95% plausible set.

Figure 13.3 also shows the nested design defined by this 95% plausible set of cladograms. The ambiguities in cladogram topology discussed previously have no impact on the nested design, as is frequently the case (Templeton and Sing, 1993). Hence, in terms of the statistical analysis of the geographical data, the cladogram is effectively completely and unambiguously resolved.

Figure 13.4 gives the results of the geographical distance analysis, and Table 13.2 gives the inferences made from these results using the key in the Appendix. As can be seen, the null hypothesis is rejected in five nesting

Table 13.1. Mitochondrial Haplotype Frequencies within Bovid Species at Five Sampling Locations in Africa

mtDNA Haplotype Number for Buffalo (*Syncerus caffer*)

Location	1	2	3	4	5	6	7	8	9	10	11	12	13	14	15	16	17	18	19	20	21	22	23	24	25	26	27	28	Sample Size
Seronera	1	7	2	1	5	1																							17
Lobo	1	8	2	1	1		2	1	1																				17
Ngorongoro										1	6	1	3	2	1	1		1	1										17
Tarangire		3			4	2					1		3				2	1	1										17
Chobe										2										2	2	1	3	2	1	1	1	1	17

mtDNA Haplotype Number for Impala (*Aepyceros melampus*)

Location	1	2	3	4	5	6	7	8	9	10	11	12	13	14	15	16	17	18	19	20	21	22	23	24	25	Sample Size
Seronera	6	2	6	2	1																					17
Lobo	6	3	3	2	1		2																			17
Ngorongoro	(Not Found)																									0
Tarangire	1					2				6	1	1	2	1	2	1										17
Chobe																	4	4	1	1	1	1	3	1	1	17

mtDNA Haplotype Number for Wildebeest (*Connochaetes taurinus*)

Location	1	2	3	4	5	6	7	8	9	10	11	12	13	14	15	16	17	18	19	20	21	22	23	24	Sample Size
Seronera	6	3	3	2	1	1	1																		17
Lobo	2	1	3	1	1			4	4	1															17
Ngorongoro		2	4	1				3	6	1															17
Tarangire													4	1	10	1	1								17
Hwange																		8	1	1	3	2	1	1	17

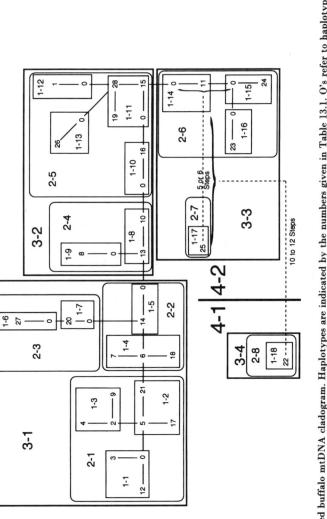

Figure 13.3. The unrooted buffalo mtDNA cladogram. Haplotypes are indicated by the numbers given in Table 13.1. O's refer to haplotype states that are intermediate between existing haplotypes but were not found in the sample. Each solid line connecting haplotypes or haplotype states indicates a single mutational change. Dashed lines indicate multiple-step branches for which the exact connections are ambiguous. Haplotypes are nested into 1-step clades as indicated by narrow line boxes and designated by "1-#" where "#" is an Arabic number assigned to the clade. Higher step clades are similarly indicated, with narrow-line boxes with rounded corners indicating 2-step clades, boxes with double-thick lines indicating 3-step clades, and a triple-thick line separating 4-step clades.

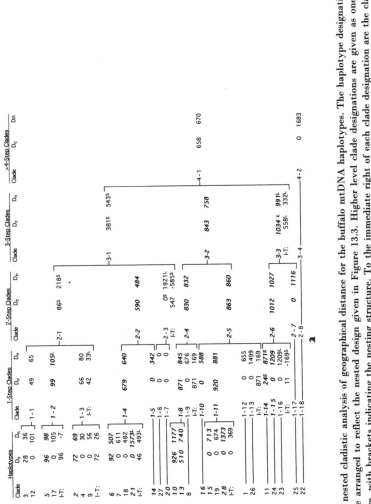

Figure 13.4. Results of the nested cladistic analysis of geographical distance for the buffalo mtDNA haplotypes. The haplotype designations are given on the far left and are arranged to reflect the nested design given in Figure 13.3. Higher level clade designations are given as one moves toward the right in the figure, with brackets indicating the nesting structure. To the immediate right of each clade designation are the clade and nested-clade distances. An "S" superscript indicates the distance is significantly small at the 5% level, and an "L" superscript indicates that it is significantly large. For nested clades in which the tip/interior status is known, the nested-clade designations and distances for interior clades are given in italics and boldface. Immediately below the nested groups with tips and interiors, the symbol "I-T" is given and indicates the average difference in distances between interior clades and tip clades within the nested group.

Table 13.2. **The Nesting Clades Containing One or More Significant Distance Measures for Buffalo mtDNA from Figure 13.4, along with the Inference Chain Obtained by Using the Key in the Appendix, and the Final Inference**

Nesting clades	Chain of inference using steps given in Appendix	Final inference
1-4	1-2-11-17-4-9-10 (NO)	Geographical sampling inadequate: Fragmentation or isolation by distance
2-1	1-2-11-17-4 (NO)	Gene flow restricted by isolation by distance
2-6	1-2-3-5-15-16-18 (YES)	Geographical sampling inadequate: Fragmentation or isolation by distance
3-1	1-2-3-5-6-7-8 (NO)	Geographical sampling inadequate: Isolation by distance or occasional long-distance dispersal
4-1	1-2-3-4 (NO)	Gene flow restricted by isolation by distance

clades. In two of these cases, the inference was that the geographical associations were caused by gene flow restricted by isolation by distance. For the significant results found in nesting clade 3-1, restricted gene flow is also indicated, but inadequacies of the geographical sampling scheme (the absence of samples between Chobe and Tanzania) prevent us from excluding the possibility of some long-distance dispersal. The other two nesting clades with significant results also suffer from partially ambiguous outcomes because of the failure to obtain samples between Chobe and Tanzania, but in both cases the possible inferences include gene flow restricted by isolation by distance. Hence, the overall pattern found in the buffalo can be explained entirely by an isolation by distance model of restricted gene flow.

Impala

Haplotypes separated by up to three restriction-site changes lie within the 95% confidence bounds of parsimony. All haplotypes are separated by three or fewer restriction-site changes from one or more other haplotypes, so only parsimonious alternatives need be considered (Figure 13.5). Within these parsimonious alternatives, two loops exist: one involving haplotypes 3, 4, 10, and 12, which can be broken in four different ways; the other involving the possible connections of haplotypes 22 and 21 to each other and to haplotypes 17 and 23, which can be broken in three different ways. Hence, there are a total of 12 cladograms in the 95% plausible set. A probability was assigned to all the ways of breaking these two loops using the procedure of Crandall et al. (1994), as shown in Table 13.3. No loop was resolved at the 95% level, so all loops were retained in the nested design, which is also shown in Figure 13.5.

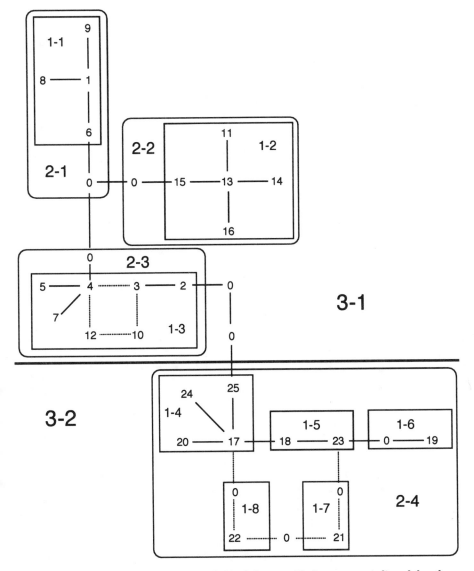

Figure 13.5. The unrooted impala mtDNA cladogram. Haplotypes are indicated by the numbers given in Table 13.1. Dotted lines indicate mutational changes forming loops for which there is uncertainty whether that mutational change actually occurred in the true evolutionary history of mtDNA. For descriptions of other features of this figure, see the caption to Figure 13.3.

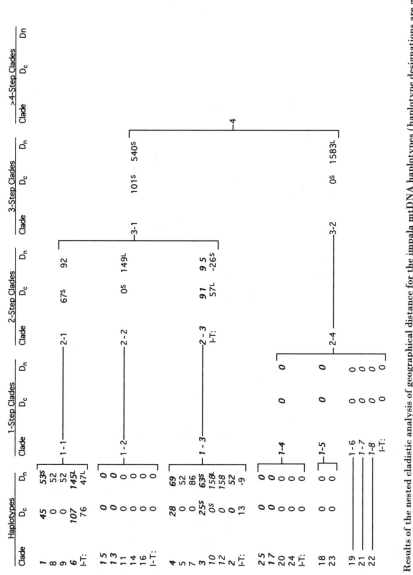

Figure 13.6. Results of the nested cladistic analysis of geographical distance for the impala mtDNA haplotypes (haplotype designations are given on the far left and are arranged to reflect the nested design given in Figure 13.5). If a clade has a high probability (≥ 0.05) of being either a tip or an interior, its designations and distances are given in italics and boldface. For descriptions of other features of this figure, see the caption to Figure 13.4.

Table 13.3. The Probabilities (Using the Algorithm of Crandall et al., 1994) of the Four Alternative Ways of Breaking the Loop Involving the Impala mtDNA Haplotypes 3, 4, 10 and 12, and the Three Alternative Ways of Breaking the Loop Involving Haplotypes 17, 21, 22 and 23

Alternatives for breaking loop	Probability
3,4,10,12 Loop:	

3,4,10,12 Loop:

```
4 — 3              0.095
|   |
12  10
```

```
4   3              0.216
|   |
12 — 10
```

```
4 — 3              0.658
    |
12 — 10
```

```
4 — 3              0.031
|
12 — 10
```

17,21,22,23 Loop:

```
17 — 18 — 23       0.702
|         |
0         0
|         |
22        21
```

```
17 — 18 — 23       0.149
          |
          0
          |
22 — 0 — 21
```

```
17 — 18 — 23       0.149
|
0
|
22 — 0 — 21
```

Figure 13.6 gives the results of the geographical distance analysis, and Table 13.4 gives the inferences made from these results using the key in the Appendix. As can be seen, the null hypothesis is rejected in four nesting clades. The pattern in clade 1-1 is that of gene flow restricted by isolation by distance. Within clade 1-3, the pattern is that of allopatric fragmentation. This means that no genetic exchange between two or more areas can be detected on the timescale marked by mutational accumulation in mtDNA at this clade level. In this case, the two areas so fragmented lie on either side of the Rift Valley in Tanzania. The next significant rejection of the null hypothesis occurs in nesting clade 3-1, with the inference being long-distance

Table 13.4. The Nesting Clades Containing One or More Significant Distance Measures for Impala mtDNA from Figure 13.6, along with the Inference Chain Obtained by Using the Key in the Appendix, and the Final Inference

Nesting clades	Chain of inference using steps given in Appendix	Final inference
1-1	1-2-11-17-4(NO)	Gene flow restricted by isolation by distance
1-3	1-2-3-5-15-16(YES)	Fragmentation
3-1	1-2-3-5-6-13-14(YES)	Long-distance colonization
4	1-2-3-5-15-16-18(NO)	Geographical sampling inadequate: Fragmentation, range expansion, or isolation by distance

colonization in this case. This means that genetic exchange between two areas has been detected, but it is unique or sporadic on the timescale marked by mutational accumulation at this clade level (as opposed to a gene flow model with recurrent genetic exchange). The two areas involved are once again the savannas on either side of the Rift Valley in Tanzania. Note from Figure 13.6 that 1-3 is ultimately nested within 3-1. Hence, there are long periods of no genetic contact between these two sides of the Rift Valley, with only unique or sporadic contact being detectable even on the timescale needed to generate all the genetic diversity found within clade 3-1 (Figure 13.5). The final significant rejection of the null hypothesis occurs at the highest level of contrast: 3-1 versus 3-2. Here, the inference is ambiguous between fragmentation and isolation by distance, once again because of the failure to sample between Chobe and Tanzania.

Wildebeest

Haplotypes separated by up to three restriction-site changes lie within the 95% confidence bounds of parsimony. The haplotypes separated by three or fewer restriction-site changes form three disjoint networks (Figure 13.7). Within the largest of these parsimonious networks, two intersecting loops are found, thereby indicating topological ambiguity due to five equally parsimonious alternatives. Because of the intersecting nature of these loops, the algorithm given in Crandall et al. (1994) cannot be used to assign probabilities to the alternative ways of breaking these loops. Nevertheless, the probability that haplotype 16 is a tip (and, therefore, 15 is also a tip) can be determined from the empirically derived priors that are justified by coalescent theory and are given in Crandall and Templeton (1993). This tip probability turns out to be 0.839.

The parsimonious haplotype network containing haplotypes 19 and 22 is minimally separated from the intermediate haplotype between 18 and 24 by

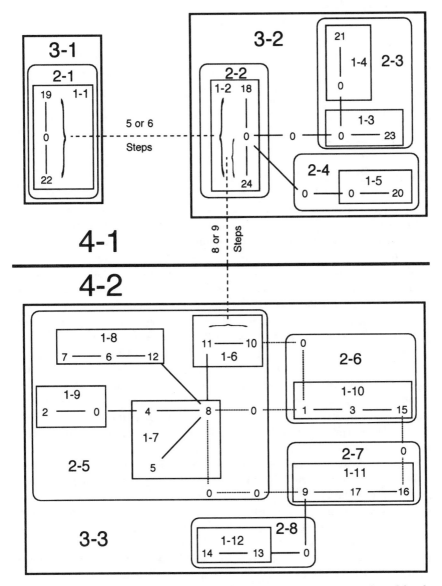

Figure 13.7. The unrooted wildebeest mtDNA cladogram. Haplotypes are indicated by the numbers given in Table 13.1. See captions to Figures 13.3 and 13.5 for further explanation.

Table 13.5. The Nesting Clades Containing One or More Significant Distance Measures for Wildebeest mtDNA from Figure 13.8, Along with the Inference Chain Obtained by Using the Key in the Appendix, and the Final Inference

Nesting clades	Chain of inference using steps given in Appendix	Final inference
1-10	1-2-3-5-15 (NO)	Fragmentation
1-11	1-2-11-17-4-9 (NO)	Fragmentation
2-5	1-2-11-17-4 (YES)	Gene flow restricted by isolation by distance
3-3	1-2-3-5-6-7-8 (YES)	Restricted gene flow with some long-distance dispersal
5	1-2-3-5-15-16-18 (YES)	Geographical sampling inadequate: Fragmentation or isolation by distance

five steps. The probability of deviating from parsimony by, at most, one additional step is 0.998, and there are a total of five possible connections (one parsimonious and four nonparsimonious) between these two networks (Figure 13.7). Haplotype 10 is minimally eight steps from haplotype 24, and the probability of deviating from parsimony by, at most, one additional step is 0.986, so only the three connections of length eight or nine are included in the 95% plausible set. Hence, there are a total of 75 cladograms in the 95% plausible set. The nested design (also shown in Figure 13.7) was affected only by the loops within the maximum parsimony network and was invariant with respect to the possible connections between the maximum parsimony loops. However, these ambiguities create uncertainties about the tip-interior status of haplotypes 1, 10, 18, 19, 22 and 24.

Figure 13.8 gives the results of the geographical distance analysis, and Table 13.5 gives the inferences made from these results using the key in the Appendix. As can be seen, the null hypothesis is rejected in five nesting clades. Both rejections of the null hypothesis by haplotypes nested within 1-step clades lead to the inference of fragmentation, and in both cases it involves fragmentation of Tanzanian populations separated by the Rift Valley. The 1-step clades nested within 2-5 indicate gene flow restricted by isolation by distance, and this pattern involves the Tanzanian populations found west of the Rift Valley. For the 2-step clades nested within 3-3, there was insufficient genetic resolution to discriminate between long-distance colonization versus restricted gene flow with some long-distance dispersal. These clades involve sets of haplotypes found in Tanzania on both sides of the Rift Valley. Note that clades 1-10 and 1-11 are both nested within 3-3, so at this deeper evolutionary level, some sort of genetic connection has finally been detected across the Rift Valley. The two 4-step clades (4-1 is found in Hwange, 4-2 in Tanzania) nested within the total cladogram also

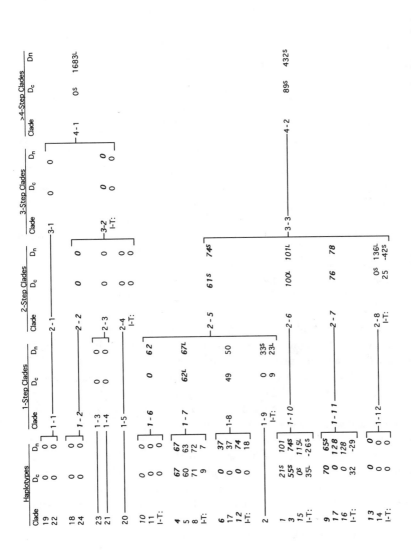

Figure 13.8. Results of the nested cladistic analysis of geographical distance for the wildebeest mtDNA haplotypes (haplotype designations are given on the far left and are arranged to reflect the nested design given in Figure 13.7). See legends to Figures 13.4 and 13.6 for further explanations.

reject the null hypothesis, but in this case the lack of sampling between Hwange and Tanzania makes the outcome inconclusive between fragmentation versus gene flow restricted by isolation by distance.

DISCUSSION AND CONSERVATION RELEVANCE

One way to identify those attributes of a landscape or greater ecosystem that have ecological and evolutionary significance is to perform studies upon more than one species (or other component of the landscape) and search for concordant patterns across the species. In this regard, much concordance exists between the inferences made from impala and from wildebeest. For both species, there is evidence for genetic fragmentation between either side of the Rift Valley in Tanzania, and perhaps, conditional upon future sampling, between Tanzania and southern Africa. This barrier to gene flow cannot be explained in terms of isolation by distance because both species show recurrent gene flow on the western side of the Rift Valley over areas of comparable geographical distance. At least for the wildebeest, the barrier to gene flow cannot be explained only in terms of geographical relief, because there is recurrent gene flow between Ngorongoro Crater and the Serengeti. The barrier for both species is associated instead with a habitat discontinuity.

Interestingly, the biological interpretation of this concordance is aided, not weakened, by the non-concordant pattern revealed by the buffalo. The geographical associations revealed by the buffalo were entirely explicable by recurrent gene flow restricted by isolation by distance, and therefore the Rift Valley and the geographical distance between Tanzania and Chobe were not significant barriers to genetic exchange in this species. Recall that buffalo have the broadest ecological niche of the three species examined in this chapter. Buffalo prefer the savanna habitat, but are also found in dense bush and montane forests. In contrast, both the impala and wildebeest are much more specialized ecologically and are found in a much narrower range of habitats. In this regard, it is critical to note that the habitat between Ngorongoro and Manyara (the geographical area associated with the statistically significant, geographically concordant fragmentation events) has traditionally consisted of upland forest, a habitat that buffalo will occupy (but not impala or wildebeest). Therefore, the three species considered together imply that the genetic population structures of the species that inhabit this landscape are determined by an interaction between the species' ecological niches and habitat patchiness at the landscape level.

The observed and statistically significant patterns of gene flow and fragmentation are not predictable from short-term studies on female dispersal behavior in these species. As pointed out earlier, buffalo females are strongly philopatric, and hence one might expect this species to show the least amount of gene flow and greatest amount of fragmentation over the sample area. However, just the opposite occurred. This perhaps is not too surprising. Population genetic theory reveals that even relatively rare events can have a major impact on gene flow patterns (Wright, 1969), and these are precisely the type of events that are not easily observable, much less quantifiable, by traditional dispersal studies. Hence, genetic surveys offer a more reliable means of inferring past and ongoing patterns of genetic exchange within a species (Templeton et al., 1990). In the species under consideration, ecological factors (niche breadth and habitat patchiness in concert) dominate behavioral philopatry in determining patterns of gene flow and fragmentation. Consequently, in contrast to Vrba (1987), observations of philopatry and mobility over ecological time are not informative about gene flow over evolutionary time. Therefore, management recommendations based upon observations of philopatry and mobility must be regarded as tentative and unreliable until confirmed by genetic survey studies. It is also important to realize that an inference of genetic fragmentation between two populations implies that they are separate demographic units and should be managed as such, but an inference of recurrent gene flow between two populations does not automatically imply that they are a single demographic unit, given that relatively rare events can result in effective gene flow (Wright, 1969).

Although genetic surveys provide a better tool for inferring patterns of gene flow and fragmentation than dispersal studies, it is essential that the genetic data utilize the natural experimental design provided by evolutionary history. Nonhistorical analyses simply lack the power to discriminate among possible alternatives (Templeton, 1994b). For example, one of the traditional, nonhistorical statistics used to investigate geographical/genetic associations are F statistics (Wright, 1969), and particularly F_{ST} which measures the degree of genetic differentiation among geographical populations (with $F_{ST} = 0$ corresponding to no genetic differentiation, and $F_{ST} = 1$ corresponding to complete differentiation). Although F_{ST} has frequently been used to estimate gene flow parameters, many factors other than restricted gene flow can create nonzero F_{ST} values (Templeton, 1994b), making the biological interpretation of these statistics difficult. For example, F_{ST} was estimated from the data given in Table 13.1 using the procedure of Davis et al. (1990) and was found to be 0.08 for the buffalo, 0.10 for the impala, and 0.16 for the wildebeest. The F_{ST} values for buffalo and impala are statistically indistinquishable. Hence, if F_{ST} were the only statistic being used to infer population structure, these two species would be regarded as having similar population structures with these data. However, the evolutionary historical analyses performed with the same dataset reveal that their patterns of genetic differentiation have different underlying causes. This illustrates that the evolutionary historical approach can discriminate among cases that appear similar with nonhistorical statistics.

Wise management decisions require more than just knowledge of genetic pattern— they also require knowledge of the underlying processes that created the pattern. For example, suppose a manager wanted to translocate individuals from one part of the species' range to another, where the native population had been extirpated or reduced to low numbers. There is no genetic rationale for not proceeding with translocation programs in the buffalo because they have recurrent genetic exchange throughout the studied range (however, before implementing a translocation program, this inference based upon mtDNA should be confirmed by genetic surveys on nuclear DNA or chromosomal variants). On the other hand, the long-term fragmentation observed in the wildebeest and impala raises the possibility of subspecies or local adaptation, so translocation should be avoided, or undertaken cautiously.

This last point raises another serious issue that appears in many conservation programs— the biological reality of *subspecies* (which along with *distinct population segments* are subject to legal protection under the U.S. Endangered Species Act). There is no consensus as to what constitutes a subspecies, but certainly one would minimally demand a subspecies to be an evolutionarily definable subset of the species; that is, a lineage fragmented from other conspecific populations. (Note that mere genetic differentiation is not a criterion for subspecies status under this view, as indicated by the discussion of F_{ST}.) Hence, by using the approach outlined in this chapter, conservation managers have explicit and statistically quantifiable criteria to identify subspecies. In the species studied here, the results of our analyses are partially consistent with the assignment of subspecies within each species as judged by morphological criteria. Morphologically distinct subspecies of wildebeest exist west (*C.t. mearnsi*) and east (*C.t. albojubatus*) of the Rift Valley in northern Tanzania, and an additional subspecies (*C.t. taurinus*) has been described in Botswana and Zimbabwe. The present analysis supports the biological reality of the first two subspecies, and is consistent with the existence of the third as well. However, additional sampling will be required in the region between Zimbabwe and Tanzania to confirm the biological reality of this third subspecies.

For the buffalo, only one subspecies exists within the sampled range, a taxonomic decision supported by the present genetic analysis. A discrepancy between morphology and molecular genetics is observed, however, for the impala. Only one subspecies of impala is recognized across the sampled range, yet the pattern of gene flow and fragmentation found in the impala is virtually identical to that of the wildebeest and discrepant with that of the buffalo. That morphologically cryptic subspecies could exist should not be surprising, given that cryptic species are found in many groups (Templeton, 1981). Before drawing firm conclusions, however, it would be best to repeat this analysis with one or more nuclear markers. It may prove that the impala has more male-mediated genetic exchange than the wildebeest across the fragmented areas.

The gene tree historical approach can be extended to the species level when the criteria for species status are made explicit and phrased as testable null hypotheses, as has been done for the cohesion species concept (Templeton, 1994b). Thus, the fundamental units of genetic management can be objectively defined at both the intraspecific and specific level using the evolutionary historical approach with rigorous statistical testing.

Although the emphasis in this chapter has been upon landscape-level conservation, in many cases additional effort will need to be focused upon species that are indeed endangered. The fact that a species' population genetic structure represents an interaction between the species-specific niche and landscape properties means that genetic studies should also be performed upon the endangered species whenever possible. Generalizations from even closely related, non-endangered species can be misleading. However, the more species studied, the more accurate these data become for making inferences about the processes responsible for the patterns, and process knowledge should yield better management recommendations for endangered species. For example, one endangered species found in this landscape is the African elephant. From a habitat-use point of view, the elephant is more similar ecologically to the buffalo than to either the wildebeest or impala. Hence, if the inference is correct that gene flow patterns are determined by the interaction of habitat patchiness with ecological-niche breadth, the elephant would be expected to display recurrent gene flow throughout this area of Africa. Direct genetic surveys upon elephants confirm that this is indeed the case (Georgiadis et al., 1994). Hence, by accumulating data on several non-endangered species, more accurate management recommendations can be made for endangered species as well.

Another recommendation stems from the inference that colonization across the Rift Valley occurred for the impala and possibly for the wildebeest. Given the rare, sporadic nature of the genetic connections between savanna habitats separated by only 50 km of forest observed for both of these species, we would predict that savanna-specialist species may not occupy all appropriate habitat patches at any given time and that the current geographical range would be limited by extrinsic habitat barriers. This implies that for an endangered savanna-specialist species, appropriate but unoccupied habitats probably exist that could serve as areas of introduction. This recommendation would not be appropriate for species such as the buffalo and elephant, which have the capacity to more thoroughly find and occupy the habitats that meet their ecological requirements.

As these recommendations show, studies on non-endangered species also can play an important role in conservation biology. By elucidating processes and not mere patterns, such studies allow us to gain greater insight into how ecosystems function over evolutionary time and to make better management recommendations for the ecosystem as a whole and for the endangered species within it. The conservation of evolutionary processes has received little attention primarily because of the difficulty of revealing or estimating the relevant processes. Modern molecular genetics combined with statistical designs and analyses that incorporate evolutionary history allow these processes to be defined and studied, thereby enhancing our ability to practice process conservation in addition to species conservation.

SUMMARY

1.   Given that intraspecific gene trees are among the essential foundations of conservation practice, it is critical that an assessment of statistical confidence about the accuracy of the estimated gene tree be made before using the gene tree to draw conservation inferences. When interpreting other data (such as geographical position) by overlaying it upon the gene tree, it is likewise critical to justify all inferences by rigorous statistical testing and explicit interpretative criteria. Otherwise, management decisions might be made from inadequate samples and/or subjective interpretations of the data.

2.   By using the gene tree as a natural statistical design, permutation testing, and explicit interpretative criteria, it is possible to separate the roles that restricted gene flow, colonization, and population fragmentation have played in determining a species' population genetic pattern over evolutionary time. In particular, the genetic pattern found in buffalo is explicable entirely by gene flow restricted by isolation by distance. In contrast, both the impala and wildebeest have mixtures of restricted gene flow, colonization and/or occasional long-distance dispersal, and population fragmentation.

3.   The genetic population structure of the surveyed bovid species that inhabit savannas in eastern and southern Africa is largely determined by an interaction between the species' ecological niche and the habitat patchiness at the landscape level.

4.   Observations of philopatry and mobility over ecological time are not informative about gene flow over evolutionary time; indeed, they are often misleading. Therefore, management recommendations based upon observations of philopatry and mobility must be regarded as tentative and unreliable until confirmed by genetic survey studies.

5. Although genetic surveys provide a better tool for inferring patterns of gene flow and fragmentation than do dispersal studies, it is essential that the genetic data utilize the natural experimental design provided by evolutionary history. Nonhistorical analytical approaches lack the power to discriminate among possible alternatives and hence are not as reliable for making management recommendations.

6. Subspecies should not be recognized upon the basis of mere genetic differentiation, but only in those cases in which the genetic differentiation is caused by fragmentation (extrinsic barriers to gene flow) resulting in stable and definable lineages over evolutionary time. By applying these criteria to mtDNA, there is only one subspecies of buffalo in the sampled area (concordant with morphologically recognized subspecies) and at least two and possibly three subspecies within both the wildebeest (concordant with morphologically recognized subspecies) and the impala (discordant with morphologically recognized subspecies).

7. It is best to perform genetic surveys on endangered species in order to formulate management guidelines, because generalization from even closely related species can be misleading. However, when genetic surveys on several species are used to infer processes rather than mere patterns, management recommendations can be made in those cases in which the endangered species is so rare that meaningful genetic surveys are impossible, or in which recent human disturbances have so altered the current state of the species as to make such surveys difficult to interpret. In particular, bovid species with broad habitat usage can be translocated over wide areas, whereas those that are savanna specialists should not (in the absence of other data). However, the specialist species can be introduced into nearby savanna habitats from which they are absent, because absence is often a matter of historical chance.

REFERENCES

AVISE, J.C. 1994. *Molecular Markers, Natural History and Evolution.* Chapman & Hall, New York.

AVISE, J.C., J. ARNOLD, R.M. BALL, E. BERMINGHAM, T. LAMB, J.E. NEIGEL, C.A. REEB, and N.C. SAUNDERS. 1987. Intraspecific phylogeography: the mitochondrial DNA bridge between population genetics and systematics. *Annu. Rev. Ecol. Syst.* 18:489–522.

BORNER, M. 1987. The increasing isolation of Tarangire National Park. *Oryx* 19:91–96.

CASTELLOE, J. and A.R. TEMPLETON. 1994. Root probabilities for intraspecific gene trees under neutral coalescent theory. *Molec. Phylo. Evol.* 3:102–113.

CRANDALL, K.A. 1994. Intraspecific cladogram estimation: accuracy at higher levels of divergence. *Syst. Biol.* 43:222–235.

CRANDALL, K.A. and A.R. TEMPLETON. 1993. Empirical tests of some predictions from coalescent theory with applications to intraspecific phylogeny reconstruction. *Genetics* 134:959–969.

CRANDALL, K.A., A.R. TEMPLETON, and C.F. SING. 1994. Intraspecific phylogenetics: Problems and solutions. In *Phylogeny Reconstruction*, eds. R.W. Scotland, D.J. Siebert, and D.M. Williams, pp. 273–297. Clarendon, Oxford.

DAVIS, S.K., J.E. STRASSMANN, C. HUGHES, L.S. PLETSCHER, and A.R. TEMPLETON. 1990. Population structure and kinship in *Polistes* (Hymenoptera, Vespidae): an analysis using ribosomal DNA and protein electrophoresis. *Evolution* 44:1242–1253.

ESTES, R. 1991. *The Behavior Guide to African Mammals*. University of California Press, Berkeley.

GEORGIADIS, N., L. BISCHOF, A. TEMPLETON, J. PATTON, W. KARESH, and D. WESTERN. 1994. Structure and history of African elephant populations: I. Eastern and Southern Africa. *J. Hered.* 85:100–104.

GRUMBINE, E. 1990. Protecting biological diversity through the greater ecosystem concept. *Nat. Areas J.* 10:114–120.

GRUMBINE, R.E. 1994. What is ecosystem management? *Conserv. Biol.* 8:27–38.

HANSEN, R., M. MUGAMBI, and S. BAUNI. 1985. Diets and trophic ranking of ungulates in northern Serengeti. *J. Wildl. Mgmt.* 49:823–829.

KARESH, W., F. SMITH, and H. FRAZIER-TAYLOR. 1989. A remote method for obtaining skin biopsy samples. *Conserv. Biol.* 1:261–262.

LAMPREY, H.F. 1963. Ecological separation of the large mammal species in the Tarangire Game Reserve, Tanganyika. *E. Afr. Wildl. J.* 1:63–92.

MANIATIS, T., E.F. FRITSCH, and J. SAMBROOK. 1989. *Molecular Cloning: A Laboratory Manual*. Cold Spring Harbor, New York.

MCNAUGHTON, S.J. 1983. Serengeti grassland ecology: the role of composite environmental factors and contingency in community organization. *Ecol. Monogr.* 53:291–320.

MURRAY, M.G. 1982. Home range, dispersal and the clan system of impala. *Afr. J. Ecol.* 20:253–269.

NEWMARK, W.D., D.N. MANYANZA, D.G. GAMASSA, and H.I. SARIKO. 1994. The conflict between wildlife and local people living adjacent to protected areas in Tanzania: human density as a predictor. *Conserv. Biol.* 8:249–255.

TEMPLETON, A.R. 1981. Mechanisms of speciation— a population genetic approach. *Annu. Rev. Ecol. Syst.* 12:23–48.

TEMPLETON, A.R. 1991. Genetics and conservation biology. In *Species Conservation: A Population–Biological Approach*, eds. A. Seitz and V. Loeschcke, pp. 15-29. Birkhäuser Verlag, Basel, Switzerland.

TEMPLETON, A.R. 1993. The "Eve" hypothesis: a genetic critique and reanalysis. *Amer. Anthropol.* 95:51–72.

TEMPLETON, A.R. 1994a. Biodiversity at the molecular genetic level: experiences from disparate macroorganisms. *Phil. Trans. Roy. Soc. Lond.* B345:59–64.

TEMPLETON, A.R. 1994b. The role of molecular genetics in speciation studies. In *Molecular Approaches to Ecology and Evolution*, eds. B. Schierwater, B. Streit, G.P. Wagner, and R. DeSalle, pp. 455–477. Birkhäuser Verlag, Basel, Switzerland.

TEMPLETON, A.R., E. BOERWINKLE, and C.F. SING. 1987. A cladistic analysis of phenotypic associations with haplotypes inferred from restriction endonuclease mapping. I. Basic theory and an analysis of alcohol dehydrogenase activity in *Drosophila*. *Genetics* 117:343–351.

TEMPLETON, A.R., K.A. CRANDALL, and C.F. SING. 1992. A cladistic analysis of phenotypic associations with haplotypes inferred from restriction endonuclease mapping and DNA sequence data. III. Cladogram estimation. *Genetics* 132:619–633.

TEMPLETON, A.R., E. ROUTMAN, and C. PHILLIPS. 1995. Separating population structure from population history: a cladistic analysis of the geographical distribution of mitochondrial DNA haplotypes in the tiger salamander, *Ambystoma tigrinum. Genetics* 140:767–782.

TEMPLETON, A.R., K. SHAW, E. ROUTMAN, and S.K. DAVIS. 1990. The genetic consequences of habitat fragmentation. *Ann. Missouri Bot. Gard.* 77:13–27.

TEMPLETON, A.R. and C.F. SING. 1993. A cladistic analysis of phenotypic associations with haplotypes inferred from restriction endonuclease mapping. IV. Nested analyses with clado-gram uncertainty and recombination. *Genetics* 134:659–669.

VRBA, E.S. 1987. Ecology in relation to speciation rates: some case histories of Miocene-Recent mammal clades. *Evol. Ecol.* 1:283–300.

WRIGHT, S. 1969. *Evolution and the Genetics of Populations, Vol. 2. The Theory of Gene Frequencies.* University of Chicago Press, Chicago.

APPENDIX

Start with haplotypes nested within a 1-step clade:

1. Are there any significant values for D_c, D_n, or I-T within the clade?
   - NO — the null hypothesis of no geographical association of haplotypes cannot be rejected (either panmixia in sexual populations, extensive dispersal in non-sexual populations, small sample size, or inadequate geographical sampling). Move on to another clade at the same or higher level.
   - YES — Go to Step 2.

2. Are the D_c values for tip or some (but not all) interior clades significantly small or is the I-T D_c distance significantly large?
   - NO — Go to Step 11.
   - YES — Go to Step 3.
   - Tip/Interior Status Cannot be Determined — **Inconclusive outcome.**

3. Are any D_n and/or I-T D_n values significantly reversed from the D_c values, and/or do one or more tip clades show significantly large D_ns or interior clades significantly small D_ns or I-T significantly small D_n with the corresponding D_c values being non-significant?
   - NO — Go to Step 4.
   - YES — Go to Step 5.

4. Do the clades (or two or more subsets of them) with restricted geographical distributions have ranges that are completely or mostly non-overlapping with the other clades in the nested group (particularly interiors), and does the pattern of restricted ranges represent a break or reversal from lower level trends within the nested series (applicable to higher level clades only)?

- NO — **Restricted gene flow with isolation by distance (restricted dispersal by distance in non-sexual species).** This inference is strengthened if the clades with restricted distributions are found in diverse locations, if the union of their ranges roughly corresponds to the range of one or more clades (usually interiors) within the same nested group (applicable only to nesting clades with many clade members or to the highest level clades regardless of number), and if the D_c values increase and become more geographically widespread with increasing clade level within a nested series (applicable to lower level clades only).
- YES — Go to Step 9.

5. Do the clades (or two or more subsets of them) with restricted geographical distributions have ranges that are completely or mostly non-overlapping with the other clades in the nested group (particularly interiors), and does the pattern of restricted ranges represent a break or reversal from lower level trends within the nested series (applicable to higher level clades only)?
   - NO — Go to Step 6.
   - YES — Go to Step 15.

6. Do clades (or haplotypes within them) with significant reversals or significant D_n values without significant D_c values define geographically concordant subsets, or are they geographically concordant with other haplotypes/clades showing similar distance patterns?
   - NO — Go to Step 7.
   - YES — Go to Step 13.
   - TOO FEW CLADES (< 2) TO DETERMINE CONCORDANCE — **Insufficient genetic resolution to discriminate between range expansion/colonization and restricted dispersal/gene flow** — Proceed to Step 7 to determine if the geographical sampling is sufficient to discriminate between short- versus long-distance movement.

7. Are the clades with significantly large D_ns (or tip clades in general when D_n for I-T is significantly small) separated from the other clades by intermediate geographical areas that were sampled?
   - NO — Go to Step 8.
   - YES — **Restricted gene flow/dispersal but with some long distance dispersal.**

8. Is the species absent in the nonsampled areas?
   - NO — **Sampling design inadequate to discriminate between isolation by distance (short-distance movements) versus long-distance dispersal.**

- YES — **Restricted gene flow/dispersal but with some long-distance dispersal over intermediate areas not occupied by the species.**

9. Are the different geographically concordant clade ranges separated by areas that have not been sampled?
   - NO — **Past fragmentation.** (If inferred at a high clade level, additional confirmation occurs if the clades displaying restricted but at least partially non-overlapping geographical distributions are mutationally connected to one another by a larger than average number of steps.)
   - YES — Go to to Step 10.

10. Is the species absent in the nonsampled areas?
    - NO — **Geographical sampling scheme inadequate to discriminate between fragmentation and isolation by distance.**
    - YES — **Allopatric fragmentation.** (If inferred at a high clade level, additional confirmation occurs if the clades displaying restricted but at least partially non-overlapping geographical distributions are mutationally connected to one another by a larger than average number of steps.)

11. Are the D_c values for some tip clades significantly large, and/or the D_cs for all interiors significantly small, and/or the I-T D_c values significantly small?
    - NO — Go to step 17.
    - YES — **Range expansion,** go to Step 12.

12. Are the D_n and/or I-T D_n values significantly reversed from the D_c values?
    - NO — **Contiguous range expansion.**
    - YES — Go to Step 13.

13. Are the clades with significantly large D_ns (or tip clades in general when D_n for I-T is significantly small) separated from the other clades by intermediate geographical areas that were sampled?
    - NO — Go to Step 14.
    - YES — **Long-distance colonization.**

14. Is the species absent in the nonsampled areas?
    - NO — **Sampling design inadequate to discriminate between contiguous range expansion and long-distance colonization.**
    - YES — **Long-distance colonization.**

15. Are the different geographically concordant areas separated by areas that have not been sampled?
    - NO — **Past fragmentation.** (If inferred at a high clade level, additional confirmation occurs if the clades displaying restricted but at

least partially non-overlapping geographical distributions are mutationally connected to one another by a larger than average number of steps.)

- YES — Go to Step 16.

16. Is the species absent in the nonsampled areas?
    - NO — Go to step 18.
    - YES — **Allopatric fragmentation.** (If inferred at a high clade level, additional confirmation occurs if the clades displaying restricted but at least partially non-overlapping geographical distributions are mutationally connected to one another by a larger than average number of steps.)

17. Are the D_n values for tip or some (but not all) interior clades significantly small, or the D_n for one or more interior clades significantly large, or is the I-T D_n value significantly large?
    - NO — **Inconclusive outcome.**
    - YES — Go to Step 4.

18. Are the clades found in the different geographical locations separated by a branch length with a larger than average number of mutational steps?
    - NO — **Geographical sampling scheme inadequate to discriminate between fragmentation, range expansion, and isolation by distance.**
    - YES — **Geographical sampling scheme inadequate to discriminate between fragmentation and isolation by distance.**

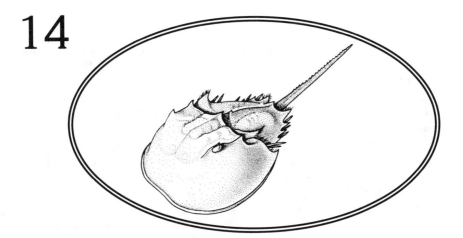

14

TOWARD A REGIONAL CONSERVATION GENETICS PERSPECTIVE: PHYLOGEOGRAPHY OF FAUNAS IN THE SOUTHEASTERN UNITED STATES

John C. Avise

INTRODUCTION

By the very nature of genetic assessment and phylogeny estimation, most molecular studies (in conservation biology or elsewhere) are directed at particular species or clades, rather than toward ecological assemblages of unrelated taxa. However, the problems of conservation biology go far beyond those of individual taxonomic groups (important though they may be) to broader concerns about environmental health and ecosystem viability. Are there additional ways (such as those developed in Chapter 13) that the data and perspectives of genetics might contribute to conservation efforts at regional or ecosystem levels?

This problem of moving from a species-based to a regional perspective finds a parallel in the distinction between the pragmatic approaches versus the ultimate goals of conservation biology under the U.S. Endangered Species Act (ESA) of 1973. The stated intent of the ESA is to "provide a means whereby the ecosystems upon which endangered species and threatened species depend may be conserved..." (16 U.S.C. Section 1531[b]); but, in practice, the means for doing so normally involve focusing preservation efforts on particular species, rather than on ecosystems or biodiversity *per se*. Although this species-based approach has been immensely useful in the conservation movement, it would be wise to remain cognizant of the final goal, which is to seek sustainable environments where ecological and evolutionary processes fostering biotic diversity are maintained. An enlightened implementation of the ESA therefore requires that more explicit concern be devoted to the viability of entire ecosystems, whereby not only keystone endangered species but also their myriad ecological associates would simultaneously benefit. There is also a practical reason for advocating this perspective. Under current convention, if the northern spotted owl or the Florida panther (for example) were to go extinct for any reason, conservationists could only hope that the legal basis for preservation of Pacific old-growth forests and the Florida Everglades (in these respective cases) would not be seriously undermined.

In any event, one extended case history does exist in which genetic perspectives explicitly have been brought to bear on regional conservation issues involving unrelated taxa. This research involves a diverse fauna in the coastal states of the southeastern United States (Box 14.1) and can be divided into three parts focusing on genetic analyses of: (1) freshwater fish across all major drainages between the Carolinas and Louisiana or Texas; (2) other freshwater and terrestrial vertebrate species in this same region; and (3) vertebrate and invertebrate species along the U.S. coastlines in the Gulf of Mexico and the Atlantic Ocean. These studies involve comparisons

---

Box 14.1. Conservation Status of Faunas in the Southeastern United States

The current chapter deals with a variety of vertebrate and invertebrate animals in coastal states bordering the Gulf of Mexico and the southern Atlantic. Most of the genetic surveys reported span the area between Texas or Louisiana to the west and the Carolinas to the east, although surveys of a few taxa also have extended inland to adjacent non-coastal states, or northward to New England along the coast. In the southeastern region, about 90 animal taxa currently are listed as threatened or endangered under the U.S. Endangered Species Act of 1973, and several of these have been the subject of molecular genetic study. However, many of the broader lessons of conservation relevance have come from genetic studies directed at non-threatened species in the region, as well as from more traditional biogeographic assessments based on species' distributions and geologic evidence.

---

across species in the geographic patterns of intraspecific population structure, as evidenced by assays of mitochondrial (and in some cases nuclear) genes.

[*Note:* The core of this synopsis is an earlier review (Avise, 1992), which along with the original data papers should be consulted for further details. The current overview updates and extends comparative trends and their conservation import.]

BACKGROUND
ON THE PHYSICAL ENVIRONMENT

As will become clear from the comparative results presented later, historical biogeographic factors must have played a significant role in shaping the genetic architectures of the freshwater, terrestrial, and maritime biotas of the southeastern United States. What follows is a brief introduction to the remarkable dynamism of the physical environment over the past few million years, the intent being to exemplify the types of physiographic factors likely to have impinged on species' distributions and phylogeographic population structures (Figure 14.1).

In the freshwater realm, about a dozen major rivers and numerous smaller streams currently traverse the southeastern region, with eastern drainages entering the Atlantic Ocean and western drainages entering the Gulf of Mexico. During the high sea-stands of the Pliocene and the moderate sea-stands of the Pleistocene interglacial periods (Figure 14.1), smaller coastal streams likely were flooded, and freshwater faunas probably were isolated in the upper reaches of the larger rivers, and perhaps in lakes and rivers of the Floridian peninsula itself. Subsequent interdrainage transfer of fish must have occurred via lateral or headwater stream capture (as, for example, between the Savannah and Apalachicola drainages that now come into close contact in the southern Appalachians). Conversely, during the low sea-stands associated with the glacial periods that dominated much of the Pleistocene, adjacent drainages occasionally may have coalesced on the broader coastal plains (as do the Tombigbee and Alabama drainages today, just before entering the Gulf of Mexico at Mobile Bay). Such histories of drainage isolation and connection (the details of which remain poorly known from geologic evidence) undoubtedly influenced the phylogenetic histories of freshwater populations. For example, if freshwater fish periodically were isolated in separate refugial regions in the southeast, the geographic strongholds involved and the subsequent range expansions (which would be governed by the historical patterns of drainage connection and by ecological conditions associated with changing landscapes) should have left assayable footprints in both the distributional and genetic records of extant faunas.

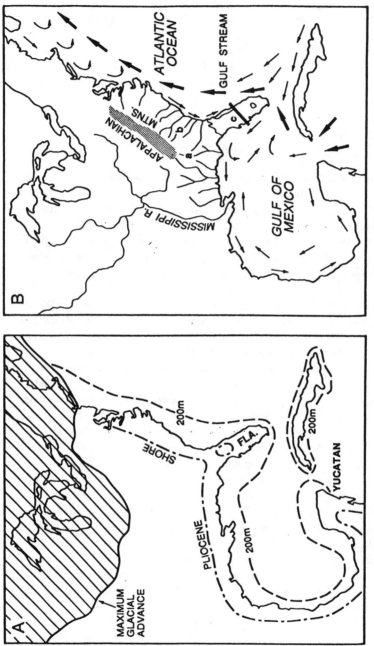

Figure 14.1. Selected physiographic features of the eastern United States, historically (A) and at the present time (B) (after Avise, 1992). Shown in A are the high sea-stands of the Pliocene (somewhat higher than any Pleistocene interglacial shorelines), the 200 m depth contour (slightly outside the exposed land areas of the Pleistocene glacial maxima), and the maximum extent of Pleistocene glacial advance. Shown in B are the Appalachian mountains, larger rivers entering the Gulf of Mexico and the Atlantic (including the biogeographically important Apalachicola drainage, labeled "a"), and some trends in the major marine currents (including the Gulf Stream indicated by heavier arrows).

In the terrestrial arena, the Floridian peninsula (particularly at its south-ern extreme) is climatically, physiographically, and ecologically distinct from the more temperate continental realms in states to the north and west. Thus, many taxa that are adapted to tropical or subtropical conditions in the southeastern United States are confined to southern and central Florida. Particularly during the Pleistocene glacial episodes, when sea levels were lower than now, the Floridian peninsula was much larger (Figure 14.1) and may have served as an important refugium for temperate species whose ranges were compressed southward by climatic changes associated with glacial advances. Furthermore, at earlier times (Miocene and perhaps Pliocene), the Ocala uplift area of west-central Florida existed as one or more large islands, cut off from the continental mainland by a shallow marine channel in the low-relief area of the present-day Okeefenokee Swamp and Suwannee River (Bert, 1986 and references therein). The net effect of these various historical and contemporary influences on biotic distributions is that a number of terrestrial and freshwater organisms endemic to the Floridian peninsula are recognizably distinct from their taxonomic counter-parts on the main body of the continent (to be discussed).

In the maritime realm, ecological and historical factors again should have left distributional and genetic footprints among extant forms. The Floridian peninsula currently protrudes southward into subtropical waters (25–28°N), separating some (but not all) temperate faunas into allopatric units on the Atlantic coast and Gulf of Mexico. During the numerous glacial advances and retreats of the Pleistocene, climatic changes and fluctuations in sea level no doubt had great impact on the coastal and marine species in the South-east. The climatic cooling associated with glacial advances must have pushed temperate populations southward, and may have increased the opportunity for contact of Atlantic and Gulf populations around south Florida. However, the glacial advances also caused drops in sea level (by as much as 150 m) and exposed great expanses of the Floridian and Yucutan peninsulas (Figure 14.1). At such times, the Floridian peninsula was more arid than it is today, and perhaps bordered by fewer of the intermediate-salinity estuaries and salt-marsh habitats favored by many coastal species. Thus, during glacial advances, an enlarged Floridian peninsula may have increased the potential for physical separation of Atlantic and western Gulf-coastal populations of some species.

Opposing influences on species' distributions also may have been at work during interglacial periods (such as the present), when sea levels were higher and the Floridian peninsula perhaps was bordered by more extensive estua-ries and salt marshes. At such times of climatic warming, some strictly temperate species may have been separated into disjunct Atlantic and Gulf populations by the tropical conditions of south Florida, while other more eurythermal and estuarine-adapted species may have expanded out of the Gulf region to regain increased contact with Atlantic populations around the southern tip of a smaller Floridian peninsula. At present, swift marine currents moving out of the Gulf of Mexico contribute to the Gulf Stream, which hugs the southeastern coast of Florida (Figure 14.1) and may facilitate transport of Gulf-derived pelagic gametes or larvae into the south Atlantic (e.g., see discussion in this chapter of genetic patterns for oysters and horseshoe crabs). The Gulf Stream veers offshore near Cape Canaveral in east-central Florida.

BIOGEOGRAPHIC PATTERNS

The following format is employed for this section. For each of three realms (freshwater, terrestrial, and maritime), general biogeographic backgrounds on species' ranges and distributional limits first are presented. These provide the traditional source of information for defining biogeographic provinces. Then, comparative molecular genetic data on phylogeographic population structures within species are considered. This information is detailed for one representative species in each realm, and the major phylogeographic patterns are then summarized for the other codistributed species similarly assayed.

Freshwater Fishes

Traditional Biogeographic Provinces.   About 250 species of freshwater fish inhabit coastal drainages of the southeastern United States, with the number of species per drainage ranging from 20 to 157. For purposes of reconstructing biogeographic histories, Swift et al. (1985) detailed the geographic ranges of these species and uncovered a great diversity of distributional patterns (as might be expected across a fauna this diverse). Nonetheless, from observed concentrations of species' distributional limits, the authors were able to define about nine faunal provinces that implicate significant historical and contemporary barriers to fish dispersal. For example, one method of data analysis involved the grouping of river drainages according to overall similarity in species composition, as determined by cluster analyses as applied to a data matrix consisting of the presence versus absence of species. In this analysis, the most basal split in the phenogram distinguished all western drainages (Apalachicola westward through Louisiana) from all drainages to the east (Atlantic coastal rivers, and those throughout the Florida peninsula west to, but not including, the Apalachicola). Thus, an unusually high proportion of species tended to have eastern or western distributional limits in the general area near the Alabama–Georgia state line. For present purposes, these two piscine provinces (Figure 14.2) are emphasized (at the risk of considerable oversimplification) because they were *the* fundamental units recognized in the conventional biogeographic data (species' ranges), and because, as shown next, similar phylogeographic patterns exist *within* several fish species that display continuous distributions across the southeastern United States.

Intraspecific Phylogeography as Registered in Genetic Assays. The bowfin (*Amia calva*) is one of the few living representatives of Holostei, an assemblage of primitive bony fish (also including gars) that was much more diverse during the Mesozoic era. The bowfin's range includes the Atlantic seaboard states from Pennsylvania south through Florida and the Gulf states, and into the Mississippi and Great Lakes drainages. Bermingham and Avise (1986) assayed mitochondrial (mt) DNA genotypes from 75 bowfin collected from 13 major river drainages in six southeastern states. Based on the restriction-site assays (14 endonucleases, mean of 54 restriction sites scored per individual), 13 different haplotypes were revealed, the phylogenetic relationships among which are summarized in Figure 14.3. Clearly, considerable mtDNA variation and differentiation exists in this species, and the genotypes are non-randomly distributed across river drainages. Most notable was a fundamental phylogeographic "break" that distinguished all assayed individuals in eastern drainages (Cooper to the Apalachicola Rivers) from those to the west (Escambia to the Mississippi). This genetic gap involved *at least* four assayed restriction-site changes, and an estimated 0.6% net sequence divergence between the eastern and western clades (after correction for within-clade genetic variation).

The mtDNA survey by Bermingham and Avise (1986) also included three sunfish species (Centrarchidae) similarly assayed across southeastern drainages. Within each species—warmouth (*Lepomis gulosus*), redear (*L. microlophus*), and spotted sunfish (*L. punctatus*)—numerous mtDNA genotypes were observed that also displayed fundamental phylogeographic splits distinguishing populations in eastern versus western drainages (Figure 14.4), albeit at considerably different magnitudes of estimated net sequence divergence. The geographic distributions of the two major mtDNA genotypic assemblages or clades observed within these respective species are summarized in Figure 14.5.

Similar mtDNA phylogeographic analyses have been conducted on two additional freshwater fishes across this region, and these provide further evidence for fundamental genetic partitions distinguishing populations in eastern versus western drainages. In the bluegill sunfish (*L. macrochirus*), mtDNA restriction sites (Avise et al., 1984) appeared to readily distinguish

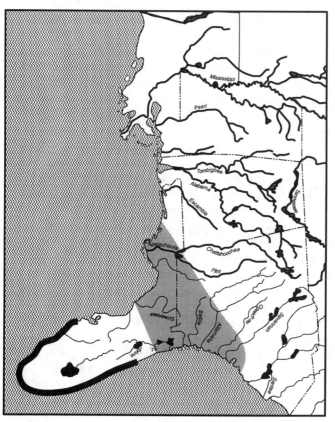

Figure 14.2. Major biogeographic provinces and their boundaries in the southeastern coastal states as identified by traditional considerations of species' ranges and concentrations of distributional limits. For freshwater fish, the darker lines indicate rivers in the distinctive western faunal zone, and the lighter lines are drainages in an eastern faunal zone, as implied by the Swift et al. (1985) phenetic summary of drainage relationships based on similarities in species' compositions (see text and Swift et al. for details and qualifications). For terrestrial organisms, the large shaded area represents Remington's (1968) classic "suture-zone," a general region of presumed secondary contact and hybridization between many endemic Floridian forms and their continental counterparts. For the maritime realm, the heavy line along the Florida coast indicates influence of the tropical "Caribbean" faunal element (Briggs, 1974), which gradually replaces representatives of the warm-temperate "Carolinean" province in a transitional area along the east-central Floridian coast (and perhaps also along Florida's Gulf coast).

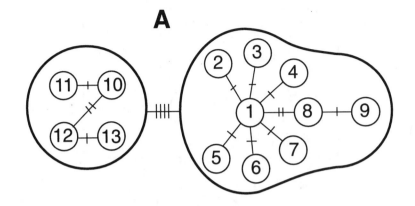

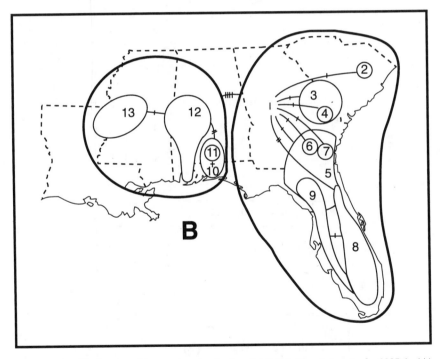

Figure 14.3. Phylogeographic patterns in the bowfin fish (after Avise et al., 1987a). (A) Parsimony network, in which slashes crossing branches indicate restriction-site changes among the 13 observed mtDNA haplotypes. The heavier lines encompass two major genotypic arrays whose members were distinguished by at least four restriction-site changes. (B) The same parsimony network as in A, superimposed over the geographic sources of the collections.

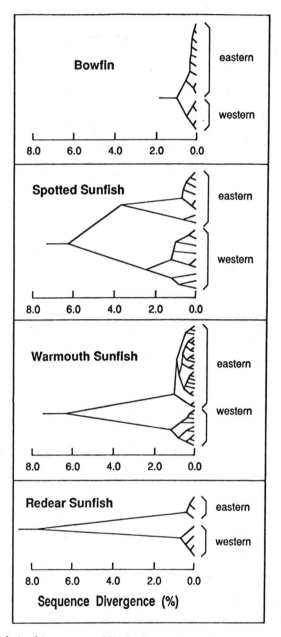

Figure 14.4. Relationships among mtDNA haplotypes observed within each of four freshwater fish species analyzed by Bermingham and Avise (1986). Note that all cluster phenograms are plotted on the same scale of estimated mtDNA sequence divergence.

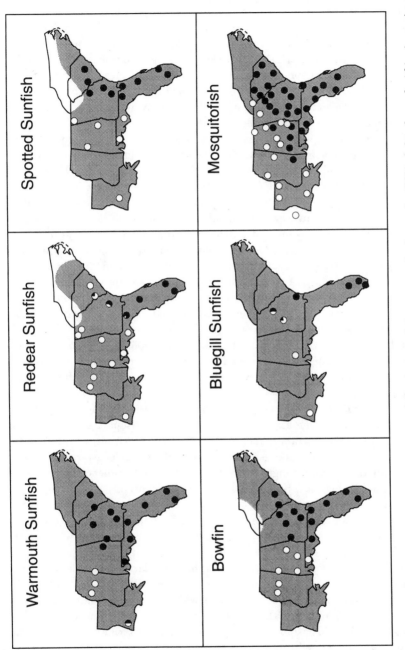

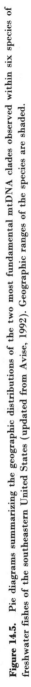

Figure 14.5. Pie diagrams summarizing the geographic distributions of the two most fundamental mtDNA clades observed within six species of freshwater fishes of the southeastern United States (updated from Avise, 1992). Geographic ranges of the species are shaded.

the two previously named subspecies— *L.m. purpurescens* from peninsular Florida and the Atlantic coast, versus *L.m. macrochirus* from the Apalachicola drainage westward (Figure 14.5); and the same can be said for the clear distinction in mtDNA between eastern populations of mosquitofish (*Gambusia holbrooki*) versus those in the western drainages (*G. affinis*; Scribner and Avise, 1993). (These forms previously had been recognized as subspecies [*G.a. holbrooki* and *G.a. affinis*], and only recently `were raised to species rank; Wooten et al., 1988.)

For three fish species, allozyme evidence has been gathered, and the results bolster and extend these phylogeographic conclusions. In both the bluegill (Avise and Smith, 1974) and the mosquitofish complex (Wooten and Lydeard, 1990; Scribner and Avise, 1993), nearly fixed differences in several allozyme markers distinguish the respective eastern from western populations, and also indicate that the taxonomic forms involved have come into secondary contact in regions of extensive introgressive hybridization across parts of South Carolina, western Georgia, and (especially in the mosquitofish) parts of southeastern Alabama, Mississippi, and the Florida panhandle (Figure 14.6). A remarkably similar protein-electrophoretic result was observed in the largemouth bass (*Micropterus salmoides*; Figure 14.6), where nearly fixed allele-frequency differences at several allozyme loci distinguish the Floridian subspecies *M.s. floridanus* from the northern (continental) subspecies *M.s. salmoides* (Philipp et al., 1983). These markers also evidence a broad zone of intergradation across northern Florida and Georgia, where the two subspecies presumably have come into secondary hybrid contact.

In summary, at least seven freshwater fish species or species-complexes have been extensively assayed for genetic composition across the major Atlantic and Gulf coastal drainages of the southeastern United States. Without exception, fundamental genetic partitions distinguish eastern from western populations, thus strongly suggesting a significant influence for historical biogeographic factors. The Floridian peninsula invariably is the geographic stronghold of the eastern forms, whereas drainages from the Alabama-Tombigbee westward almost invariably house exclusively the respective western forms (for which the Mississippi and/or the Alabama-Tombigbee drainages may be the historical [preglacial] biogeographic stronghold; see Mayden, 1988). The exact positions of the geographic boundaries otherwise appear to shift somewhat from one species to the next, as do the magnitudes of the genetic separations. In three species for which allozyme data are available (alone or in conjunction with mtDNA information), secondary contacts and introgressive hybridization between the respective eastern and western forms are strongly implicated in geographically intermediate areas.

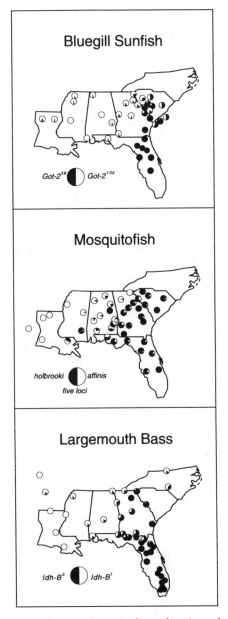

Figure 14.6. Geographic distributions of protein-electrophoretic markers in three species of freshwater fish (after Avise and Smith, 1974; Scribner and Avise, 1993; and Philipp et al., 1983, respectively). Although these diagrams are for particular allozyme loci (*Got-2* in bluegill, mean allele frequencies at five loci in mosquitofish, and *Idh-B* in largemouth bass), they faithfully reflect the major genetic subdivisions revealed in the broader allozyme data for each species.

Terrestrial and Freshwater Tetrapods

Traditional Biogeographic Provinces. Remington (1968) was one of
the first to appreciate the taxonomically widespread occurrence of distinctive forms of endemic plants and animals in the Floridian peninsula, *vis-à-vis*
their respective continental near-relatives. Primarily, the evidence consisted
of geographically concordant contact regions of known or suspected hybridization between numerous Floridian forms (50 in his list) and their continental counterparts along a "suture-zone," usually situated in north Florida
and sometimes extending northeastward along the Atlantic coast into the
Carolinas and/or westward along the Gulf coast into Alabama and Mississippi (Figure 14.2). So important was this secondary contact region deemed
to be that Remington afforded it a status equal to that of only five other
major suture zones recognized across the entire North American continent.
This suture zone therefore demarcates a distinctive peninsular-Florida biotic
province from that of the continental mainland.

Intraspecific Phylogeography as Registered in Genetic Assays. At
least two extant species of pocket gophers (fossorial rodents in Geomyidae)
have been recognized in the southeastern United States. The southeastern
pocket gopher (*Geomys pinetis*) is widely distributed across the northern half
of Florida and the southern halves of Alabama and Georgia. The endangered
or extinct colonial pocket gopher (*G. colonus*), whose taxonomic status
recently has been called into question from genetic evidence (Box 14.2), had
a range confined to Camden County in southeastern Georgia (Figure 14.7).
   Avise et al. (1979b) assayed mtDNA genotypes from 87 specimens of *G.*
pinetis collected across the species' range. Based on the restriction assays
(six endonucleases, 29 restriction fragments scored on average per individual), 23 different haplotypes were revealed, the phylogenetic relationships
among which are summarized in Figures 14.7 and 14.8. Considerable mtDNA
variation and differentiation clearly exist in this species, and the genotypes
were distributed non-randomly across locales. Furthermore, a fundamental
phylogeographic "break" distinguished all assayed individuals in the eastern
half of the species' range from those to the west (Figure 14.7). This genetic
gap involved *at least* nine assayed restriction-site changes, and an estimated
3.1% net sequence divergence. The genetic survey also included protein-
electrophoretic assays, which revealed a nearly identical pattern of distinction between eastern and western populations, including essentially fixed
differences in allelic frequency at an albumin locus (Avise et al., 1979b).
   The gopher tortoise (*Gopherus polyphemus*) has a nearly identical distribution to that of *Geomys pinetis*, and also displays a remarkably similar
phylogeographic pattern as registered in mtDNA (Figures 14.8 and 14.9).

A

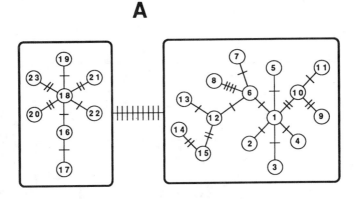

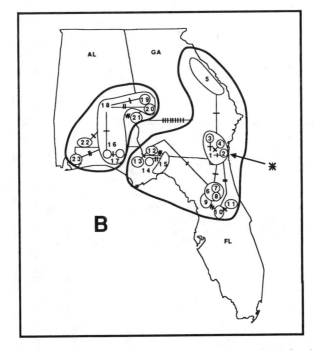

Figure 14.7. Phylogeographic patterns in the southeastern pocket gopher (after Avise et. al., 1979b). (A) Parsimony network, in which slashes crossing branches indicate inferred restriction-site changes among the 23 observed mtDNA haplotypes. The larger boxes encompass two major genotypic arrays whose members were distinguished by at least nine restriction-site changes. (B) The same parsimony network as in A, superimposed over the geographic sources of the collections in Alabama, Georgia, and Florida. The star indicates the locale of the colonial pocket gopher (see Box 14.2).

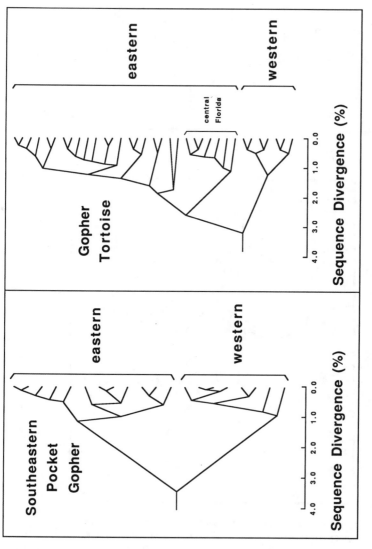

Figure 14.8. Relationships among mtDNA haplotypes observed within the southeastern pocket gopher (Avise et al., 1979b; three haplotypes are not shown because they were assayed for only a subset of endonucleases) and the gopher tortoise (after Osentoski, 1993). Although both cluster phenograms are plotted on the same scale of mtDNA sequence divergence, note that the assays differed with respect to which portions of the mtDNA molecule were monitored (see text).

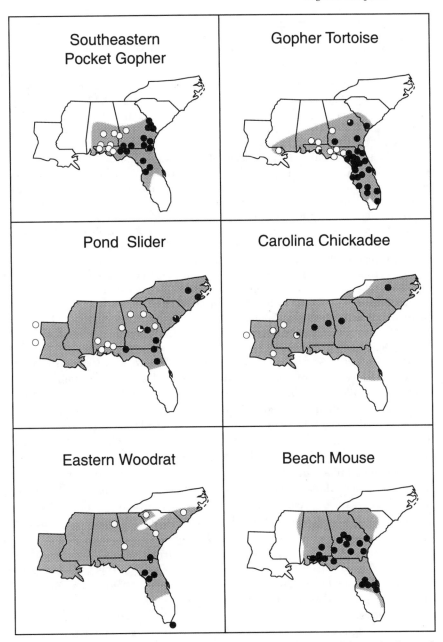

Figure 14.9.   Pie diagrams summarizing the geographic distributions of the most fundamental mtDNA clades observed within each of six species of freshwater and terrestrial tetrapods across the southeastern United States (see text). Geographic ranges of the species are shaded.

Box 14.2. Taxonomic and Conservation Status of the Colonial Pocket Gopher

Geomys colonus was described in the 1800s (Bangs, 1898) on the basis of a darker pelage and minor cranial differences from its common and widespread congener *G. pinetis*. The original identified range of *G. colonus* encompassed about 16 km^2 of coastal plain in Camden County, Georgia (Figure 14.7), but the population remained essentially unnoticed and unstudied until 1967, when a pocket gopher population within this historical range was "rediscovered." This modern population consisted of less than 100 individuals on 200 hectares, and was listed as endangered by the State of Georgia. Subsequently, a multifaceted genetic survey was conducted based on allozymes, mtDNA, and karyotypes (Laerm et al., 1982). None of the assays detected any consistent distinction between "*G. colonus*" and geographically adjacent populations of *G. pinetis*. This result could not be attributed to a lack of sensitivity of the methods employed, because each revealed dramatic genetic differences among a broader geographic array of *G. pinetis* populations (e.g., Figure 14.7). The conclusion was clear: "*G. colonus*" did not warrant recognition as a distinct species. Either the original description in 1898 was inappropriate, or a valid *G. colonus* species had gone extinct in this century, replaced by recent *G. pinetis* immigrants into Camden County (Avise, 1989).

Osentoski (1993) assayed mtDNA restriction sites in blood samples taken from 112 individuals of this federally threatened species. Examination of tortoises from more than 50 locales revealed 35 mtDNA haplotypes that grouped into two distinct genetic clusters whose distributions are summarized in Figure 14.9. Again, an eastern phylogenetic assemblage differed from a western assemblage (estimated sequence divergence ca. 3.2%, based on assays primarily of the rapidly evolving control region of the molecule), with the geographic boundary area between the two forms occurring in the vicinity of the Alabama–Georgia border and the panhandle of Florida. Within the eastern "clade," a pronounced phylogeographic subcluster of gopher tortoises was observed in the west-central highlands of peninsular Florida (sequence divergence ca. 2.7%; Figure 14.8).

   In a freshwater turtle, the pond slider (*Trachemys scripta*), analyses of 55 individuals from 19 locales revealed two mtDNA haplotypes (differing at three restriction sites; estimated sequence divergence 0.6%), whose distributions are shown in Figure 14.9 (Avise et al., 1992). In another freshwater turtle species, the musk turtle *Sternotherus minor*, mtDNA analyses (both restriction fragment length polymorphism [RFLP] assays and control-region sequences) of 52 specimens from throughout the species' range uncovered extensive mtDNA variation among locales (Walker et al., 1995), with a fundamental feature of the data again being a strong phylogenetic subdivi-

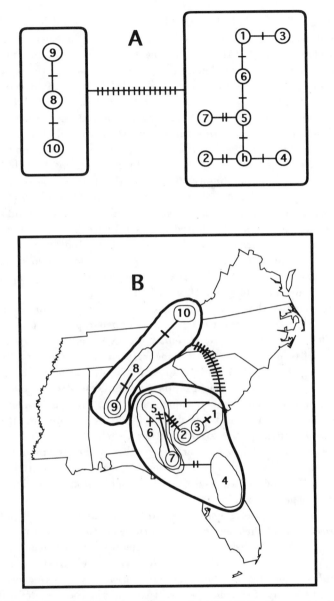

Figure 14.10. Phylogeographic patterns in the musk turtle. (A) Parsimony network, in which slashes crossing branches indicate inferred restriction-site changes among 10 observed (and one hypothetical, h) mtDNA haplotypes. The heavier lines encompass two major genotypic arrays whose members were distinguished by at least 16 restriction-site changes. (B) The same parsimony network as in A, superimposed over the geographic sources of the collections from the southeastern United States.

sion distinguishing populations in Florida and Georgia from those to the north and west in Alabama, northwest Georgia, Tennessee, and western Virginia (Figure 14.10). Although the geographic range of the musk turtle extends farther north than that of the gopher tortoise (or the southeastern pocket gopher), the geographic orientation of the major phylogenetic subdivision in that species is similar to those presented in Figure 14.9.

Geographic surveys of mtDNA have been reported for at least four other tetrapod species sampled across the southeastern coastal states. In a bird, the Carolina chickadee *Parus carolinensis*, analyses of 55 individuals from six southeastern coastal states revealed 14 different mtDNA haplotypes, which grouped into two highly distinct clades (net sequence divergence 3.7%) whose geographic distributions are shown in Figure 14.9 (Gill et al., 1993). In the woodrat, *Neotoma floridanus*, analyses of 114 individuals from 33 locales across the eastern United States revealed 28 mtDNA genotypes, most of which were localized geographically (Hayes and Harrison, 1992). If attention is confined to the southeastern samples alone, two major clades (net sequence divergence ca. 1.6%) were revealed with geographic distributions shown in Figure 14.9 (see Hayes and Harrison for further details). The authors concluded that "vicariance events, including the insularization of Florida, have been important determinants of geographic variation." In the white-tailed deer, *Odocoileus virginianus*, mtDNA analyses based on RFLP patterns (Ellsworth et al., 1994) documented a fundamental phylogeographic split distinguishing most Floridian and southern Atlantic coast populations from those in the Florida panhandle westward to Mississippi (map not shown). The authors concluded that "the pattern of mtDNA variation observed in white-tailed deer in this region is concordant spatially with those identified previously in... other vertebrates" (including some of those in Figure 14.9). They went on to hypothesize that the phylogeographic distinctiveness of the Floridian/East Coast assemblage was attributable in this case to Pleistocene dispersal from Neotropical source populations via a Gulf Coast dispersal corridor.

In the beach mouse, *Peromyscus polionotus*, analyses of 68 individuals from across most of the species' range in three southeastern states (Avise et al., 1979a, 1983) revealed a total of 22 different mtDNA genotypes, all of which were geographically localized. However, in contrast to the species considered previously, no fundamental "break" was evidenced in the mtDNA phylogeny for this species. Instead, the poorly defined clades showed no obvious geographic or phylogenetic disjunctions. However, when viewed from a broader phylogeographic perspective, the confinement of *P. polionotus* to the southeastern states (Figure 14.9) may itself register a historical vicariant event (or perhaps a dispersal episode) involving the Floridian peninsula. From genetic and other evidence, the beach mouse shares a recent ancestor with the deer mouse (*P. maniculatus*), an abundant species across the remainder of the continent that exhibits a paraphyletic relationship to *P. polionotus* with respect to mtDNA lineages (Avise et al., 1983; Lansman et al., 1983).

In summary, phylogeographic datasets for a number of terrestrial and freshwater tetrapods in the southeastern United States evidence pronounced southeast–northwest phylogenetic discontinuities. Although these patterns appear somewhat varied in terms of evolutionary depth and precise spatial distribution, available genetic data are consistent with strong historical–biogeographic influences on population structures, including a probable insularization role for peninsular Florida. In the various taxa, this biogeographic influence may be reflected in significant phylogeographic breaks within species (as in the eastern woodrat), in species' ranges *per se* (as in the beach mouse), or in both of these aspects of evolutionary genetic structure (as in the gopher tortoise and pocket gopher, which have primary, intraspecific genetic subdivisions within the southeastern United States but also differ from phylogenetically allied species in western regions of North America). The differing magnitudes of genetic divergence suggested by the molecular data might be attributable in part to population separations that date to different times in the cyclical nature of biogeographic changes in the southeastern United States (to be discussed), and/or to lineage heterogeneities in the rates with which genetic differences accumulate (e.g., Avise et al., 1992).

Coastal Vertebrates and Invertebrates

Traditional biogeographic provinces. Several marine zoogeographers and faunal specialists have summarized distributional records for various marine taxa in the western Atlantic (review in Briggs, 1974). With regard to the coastal fauna of the southeastern United States, the most striking feature of these distributions is the north–south transition between the tropical marine faunas of southern Florida (and the Caribbean), and the warm-temperate faunas farther north along the Atlantic and Gulf coasts (Figure 14.2). For example, along the inner continental shelf of the Atlantic coast, a temperate "Appalachian Province" for mollusks extends down the northern third of Florida, where a gradual replacement by more tropical-adapted species occurs; a faunal break for octocorals similarly exists in east-central Florida; and a large number of fish species have northern or southern distributional limits along the east-central Floridian coastline (Briggs, 1958). A general consensus is that the area around Cape Canaveral (Figure 14.2) is more or less the center of a pronounced transitional boundary between highly distinctive maritime faunal provinces. A similar (although perhaps less well defined) transitional area between tropical and warm-temperate faunas occurs along the Gulf coast of Florida (Figure 14.2), although some authors have considered the whole of the Gulf of Mexico to be primarily tropical in biotic composition (see references in Briggs, 1974).

Intraspecific Phylogeography as Registered in Genetic Assays. The seaside sparrow assemblage (*Ammodramus maritimus*) has had a complex taxonomic history, with varying numbers of species and subspecies described. Currently, about nine subspecies are recognized, with more or less abutting ranges along the species' coastal-marsh habitat from New England to south Texas. One morphologically distinctive form (the dusky seaside sparrow) had been listed as an endangered species prior to its extinction (Box 14.3), and another subspecies confined to the western

---

Box 14.3. Taxonomic and Conservation Status of the Dusky Seaside Sparrow

In 1872, a melanistic form of seaside sparrow was discovered in Brevard County, Florida (Figure 14.11), and subsequently described as a distinct species, *Ammodramus nigrescens*. In the 1960s, the population (by then reduced to subspecific status, *A. maritimus nigrescens*) was in severe decline due to habitat alterations by humans, and the dusky seaside sparrow was placed on the U.S. Endangered Species List. By 1980, the few remaining birds (all of which were males) were brought into captivity and mated to individuals from a Gulf coast subspecies, the intent being to produce F_1 hybrids and backcross progeny (the latter carrying primarily dusky [nuclear] genes) for eventual reintroduction into Brevard County. The breeding program later was discontinued.

The molecular genetic survey by Avise and Nelson (1989; see text) included tissue from the last available dusky male, which had died in captivity in 1987. With respect to the mtDNA restriction sites assayed, the dusky proved indistinguishable from the most common genotype (haplotype "1"; Figure 14.11) observed in other Atlantic coast seaside sparrows, yet was highly distinct from all Gulf coast birds. Thus, the traditional taxonomy for seaside sparrows (from which conservation priorities understandably were derived) probably had been a misleading guide to evolutionary relationships in the complex in two respects: (1) in failure to recognize the fundamental phylogenetic dichotomy between Atlantic and Gulf populations; and (2) in taxonomic emphasis on distinctions within both coastal regions that appear evolutionarily minor compared to the between-region genetic differences.

One caveat should be added. The genetic assays in this case involved only mtDNA, and thus different phylogeographic outcomes might be envisioned for nuclear genes. The principle counterarguments currently available against this possibility are that an Atlantic–Gulf phylogenetic dichotomy: (1) makes historical and zoogeographic "sense" (and indeed was predicted by Funderburg and Quay [1983] from other lines of evidence); and (2) is displayed concordantly by a number of other coastal-restricted species (see text).

Everglades (the Cape Sable seaside sparrow) remains on the federal list as endangered.

Avise and Nelson (1989) assayed mtDNA genotypes from 40 seaside sparrows representing seven named subspecies. Based on the restriction-site assays (18 endonucleases, 89 sites scored on average per individual), 11 different haplotypes were revealed, with phylogenetic relationships summarized in Figure 14.11. The most noteworthy finding was a fundamental phylogeographic "break" that distinguished all Atlantic coast birds from all specimens collected along the Gulf of Mexico. This genetic gap involved *at least* five assayed restriction-site changes, and an estimated 1.0% net sequence divergence. Such a phylogeographic outcome had been anticipated by Funderburg and Quay (1983), who speculated from distributional and other historical zoogeographic information that populations in the seaside sparrow complex were phylogenetically split into Atlantic versus Gulf forms.

Currently, there is a pronounced hiatus in the geographic distribution of seaside sparrows, with no extant populations in southeastern Florida. Another coastal species with disjunct Atlantic and Gulf populations that has been assayed for mtDNA is the black sea bass, *Centropristis striata* (Bowen and Avise, 1990). Here too, Atlantic and Gulf forms (conventionally recognized as different subspecies) proved to be quite distinct genetically, with an estimated net nucleotide sequence divergence of 0.7% (Figure 14.12).

Two other coastal vertebrate species (or species-complexes) that show evidence for a fundamental Atlantic–Gulf disjunction in mtDNA phylogeny involve the toadfish complex *Opsanus* (Avise et al., 1987b) and the diamondback terrapin *Malaclemys terrapin* (Lamb and Avise, 1992). In the former assemblage, two related species (*O. tau* and *O. beta*) essentially are confined to the Atlantic and Gulf coasts, respectively, and differ by an estimated 9.6% mtDNA sequence divergence. In the latter species, mtDNA variation was limited, but the one fixed restriction-site difference observed did distinguish Gulf samples from most of those in the Atlantic. In this case, the range of the "Gulf form" also extended into southeast Florida, in a pattern (Figure 14.13) nearly identical to those of some other coastal species, as described next.

Three coastal invertebrates with more or less continuous distributions along the Atlantic and Gulf of Mexico have been assayed for mtDNA. In the tiger beetle *Cicindela dorsalis* (Vogler and DeSalle, 1993), horseshoe crab *Limulus polyphemus* (Saunders et al., 1986) and American oyster *Crassostrea virginica* (Reeb and Avise, 1990), Atlantic and Gulf populations again proved highly distinct genetically (Figures 14.12 and 14.14), although in the latter two species the ranges of the "Gulf forms" extended into southeast Florida (Figure 14.13). In the case of the American oyster, subsequent restriction-site assays of anonymous nuclear loci added support for this

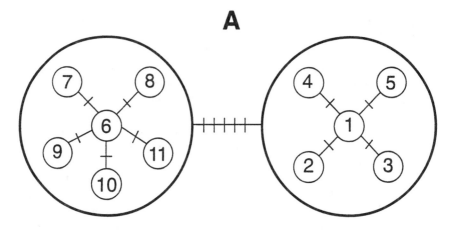

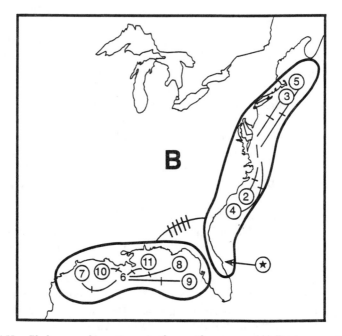

Figure 14.11. Phylogeographic patterns in the seaside sparrow. (A) Parsimony network, in which slashes crossing branches indicate inferred restriction-site changes among the 11 observed mtDNA haplotypes. The heavier lines encompass two major genotypic arrays whose members were distinguished by at least five restriction-site changes. (B) The same parsimony network as in A, superimposed over the geographic sources of the collections. The star indicates the locale of the dusky seaside sparrow (see Box 14.3).

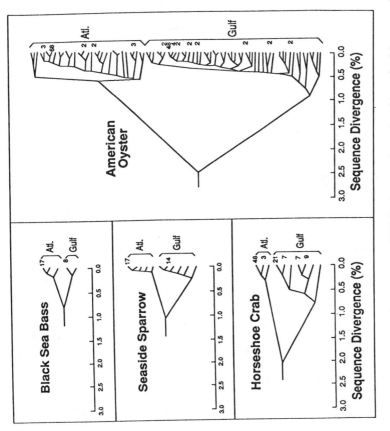

Figure 14.12. Relationships among mtDNA haplotypes observed within each of four coastal-restricted species in the southeastern U.S. (after Avise, 1992). Numbers of individuals showing the various mtDNA genotypes are indicated to the right; terminal branches without numbers were represented by single individuals. Note that all cluster phenograms are plotted on the same scale of sequence divergence.

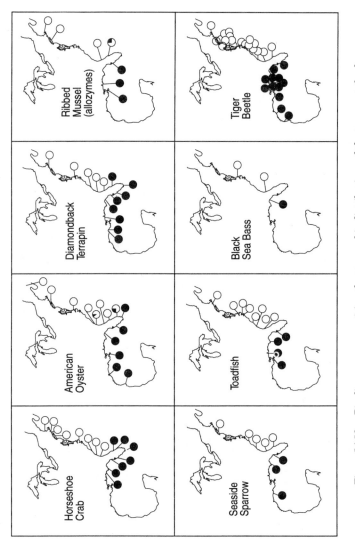

Figure 14.13. Pie diagrams summarizing the geographic distributions of the two most fundamental mtDNA clades observed within each of several species of coastal-restricted vertebrates and invertebrates in the southeastern U.S. (after Avise, 1992 and Vogler and DeSalle, 1993). Also shown, in the upper right-hand corner, are pie diagrams summarizing the geographic distributions of two distinct groups of ribbed mussels as identified by allozymic (and morphological) markers (from data in Sarver et al., 1992). A nearly identical pattern of allozyme differentiation (not shown) also characterizes the *Fundulus majalis–F. similis* fish complex (Duggins et al., 1995).

Atlantic–Gulf distinction (Karl and Avise, 1992), which had not, however, been apparent in earlier allozyme assays (Buroker, 1983). In all three invertebrates, the level of genetic divergence between Atlantic and Gulf populations far surpassed observed differences within either region.

Population genetic studies based on allozymes have uncovered further evidence for Atlantic–Gulf distinctions in some other coastal invertebrates and vertebrates. In the ribbed mussel *Geukensia demissa* (Sarver et al., 1992), and in the *Fundulus majalis–F. similis* fish complex (Duggins et al., 1995), pronounced genetic discontinuities (fixed or nearly fixed allelic frequency differences at several loci) distinguished the respective Atlantic from Gulf populations (which the authors in both cases recommend be considered separate species). The respective Gulf forms again extended along the southeastern coastline of Florida in patterns (Figure 14.13) reminiscent of that for the American oyster, horseshoe crab, and diamondback terrapin. In the stone crab, *Menippe mercenaria* (Bert, 1986), allozyme data documented a pronounced genetic distinction between Floridian and western Gulf forms (now considered distinct species), although in this case a primary boundary between the two populations appeared to be on the west coast of Florida, with evidence of interregional hybridization along both the northeastern Gulf and the south Atlantic coast (Bert and Harrison, 1988).

Not all coastal species genetically assayed have shown clear evidence for fundamental phylogenetic dichotomies between populations in the Atlantic versus the Gulf of Mexico (Figure 14.15). Five assayed species or species complexes of fish whose populations appear to lack such obvious genetic distinctions are the red drum *Sciaenops ocellatus* (Gold and Richardson, 1991), American eel *Anguilla rostrata* (Avise et al., 1986), hardhead catfish *Arius felis* (Avise et al., 1987b), the menhaden complex *Brevoortia*, and (to an arguable extent) the Atlantic sturgeon *Acipenser oxyrhynchus* (Bowen and Avise, 1990). An absence of regional genetic differentiation in the American eel no doubt is attributable to its catadromous life cycle, which involves spawning in the tropical mid-Atlantic ocean (Sargasso Sea), and subsequent larval transport by ocean currents. The larvae may be deposited more or less at random to coastal streams, where maturation to adulthood takes place (and where the samples were collected for the genetic analyses). Lack of pronounced regional differentiation in the other groups may be due to contemporary gene flow around south Florida, and/or to retention of ancestral polymorphisms by extant Atlantic and Gulf populations that have been in contact in relatively recent evolutionary time (Bowen and Avise, 1990).

In summary, large-scale molecular phylogeographic surveys have been conducted on more than a dozen coastal-restricted species or species complexes in the southeastern United States. Across a surprising diversity of

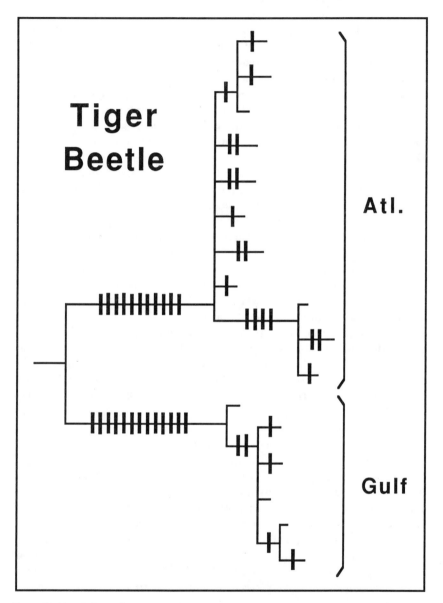

Figure 14.14.   Relationships among the 17 mtDNA haplotypes observed in the tiger beetle, a species confined to coastal beach habitat (after Vogler and DeSalle, 1993). Shown is a rooted, strict consensus parsimony tree, with inferred character state changes (nucleotide substitutions) indicated by slashes crossing tree branches.

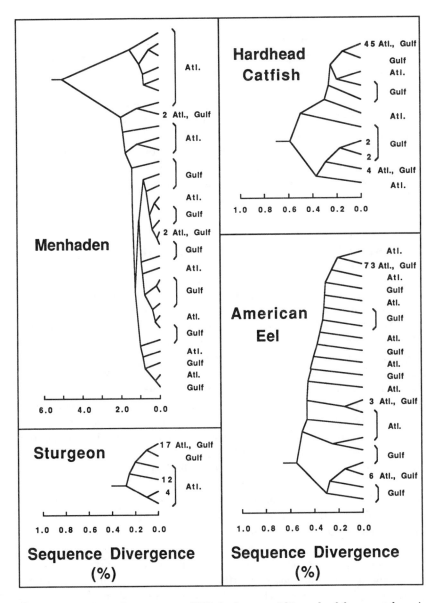

Figure 14.15. Relationships among mtDNA haplotypes within each of four coastal species whose populations in the Atlantic versus Gulf of Mexico appear to lack pronounced genetic differentiation (after Avise, 1992). Numbers of individuals showing the various mtDNA genotypes are indicated to the right; terminal branches without numbers were represented by single individuals. Note that the cluster phenogram for the menhaden is plotted on a different scale from that of the other species.

taxa, ranging from several invertebrates to fish and salt-marsh tetrapods, genetic data revealed fundamental phylogenetic discontinuities between populations along the Gulf of Mexico versus most of the south-Atlantic coastline, yet little evidence for dramatic historical genetic partitions within either region. Some of these studies involved species with disjunct Gulf and Atlantic populations. Others involved species continuously distributed around southern Florida at the present time. In some of these latter groups, genotypes normally characteristic of the Gulf of Mexico also occur along Florida's southeastern coast. Although the Atlantic–Gulf populational distinction is far from universal, the level of concordance in patterns of genetic subdivision across multiple taxa strongly implicates shared biogeographic influences on the genetic architectures of a substantial fraction of this maritime fauna.

GENEALOGICAL DEPTH, CONCORDANCE, AND BIOGEOGRAPHY

Geographic populations can be genetically structured at a variety of ecological and evolutionary levels, such that not all population genetic structures that are "statistically significant" can be considered equal. In any species in which single-generation dispersal distances are much less than the total geographic range occupied by the species, at least some newly arisen mutations will have circumscribed geographic distributions (Neigel et al., 1991; Neigel and Avise, 1993). With sufficient laboratory effort (or by serendipity), some of these mutations likely will be uncovered in various molecular genetic assays, yielding statistically significant differences in allelic frequencies among geographic sites. However, for such populations isolated by distance over short (ecological) timescales, relatively few mutational steps normally will be registered in any gene (such as mtDNA). Furthermore, due to the vagaries of the origin and survival of mutations, the particular geographic units that happen to be recognizable likely will differ from one unlinked gene to the next within a species, as well as from one species to the next among a series of codistributed taxa. In the faunas of the southeastern United States, many examples of such "shallow" or "idiosyncratic" population genetic structures exist, particularly for sedentary or habitat-restricted species. For example, *within* either the eastern or the western forms of the low-vagility pocket gopher (Figure 14.7), or of the musk turtle (Figure 14.10), mtDNA markers distinguished various local demes, but the numbers of mutational steps typically were small. Furthermore, in the pocket gopher an identical (overlaid) demic structure was not registered concordantly by nuclear (allozyme) or morphological markers (Avise et al., 1979b).

On the other hand, when populations have experienced limited or no genetic contact over longer (evolutionary) timescales, geographic population structures will be historically "deep." In such cases, populations (or species) will tend to accumulate additional mutational differences in particular genes, as well as to become concordantly recognizable in assays of multiple independent (unlinked and non-epistatic) loci and the traits they encode. Furthermore, if significant historical biogeographic barriers to dispersal were responsible for the population separations, codistributed species with similar ecological requirements should tend to become genetically structured in a geographically similar fashion. As emphasized earlier, such geographic concordance does in fact exist with respect to the major phylogenetic discontinuities within several groups of codistributed taxa in the southeastern United States.

These considerations have led to suggestions that the degree of phylogenetic concordance *per se* across independent genetic markers should be a useful if not necessary criterion for distinguishing deep population genetic structures from those that are evolutionarily shallow (Avise and Ball, 1990). Any two populations isolated from one another for a sufficiently long period of time (relative to effective population size, and to the mutation rate at marker loci) should tend to evolve to a situation in which multiple genetic markers concordantly distinguish them. Such genealogical concordance has several aspects (Table 14.1): (1) coincidence in the population units identified by multiple independent mutations in a given gene (such as mtDNA); (2) coincidence in the population units identified by multiple independent genes within a species' pedigree (such as mtDNA and various unlinked nuclear loci); and (3) coincidence in the geographic patterns of

Table 14.1. Four Aspects of Genealogical Concordance in Phylogeographic Inference

1. *Concordance across sequence or other characters within a non-recombining segment of DNA.*
   Relevance: contributes to the statistical significance of putative clades within a gene tree.

2. *Concordance in significant genealogical partitions across multiple independent loci within a species.*
   Relevance: establishes that phylogenetic partitions in gene trees register significant partitions in organismal phylogeny.

3. *Concordance in the geographic positions of gene-tree partitions across multiple codistributed species.*
   Relevance: strongly implicates shared historical biogeographic factors in shaping intraspecific phylogenies.

4. *Concordance of gene-tree partitions with geographic boundaries between traditionally recognized biogeographic provinces.*
   Relevance: strongly implicates shared historical biogeographic factors in shaping intraspecific organismal phylogenies and distributional patterns.

genetic structure across multiple codistributed species whose populations had been concordantly shaped by historical biogeographic forces. All three aspects of concordance were abundantly evidenced in the southeastern fauna.

Although the geographic patterns of several of the deeper phylogenetic disjunctions were remarkably concordant across various southeastern taxa (Figures 14.5, 14.9, and 14.13), the absolute magnitudes of the population genetic separations (e.g., in mtDNA) sometimes differed considerably across species (Figures 14.4 and 14.12). Assuming that the shared geographic patterns reflect shared historical biogeographic influences, several explanations might account for the differing magnitudes of genetic divergence: (1) heterogeneity in rates of molecular evolution across taxa (independent evidence does exist for several-fold differences in evolutionary rate of mtDNA [review in Rand, 1994], including rate differences among some of the taxa included in this report [Avise et al., 1992]); (2) differences in the levels of ancestral genetic variability that were available for conversion to between-regional differences subsequent to the vicariant or dispersal event(s); and/or (3) different timings of the relevant historical episodes.

As to the latter possibility, physiographic changes that likely have molded the genetic architectures of southeastern biotas certainly have been cyclical. For example, the Pleistocene was characterized by numerous glacial advances and retreats, such that the separations between extant eastern versus western (or Gulf vs. Atlantic) populations in various species might trace genealogically to different glacial or interglacial episodes (depending in part on the patterns of extinction of former regional isolates). Furthermore, some of the biogeographic footprints on population structure might trace to earlier (Miocene or Pliocene) times, when (for example) the central highlands of Florida existed periodically as one or more large islands divorced from the mainland. Indeed, *within* particular species and species groups, different depths for multiple genetic disjunctions may be registered. For example, the east–west disjunctions within *Geomys pinetis* and *Gopherus polyphemus* (Figure 14.8) might reflect relatively recent barriers to gene flow, whereas the phylogenetic separations of these species from their respective congeners in the western United States probably trace to earlier vicariant episodes (Lamb et al., 1989).

A fourth class of concordance also should be emphasized (Table 14.1). In the freshwater and maritime realms (and perhaps also in the terrestrial realm), the major phylogeographic disjunctions registered in the genetic assays (Figures 14.5, 14.9, and 14.13) tend to generally coincide with the boundaries between the respective zoogeographic provinces as previously identified by concentrations of species' distributional limits (Figure 14.2). Similar historical–biogeographic processes probably are evidenced by these intra- and interspecific phenomena. For example, if either of the two regional genetic forms within a species (such as the bluegill sunfish, or the horseshoe crab) were to have gone extinct (for whatever reason), the remaining taxon then would display a distributional boundary contributing to the recognition of zoogeographic provinces as defined by species' ranges.

CONSERVATION RAMIFICATIONS

It is now abundantly clear that most species should not be viewed as monotypic entities, but rather as consisting of a series of genetically differing populations with a rich fabric of phylogeographic structure. This structure can range from evolutionarily shallow (evidenced by relatively minor genetic differences idiosyncratic to particular genes and geographic populations) to deep (evidenced by pronounced genetic differences often concordant across genes, across geography, and sometimes across the geographic populations of codistributed species). The molecular genetic observations described in this chapter evidence both shallow and deep population structures within a diverse regional fauna. Only recently have concepts of varying "depth" of phylogeographic structure been incorporated into discussions of population genetics and conservation biology (e.g., Avise, 1987, 1994; Dizon et al., 1992).

In this latter context, Moritz (1994a, 1994b) makes an important distinction between *management units* (MUs, or "stocks") and *evolutionarily significant units* (ESUs; see also Ryder, 1986). MUs are proposed to reflect the shallower genetic subdivisions within a species—populations connected by little or no contemporary gene flow, but not separated historically for long periods of time. Shallow population units (MUs) can be of conservation relevance because they reflect low contemporary dispersal and gene flow within a species, perhaps sufficiently so that demes overexploited or extirpated by humans (or other causes) are unlikely to recover via natural exogenous recruitment, at least over ecological timescales relevant to immediate management interests (Avise, 1995). ESUs represent major (deep) phylogenetic subdivisions within a species—populations or groups of populations long-isolated from one another, and therefore collectively encompassing a substantial fraction of a species' genetic diversity. Knowledge concerning deeper population structures (evidenced, e.g., by concordant genetic patterns within and among species) is also important for conservation biology, as, for example, in the following areas.

Recognition of Endangered Taxa

Taxonomic assignments inevitably shape perceptions of biological diversity, so it is somewhat disconcerting to realize that many subspecies' and species' descriptions trace to limited information, often gathered in the last century, on the geographic distributions of a small number of (usually morphological) traits with unknown genetic bases. For most of the species considered in this chapter, the subspecies' designations previously available failed to register the fundamental phylogeographic splits that were uncovered in the molecular genetic assays. To cite but one example, four subspecies of the tiger beetle (two in the Atlantic and two in the Gulf) had been recognized by morphological criteria, but this taxonomy at face value gave no clear clue to the basal Atlantic–Gulf dichotomy revealed in the mtDNA assays. Furthermore, within either coastal region, "the geographic distribution of haplotypes did not coincide with the distribution of morphological subspecies" (Vogler and DeSalle, 1993), one of which (*C.d. dorsalis*) is among the few insects currently listed for protection under the U.S. Endangered Species Act (Vogler et al., 1993).

Once a Latin binomial or trinomial is in the literature, the group of organisms to which it refers almost automatically assumes an aura of reality that may or may not be commensurate with the taxon's true evolutionary distinctiveness. Thus, given the overriding importance of taxonomic assignments on the recognition and management of biodiversity, increased attention should be devoted to taxonomic assessments (from molecular as well as other data). This may be especially true at the "intraspecific" level, where a greater appreciation of the different patterns and levels of population structure (and of the relevance of genealogical concordance principles in interpreting these patterns) can only enlighten intraspecific systematics. Conservationists should keep open minds regarding possible taxonomic realignments on the basis of new information. In the southeastern fauna, examples of taxonomic assignments that appear to have been misleading as guides to "endangered species" management (in the sense of not reflecting ESUs) are summarized in Boxes 14.2 and 14.3.

Design and Maintenance of Biodiversity Sanctuaries

In the final analysis, concerns about the conservation of biodiversity represent concerns about the conservation of genetic diversity (Avise, 1994). Genetic studies of the southeastern fauna indicate that major components of evolutionary genetic diversity within and among species are strongly organized on a regional basis, presumably because of shared historical-biogeographic influences. Such findings, if they should prove generalizable to other biotic provinces, suggest several guidelines for recognizing and maintaining the integrity of genetically distinct regional biotas (ESUs), and thereby conserving significant fractions of extant genetic diversity both within and among species.

One management element in maintaining the sanctity of distinct regional biotas should involve limiting unnecessary interregional translocations. However, many public and private management agencies actively *promote* geographic translocations, for purposes such as bolstering local population sizes, introducing "desirable" genetic traits into a region, or increasing local heterozygosity. Unfortunately, several undesirable consequences also may arise from such translocation programs, including: (1) the possibility of disease or parasite spread; (2) irretrievable loss of the rich phylogeographic records of species (an analogy in the field of anthropology would be the uncontrolled mixing of artifacts between archaeological sites); and (3) inevitable erosion of overall genetic diversity within a species, much of which we now recognize to be generated and maintained by geographic isolation (both contemporary and historical). Translocations sometimes may be justified, as in the reintroduction of a native species to its former range where it had been extirpated by humans. In general, however, the burden of proof in proposed translocation programs should rest on proponents of this strategy, rather than on the opponents (as is currently the case). Many biologists recognize that introductions of exotic species can cause irreparable harm to regional biodiversity by forcing extinctions of native species, but they have been slower to appreciate the similar kinds of problems that can stem from translocations among regional ESUs within species.

Another management approach in conservation biology might involve the explicit design of biogeographic reserves that would "capture" significant fractions of regional biotic diversity. In the freshwater and terrestrial realms of the southeastern United States, for example, these biogeographic reserves might include portions of the central Floridian peninsula and Atlantic coastal states, as distinct from other reserves in the more westerly Gulf coastal states. In the maritime realm, distinct biogeographic sanctuaries could be envisioned within the Gulf of Mexico, and within the Atlantic. Additional sanctuaries also might be designed to encompass major boundary regions between provinces (as in the Cape Canaveral region of Florida), where active zones of species contact and hybridization appear to be unusually common. On a broader geographic scale, one exciting conservation vision would be the development of a series of national and international Biogeographic or Biodiversity Parks, perhaps analogous to the current National Park system in the United States that focuses preservation efforts on exceptionally beautiful or unique features of the geological landscape.

From a pragmatic conservation viewpoint, perhaps the most important finding from genetic studies of the southeastern fauna is the concordance in geographic position between genetically defined biotic provinces at the intraspecific level, and zoogeographically defined biotic provinces as traditionally identified through species' distributional limits and considerations of historical physiography (Figure 14.2). Because of the time and expense required to conduct multifaceted genetic studies on regional biotas, reasonably comprehensive molecular examinations of the sort reported in this chapter can realistically be contemplated in only a few model circumstances. Nonetheless, if studies from the southeastern fauna can be generalized, then a highly encouraging message is that much of the biodiversity information needed to design regional biogeographic reserves already may be available— in the form of traditionally recognized biogeographic provinces and subprovinces (on which there is a large literature). Thus, we need not wait for complete genetic reexaminations of the biological world before embracing regional conservation perspectives.

SUMMARY

1.   Most conservation genetic studies inherently adopt a taxonomic (rather than ecosystem) orientation, but many of the broader problems in conservation biology involve concerns about regional biodiversity. The case histories presented in this chapter summarize regional population genetic findings on the comparative phylogeographic patterns observed within numerous taxonomically unrelated species in the freshwater, terrestrial, and maritime realms of the southeastern United States.

2.   As gauged by assays of mtDNA and some nuclear markers, including allozymes, conspecific populations of most southeastern taxa display phylogeographic structures that range from evolutionarily shallow to deep. The shallow separations are evidenced by molecular differences that are idiosyncratic to particular genetic markers or populations, and presumably reflect the limited vagilities of most individuals relative to the respective species' ranges. The deeper phylogeographic separations are evidenced by genetic differences that tend to be concordant across multiple markers within a gene, across multiple genes within a species' pedigree, and across the geographic populations of multiple codistributed species. These deeper population structures presumably reflect long-term historical biogeographic influences.

3.   The phylogeographic concordances across conspecific populations of several surveyed taxa suggest that particular regions in the Southeast are centers for a substantial proportion of extant genetic diversity at the intraspecific level. In the freshwater realm, eastern populations (in drainages of peninsular Florida and along the Atlantic coast) often tend to be highly distinct from populations to the west. In the terrestrial realm, populations in the Floridian peninsula and adjoining areas frequently tend to be highly

distinct from those to the north and west. In the maritime realm, large genetic differences frequently characterize Atlantic coastal populations from those along the Gulf of Mexico. In only a fraction of cases have these fundamental phylogeographic splits been adequately registered in the previously available species' or subspecies' taxonomies upon which conservation plans are built.

4. These regional centers of intraspecific genetic diversity also coincide quite well with the major zoogeographic provinces in the southeastern United States as identified previously by considerations of species' ranges, concentrations of species' distributional limits, and historical physiography. If such agreement proves generalizable to regional biotas elsewhere, it suggests that similar classes of historical biogeographic factors may be responsible for the intra- and interspecific patterns. It also suggests that the traditional data of biogeography will be of great utility in identifying regional centers of biodiversity.

5. The comparative findings on regional phylogeography at intra- and interspecific levels carry several management implications, ranging from taxonomic recognition in endangered species complexes, to translocation strategies, to the design of regional biogeographic sanctuaries for the preservation of extant biodiversity.

REFERENCES

AVISE, J.C. 1987. Identification and interpretation of mitochondrial DNA stocks in marine species. In *Proceedings of Stock Identification Workshop*, eds. H. Kumpf and E.L. Nakamura, pp. 105–136. Publication of the National Oceanographic and Atmospheric Administration, Panama City, FL.

AVISE, J.C. 1989. A role for molecular genetics in the recognition and conservation of endangered species. *Trends Ecol. Evol.* 4:279–281.

AVISE, J.C. 1992. Molecular population structure and the biogeographic history of a regional fauna: a case history with lessons for conservation biology. *Oikos* 63:62–76.

AVISE, J.C. 1994. *Molecular Markers, Natural History and Evolution*. Chapman & Hall, New York.

AVISE, J.C. 1995. Mitochondrial DNA polymorphism and a connection between genetics and demography of relevance to conservation. *Conserv. Biol.* 9:686–690.

AVISE, J.C., J. ARNOLD, R.M. BALL, E. BERMINGHAM, T. LAMB, J.E. NEIGEL, C.A. REEB, and N.C. SAUNDERS. 1987a. Intraspecific phylogeography: the mitochondrial DNA bridge between population genetics and systematics. *Annu. Rev. Ecol. Syst.* 18:489–522.

AVISE, J.C. and R.M. BALL., Jr. 1990. Principles of genealogical concordance in species concepts and biological taxonomy. *Oxford Surv. Evol. Biol.* 7:45–67.

AVISE, J.C., E. BERMINGHAM, L.G. KESSLER, and N.C. SAUNDERS. 1984. Characterization of mitochondrial DNA variability in a hybrid swarm between subspecies of bluegill sunfish (*Lepomis macrochirus*). *Evolution* 38:931–941.

AVISE, J.C., B.W. BOWEN, T. LAMB, A.B. MEYLAN, and E. BERMINGHAM. 1992. Mitochondrial DNA evolution at a turtle's pace: evidence for low genetic variability and reduced microevolutionary rate in the Testudines. *Molec. Biol. Evol.* 9:457–473.

AVISE, J.C., C. GIBLIN-DAVIDSON, J. LAERM, J.C. PATTON, and R.A. LANSMAN. 1979b. Mitochondrial DNA clones and matriarchal phylogeny within and among geographic populations of the pocket gopher, *Geomys pinetis. Proc. Natl. Acad. Sci. USA* 76:6694–6698.

AVISE, J.C., G.S. HELFMAN, N.C. SAUNDERS, and L.S. HALES. 1986. Mitochondrial DNA differentiation in North Atlantic eels: population genetic consequences of an unusual life history pattern. *Proc. Natl. Acad. Sci. USA* 83:4350–4354.

AVISE, J.C., R.A. LANSMAN, and R.O. SHADE. 1979a. The use of restriction endonucleases to measure mitochondrial DNA sequence relatedness in natural populations: I. Population structure and evolution in the genus *Peromyscus. Genetics* 93:279–295.

AVISE, J.C. and W.S. NELSON. 1989. Molecular genetic relationships of the extinct dusky seaside sparrow. *Science* 243:646–648.

AVISE, J.C., C.A. REEB, and N.C. SAUNDERS. 1987b. Geographic population structure and species differences in mitochondrial DNA of mouthbrooding marine catfishes (Ariidae) and demersal spawning toadfishes (Batrachoididae). *Evolution* 41:991–1002.

AVISE, J.C., J.F. SHAPIRA, S.W. DANIEL, C.F. AQUADRO, and R.A. LANSMAN. 1983. Mitochondrial DNA differentiation during the speciation process in *Peromyscus. Molec. Biol. Evol.* 1:38–56.

AVISE, J.C. and M.H. SMITH. 1974. Biochemical genetics of sunfish: I. Geographic variation and subspecific intergradation in the bluegill, *Lepomis macrochirus. Evolution* 28:42–56.

BANGS, O. 1898. Land mammals of Florida and Georgia. *Proc. Boston Soc. Nat. Hist.* 28:157.

BERMINGHAM, E. and J.C. AVISE. 1986. Molecular zoogeography of freshwater fishes in the southeastern United States. *Genetics* 113:939–965.

BERT, T.M. 1986. Speciation in western Atlantic stone crabs (genus *Menippe*): the role of geological processes and climatic events in the formation and distribution of species. *Mar. Biol.* 93:157–170.

BERT, T.M. and R.G. HARRISON. 1988. Hybridization in western Atlantic stone crabs (genus *Menippe*): evolutionary history and ecological context influence species interactions. *Evolution* 42:528–544.

BOWEN, B.W. and J.C. AVISE. 1990. Genetic structure of Atlantic and Gulf of Mexico populations of sea bass, menhaden, and sturgeon: influence of zoogeographic factors and life-history patterns. *Mar. Biol.* 107:371–381.

BRIGGS, J.C. 1958. A list of Florida fishes and their distribution. *Bull. Fla. State Mus. Biol. Sci.* 2:223–318.

BRIGGS, J.C. 1974. *Marine Zoogeography.* McGraw-Hill, New York.

BUROKER, N.E. 1983. Population genetics of the American oyster *Crassostrea virginica* along the Atlantic coast and the Gulf of Mexico. *Mar. Biol.* 75:99–112.

DIZON, A.E., C. LOCKYER, W.F. PERRIN, D.P. DEMASTER, and J. SISSON. 1992. Rethinking the stock concept: a phylogeographic approach. *Conserv. Biol.* 6:24–36.

DUGGINS, C.F., Jr., A.A. KARLIN, T.A. MOUSSEAU, and K.G. RELYEA. 1995. Analysis of a hybrid zone in *Fundulus majalis* in a northeastern Florida ecotone. *Heredity* 74:117–128.

ELLSWORTH, D.L., R.L. HONEYCUTT, N.J. SILVY, J.W. BICKHAM, and W.D. KLIMSTRA. 1994. Historical biogeography and contemporary patterns of mitochondrial DNA variation in white-tailed deer from the southeastern United States. *Evolution* 48:122–136.

FUNDERBURG, J.B., Jr. and T.L. QUAY. 1983. Distributional evolution of the seaside sparrow. In *The Seaside Sparrow, Its Biology and Management*, eds. T.L. Quay, J.B. Funderburg, Jr., D.S. Lee, E.F. Potter, and C.S. Robbins, pp. 19–27. North Carolina State Museum of Natural History, Raleigh, NC.

GILL, F.B., A.M. MOSTROM, and A.L. MACK. 1993. Speciation in North American chickadees: I. Patterns of mtDNA genetic divergence. *Evolution* 47:195–212.

GOLD, J.R. and L.R. RICHARDSON. 1991. Genetic studies in marine fishes: IV. An analysis of population structure in the red drum (*Sciaenops ocellatus*) using mitochondrial DNA. *Fish. Res.* 12:213–241.

HAYES, J.P. and R.G. HARRISON. 1992. Variation in mitochondrial DNA and the biogeographic history of woodrats (*Neotoma*) of the eastern United States. *Syst. Biol.* 41:331–344.

KARL, S.A. and J.C. AVISE. 1992. Balancing selection at allozyme loci in oysters: implications from nuclear RFLPs. *Science* 256:100–102.

LAERM, J., J.C. AVISE, J.C. PATTON, and R.A. LANSMAN. 1982. Genetic determination of the status of an endangered species of pocket gopher in Georgia. *J. Wildl. Mgmt.* 46:513–518.

LAMB, T. and J.C. AVISE. 1992. Molecular and population genetic aspects of mitochondrial DNA variability in the diamondback terrapin, *Malaclemys terrapin. J. Hered.* 83:262–269.

LAMB, T., J.C. AVISE, and J.W. GIBBONS. 1989. Phylogeographic patterns in mitochondrial DNA of the desert tortoise (*Xerobates agassizi*), and evolutionary relationships among the North American gopher tortoises. *Evolution* 43:76–87.

LANSMAN, R.A., J.C. AVISE, C.F. AQUADRO, J.F. SHAPIRA, and S.W. DANIEL. 1983. Extensive genetic variation in mitochondrial DNA's among geographic populations of the deer mouse, *Peromyscus maniculatus. Evolution* 37:1–16.

MAYDEN, R.L. 1988. Vicariance biogeography, parsimony, and evolution in North American freshwater fishes. *Syst. Zool.* 37:329–355.

MORITZ, C. 1994a. Applications of mitochondrial DNA analysis in conservation: a critical review. *Molec. Ecol.* 3:401–411.

MORITZ, C. 1994b. Defining "evolutionarily significant units" for conservation. *Trends Ecol. Evol.* 9:373–375.

NEIGEL, J.E. and J.C. AVISE. 1993. Application of a random walk model to geographic distributions of animal mitochondrial DNA variation. *Genetics* 135:1209–1220.

NEIGEL, J.E., R.M. BALL, Jr., and J.C. AVISE. 1991. Estimation of single generation migration distances from geographic variation in animal mitochondrial DNA. *Evolution* 45:423–432.

OSENTOSKI, M.F. 1993. MtDNA variation in the gopher tortoise, *Gopherus polyphemus*. MS thesis, East Carolina University, Greenville, NC.

PHILIPP, D.P., W.F. CHILDERS, and G.S. WHITT. 1983. A biochemical genetic evaluation of the northern and Florida subspecies of largemouth bass. *Trans. Amer. Fish. Soc.* 112:1–20.

RAND, D.M. 1994. Thermal habit, metabolic rate and the evolution of mitochondrial DNA. *Trends Ecol. Evol.* 9:125–130.

REEB, C.A. and J.C. AVISE. 1990. A genetic discontinuity in a continuously distributed species: mitochondrial DNA in the American oyster, *Crassostrea virginica. Genetics* 124:397–406.

REMINGTON, C.L. 1968. Suture-zones of hybrid interaction between recently joined biotas. *Evol. Biol.* 2:321–428.

RYDER, O.A. 1986. Species conservation and the dilemma of subspecies. *Trends Ecol. Evol.* 1:9–10.

SARVER, S.K., M.C. LANDRUM, and D.W. FOLTZ. 1992. Genetics and taxonomy of ribbed mussels (*Geukensia* spp). *Mar. Biol.* 113:385–390.

SAUNDERS, N.C., L.G. KESSLER, and J.C. AVISE. 1986. Genetic variation and geographic differentiation in mitochondrial DNA of the horseshoe crab, *Limulus polyphemus. Genetics* 112:613–627.

SCRIBNER, K.T. and J.C. AVISE. 1993. Cytonuclear genetic architecture in mosquitofish populations and the possible roles of introgressive hybridization. *Molec. Ecol.* 2:139–149.

SWIFT, C.C., C.R. GILBERT, S.A. BORTONE, G.H. BURGESS, and R.W. YERGER. 1985. Zoogeography of the southeastern United States: Savannah River to Lake Ponchartrain. In *Zoogeog-*

raphy of the North American Freshwater Fishes, eds. C.H. Hocutt and E.O. Wiley, pp. 213–265. Wiley, New York.

VOGLER, A.P. and R. DESALLE. 1993. Phylogeographic patterns in coastal North American tiger beetles (*Cicindela dorsalis* Say) inferred from mitochondrial DNA sequences. *Evolution* 47:1192–1202.

VOGLER, A.P., R. DESALLE, T. ASSMANN, C.B. KNISLEY, and T.D. SCHULTZ. 1993. Molecular population genetics of the endangered tiger beetle *Cicindela dorsalis* (Coleoptera: Cicindelidae). *Ann. Entomol. Soc. Amer.* 86:142–152.

WALKER, D., V.J. BURKE, I. BARÁK, and J.C. AVISE. 1995. A comparison of mtDNA restriction sites vs. control region sequences in phylogeographic assessment of the musk turtle (*Sternotherus minor*). *Molec. Ecol.* 4:365–373.

WOOTEN, M.C. and C. LYDEARD. 1990. Allozyme variation in a natural contact zone between *Gambusia affinis* and *Gambusia holbrooki*. *Biochem. Syst. Ecol.* 18:169–173.

WOOTEN, M.C., K.T. SCRIBNER, and M.H. SMITH. 1988. Genetic variability and systematics of *Gambusia* in the southeastern United States. *Copeia* 1988:283–289.

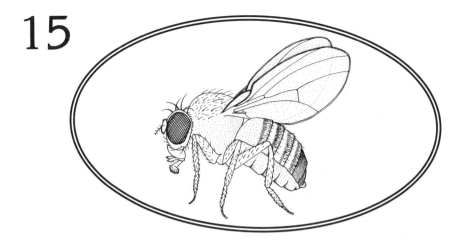

15

A QUANTITATIVE-GENETIC PERSPECTIVE ON CONSERVATION ISSUES

Michael Lynch

INTRODUCTION

The multiplicity of populations, species, and higher taxonomic levels that we call *biodiversity* is a product of thousands to billions of years of evolution. Essentially everyone agrees that the short-term key to maintaining this diversity is the protection of critical habitat upon which species have depended historically. However, comprehensive strategies for preserving biodiversity require both prospective and retrospective views of the issues. Looking to the future, a primary consideration is the preservation of intrinsic genetic features that enable evolutionary lineages to cope with challenges from changing environments.

Whether intentional or a matter of chance, the current policy of the United States government in decisions to formally list species as endangered or threatened under the Endangered Species Act (U.S. Fish and Wildlife Service, 1992) has been to forego listing until a candidate population has dwindled to about 1,000 individuals in the case of animals, and 100 individuals in the case of plants (Wilcove et al., 1993). Based on our knowledge of the vulnerability of small populations to demographic and environmental stochasticity alone, such target sizes for listing are well within the range in which the short-term risk of extinction is of significant concern (Ludwig, 1976; Leigh, 1981; Shaffer, 1981; Ginzburg et al., 1982; Goodman, 1987; Burgman et al., 1992; Lande, 1993; Foley, 1994). The results reviewed in this chapter provide a genetic perspective on the issues.

Most of the features of organisms that we value involve their outward appearances—morphology, behavior, and in some cases, physiology, the very characters upon which natural selection operates. Most characters of this nature appear to be products of tens to perhaps hundreds of genetic loci, and their expression can be modified by numerous environmental factors (Wright, 1968; Falconer, 1981; Lande, 1981), although there are exceptions (Gottlieb, 1984; Orr and Coyne, 1992). This conclusion is based on indirect statistical inference, rather than on direct observation of specific gene products. Nevertheless, over the past half century, a rich theory has developed to explain the emergent genetic and evolutionary properties of complex characters based on the assumption of multifactorial inheritance. This field of study, known as *quantitative genetics*, has been subject to considerable empirical scrutiny, and its long history of successful application in plant and animal breeding programs testifies to its practical utility (Falconer, 1981; Hallauer and Miranda, 1981; Pirchner, 1983; Wricke and Weber, 1986). There seems to be little question that the field of conservation genetics could profit from an influx of quantitative-genetic thinking.

When phenotypes are a function of multiple genetic and environmental effects, it is essentially impossible to ascertain an individual's genotype from its outward appearance. However, quantitative-genetic studies provide a basis for identifying sources of variation contributing to individual differences. Information on the relative contributions of genes and environment to the total variation in a population is acquired by comparing phenotypes of relatives. For example, a parent passes on a single gene per genetic locus to each of its offspring, but except in the case of cultural inheritance or strong maternal effects, most environmental effects are not transmitted. Consequently, the resemblance between parents and their offspring is largely, or entirely, a function of variable genes in the population. If all parents have identical genotypes, their phenotypic differences are due entirely to environmental effects, and their phenotypes are uncorrelated with those of their progeny.

One of the most familiar concepts in quantitative genetics is the heritability of a trait, h^2, which is the fraction of the phenotypic variance for the trait that has an additive genetic basis. As a first-order approximation, the heritability of a trait is equivalent to the slope of a regression between midparent (mean value of the two parents) and offspring phenotypes. The heritability concept is important because it relates in a simple way to the ability of a population to evolve in response to directional selection.

A fundamental law of evolutionary biology is expressed by $\Delta\bar{z} = h^2 S$, which states that the rate of evolution of the mean phenotype $\Delta\bar{z}$ is equal to the product of the selection differential S (the change in the mean phenotype in the population caused by selection prior to reproduction) and the heritability of the trait h^2 (Falconer, 1981). Often referred to as the breeders' equation, this formula provides the theoretical basis for most selection programs in plant and animal breeding. The breeders' equation nicely separates evolution into two components—selection and inheritance. Selection favors individuals on the basis of their phenotypes, regardless of whether the favorable features are due to genetic or environmental effects. However, the degree to which the selective change in the mean phenotype is transmitted across generations depends on the heritability of the trait. Since heritability has a zero-to-one range, it can be viewed as the efficiency with which a population responds to natural selection.

The goal of most conservation-genetic programs is to preserve significant pools of heritable variation, while simultaneously preventing the chance fixation of deleterious alleles (Foose et al., 1986; Falk and Holsinger, 1991; Hedrick, 1992; Hedrick and Miller, 1992; Lande, 1995). As testified by numerous chapters in this volume, progress toward these goals is generally monitored by use of molecular markers. Implicit in this approach is the assumption that molecular techniques do, in fact, serve as suitable surrogates for estimating adaptive genetic diversity and population-genetic structure. In the following section, I point out some of the limitations of molecular genetic approaches to conservation biology (see also Hedrick et al., 1986), and some reasons why molecular surveys might sometimes be misleading. That section motivates the remainder of the chapter, in which I advocate the use of quantitative-genetic principles as supplemental guides to developing management programs in conservation biology.

MOLECULAR VERSUS QUANTITATIVE-GENETIC APPROACHES TO CONSERVATION ISSUES

Several attributes of molecular markers, especially those that are DNA-based, have led to their widespread use in conservation studies. Molecular methods have the advantage of sampling broadly over the genome, and those that are based on the polymerase chain reaction can be applied to live organisms with little disturbance, and even to museum specimens, providing a historical perspective of gene turnover. DNA differences among individuals and/or populations have an unambiguous genetic basis, and the fact that most molecular polymorphisms behave in an essentially neutral fashion makes them very useful for ascertaining pedigrees, reconstructing phylogenies, identifying phylogeographic patterns, and estimating patterns of gene flow (Avise, 1994; Milligan et al., 1994; Moritz, 1994).

Molecular studies in conservation biology often take the data a step further, using them to infer *adaptive* features of population-genetic structure. For example, a lack of molecular variation in the cheetah has been taken to imply an absence of genetic variation for adaptive quantitative characters and, by extrapolation, enhanced risk of extinction due to genetic homogeneity (O'Brien et al., 1985; see Chapter 3). The confidence that can be attached to such extrapolations is limited (Caughley, 1994). Although empirical data on the matter are in short supply, there are several good theoretical reasons to doubt that a strong connection will normally be found between levels of molecular and quantitative-genetic diversity within populations:

1. Variation at the molecular level (heterozygosity) is introduced to a population at the per-locus rate of mutation for the molecular marker, typically on the order of 10^{-8} to 10^{-5} per year (Kimura, 1983), whereas variation for quantitative traits (heritability) is introduced at a rate of approximately 10^{-3} to 10^{-2} per generation (Lynch, 1988a). Consequently, populations that go through significant-enough bottlenecks to lose most of their genetic variation will exhibit the molecular signature of such an event for tens to hundreds of thousands of years, while having ample time to recover normal levels of heritable variation (Lande and Barrowclough, 1987). A possible example of this situation is provided by a recent study of the highly endangered cotton-top tamarin. This species has a very low level of molecular heterozygosity, nearly identical to that in the cheetah, yet it exhibits a rather high level of heritability for body weight (Cheverud et al., 1994).

2. Although the expected level of heterozygosity at neutral molecular markers declines linearly with the inbreeding coefficient, when significant sources of variation for quantitative traits are due to nonadditive gene action (dominance and epistasis), substantial departures from this behavior are seen. For example, due to the changes in the average effects of genes that occur as the frequencies of interacting genes are altered by genetic drift, it is possible for the expected additive genetic variance to increase with a population bottleneck (Robertson, 1952; Goodnight, 1988; Cockerham and Tachida, 1988; Whitlock et al., 1993; Willis and Orr, 1993). Such increases have been observed in bottlenecked populations of house flies (Bryant et al., 1986; Bryant and Meffert, 1993) and fruit flies (López-Fanjul and Villaverde, 1989), although interpretation of the adaptive significance of these results is complicated by the fact that the inflation of the genetic variance is typically accompanied by a reduction in mean fitness.

3. For reasons purely related to statistical sampling, the relationship between molecular and quantitative-genetic variation is expected to be weak. Even for a quantitative trait with a purely additive genetic basis, the variation in quantitative-genetic parameters that arises among small populations can be enormous (Avery and Hill, 1977; Weir et al., 1980; Lynch, 1988b). Components of quantitative-genetic variance can drift substantially above or below their expected values, as López-Fanjul et al. (1989) have demonstrated in replicate populations of *Drosophila*. In addition, empirical estimates of quantitative-genetic parameters are subject to substantial sampling error.

4. Similar sampling problems exist in the estimation of heterozygosity with molecular markers—unless a substantial number of loci and individuals has been assayed, the sampling variance of measures of molecular variation can be very high (Nei and Roychoudhury, 1974; Mitton and Pierce, 1980; Chakraborty, 1981; Lynch and Crease, 1990; Lynch and Milligan, 1994). This point is well illustrated by an experiment that put replicate populations of mosquitofish through small bottlenecks (Leberg, 1992). Although the average response of the replicates was a reduction in allozyme heterozygosity, some replicates exhibited increases. Even when a large enough survey is done that one can be relatively confident that an accurate assessment of the molecular heterozygosity has been acquired, the molecular-marker loci will provide little insight into conditions at loci underlying adaptive variation unless a large fraction of the former are tightly linked to the relevant quantitative-trait loci. This seems unlikely except in species with very small chromosome numbers. In the absence of such linkage, marker data will provide little information as to whether past selective conditions might have led to abnormally low or high levels of genetic variation for adaptive quantitative traits.

All of the above caveats can also be used as arguments against the indiscriminate use of *interdemic* measures of molecular divergence to derive inferences about *adaptive divergence*. Certainly, significant molecular divergence provides strong evidence that the *opportunity* for adaptive divergence has existed. However, a lack of molecular divergence is uninformative, an issue to which we will return later.

Not many empirical studies have been done on the relationship between molecular variation and quantitative-trait variation, and only two of these have actually considered the genetic component of phenotypic variance. In a comparison of several laboratory populations of *Drosophila melanogaster*, Briscoe et al. (1992) found a strong, positive correlation between allozyme heterozygosity and the heritability of bristle number. Since this character is known to have an additive genetic basis and to be under only weak natural selection, it is much more likely to reflect the patterns of variation of neutral molecular markers than characters more closely related with fitness, which often exhibit a large nonadditive component (Falconer, 1981). In our studies of several *Daphnia* populations, we have yet to find an association between allozyme heterozygosity and the total genetic variance for fitness characters (Lynch and Spitze, in prep.)

Studies that have focused on phenotypic variation have yielded mixed results. Yezerinac et al. (1992) found a very weak, but positive, correlation between allozyme heterozygosity and morphological variation within populations of rufous-collared sparrows (*Zonotrichia capensis*), and Strauss (1991) found a strong positive correlation between allozyme heterozygosity and phenotypic variation within populations of freshwater sculpins in the genus *Cottus*. On the other hand, in humans (Kobyliansky and Livshits, 1983), fox sparrows, and pocket gophers (Zink et al., 1985), there appears to be a weak, inverse relationship between the enzyme heterozygosity in a population and its morphological variation. Interpretation of these kinds of results is difficult. When the measure of morphological variation is purely phenotypic, it is unclear whether the fit would be better or worse if it were based purely on the genetic component of variation. Arguments have been made, with a fair amount of supporting evidence, that homozygous individuals are more developmentally unstable, and hence exhibit more phenotypic variation for purely environmental reasons than heterozygous individuals (Lerner, 1954; Soulé, 1982; Palmer and Strobeck, 1986). If this is generally true, we should expect a negative correlation between molecular variation and the environmental component of phenotypic variation to obscure any positive relationship between molecular and quantitative-genetic variation.

In summary, some major challenges confront a science of conservation genetics that relies largely on molecular markers— most notably, to document a connection between molecular and adaptive quantitative-genetic variation, and to develop methods for rapidly assaying the latter in natural populations. Although the methods of quantitative genetics have almost never been applied to problems in conservation biology, in one sense, intensively studied populations of endangered species are ideally suited to such analysis. For captive populations with controlled breeding programs and for marked field populations, substantial information on genealogical relationships is often available, in which case the only additional labor required for a quantitative-genetic analysis is the measurement of the traits of interest. This, however, is where a difficult and perhaps unanswerable question arises. What characters should be measured? In almost all cases, it will be a judgment call as to which traits are currently most critical to survival and reproduction, and even more so as to which ones are most likely to be confronted with future selective challenges. As will be argued later, quantitative genetics certainly has much to contribute to endangered species management, but as in the case of molecular analysis, there are limits to what can be accomplished.

Rather than focusing on the practical methodology of quantitative genetics, which is amply covered in many textbooks, the remainder of this chapter outlines several basic theoretical principles that have important and general implications for the management of species for long-term survival. Particular attention is given to the genetic consequences of small population size, one attribute that all endangered species have in common, with a goal of identifying the minimum effective population size necessary to secure the genetic integrity of a population. Much of the literature cited is quite technical mathematically, and an attempt has been made to reduce the basic results to simple qualitative relationships that may be useful to those involved in management decisions.

MAINTENANCE OF ADAPTIVE VARIATION

Most populations, even those undisturbed by human activity, are exposed regularly to temporal and spatial variation in physical and biotic features of the environment. Although most organisms are endowed with an array of behavioral and/or physiological mechanisms for coping with short-term environmental changes, the range of environments within which such homeostatic mechanisms are operative is normally confined to the conditions experienced over recent evolutionary time. In principle, some species can cope with new selective challenges by simply migrating to suitable habitat (Pease et al., 1989). However, endangered species typically live in highly fragmented habitats with inhospitable barriers to movement. This leaves adaptive evolutionary change as the primary means of responding to selective challenges.

There is an additional reason why the maintenance of evolutionary flexibility may be especially crucial for the long-term survival of endangered species. The same human activities that threaten their demographic stability may also impose selective pressures that are substantially greater than those typically experienced in natural settings. For many species, habitat degradation, environmental pollution, global climatic change, species introductions, harvesting, and so on, impose an array of conditions that have never been experienced before. As noted previously, the ability to respond to novel selective challenges is proportional to the additive genetic variance for the selected trait. Thus, a common goal of genetic conservation programs, the maintenance of adaptive genetic variation, is well founded.

In small isolated populations, three factors interact to determine standing levels of genetic variation for characters associated with morphology, physiology, and behavior. First, most forms of natural selection cause a reduction in the genetic variance by eliminating extreme genotypes, the exact amount depending on the intensity of selection and the form of the fitness function (Shnol and Kondrashov, 1993). Second, small populations with an effective number of breeding adults N_e also lose an expected fraction $1/(2N_e)$ of their genetic variance each generation by random genetic drift. Finally, genetic variation is added to a population each generation by mutation, at the rate σ^2_m. When populations are kept at a constant size and under constant selective pressures, a quasi-equilibrium level of genetic variance eventually evolves, at which point the loss due to selection and drift is approximately balanced by mutational input.

For populations with effective sizes smaller than a few hundred individuals, the expected amount of variation for a typical quantitative character is nearly independent of the strength of selection and is largely a result of mutation-drift balance (Keightley and Hill, 1988; Barton, 1989; Bürger et al., 1989; Houle, 1989; Foley, 1992). This situation arises with polygenic characters because the forces of selection are distributed over large numbers of loci, rendering the selective pressures on specific loci small enough to be overwhelmed by random genetic drift. Under these conditions, characters with an additive genetic basis have an expected genetic variance equal to $2N_e\sigma^2_m$. This result, which is also the neutral expectation (Lynch and Hill, 1986), is fairly general. It depends very little on whether gene action is nonadditive or on the linkage relationships of the constituent loci. Thus, for small populations, a doubling in population size effectively doubles the evolutionary potential of the population.

As the effective population size increases, random genetic drift becomes less significant as an evolutionary force, until with large populations, the equilibrium level of genetic variance is due entirely to selection–mutation balance. There is still some debate as to the magnitude of the genetic variance resulting from selection-mutation balance, as it depends on the gametic mutation rate and the distribution of mutational effects, neither of which are very well understood (Turelli, 1984; Barton and Turelli, 1989). However, there seems to be general agreement that the *average* genetic variance is essentially independent of population size once N_e exceeds 1,000 or so individuals. This does not mean that a population with $N_e > 1,000$ is genetically equivalent to an effectively infinite population.

Except in populations containing many thousands of individuals, random genetic drift causes the actual genetic variance in a population to wander around its expected value from generation to generation (Keightley and Hill, 1989; Zeng and Cockerham, 1991). The variance of the genetic variance is inversely related to the number of mutations entering the population per generation, such that the coefficient of variation of the additive genetic variance is approximately $(2\mu N_e)^{-1/2}$, where μ is the genomic mutation rate for the character (the sum of the mutation rates for the constituent loci; Lynch and Hill, 1986; Barton, 1989; Bürger and Lande, 1994). Thus, if μ were 0.05, which appears to be reasonable for a typical quantitative trait (Lande, 1975; Turelli, 1984; Lynch, 1988a), the coefficient of variation of the genetic variance would be on the order of 0.3 and 0.1 for populations with $N_e = 100$ and 1,000. Due to inheritance, any random declines in the genetic variance are likely to persist for several generations. For N_e less than a thousand or so, the expected correlation of the genetic variance in the same population at times separated by t generations is approximately $e^{-t/(2N_e)}$ (Bürger and Lande, 1994), which declines only to 0.5 after $t = 1.4\,N_e$ generations.

The practical implication of these results is that by chance, even in the absence of a population bottleneck, in a population of moderate size the genetic variance for a quantitative trait can essentially disappear temporarily and remain at a low level for many generations until mutation has had an opportunity to replenish it. Such temporary excursions to low levels of genetic variance can jeopardize the survival of populations inhabiting changing environments (Bürger and Lynch, 1994).

One important caveat to the aforementioned results is that they apply largely to characters with a purely additive genetic basis, which may not include many fitness-related characters. Limited work has been done on the maintenance of genetic variance by drift–selection–mutation balance when genes exhibit dominance, and the results suggest that the predictions of the purely additive model are still reliable as a first approximation (Lynch and Hill, 1986; Caballero and Keightley, 1994). Whether this is true when epistatic interactions are important remains to be seen.

The Minimum Effective Size
for an Adaptively Secure Population

Some past attempts to identify a critical minimum population size from a genetic perspective have focused on goals such as the maintenance of 90% of the genetic variability present in the ancestral (predisturbance) population for 200 years (Franklin, 1980; Soulé, 1980; Soulé et al., 1986; Lande, 1995). Short-term goals of this nature take into consideration the fact that populations that are dwindling in size cannot be in equilibrium. However, such goals are rather arbitrary with respect to choice of acceptable loss and time span.

For long-term planning, an alternative approach to defining a genetically secure population is to consider the equilibrium conditions outlined previously. Beyond an effective population size of approximately 1,000 individuals, any further increase in N_e will not usually enhance the expected amount of genetic variance maintained in a population. For many species, the effective population size is often on the order of one-tenth to one-third of the actual number of breeding adults (Heywood, 1986; Lande and Barrowclough, 1987; Briscoe et al., 1992). Thus, from the standpoint of the maintenance of adaptive variation, the $N_e > 1,000$ criterion translates into the need for stable persistence of 5,000 to 10,000 breeding adults each generation. Programs to manage populations at a larger size solely to enhance their average evolutionary potential would be unlikely to achieve that objective, although a further reduction in the fluctuations of the genetic variance would be obtained. Additional considerations on the maintenance of adaptive variation in managed populations, including population subdivision and migration, are provided in Lande (1995).

The Maximum Rate of Sustainable Evolution

In the development of policies for species protection, the major focus is usually on short-term local issues such as dam and road building, logging and mining, grazing, housing development, and so on. However, the evidence is mounting that human activity is causing, and will continue to cause, global changes in the temperature and chemical composition of the atmosphere (Abrahamson, 1989; Wyman, 1991; Kareiva et al., 1993). These types of changes, particularly when combined with habitat fragmentation, pose threats to all species, even those that are currently abundant, and raise a fundamental question: What is the maximum rate of environmental change that can be tolerated by a species?

Laboratory studies with several species of plants and animals have shown that continuous programs of intense directional selection on quantitative traits often lead to approximately linear changes in mean phenotypes for several dozens of generations (Jones et al., 1968; Kress, 1975; Dudley, 1977; Eisen, 1980; Weber and Diggins, 1990). Although the effective sizes of the populations involved in these studies are generally on the order of only a few dozen individuals, it is not unusual for the evolutionary change of the mean phenotype to exceed five to ten phenotypic standard deviations before the response to selection begins to diminish. Such results clearly indicate that even small isolates of natural populations typically harbor enough quantitative-genetic variation that, when confronted with a novel environment, mean phenotypes well outside of the observed range of variation in the initial population can evolve rapidly.

After a few dozen generations of fairly strong selection, laboratory experiments usually exhibit a selection plateau due to the evolutionary advancement of genes with major effects on the selected trait and negative pleiotropic effects on fitness (i.e., to a conflict between artificial and natural selection). Such antagonisms are less likely to arise in more natural settings, where genes are selected for their total effects on fitness. Nevertheless, in small populations, the rate of phenotypic evolution can become limited by the availability of additive genetic variance.

Once the favorable genes in a founder population have been advanced to fixation by natural selection or lost by random genetic drift, all future evolutionary change will be dependent upon the input of new variation by mutation. This transition certainly takes on quantitative significance by the time N_e generations have elapsed (Clayton and Robertson, 1955; Lynch and Hill, 1986). For example, even in the absence of selection, the fraction of initial additive genetic variation surviving after t generations of isolation is approximately $e^{-t/(2N_e)}$, whereas the variation resulting from mutations subsequent to the founder event has expected value $2N_e \sigma_m^2 (1 - e^{-t/(2N_e)})$. Thus, after approximately $1.4 N_e$ generations have elapsed, 50% of the genetic variance in the founder population is expected to have been lost, while the new variation due to mutation will have increased to 50% of its equilibrium value. After approximately $4.6 N_e$ generations, 90% of the variation in the founder individuals is expected to have been lost, while the new variation from mutations will have increased to within 10% of its equilibrium value.

Consider a population confronted with a gradual long-term change in a critical environmental parameter (e.g., temperature, humidity, prey size, etc.). If the rate of environmental change is sufficiently slow and the amount of genetic variance for the characters needed to cope with it is sufficiently high, then the population will be able to slowly track the environmental change evolutionarily, without a major reduction in population size. If, however, the rate of environmental change is too high, the selective load (reduced viability and fecundity) on the population will exceed the population's capacity to assimilate new genetic variation and maintain a positive rate of growth. In this case, although the population may respond evolutionarily, it will go extinct in the process. Thus, for any population, there must be a critical rate of environmental change that allows the population to evolve just fast enough to maintain a stable size.

Theoretical studies on the response to long-term directional selection show that as populations settle into a new equilibrium level of genetic variance, they also settle into an evolutionary trajectory that is parallel to the moving optimum for the selected trait, provided the optimum is not out of the range of possible adaptive variation (Lynch et al., 1991; Lynch and Lande, 1993; Bürger and Lynch, 1994). Thus, the critical rate of environmental change is equivalent to the maximum rate of sustainable evolution. The amount by which the mean phenotype lags behind the moving optimum is determined by the strength of selection and the magnitude of genetic variance for the trait. The magnitude of the lag in turn determines the degree of maladaptation of the population and hence its demographic potential.

To gain some appreciation of the magnitude of environmental change that can be tolerated for a sustained period of time, consider the situation in which a population is exposed to a Gaussian (bell-shaped) fitness function with a constant width but moving optimum for the selected trait. In units of phenotypic standard deviations, the critical rate of environmental change (rate of movement of the optimum) beyond which such a population is incapable of replacing itself is less than $\phi \tilde{h}^2 [2r_{max} - (1/(2N_e))]^{1/2}$, where $\phi = \sigma_z/\sigma_w$ is the ratio of the phenotypic standard deviation to the width of the fitness function, $\tilde{h}^2$ is the equilibrium heritability of the trait, and r_{max} is the rate of population growth in a constant environment (assuming in that case that the population is fully adapted, such that the mean phenotype coincides with the optimum; Lynch and Lande, 1993; Bürger and Lynch, 1994). What does this expression imply about the maximum sustainable rate of evolution?

First, we note that if the effective population size (N_e) is less than $1/(4r_{max})$, a population has no long-term viability even in a constant environment. With such small population sizes, the mean phenotype drifts to a large enough extent from the optimum phenotype that the average selective load exceeds the demographic potential.

For populations with larger size (i.e., $N_e \gg 1/(4r_{max})$), the critical rate of environmental change depends on the product $\phi \tilde{h}^2$. Only limited work has been done on the equilibrium heritability of quantitative traits experiencing directional selection. However, it appears that for large populations, $\tilde{h}^2$ is elevated somewhat above $2\mu/\phi^2$, where μ is the zygotic mutation rate for the trait (Bürger and Lynch, 1994). Thus, even for populations with effectively infinite size, the critical rate of environmental change is less than $2\mu[2r_{max}]^{1/2}/\phi$. As noted previously, estimates of μ for typical quantitative traits appear to be on the order of 0.05 or smaller. ϕ appears often to be on the order of 0.1 to 0.4 (Turelli, 1984; Endler, 1986), and on time scales of generations, r_{max} is usually less than 1.0. Since the preceding expression overestimates the critical rate of environmental change if the directional trend in the environment contains a stochastic component (Lynch and Lande, 1993), which is essentially always the case (Boag and Grant, 1981; Kalisz, 1986; Hairston and Dillon, 1990; Weis et al., 1992), our results imply that even large populations are unlikely to sustain rates of environmental change that exceed a few percent of a phenotypic standard deviation per generation. The situation is even more stringent in small populations (Lynch and Lande, 1993; Bürger and Lynch, 1994).

The practical implication of these results is that sustained periods of environmental change of fairly small magnitudes can eventually cause the selective mortality of a population to exceed its demographic potential. This conclusion applies to both common and rare species, and there is reason to believe that the demographic consequences of environmental change may be threshold in nature. The extent to which current long-term trends of global warming and atmospheric change will translate into a major extinction episode will depend on the degree to which such changes exacerbate the intensity of natural selection. In principle, climatic changes may induce shifts in the optima for multiple physiological characters required for coping directly with such changes, as well as generate secondary changes in species' ecological settings (abundances and phenotypes of predators, competitors, and prey items). Thus, for isolated populations in particular, it is conceivable that global climate change may impose as substantial a threat to long-term survival as habitat destruction currently poses for short-term survival.

POPULATION BOTTLENECKS: PURGING VERSUS FIXATION OF DELETERIOUS GENES

For many quantitative traits, the fitness effects of mutations may be context dependent (i.e., a mutation for smaller body size may be beneficial or detrimental depending on the current direction of selection for body size). However, it has long been thought that a large fraction of mutations is unconditionally deleterious. Such mutations influence the mean fitness of an isolated population in three ways. First, segregating mutations inherited from the ancestral population may either be purged by natural selection or they may rise to fixation by random genetic drift. Second, mutations arising in the isolated population lead to the establishment of a new segregational load defined by the balance between the present forces of drift, mutation, and selection. Third, a fraction of the mutations entering the population each generation becomes fixed by random genetic drift at some future time. Unlike the first two sources of mutation load, the fixation of recurrent deleterious mutations leads to a progressive loss in fitness. In the following sections, I consider how the accumulation of deleterious mutations can influence the mean fitness and risk of extinction of small populations.

Deleterious Genes in the Ancestral Population

Most attempts to model the genetic risk of extinction for small populations have focused entirely on rare deleterious genes carried in the founder individuals (Senner, 1980; Barrett and Charlesworth, 1991; Lacy, 1992, 1993; Halley and Manasse, 1993; Hedrick, 1994; Mills and Smouse, 1994). The primary concern of these studies has been to ascertain the expected buildup of inbreeding depression due to the random increase in frequency of deleterious recessive genes. However, with the exception of Hedrick (1994), who confined his analysis to full-sib mated lines, all of these studies have used rather *ad hoc* genetic models. For example, Senner (1980) simply assumed that the load due to inbreeding depression increases linearly with the average coefficient of inbreeding at neutral loci. Whereas Mills and Smouse (1994) allowed for a nonlinear response, their approach, as well as that of Senner, is phenomenological, ignoring the underlying genetic basis of inbreeding depression and failing to take selection against deleterious alleles into account. Lacy (1992, 1993) modeled inbreeding depression with a genetically explicit model, but he assumed that all of the recessive load is due to lethals at a single locus, with each founder individual carrying a unique allele. Such a scenario is inconsistent with extensive observations that the load in populations due to recessive deleterious genes is spread over many loci, a large fraction of which have small individual effects (Charlesworth and Charlesworth, 1987).

A popular idea in the design of captive breeding programs is that intentional inbreeding, combined with rapid population expansion, can lead to a purging of the deleterious genes from a population (Templeton and Read, 1983, 1984). Such a purging of inbreeding depression has, in fact, been recorded in short-term studies with flies (Bryant et al., 1990), plants (van Treuren et al., 1993), and mice (Bowman and Falconer, 1960; C.B. Lynch, 1977; Connor and Bellucci, 1979), although in the latter case it has only been accomplished at the expense of extreme selection (extinction) among replicate lines. Were this procedure to be generally successful, it would provide a powerful way for managing captive populations for viability. However, a close look at the issues raises questions about the utility of such a treatment.

The effects of small population size on the evolution of inbreeding depression need to be considered from two frames of reference. By eliminating the genetic variation within an isolated population, long-term inbreeding can lead to a situation in which any further inbreeding has essentially no effect on the number of homozygous recessive genes expressed per individual. Once this situation has been reached, from the standpoint of the current population, there is no inbreeding depression. However, this is certainly not the case from the standpoint of the ancestral population if, during the inbreeding process, deleterious genes have become fixed. This point is nicely illustrated by a study of self-fertilized lines of the normally outcrossing aquatic plant *Eichhornia paniculata* (Barrett and Charlesworth, 1991). Inbreeding caused an immediate depression in fitness, but after only two generations of selfing there was no further decline, suggesting that the vast majority of loci affecting fitness had become homozygous within lines. Nevertheless, despite the absence of inbreeding depression within the derived lines, crosses between lines exhibited a substantial increase in mean fitness, as expected if the different lines had become fixed for different deleterious recessives. Similar results have been obtained for other species of habitually self-fertilizing plants (Charlesworth et al., 1990; Holtsford and Ellstrand, 1990; Ågren and Schemske, 1993). The salient point here is that the absence of local inbreeding depression does not eliminate the possibility that a population harbors a substantial mutation load.

A general understanding of the consequences of deleterious genes contained in founder individuals can be achieved as follows. Consider an ancestral population containing N individuals, each of which incurs an expected μ new deleterious mutations per generation. Let the relative fitnesses at a genetic locus for mutation-free homozygotes and mutant heterozygotes and homozygotes be, respectively, 1, $(1 - 2hs)$, and $(1 - 2s)$, where h is a measure of the degree of dominance ($h = 0.5$ denoting additivity). For large populations with reasonably stable sizes, there is a mutation–selection balance between the number of deleterious mutations arising each generation ($N\mu$) and the number eliminated by natural selection. Assuming that the mutant alleles do not have epistatic effects on fitness, this equilibrium load of mutations causes the mean fitness of individuals to be reduced to approximately $e^{-\mu}$ of that expected for an individual free of deleterious mutations (Haldane, 1937). Haldane's result has some remarkable features. First, it applies to all populations with effective sizes greater than $5/s$ (Kimura et al., 1963; Bürger and Hofbauer, 1994). Even with s as small as 0.001, this requires only that the population contain at least a few thousand individuals. Second, it depends only on the genomic deleterious mutation rate, not on the effects (s or h) of the individual mutations. This is because there is an inverse relationship between the effect of a mutation and its equilibrium frequency.

Haldane's result allows a prediction about the number of deleterious alleles that are likely to be inherited by a founder population. Let $\bar{n}$ be the mean number of deleterious genes per individual in the ancestral population. Making the reasonable assumption that most deleterious alleles have low enough frequencies in the ancestral population that they are almost always in the heterozygous state, and assuming that mutations at different loci influence individual fitness independently, the mean fitness of an ancestral individual is approximately $(1 - 2hs)^{\bar{n}} \simeq e^{-2hs\bar{n}}$. Setting this expression equal to Haldane's $e^{-\mu}$, the expected number of deleterious genes carried by a founder individual is found to be approximately $\bar{n} = \mu/(2hs)$.

The magnitude of $\bar{n}$ depends on the mutational properties μ, s, and h. Unfortunately, estimates of these parameters are only available for a single higher organism, the fruitfly, although in this case the data are extensive (Mukai, 1964, 1969, 1979; Mukai et al., 1965; Mukai and Yamazaki, 1968; Ohnishi, 1977; Houle et al., 1992). A survey of the available data suggests that $\mu \simeq 1.5$, $\bar{s} \simeq 0.01$, and $\bar{h} \simeq 0.36$ (Lynch et al., 1995). These estimates exclude lethal mutations which, in *Drosophila*, appear to arise at the rate of approximately 0.02 per genome per generation. These data imply that the average number of mildly deleterious segregating genes per individual is on the order of $\bar{n} \simeq \mu/(2\bar{h}\bar{s}) \simeq 140$. The effects of individual mutations are, of course, variable. Some indirect evidence supports the idea that the distribution of deleterious mutational effects is approximately exponential (Gregory, 1965; Edwards et al., 1987; Mackay et al., 1992; Santiago et al., 1992; Keightley, 1995), with the frequency of a mutation decreasing with increasing effect.

Research on the fitness effects of spontaneous mutations in other organisms is necessary before we can be certain about the generality of the *Drosophila* results, but there is no reason to expect that they will be highly unusual. Thus, it seems reasonable to conclude that, since most individuals are expected to carry unique sets of rare deleterious genes, even small founder populations are likely to harbor segregating deleterious genes at several hundred loci. All such genes are subject to eventual chance fixation in a small founder population.

The worst-case scenario is realized when the size of the founder population is so small that random genetic drift completely overwhelms the power of natural selection. Roughly speaking, this requires that the effective size of the isolated population, N_0, be $<1/(2s)$, or using the *Drosophila* data, that $N_0 < 33$. Under these circumstances, the probability that a deleterious mutation goes to fixation is close to its initial frequency in the founder population. Thus, for deleterious mutations with additive effects, the *average* impact of a population bottleneck on mean fitness is essentially zero, although in any particular population, random genetic drift will cause the realized mean fitness to be above or below this expectation. For populations with sizes $>1/(2s)$, some or all of the ancestral additive deleterious mutation load will be purged, and when $N_0 > 10/(2s)$, essentially all of it will be eliminated (Lynch et al., 1995).

Suppose now that the ancestral mutations are partially recessive, rather than additive, again averaging $\mu/(2hs)$ in number per individual. Assuming that these mutations are distributed over a large number of loci, then there are approximately $N_0\mu/(2hs)$ deleterious genes in the founder population, each with approximate frequency $1/(2N_0)$. Thus, if selection is completely ineffective (so that the fixation probability equals the initial frequency), the number of deleterious mutations expected to become fixed in a permanently bottlenecked population is approximately $\mu/(4hs)$. In this extreme situation, the mean fitness would be $(1 - 2s)^{\mu/(4hs)}$, compared to $(1 - 2hs)^{\mu/(2hs)}$ in the ancestral population. The ratio of these two quantities is approximately $e^{\mu[1 - 1/(2h)]}$, which with the *Drosophila* parameters equals 0.56.

These simple qualitative arguments validate the idea that population bottlenecks can lead to a reduction in the average fitness of individuals. However, they also suggest that the depression in mean fitness due to the fixation of unconditionally deleterious mutations *inherited from the ancestral population* is unlikely to be more than 50% or so. This conclusion still needs some major qualification. First, the results cited indicate only what is expected to happen on average; in any particular case, random genetic drift will result in a situation that is somewhat better or somewhat worse than the expectation. Second, we have yet to consider the fate of the substantial number of deleterious mutations that arise subsequent to the founder event.

New Deleterious Mutations in the Isolated Population

As noted previously, for populations of moderate size, one may reasonably expect natural selection to lead to a substantial purging of the deleterious genes carried in the founder individuals. However, this gives a rather distorted picture of the true mutation load. At the same time the deleterious mutations inherited from the ancestral population are being eliminated, new ones are appearing. Indeed, so long as $N_0 > (5/s)$, a population will behave as though it is effectively infinite— the loss of old mutations by selection will be balanced by the input by mutation each generation, and the fitness due to segregating mutations will remain stable at $e^{-\mu}$, as predicted by Haldane (1937).

For populations with effective sizes smaller than $5/s$ breeding adults per generation, the fitness associated with *segregating* deleterious mutations can decline to a value less than $e^{-\mu}$, the minimum occurring when the effective population size is approximately $1/(2s)$ when mutations have additive effects (Lynch et al., 1995). With the *Drosophila* parameters, this result implies that populations with effective sizes of about 33 individuals will exhibit the highest load of segregating mutations. It may seem counterintuitive that the segregational load declines in populations below this point. However, as noted, in very small populations, the mutations behave as though they are effectively neutral. Under drift–mutation equilibrium, the expected heterozygosity per locus is $4N_e\mu$, where μ is the genic mutation rate, and the expected number of heterozygous loci per individual is $2N_e\mu$. Thus, for very small populations, the expected fitness associated with segregating mutations is approximately $(1 - 2hs)^{2N_e\mu} \simeq e^{-4N_e\mu hs}$. This value is greater than the expectation for an effectively infinite population, $e^{-\mu}$, if $N_e < 1/(4hs)$.

At first sight, this result might suggest that populations could be managed for a minimal mutation load by simply maintaining them in small isolates. However, the fact that very small populations have a reduced load due to segregating mutations is again misleading. In large populations, the mutation load results almost entirely from a balance between the input of new deleterious genes by mutation and their removal by selection. In small populations, selection is less effective, and the input of new mutations is balanced by some selective removal *and* fixation by random genetic drift. The progressive and permanent decline in fitness due to fixation of recurrent mutations in small populations is not inconsequential. With μN new deleterious mutations entering a population each generation, each with initial frequency $1/(2N)$, for populations with effective sizes small enough that selection is overwhelmed by random genetic drift, $\mu/2$ fixations are expected to occur per generation. This implies a decline in fitness per generation of $[1 - (1 - 2s)^{\mu/2}] \simeq (1 - e^{-\mu s})$, or approximately 2.2% with the *Drosophila* parameters.

The preceding discussion provides a qualitative framework for understanding the consequences of deleterious mutation accumulation for the viability of populations. Suppose that the maximum number of progeny produced per adult in the ancestral population is R. If the fitness load due to deleterious mutations increases to the point at which the probability of survival to maturity is less than $1/R$, the per capita reproductive rate will be less than one. At this point, the population is no longer capable of maintaining a stable population size, and as it begins to decline, a synergistic interaction between random genetic drift and mutation accumulation is set in motion (Lynch and Gabriel, 1990; Lande, 1994; Lynch et al., 1994, 1995). As the population size declines, random genetic drift becomes a more significant evolutionary force and the rate of accumulation of deleterious mutations increases, causing a further decline in population size. We refer to this extinction phenomenon, which can be quite rapid, as a *mutational meltdown*.

We have modeled the mutational meltdown using the *Drosophila* mutational parameters. For populations with effective sizes smaller than several dozen individuals, the mean time to extinction due to the buildup of mutation load is typically on the order of a few tens to a few hundreds of generations *in the absence of any demographic or environmental stochasticity* (Lynch et al., 1995). Of course, virtually all populations continuously experience both kinds of stochasticity (in the form of fluctuations in sex ratios, carrying capacities, and survival and reproductive rates). Such conditions cause a reduction in the long-term effective population size, which under realistic conditions can reduce the time to entry to a mutational meltdown by an order of magnitude or more (Lynch et al., 1994, 1995).

There are additional reasons why the preceding results probably underestimate the threat of deleterious mutations to population survival. First, we have relied upon the mathematically tractable, but biologically implausible, assumption that mutations have constant effects s and h. However, it is primarily the effectively neutral fraction of mutations, that is, those with selection coefficients smaller than $1/(2N_e)$, that contributes most to the mutation load. If significant numbers of deleterious mutations have effects that are smaller than $1/(2N_e)$, as suggested previously, the rate of mutation accumulation will be substantially greater, and the mean time to extinction substantially smaller, than anticipated on the basis of the *average* mutational effect (Lande, 1994).

Second, the problem of deleterious mutation accumulation may be exacerbated in endangered species that are confined to breeding facilities. Since captive environments are usually quite benign (including services from dieticians, veterinarians, artificial inseminators, etc.), a real possibility exists that mutations that are significantly deleterious in nature are rendered nearly neutral. If that were the case, regardless of the population size, deleterious mutations would accumulate at nearly the neutral rate, $\mu/2$ per generation, although their effects would go undetected until the population was reintroduced into the wild. At that point, the population might no longer be capable of sustaining itself without continued human intervention.

Third, empirical evidence supports the idea that the expression of deleterious mutations is magnified in harsh environments. Relative to controls, *Drosophila* lines that have accumulated mildly deleterious mutations have substantially reduced fitness when raised under stressful conditions ($< 10\%$ of that in a benign environment; Kondrashov and Houle, in preparation). Moreover, numerous studies have shown that the expression of inbreeding depression, a consequence of deleterious recessive genes, is exacerbated in extreme environments (Parsons, 1971; Barlow, 1981; Dudash, 1990; Jiménez et al., 1994; Pray et al., 1994).

In summary, what little we know about deleterious mutations raises the real concern that their recurrent introduction can threaten the persistence of even moderately large populations over timescales of several dozens of generations. The issues are complex, and there is a serious need for more data in other species so that firmer quantitative statements can be made. However, for captive breeding programs in particular, there seems to be little question that management programs whose genetic focus is entirely on the deleterious genes contained in founder individuals are misguided.

The Threat from Elevated Mutation Rates

An important issue relevant to long-term species survival is the presence of environmental mutagens. Over the past century, the release of environmental pollutants through human activities has almost certainly increased the mutagenic potential of the environments of many species beyond our own. For example, concern over the depletion of the ozone layer is largely motivated by the resultant increase in the intensity of ultraviolet radiation, a significant source of somatic mutations, and presumably of germline mutations in organisms with exposed gametes and embryos. Although a massive amount of bioassay research has been done on the short-term effects of various pollutants on survival and reproduction, almost no quantitative information exists on the consequences of environmental mutagens for the genomic mutation rate and spectrum of mutational effects in natural populations.

A simple extrapolation of Haldane's result provides some insight into the long-term consequences of an increase in the genomic mutation rate. For a large population, an increase in the genomic mutation rate by a factor of x will reduce the fitness from segregating mutations from $e^{-\mu}$ to $e^{-\mu x}$. Even disregarding the fixation of new mutations, it is clear that populations for which $x > \ln(R)/\mu$ are doomed to rapid extinction. If the genomic mutation rate estimate from *Drosophila* is reasonably representative of that in other organisms, the critical value of x is not very large. For example, with $R = 20$, a generous situation for many birds and mammals, and $\mu = 1.5$, it is only 2.0, and even with $R = 1,000$, it is 4.6. Thus, if the *Drosophila* estimate for μ is correct, the long-term consequence of a doubling of the genomic mutation rate would be extinction for many species. It is worth emphasizing that this argument applies to all species, regardless of their current population size, including our own. Although most governmental policies on the disposal of toxic pollutants are guided by human concerns about carcinogenesis, it is clear that the long-term effects of environmental mutagens on ecosystem stability warrant close attention.

POPULATION AUGMENTATION

An increasingly common strategy for maintaining wild populations of endangered species is augmentation with stock from breeding facilities. An implicit assumption of such procedures is that recipient populations, when they still exist, actually derive some benefit from an artificial boost in population size. There are, however, several reasons why the long-term deleterious consequences of supplementation may outweigh the short-term advantage of increased population size.

First, over evolutionary time, successful populations are expected to become morphologically, physiologically, and behaviorally adapted to their local environments. This principle has been most convincingly documented through reciprocal transplant experiments with plants, which have often shown adaptive divergence on spatial scales as small as a few meters (Schemske, 1984; Waser and Price, 1985; McGraw, 1987; Schmitt and Gamble, 1990; Galen et al., 1991). Thus, there is little question that the introduction of non-native stock has the potential to disrupt local adaptations.

This type of problem takes on added significance when the population employed in stocking has been maintained in captivity. As noted, captive environments are often radically different from those in the wild. Given the amounts of genetic variation that exist for most quantitative characters, over a period of several generations captive populations will naturally evolve behavioral and/or morphological phenotypes that perform best in the novel environment. Such "domestication" selection has been clearly documented with invertebrates and fish (Doyle and Hunte, 1981; Doyle, 1983; Frankham et al., 1986; Ruzzante and Doyle, 1991; Frankham and Loebel, 1992; Reisenbichler, in preparation). To the extent that the genes advanced by selection in captivity are actually deleterious in nature, as seems likely, given that they had previously been kept at low frequencies, domestication selection can only further detract from the potential success of a reintroduction program. Fairly convincing circumstantial evidence now exists that phenotypic changes that evolve in hatchery-reared salmonids are deleterious in nature (Reisenbichler, in preparation; see also Chapter 8). On the other hand, as noted previously, an overprotective captive breeding program may simply result in a relaxation of natural selection and the gradual accumulation of deleterious genes. For example, for hatchery-raised salmonids, egg to smolt survivorship is typically 50% or greater, as compared to 10% or less in natural populations (Waples, 1991).

Second, local gene pools can be coadapted intrinsically (Dobzhansky, 1948; Templeton, 1986; Lynch, 1991). Just as the external environment molds the evolution of local adaptations by natural selection, the internal genetic environment of populations is expected to lead to the evolution of local complexes of genes that interact in a mutually favorable manner. The particular gene combinations that evolve in any local population may be largely fortuitous, depending in the long run on the chance variants that mutation provides for natural selection. By breaking up coadapted gene complexes, hybridization can lead to the production of individuals that have lower fitness than either parental type, and can even occur between populations that appear to be adapted to identical extrinsic environments. Burton (1987, 1990a, 1990b) has provided extensive evidence for the breakdown of physiological competence in crosses between populations of the intertidal marine copepod *Tigriopus californicus* separated by only tens of kilometers. Other dramatic evidence of outbreeding depression comes from observations of reduced fitness in crosses of inbred lines of flies (Templeton et al., 1976) and plants (Parker, 1992) adapted to identical environments. Crosses between outbreeding plants separated by several tens of meters can exhibit reduced fitness (Waser and Price, 1989), as can crosses between fish derived from different sites in the same drainage basin (Leberg, 1993) and crosses between clones of *Daphnia* from the same pond (Lynch and Deng, 1994). Populations that exhibit outbreeding depression upon crossing are clearly on different evolutionary trajectories.

Third, augmentation of wild populations with stock from captive breeding programs can have ecological consequences that have both immediate demographic effects and long-term evolutionary implications. For example, high-density hatchery populations of fish are prone to epidemics involving diseases that are uncommon in the natural environment. Such events provide strong selection for disease-resistant varieties of hatchery-reared fish, which subsequently can act as vectors to the wild population. The Norwegian Atlantic salmon is now threatened with extinction because of a parasite brought to Atlantic drainages by resistant stock from the Baltic (Johnsen and Jensen, 1986).

These three arguments strongly suggest that augmentation programs be used only as a last resort in the recovery and/or management of endangered species, even when both phenotypic and molecular data indicate strong similarities among demes. The studies cited here provide ample evidence that the absence of any obvious molecular divergence gives little, if any, insight into the potential for negative impacts of outcrossing. Even when the population subdivision of an endangered species is obviously due to human-induced fragmentation, a decision to manage the isolates as a single genetic entity by encouraging gene flow among demes can be shortsighted. Reproductively isolated groups created (or exaggerated) by human disturbance might reflect more natural units of genetic diversity that had arisen from long-term ecological and evolutionary forces, including isolation by distance. Alternatively, isolation induced by habitat loss may directly promote adaptive differentiation of demes to local conditions created by human activity. In either case, a management scheme that encourages interdemic migration might actually have the negative consequence of eroding local adaptation.

A particularly difficult issue underlying practical assessments of the potential for outbreeding depression concerns the timescale over which outbreeding depression is revealed. It is extremely common for the F_1 progeny of interpopulation crosses to exhibit enhanced fitness relative to their parents (Lerner, 1954; Thornhill, 1993), only to have a dramatic reduction in fitness occur in the F_2 generation (Wu and Palopoli, 1994). In some situations, the development of negative consequences of mixing coadapted gene complexes can be more subtle, taking several generations to emerge fully (Lynch, 1991). Thus, what might initially appear to be a successful management decision, encouraging further augmentation, may later become a liability to the wild population (Reisenbichler, in preparation).

Arguments have been made that populations suffering from outbreeding depression will eventually recover by evolving new coadapted gene complexes out of the hybrid gene pool, and perhaps even benefit from the influx of genetic variation (Templeton, 1986; Templeton et al., 1990). However, given that we know essentially nothing about the timescale over which such recovery might occur, subjecting a small population to even a few generations of increased genetic risks of extinction seems hazardous. Once an augmentation program has proceeded to the point at which F_2 individuals have begun to appear, it will often be next to impossible to erase the process of introgression.

SUMMARY

1. The vast majority of research in the field of conservation genetics has been focused on the use of molecular markers to elucidate patterns of variation. Such information provides a valuable perspective on numerous issues such as phylogenetic and pedigree relationships, patterns of gene flow, and phylogeographic domains. However, molecular genetics has yet to provide much insight into the fundamental genetic problems confronting most small populations of endangered species: the accumulation of deleterious mutations, the loss of adaptive potential, and the negative effects of population augmentation. Quantitative genetics, the branch of genetics that deals directly with the evolutionary properties of morphological and behavioral traits, provides a foundation for understanding the relevance of these issues to conservation biology.

2. An overview of theoretical and empirical results in quantitative genetics provides some insight into the critical population sizes below which species begin to experience genetic problems that exacerbate the risk of extinction. Populations that regularly contain fewer than 100 individuals are *extremely* vulnerable to *both* deleterious mutation accumulation and loss of adaptive potential. Security from long-term deleterious-mutation degradation requires a harmonic mean population size of at least 1,000 reproductive adults. Moreover, populations with fewer than 10,000 reproductive adults are likely to be limited with respect to adaptive genetic variation. It is perhaps not until 10^5 or so adults are regularly present that a population begins to behave genetically as if it were effectively infinite.

3.   Current national and international policies are such that endangered species are usually only endowed with formal protection after their total census number has dwindled to several hundred or fewer individuals. Based on the aforementioned, such population sizes are two to three orders of magnitude below the point at which the genetic integrity of species begins to be at risk. Thus, on genetic grounds alone, there is a need for much higher standards in the protection of species.

4.   Although most of the focus of genetic conservation is on the immediate deleterious consequences of small population size, human activities also potentially threaten the long-term persistence of very large populations. Due to the fact that natural selection imposes a load on a population through reduced individual survivorship and/or fecundity, there is a maximum rate of sustainable evolution beyond which the demographic cost of selection exceeds the ability of a population to replace itself. Rates of environmental change (such as global warming) that exceed a critical value can cause the extinction of any population. This problem is particularly acute for natural populations confined to fragmented habitat, because this eliminates migration as an alternative strategy for coping with environmental change. The critical rate of environmental change beyond which extinction is inevitable is quite low, even for populations that are effectively infinite in size from a genetic perspective. In some cases, increased deleterious mutation rates induced by human activities may impose an additional load on species, especially those with low demographic potential. Thus, the genetic consequences of human activities extend well beyond those species that currently appear to have dwindled to small population sizes.

REFERENCES

ÅGREN, J. and D.W. SCHEMSKE. 1993. Outcrossing rate and inbreeding depression in two annual monoecious herbs, *Begonia hirsuta* and *B. semiovata*. *Evolution* 47:125–135.

ABRAHAMSON, D.E. (ed.). 1989. *The Challenge of Global Warming*. Island Press, Washington, DC.

AVERY, P.J. and W.G. HILL. 1977. Variability in genetic parameters among small populations. *Genet. Res.* 29:193–213.

AVISE, J.C. 1994. *Molecular Markers, Natural History and Evolution*. Chapman & Hall, New York.

BARLOW, R. 1981. Experimental evidence for interaction between heterosis and environment in animals. *Anim. Breed. Abstr.* 49:715–737.

BARRETT, S.C.H. and D. CHARLESWORTH. 1991. Effects of a change in the level of inbreeding on the genetic load. *Nature* 352:522–524.

BARTON, N.H. 1989. Divergence of a polygenic system subject to stabilizing selection, mutation and drift. *Genet. Res.* 54:59–77.

BARTON, N.H. and M. TURELLI. 1989. Evolutionary quantitative genetics: how little do we know? *Annu. Rev. Genet.* 23:337–370.

BOAG, P.T. and P.R. GRANT. 1981. Intense natural selection in a population of Darwin's finches (Geospizinae). *Science* 214:82–85.

BOWMAN, J.C. and D.S. FALCONER. 1960. Inbreeding depression and heterosis of litter size in mice. *Genet. Res.* 1:262–274.

BRISCOE, D.A., J.M. MALPICA, A. ROBERTSON, G.J. SMITH, R. FRANKHAM, R.G. BANKS, and J.S.F. BARKER. 1992. Rapid loss of genetic variation in large captive populations of *Drosophila* flies: implications for the genetic management of captive populations. *Conserv. Biol.* 6:416–425.

BRYANT, E.H., S.A. McCOMMAS, and L.M. COMBS. 1986. The effect of an experimental bottle-neck upon quantitative genetic variation in the housefly. *Genetics* 114:1191–1211.

BRYANT, E.H. and L.M. MEFFERT. 1993. The effect of serial founder-flush cycles on quantitative genetic variation in the housefly. *Heredity* 70:122–129.

BRYANT, E.H., L.M. MEFFERT, and S.A. McCOMMAS. 1990. Fitness rebound in serially bottle-necked populations of the house fly. *Amer. Nat.* 136:542–549.

BÜRGER, R. and J. HOFBAUER. 1994. Mutation load and quantitative genetic traits. *J. Math. Biol.* 32:193–218.

BÜRGER, R. and R. LANDE. 1994. On the distribution of the mean and variance of a quantitative trait under mutation-selection-drift balance. *Genetics* 138:901–912.

BÜRGER, R. and M. LYNCH. 1995. Evolution and extinction in a changing environment: a quantitative-genetic analysis. *Evolution* 49:151–163.

BÜRGER, R., G.P. WAGNER, and F. STETTINGER. 1989. How much heritable variation can be maintained in finite populations by a mutation–selection balance? *Evolution* 43:1748–1766.

BURGMAN, M.A., S. FERSON, and H.R. AKCAKAYA. 1992. *Risk Assessment in Conservation Biology.* Chapman & Hall, New York.

BURTON, R.S. 1987. Differentiation and integration of the genome in populations of *Tigriopus californicus. Evolution* 41:504–513.

BURTON, R.S. 1990a. Hybrid breakdown in physiological response: a mechanistic approach. *Evolution* 44:1806–1813.

BURTON, R.S. 1990b. Hybrid breakdown in developmental time in the copepod *Tigriopus californicus. Evolution* 44:1814–1822.

CABALLERO, A. and P.D. KEIGHTLEY. 1994. A pleiotropic nonadditive model of variation in quantitative traits. *Genetics* 138:883–900.

CAUGHLEY, G. 1994. Directions in conservation biology. *J. Anim. Ecol.* 63:215–244.

CHAKRABORTY, R. 1981. The distribution of the number of heterozygous loci in an individual in natural populations. *Genetics* 98:461–466.

CHARLESWORTH, B., D. CHARLESWORTH, and M.T. MORGAN. 1990. Genetic loads and estimates of mutation rates in highly inbred plant populations. *Nature* 347:380–382.

CHARLESWORTH, D. and B. CHARLESWORTH. 1987. Inbreeding depression and its evolutionary consequences. *Annu. Rev. Ecol. Syst.* 18:237–268.

CHEVERUD, J., E. ROUTMAN, C. JAQUISH, S. TARDIF, G. PETERSON, N. BELFIORE, and L. FORMAN. 1994. Quantitative and molecular genetic variation in captive cotton-top tamarins (*Saguinus oedipus*). *Conserv. Biol.* 8:95–105.

CLAYTON, G.A. and A. ROBERTSON. 1955. Mutation and quantitative variation. *Amer. Nat.* 89:151–158.

COCKERHAM, C.C. and H. TACHIDA. 1988. Permanency of response to selection for quantitative characters in finite populations. *Proc. Natl. Acad. Sci. USA* 85:1563–1565.

CONNOR, J.L. and M.J. BELLUCCI. 1979. Natural selection resisting inbreeding depression in captive wild housemice (*Mus musculus*). *Evolution* 33:929–940.

DOBZHANSKY, T. 1948. Genetics of natural populations. XVIII. Experiments on chromosomes of *Drosophila pseudoobscura* from different geographical regions. *Genetics* 33:588–602.

DOYLE, R.W. 1983. An approach to the quantitative analysis of domestication selection in aquaculture. *Aquaculture* 33:167–185.

DOYLE, R.W. and W. HUNTE. 1981. Demography of an estuarine amphipod (*Gammarus lawrencianus*) experimentally selected for high "r": a model of the genetic effects of environmental change. *Can. J. Fish. Aquat. Sci.* 38:1120–1127.

DUDASH, M.R. 1990. Relative fitness of selfed and outcrossed progeny in a self-compatible, protandrous species, *Sabatia angularis* L. (Gentianaceae): a comparison in three environments. *Evolution* 44:1129–1139.

DUDLEY, J.W. 1977. Seventy-six generations of selection for oil and protein percentage in maize. In *Proceedings of the International Conference on Quantitative Genetics*, eds. E. Pollak, O. Kempthorne, and T.B. Bailey, Jr., pp. 186–200. Iowa State University Press, Ames.

EDWARDS, M.D., C.W. STUBER, and J.F. WENDEL. 1987. Molecular-marker facilitated investigations of quantitative-trait loci in maize: I. Numbers, genomic distribution and types of gene action. *Genetics* 116:113–125.

EISEN, E.J. 1980. Conclusions from long-term selection experiments with mice. *Z. Tierzüchtg. Züchtgsbiol.* 97:305–319.

ENDLER, J.A. 1986. *Natural Selection in the Wild*. Princeton University Press, Princeton, NJ.

FALCONER, D.S. 1981. *Introduction to Quantitative Genetics*, 2nd ed. Longman, London.

FALK, D.A. and K.E. HOLSINGER (eds.). 1991. *Genetics and Conservation of Rare Plants*. Oxford University Press, New York.

FOLEY, P. 1992. Small population genetic variability at loci under stabilizing selection. *Evolution* 46:763–774.

FOLEY, P. 1994. Predicting extinction times from environmental stochasticity and carrying capacity. *Conserv. Biol.* 8:124–137.

FOOSE, T.J., R. LANDE, N.R. FLESNESS, G. RABB, and B. READ. 1986. Propagation plans. *Zoo Biol.* 20:139–146.

FRANKHAM, R., H. HEMMER, O.A. RYDER, E.G. COTHRAN, M.E. SOULÉ, N.D. MURRAY, and M. SNYDER. 1986. Selection in captive populations. *Zoo Biol.* 5:127–138.

FRANKHAM, R. and D.A. LOEBEL. 1992. Modeling problems in conservation genetics using captive *Drosophila* populations: rapid genetic adaptation to captivity. *Zoo Biol.* 11:333–342.

FRANKLIN, I.R. 1980. Evolutionary changes in small populations. In *Conservation Biology: An Evolutionary-ecological Perspective*, eds. M.E. Soulé and B.A. Wilcox, pp. 135–149. Sinauer, Sunderland, MA.

GALEN, C., J.S. SHORE, and H. DEYOE. 1991. Ecotypic divergence in alpine *Polemonium viscosum*: genetic structure, quantitative variation, and local adaptation. *Evolution* 45:1218–1228.

GINZBURG, L.R., L.B. SLOBODKIN, K. JOHNSON, and A.G. BINDMAN. 1982. Quasiextinction probabilities as a measure of impact on population growth. *Risk Anal.* 2:171–181.

GOODMAN, D. 1987. The demography of chance extinction. In *Viable Populations for Conservation*, ed. M.E. Soulé, pp. 11–43. Cambridge University Press, New York.

GOODNIGHT, C.J. 1988. Epistasis and the effect of founder events on the additive genetic variance. *Evolution* 42:441–454.

GOTTLIEB, L.D. 1984. Genetics and morphological evolution in plants. *Amer. Nat.* 123:681–709.

GREGORY, W.C. 1965. Mutation frequency, magnitude of change and the probability of improvement in adaptation. *Radiat. Bot.* 5 (Suppl.):429–441.

HAIRSTON, N.G., Jr. and T.A. DILLON. 1990. Fluctuating selection and response in a population of freshwater copepods. *Evolution* 44:1796–1805.

HALDANE, J.B.S. 1937. The effect of variation on fitness. *Amer. Nat.* 71:337–349.

HALLAUER, A.R. and J.B. MIRANDA. 1981. *Quantitative Genetics in Maize Breeding.* Iowa State University Press, Ames.

HALLEY, J.M. and R.S. MANASSE. 1993. A population dynamics model subject to inbreeding depression. *Evol. Ecol.* 7:15–24.

HEDRICK, P.W. 1992. Genetic conservation in captive populations and endangered species. In *Applied Population Biology*, eds. S.K. Jain and L.W. Botsford, pp. 45–68. Kluwer Academic, Netherlands.

HEDRICK, P.W. 1994. Purging inbreeding depression and the probability of extinction: full-sib mating. *Heredity* 73:363–372.

HEDRICK, P.W., P.F. BRUSSARD, F.W. ALLENDORF, J.A. BEARDMORE, and S. ORZACK. 1986. Protein variation, fitness, and captive propagation. *Zoo Biol.* 5:91–99.

HEDRICK, P.W. and P.S. MILLER. 1992. Conservation genetics: techniques and fundamentals. *Ecol. Applic.* 2:30–46.

HEYWOOD, J. 1986. The effect of plant size variation on genetic drift in populations of annuals. *Amer. Nat.* 127:851–861.

HOLTSFORD, T.P. and N.C. ELLSTRAND. 1990. Inbreeding effects in *Clarkia tembloriensis* (Onagraceae) populations with different natural outcrossing rates. *Evolution* 44:2031–2046.

HOULE, D. 1989. The maintenance of polygenic variation in finite populations. *Evolution* 43:1767–1780.

HOULE, D., D.K. HOFFMASTER, S. ASSIMACOPOULOS, and B. CHARLESWORTH. 1992. The genomic mutation rate for fitness in *Drosophila*. *Nature* 359:58–60.

JIMÉNEZ, J.A., K.A. HUGHES, G. ALAKS, L. GRAHAM, and R.C. LACY. 1994. An experimental study of inbreeding depression in a natural habitat. *Science* 266:271–273.

JOHNSEN, B.O. and A.J. JENSEN. 1986. Infestations of Atlantic salmon, *Salmo salar*, by *Gyrodactylus salaris* in Norwegian rivers. *J. Fish Biol.* 29:233–241.

JONES, L.P., R. FRANKHAM, and J.S.F. BARKER. 1968. The effects of population size and selection intensity in selection for a quantitative character in *Drosophila*. *Genet. Res.* 12:249–266.

KALISZ, S. 1986. Variable selection on the timing of *Collinsia verna* (Scrophulariaceae). *Evolution* 40:479–491.

KAREIVA, P.M., J.G. KINGSOLVER, and R.B. HUEY (eds.). 1993. *Biotic Interactions and Global Change.* Sinauer, Sunderland, MA.

KEIGHTLEY, P.D. 1995. The distribution of mutation effects on viability in *Drosophila melanogaster*. *Genetics* 138:1315–1322.

KEIGHTLEY, P.D. and W.G. HILL. 1988. Quantitative genetic variation maintained by mutation-stabilizing selection balance in finite populations. *Genet. Res.* 52:33–43.

KEIGHTLEY, P.D. and W.G. HILL. 1989. Quantitative genetic variability maintained by mutation-stabilizing selection balance: sampling variation and response to subsequent directional selection. *Genet. Res.* 54:45–57.

KIMURA, M. 1983. *The Neutral Theory of Molecular Evolution.* Cambridge University Press, Cambridge, UK.

KIMURA, M., T. MARUYAMA, and J.F. CROW. 1963. The mutation load in small populations. *Genetics* 48:1303–1312.

KOBYLIANSKY, E. and G. LIVSHITS. 1983. Relationship between levels of biochemical heterozygosity and morphological variability in human populations. *Ann. Hum. Genet.* 47:215–223.

KRESS, D.D. 1975. Results from long-term selection experiments relative to selection limits. *Genet. Lect.* 4:253–271.

LACY, R.C. 1992. The effects of inbreeding on isolated populations: are minimum viable population sizes predictable? In *Conservation Biology*, eds. P.L. Fiedler and S.K. Jain, pp. 277–296. Chapman & Hall, New York.

LACY, R.C. 1993. VORTEX: a computer simulation model for population viability analysis. *Wildl. Res.* 20:45–65.

LANDE, R. 1975. The maintenance of genetic variability by mutation in a polygenic character with linked loci. *Genet. Res.* 26:221–235.

LANDE, R. 1981. The minimum number of genes contributing to quantitative variation between and within populations. *Genetics* 99:541–553.

LANDE, R. 1993. Risks of population extinction from demographic and environmental stochasticity, and random catastrophes. *Amer. Nat.* 142:911–927.

LANDE, R. 1994. Risk of population extinction from new deleterious mutations. *Evolution* 48:1460–1469.

LANDE, R. 1995. Breeding plans for small populations, based on the dynamics of quantitative genetic variance. In *Population Management For Survival and Recovery*, eds. J.D. Ballou, M. Gilpin, and T.J. Foose, pp. 318–340. Columbia University Press, New York.

LANDE, R. and G.F. BARROWCLOUGH. 1987. Effective population size, genetic variation, and their use in population management. In *Viable Populations for Conservation*, ed. M.E. Soulé, pp. 87–123. Cambridge University Press, New York.

LEBERG, P.L. 1992. Effects of population bottlenecks on genetic diversity as measured by allozyme electrophoresis. *Evolution* 46:477–494.

LEBERG, P.L. 1993. Strategies for population reintroduction: effects of genetic variability on population growth and size. *Conserv. Biol.* 7:194–199.

LEIGH, E.G., Jr. 1981. The average lifetime of a population in a varying environment. *J. Theor. Biol.* 90:213–239.

LERNER, I.M. 1954. *Genetic Homeostasis.* Oliver & Boyd, London.

LÓPEZ-FANJUL, C., J. GUERRA, and A. GARCIA. 1989. Changes in the distribution of the genetic variance in a quantitative trait in small populations of *Drosophila melanogaster*. *Génét. Sél. Evol.* 21:159–168.

LÓPEZ-FANJUL, C. and A. VILLAVERDE. 1989. Inbreeding increases genetic variation for viability in *Drosophila melanogaster*. *Evolution* 43:1800–1804.

LUDWIG, D. 1976. A singular perturbation problem in the theory of population extinction. Society for Industrial and Applied Mathematics– *Amer. Math. Soc. Proc.* 10:87–104.

LYNCH, C.B. 1977. Inbreeding effects upon animals derived from a wild population of *Mus musculus*. *Evolution* 31:526–537.

LYNCH, M. 1988a. The rate of polygenic mutation. *Genet. Res.* 51:137–148.

LYNCH, M. 1988b. Design and analysis of experiments on random drift and inbreeding depression. *Genetics* 120:791–807.

LYNCH, M. 1991. The genetic interpretation of inbreeding depression and outbreeding depression. *Evolution* 45:622–629.

LYNCH, M., J. CONERY, and R. BÜRGER. 1995. Mutational meltdowns in sexual populations. *Evolution, in press*.

LYNCH, M., J. CONERY, and R. BÜRGER. 1995. Mutation accumulation and the extinction of small populations. *Amer. Nat., in press*.

LYNCH, M. and T.J. CREASE. 1990. The analysis of population survey data on DNA sequence variation. *Molec. Biol. Evol.* 7:377–394.

LYNCH, M. and H.-W. DENG. 1994. Genetic slippage in response to sex. *Amer. Nat.* 144:242–261.

LYNCH, M. and W. GABRIEL. 1990. Mutation load and the survival of small populations. *Evolution* 44:1725–1737.

LYNCH, M., W. GABRIEL, and A.M. WOOD. 1991. Adaptive and demographic responses of plankton populations to environmental change. *Limnol. Oceanogr.* 36:1301–1312.

LYNCH, M. and W.G. HILL. 1986. Phenotypic evolution by neutral mutation. *Evolution* 40:915–935.

LYNCH, M. and R. LANDE. 1993. Evolution and extinction in response to environmental change. In *Biotic Interactions and Global Change*, eds. P.M. Kareiva, J.G. Kingsolver, and R.B. Huey, pp. 234–250. Sinauer, Sunderland, MA.

LYNCH, M. and B.G. MILLIGAN. 1994. Analysis of population genetic structure with RAPD markers. *Molec. Ecol.* 3:91–99.

MACKAY, T.F.C., R.F. LYMAN, and M.S. JACKSON. 1992. Effects of *P* element insertions on quantitative traits in *Drosophila melanogaster*. *Genetics* 130:315–332.

McGRAW, J.B. 1987. Experimental ecology of *Dryas octapetala* ecotypes: IV. Fitness response to reciprocal transplanting in ecotypes with differing plasticity. *Oecologia* 73:465–468.

MILLIGAN, B.G., J. LEEBENS-MACK, and A.E. STRAND. 1994. Conservation genetics: beyond the maintenance of marker diversity. *Molec. Ecol.* 3:423–435.

MILLS, L.S. and P.E. SMOUSE. 1994. Demographic consequences of inbreeding in remnant populations. *Amer. Nat.* 144:412–431.

MITTON, J.B. and B.A. PIERCE. 1980. The distribution of individual heterozygosity in natural populations. *Genetics* 95:1043–1054.

MORITZ, C. 1994. Applications of mitochondrial DNA analysis in conservation: a critical review. *Molec. Ecol.* 3:401–412.

MUKAI, T. 1964. The genetic structure of natural populations of *Drosophila melanogaster*. I. Spontaneous mutation rate of polygenes controlling viability. *Genetics* 50:1–19.

MUKAI, T. 1969. The genetic structure of natural populations of *Drosophila melanogaster*. VIII. Natural selection on the degree of dominance of viability polygenes. *Genetics* 63:467–478.

MUKAI, T. 1979. Polygenic mutations. In *Quantitative Genetic Variation*, eds. J.N. Thompson, Jr. and J.M. Thoday, pp. 177–196. Academic Press, New York.

MUKAI, T., S. CHIGUSA, and I. YOSHIKAWA. 1965. The genetic structure of natural populations of *Drosophila melanogaster*. III. Dominance effect of spontaneous mutant polygenes controlling viability in heterozygous genetic backgrounds. *Genetics* 52:493–501.

MUKAI, T. and T. YAMAZAKI. 1968. The genetic structure of natural populations of *Drosophila melanogaster*. V. Coupling-repulsion effect of spontaneous mutant polygenes controlling viability. *Genetics* 59:513–535.

NEI, M. and A.K. ROYCHOUDHURY. 1974. Sampling variances of heterozygosity and genetic distance. *Genetics* 76:379–390.

O'BRIEN, S.J., M.E. ROELKE, L. MARKER, A. NEWMAN, C.A. WINKLER, D. MELTZER, L. COLLY, J.F. EVERMANN, M. BUSH, and D.E. WILDT. 1985. Genetic basis for species vulnerability in the cheetah. *Science* 227:1428–1434.

OHNISHI, O. 1977. Spontaneous and ethyl methanesulfonate-induced mutations controlling viability in *Drosophila melanogaster*. II. Homozygous effect of polygenic mutations. *Genetics* 87:529–545.

ORR, H.A. and H.A. COYNE. 1992. The genetics of adaptation: a reassessment. *Amer. Nat.* 140:725–742.

PALMER, A.R. and C. STROBECK. 1986. Fluctuating asymmetry: measurement, analysis, patterns. *Annu. Rev. Ecol. Syst.* 17:391–421.

PARKER, M.A. 1992. Outbreeding depression in a selfing annual. *Evolution* 46:837–841.

PARSONS, P.A. 1971. Extreme environment heterosis and genetic loads. *Heredity* 26:479–483.

PEASE, C.M., R. LANDE, and J.J. BULL. 1989. A model of population growth, dispersal and evolution in a changing environment. *Ecology* 70:1657–1664.

PIRCHNER, F. 1983. *Population Genetics in Animal Breeding*, 2nd ed. Plenum, New York.

PRAY, L.A., J.M. SCHWARTZ, C.J. GOODNIGHT, and L. STEVENS. 1994. Environmental dependency of inbreeding depression: implications for conservation biology. *Conserv. Biol.* 8:562–568.

ROBERTSON, A. 1952. The effect of inbreeding on variation due to recessive genes. *Genetics* 37:189–207.

RUZZANTE, D.E. and R.W. DOYLE. 1991. Rapid behavioral changes in medaka (*Oryzias latipes*) caused by selection for competitive and noncompetitive growth. *Evolution* 45:1936–1946.

SANTIAGO, E., J. ALBORNOZ, A. DOMINGUEZ, M.A. TORO, and C. LÓPEZ-FANJUL. 1992. The distribution of effects of spontaneous mutations on quantitative traits and fitness in *Drosophila melanogaster*. *Genetics* 132:771–781.

SCHEMSKE, D.W. 1984. Population structure and local selection in *Impatiens* (Balsaminaceae), a selfing annual. *Evolution* 37:523–539.

SCHMITT, J. and S.E. GAMBLE. 1990. The effect of distance from the parental site on offspring performance and inbreeding depression in *Impatiens capensis*: a test of the local adaptation hypothesis. *Evolution* 44:2022–2030.

SENNER, J.W. 1980. Inbreeding depression and the survival of zoo populations. In *Conservation Biology*, eds. M.E. Soulé and B.A. Wilcox, pp. 209–224. Sinauer, Sunderland, MA.

SHAFFER, M.L. 1981. Minimum population sizes for species conservation. *BioScience* 31:131–134.

SHNOL, E.E. and A.S. KONDRASHOV. 1993. The effect of selection on the phenotypic variance. *Genetics* 134:995–996.

SOULÉ, M.E. 1980. Thresholds for survival: maintaining fitness and evolutionary potential. In *Conservation Biology: An Evolutionary–ecological Perspective*, eds. M.E. Soulé and B.A. Wilcox, pp. 151–169. Sinauer, Sunderland, MA.

SOULÉ, M.E. 1982. Allomeric variation. 1. The theory and some consequences. *Amer. Nat.* 120:751–764.

SOULÉ, M.E., M. GILPIN, N. CONWAY, and T. FOOSE. 1986. The millenium ark: how long a voyage, how many staterooms, how many passengers? *Zoo Biol.* 5:101–114.

STRAUSS, R.E. 1991. Correlations between heterozygosity and phenotypic variability in *Cottus* (Teleostei: Cottidae): character components. *Evolution* 45:1950–1956.

TEMPLETON, A.R. 1986. Coadaptation and outbreeding depression. In *Conservation Biology: The Science of Scarcity and Diversity*, ed. M.E. SOULÉ, pp. 105–116. Sinauer, Sunderland, MA.

TEMPLETON, A.R. and B. READ. 1983. The elimination of inbreeding depression in a captive herd of Speke's gazelle. In *Genetics and Conservation: A Reference for Managing Wild Animal and Plant Populations*, eds. C.M. Schonewald-Cox, S.M. Chambers, B. MacBryde, and L. Thomas, pp. 241–261. Addison-Wesley, Reading, MA.

TEMPLETON, A.R. and B. READ. 1984. Factors eliminating inbreeding depression in a captive herd of Speke's gazelle. *Zoo Biol.* 3:177–199.

TEMPLETON, A.R., K. SHAW, E. ROUTMAN, and S.K. DAVIS. 1990. The genetic consequences of habitat fragmentation. *Ann. Missouri Bot. Gard.* 77:13–27.

TEMPLETON, A.R., C.F. SING, and B. BROKAW. 1976. The unit of selection in *Drosophila mercatorum*. I. The interaction of selection and meiosis in parthenogenetic strains. *Genetics* 82:349–376.

THORNHILL, N.W. (ed.). 1993. *The Natural History of Inbreeding and Outbreeding*. University of Chicago Press, Chicago, IL.

TURELLI, M. 1984. Heritable genetic variation via mutation-selection balance: Lerch's zeta meets the abdominal bristle. *Theor. Pop. Biol.* 25:138–193.

UNITED STATES FISH AND WILDLIFE SERVICE. 1992. *Endangered Species Act of 1973 as Amended through the 100th Congress.* U.S. Dept. of Interior, Washington, DC.

VAN TREUREN, R., R. BIJLSMA, N.J. OUBORG, and W. VAN DELDEN. 1993. The significance of genetic erosion in the process of extinction. IV. Inbreeding depression and heterosis effects caused by selfing and outcrossing in *Scabiosa columbaria. Evolution* 47:1669–1680.

WAPLES, R.S. 1991. Genetic interactions between hatchery and wild salmonids: lessons from the Pacific Northwest. *Can. J. Fish. Aquat. Sci.* 48:124–133.

WASER, N.M. and M.V. PRICE. 1985. Reciprocal transplant experiments with *Delphinium nelsonii* (Ranunculaceae): evidence for local adaptation. *Amer. J. Bot.* 72:1726–1732.

WASER, N.M. and M.V. PRICE. 1989. Optimal outcrossing in *Ipomopsis aggregata*: seed set and offspring fitness. *Evolution* 43:1097–1109.

WEBER, K.E. and L.T. DIGGINS. 1990. Increased selection response in larger populations. II. Selection for ethanol vapor resistance in *Drosophila melanogaster* at two population sizes. *Genetics* 125:585–597.

WEIR, B.S., P.J. AVERY, and W.G. HILL. 1980. Effect of mating structure on variation in inbreeding. *Theor. Pop. Biol.* 18:396–429.

WEIS, A.E., W.G. ABRAHAMSON, and M.C. ANDERSEN. 1992. Variable selection on *Eurosta*'s gall size, I: the extent and nature of variation in phenotypic selection. *Evolution* 46:1674–1697.

WHITLOCK, M.C., P.C. PHILLIPS, and M.J. WADE. 1993. Gene interaction affects the additive genetic variance in subdivided populations with migration and extinction. *Evolution* 47:1758–1768.

WILCOVE, D.S., M. MCMILLAN, and K.C. WINSTON. 1993. What exactly is an endangered species? An analysis of the endangered species list, 1985–91. *Conserv. Biol.* 7:87–93.

WILLIS, J.H. and H.A. ORR. 1993. Increased heritable variation following population bottlenecks: the role of dominance. *Evolution* 47:949–957.

WRICKE, G. and W.E. WEBER. 1986. *Quantitative Genetics and Selection in Plant Breeding.* Walter de Gruyter, New York.

WRIGHT, S. 1968. *Evolution and the Genetics of Populations. Vol. 1: Genetic and Biometric Foundations.* University of Chicago Press, Chicago, IL.

WU, C.-I. and M.F. PALOPOLI. 1994. Genetics of postmating reproductive isolation in animals. *Annu. Rev. Genet.* 28:283–308.

WYMAN, R.L. (ed.). 1991. *Global Climate Change and Life on Earth.* Chapman & Hall, New York.

YEZERINAC, S.M., S.C. LOUGHEED, and P. HANDFORD. 1992. Morphological variability and enzyme heterozygosity: individual and population level correlations. *Evolution* 46:1959–1964.

ZENG, Z.-B. and C.C. COCKERHAM. 1991. Variance of neutral genetic variances within and between populations for a quantitative character. *Genetics* 129:535–553.

ZINK, R.M., M.F. SMITH, and J.L. PATTEN. 1985. Association between heterozygosity and morphological variance. *J. Hered.* 76:415–420.

EPILOGUE

Thanks to the research programs summarized in this book, and to others like them, we now have provisional if not definitive answers to many of the empirical and conceptual questions raised in Table 1.1 of the Introduction. The following are examples that illustrate the breadth and richness of current scientific inquiry in conservation genetics.

Yes, levels of molecular genetic variation are exceptionally low in populations of some endangered species (e.g., the cheetah, Chapter 3; Channel Island fox, Chapter 4) but not others (e.g., the red-cockaded woodpecker, Chapter 6). Reduced fitness associated with grossly diminished molecular heterozygosity has been inferred in some studies (e.g., big cats, Chapter 3) and experimentally confirmed on occasion (e.g., Sonoran topminnow, Chapter 12). However, the universality of the relationship between assayable molecular variation on the one hand, and ecological viability and/or evolutionary potential on the other, remains far from certain (Chapter 5). There are both theoretical and empirical reasons for questioning whether molecular assessments faithfully reflect variation in the quantitative morphological and physiological characters upon which natural selection is known to operate (Chapter 15).

Yes, molecular genetic approaches can help to reveal the kinship structures, socioecologies and reproductive systems of endangered species (e.g., cetaceans, Chapter 2; gray wolf, Chapter 4; monkeys, Chapter 5; Santa Cruz Island bush mallow, Chapter 10), and the data employed to assist in establishing breeding priorities for population management (e.g., Micronesian kingfisher, Chapter 6).

Molecular genetic approaches have been extremely informative in revealing conservation-relevant demographic and natural-history features of organisms (e.g., natal homing in marine turtles, Chapter 7; alternative life-history strategies of salmonid fishes, Chapter 8; stock structure in commercially harvested pelagic fishes, Chapter 11). Indeed, it is now abundantly clear that geographic populations of many species display pronounced genetic heterogeneity (e.g., foxes, Chapter 4; primates, Chapter 5; plants, Chapter 9; desert fishes, Chapter 12; etc.), that much of this heterogeneity reflects historical demographic factors, and that genealogical separations at the intraspecific level can range from evolutionarily shallow to deep (Chapter 14). Remarkably, population genetic structures sometimes are oriented in intelligible cross-species patterns, presumably reflective of the landscape ecologies and congruent phylogeographic histories of the surveyed regional biotas (e.g., African bovids, Chapter 13; freshwater, terrestrial, and maritime faunas in the southeastern United States, Chapter 14). Conservation ramifications include the development of guidelines for translocation or augmentation programs, as well as designs for regional biogeographic preserves. In general, knowledge of pattern and depth in the partitioning of genetic variation within and among populations and related species is highly relevant to virtually all taxonomic judgments and management decisions in conservation biology.

Molecular genetic approaches also have proven extremely powerful for addressing hybridization and introgression phenomena in endangered species (e.g., canids, Chapter 4) or those that are commercially important (e.g., salmonids, Chapter 8), and for estimating higher level phylogenies in endangered species complexes (e.g., felids, Chapter 3; marine turtles, Chapter 7; Hawaiian silverswords, Chapter 10). Finally, molecular genetic techniques have cogent forensic applications, ranging from illumination of the populational sources of migratory individuals (e.g., marine turtles on foraging grounds and migration pathways, Chapter 7; anadromous salmon harvested at sea, Chapter 8) to the genetic identification of biological material confiscated from illegal trade in endangered species' products (e.g., whale meat in commercial markets, Chapter 2).

Perhaps what is most important to realize is that many of the questions now routinely addressed in conservation biology would not likely have been raised (let alone answered) prior to the incorporation of genetic methods and perspectives to the field. In the absence of molecular data and genetic outlooks (including those from quantitative genetics), it seems doubtful, for example, that heterozygosity issues and the topic of inbreeding depression would be widely discussed in wildlife management; that the phylogeographic nature of populational differences would be understood or widely appreciated; that genetics, demography, behavior, and natural history would be recognized to be so intertwined and mutually illuminating; that the semipermeable nature of species' boundaries to introgressive hybridization would have been as well understood as it is today; that the level of evolutionary and phylogenetic distinctiveness of endangered populations or other taxa could be critically documented; or that molecular forensic tools would become available for determining the source of otherwise unidentified wildlife products.

These advances all have taken place within the last few years, following the introduction of molecular tools that permit direct access to the genetic properties of organisms in nature, and of conceptual tools to deal with the new classes of information. Given the youth of the field of conservation

genetics as applied to populations and species in nature, it may be anticipated that many additional empirical applications (as well as contributions
to conservation genetic theory) will soon be forthcoming. Among these will
be a further integration of the new classes of genetic information with the
concepts and data of population biology, demography, quantitative genetics,
natural-history study, ethology, ecology, phylogenetics and systematics, all
in the service of conservation efforts.

JOHN C. AVISE
J.L. HAMRICK
Athens, Georgia, December 1994

INDEX